IET CONTROL, ROBOTICS AND SENSORS SERIES 132

Human Machine Collaboration and Interaction for Smart Manufacturing

Human Machine Collaboration and Interaction for Smart Manufacturing

Automation, robotics, sensing, artificial intelligence, 5G, IoTs and Blockchain

Edited by
Wai Yie Leong

The Institution of Engineering and Technology

Published by The Institution of Engineering and Technology, London, United Kingdom

The Institution of Engineering and Technology is registered as a Charity in England & Wales (no. 211014) and Scotland (no. SC038698).

First published 2022

The Institution of Engineering and Technology
Futures Place
Kings Way, Stevenage
Hertfordshire, SG1 2UA, United Kingdom

www.theiet.org

British Library Cataloguing in Publication Data
A catalogue record for this product is available from the British Library

ISBN 978-1-83953-414-0 (hardback)
ISBN 978-1-83953-415-7 (PDF)

Typeset in India by MPS Ltd

Contents

About the editor

Wai Yie Leong is the Pro Vice Chancellor of INTI International University, Malaysia. Her specialisms are Industry 4.0 (IR4.0), wireless sensor networks, Ultra-Wideband and wireless communications, and brain signal processing for signal conditioning and classification. She is the Vice President of The Institution of Engineers Malaysia (IEM) 2020/2022, Vice President of the International Network of Women Engineers and Scientists (INWES) 2020/2023, Chairman of The Institution of Engineering and Technology (IET) Malaysian Local Network 2018/2021, IET Council Member and Committee Member of the World Federation of Engineering Organisation (Women in Engineering Committee). She holds a PhD in Electrical Engineering from The University of Queensland (UQ), Brisbane, Australia.

Human–Machine Collaboration and Interaction for Smart Manufacturing

Advances in science and innovation persistently support the improvement of industrialisation from one side of the planet to the other. The First Industrial Revolution utilised water and steam ability to motorise creation; the Second utilised electric ability to make large scale manufacturing, the Third utilised robotisation for the assembling line. The Fourth Industrial Revolution, the computerised insurgency, is portrayed by a combination of advances that is obscuring the lines between the physical, advanced, and organic circles. Fourth Industrial Revolution, 4IR, or Industry 4.0, conceptualises rapid change to technology, industries, and societal patterns and processes in the 21st century due to increasing interconnectivity and smart automation. The expansiveness and profundity of these progressions proclaimed the change of whole frameworks of creation, the board, and administration. The Fifth Industrial Revolution will be centered around the co-activity among human and machine, as human insight works in amicability back into modern creation with communitarian robots. Laborers will be upskilled to offer some benefit added errands underway, prompting mass customisation and personalisation for clients.

In a Smart Manufacturing plant, human and Artificial Intelligent (AI) intently work together and carry gigantic advantages to different capacities like Assembly, Quality Assurance, Logistics and Maintenance. In all capacities, the collaboration among human and AI and spotlight on productive human–machine interfaces to associate human specialists with AI-based frameworks.

Utilising robots is now cutting edge in practically all businesses and smart factories. Particularly for undertakings that require truly difficult work or monotonous work, robots decrease expenses and speed up, subsequently boosting efficiency. With the rising requirement for individualized items and part size, the intricacy of creation is expanding and thats only the tip of the iceberg and greater adaptability is required. Rule-based robots with fixed programming plans will miss the mark in adapting to the intricacy of creation. They will be supplanted with more adaptable AI-based frameworks. Those frameworks are a blend of AI and adaptable synergistic robots – or cobots. The frameworks will become self-streamlined frameworks that conform to changing creation prerequisites continuously. They permit direct communication between human laborers and machines. For such cooperation, portable AIs are expected to permit direct association among cobots and human laborers. Robotics builds high-performance industrial robots that work collaboratively with human workers to enable much more flexible, productive, and efficient manufacturing work cells. The Human–Machine Collaboration solution allows to create efficient assistant workflows, e.g. displaying step-by-step work instructions on tablets.

Computer-based intelligence can be utilised to lessen machine breakdowns and increment use of creation resources. In that capacity, prescient support calculations keep away from breakdowns by constantly dissecting parts and supplanting them before episodes happen. As an essential, machine information and human work process, information should be gathered consistently, so that the AI can ceaselessly gain from constant creation information. Numerous innovation suppliers currently offer frameworks that interface apparatus with modern web of things backend applications. Such frameworks can be introduced on premise or in private mists.

In this book series on Human–Machine Collaboration and Interaction for Smart Manufacturing, various technologies and application on human-machine were highlighted. The networked sensors for data interoperability and real-time production tracking, blockchain technology, 5G technology, artificial intelligence, additive manufacturing, 3D processing, drone technology, extended reality, virtual reality, fiber optics, Internet of Things, and robotics. Many case studies and projects were discussed based on medical industry, manufacturing, waste processing, ICT industry, environmental industry, and banking system.

The principle challenge and restriction in all cases is the harmony among human and computerisation. Through computerisation, man-made reasoning, and novel sensors, Industry 4.0 and Industry 5.0 will make another assembling and dispersion measure. By this, measures are digitised, robots are doing work (or team up), and the human is in control to administer fabricating and other quality guaranteeing measures. Spots, where such assembling is executed, are likewise called keen manufacturing plants. These are savvy such that their creation machines are for all time facilitated with one another and they poise while guaranteeing the most noteworthy proficiency (shrewd assembling). To accomplish a significant degree of robotisation joined with adaptability as presented above, human abilities should be depicted and moved to computerised machines (e.g., modern robots).

Editor: Wai Yie Leong
The Institution of Engineering and Technology

Chapter 1

Introduction to HMI—current and future, systems, features, and benefits Human–machine interfaces in smart manufacturing

Kamalakannan R[1], Satheesh Kumar S[2], Tan Koon Tatt[3] and Siva Sundar S[4]

Engineering technology exists in an interconnected world. The Internet of Things (IoT) is a popular movement that connects industrial and consumer goods to the Internet. Human–machine interfaces (HMIs) are becoming classier to operate those devices. HMI technology includes any software or device that takes to interact with the machine. This can be further defined as a component or feature of certain software or device application that makes humans engage and interact with several machines. With an improvement in the implementation of "Industry 4.0" in the factories of today, the concept of HMI has garnered even more importance. Adopting this technology in a factory ensures that production is closely monitored and changing production demands are met. This process ensures that efficiency is improved and the level of downtime reduces. The HMI technology also offers a glimpse into the system's operations and in case of any defect or problem; an alarm is signaled so that the operator can rectify the issue at the earliest without wasting valuable production time.

The HMI is implemented in automobiles too. For instance, initially, only luxury automobiles could afford the innovative technology in their cars but with time, this technology was also incorporated into medium-cost cars. Experts suggest that the HMI in automobiles is expected to witness significant growth in the long term. The number of consumers in HMI using augmented reality (AR) and virtual reality (VR) are discussed.

1.1 HMI on a growth drive

As a result of the manufacturing industry's adoption of Industry 4.0 in their production lines, it's not surprising that the HMI industry's growth is on the rise.

[1]Department of Mechanical Engineering, M. Kumarasamy College of Engineering, Tamil Nadu, India
[2]Department of Electrical and Electronics, University of Technology and Applied Sciences, Muscat, Oman
[3]School of Science and Technology, Wawasan Open University, Penang, Malaysia
[4]Department of Mechanical Engineering, Sethu Institute of Technology, Tamil Nadu, India

According to a report by Research and Markets, the global HMI market is predicted to earn more than $8 billion in revenue by 2023. Important aspects such as safety, energy efficiency, increased productivity, and sustainability will also contribute to the growth of the HMI market, according to the report [1].

In the workshop of the future, people will have to deal with a growing complication of procedures, machinery, and parts. New operational paradigms wish to be necessary for optimal human–machine interaction. Production processes that are nimble, flexible, and intelligent will be critical to success. By merging "practical" and "actual" to acquire a complete vision of the entire chain value, manufacturers will be able to make items faster, more efficiently, and with higher returns while using fewer resources [2]. The possible global demands, advantages, ideas, and conditions for the workshop of upcoming will be examined in this white paper. It examines how linked technologies and marketplaces are transforming the business world, as well as the implications for information, people, technology, and principles. The white paper was produced in collaboration with the Fraunhofer Institute for Production and Automation IPA by the Market Strategy Board factory of the upcoming development team. The use of artificial information in manufacturing has become a hot issue among academics and businesses all over the world. Certain academics and practitioners, on the other hand, have coined terms like intelligent manufacturing (IM) and smart manufacturing (SM) to describe what is essentially the same paradigm. While SM and IM have a lot in common, they are not the same. Little thought has been given to whether the meaning, idea, implication, and technological improvement of the notions of IM and SM are reliable across literature when using an evolutionary approach [3].

1.2 Origins of smart manufacturing

The idea of smart production appears to have arisen in the late 1990s, as indicated by "Synthetic brainpower: a means for SM," which, to our knowledge, is the first publication to link expert system AI to SM. SM with artificial intelligence (AI), released in 1987, examined how AI may help manufacturers increase efficiency and profitability [4]. AI, robots, expert systems, computer-aided process planning, vision, flexible manufacturing systems, process control, and inspection are all covered in this book. Next nearly two eras of silence, the new notion of SM has resurfaced, usually in conjunction with the advent in Industry 4.0. The basic idea of today's SM has been summarized by the General Institution of Guidelines and Technology and the Smart Manufacturing Leadership Coalition (SMLC). SM is the use of modern smart technology to allow fast and steady manufacturing of unique goods, an energetic response to individualized manufactured goods demands, and actual-time optimization of construction and resource chain networks from an engineering aspect [5]. Design, commodities, operations, and business systems can be integrated across shop floors, centers, factories, companies, and entire supply chains using SM platforms. SM is the use of IoT, CPS, and Industrial IoT (IIoT), supported by devices and transmission technologies that take data at all stages and phases of production, from a networking aspect. SM becomes smarter over time as productivity rises owing to fewer errors and waste. SM makes advantage of domain data's accessibility and availability in decision-making to help manufacturing companies improve forecast and sustainable production processes and procedures, resulting in increased production. Schedule planning, diagnostics, prognostic supply, and evaluation are just a few of the activities that SM employs big data analytics to better (BDA) [6].

1.3 HMIs

The emergence of energetic, extremely complicated human–machine systems swiftly shifted the attention to adaptive mechanisms, and early HMI research concentrated on fully controlled systems. Computerized systems have evolved into semiautonomous agents in vibrant settings and unclear surroundings, necessitating the incorporation of human-to-human interactions into HMI and human–machine teamwork. This implies that in modern human–machine systems, human and machine agents should no longer be thought of as separate entities, but rather as members of a dynamic unit or team working toward a common goal, which includes dynamic task allocation among participants. To enable the human–machine system's malleability and dynamism, full system-side determinism must be abandoned in order to provide degrees of flexibility to the automated system to react to events that could not have been precisely predicted and modeled by system creators. As a result, the machine interface's individual operator's function shifts from full control to monitoring and limited control [7]. The types and features of executed borders remain a dominant feature in the examination of HMI processes, especially in terms of ergonomic properties like usability and transparency, in order to ensure that the human operator maintains an overview and understanding of the processes and dynamics in this complex human–machine system. During the course of this essay, VR and AR will be used as current and advanced examples of customer interfaces in the study of HMI purposes in I4.0 (UIs). As a result, there are short definitions for a number of technologies. According to Paelke, AR applications are the adding of computer-generated information to a real-world setting. This involves combining numerical objects and data into a human user's vision, permitting direct engagement with data that has a straight spatial link to the real world and is easier to comprehend in that context. Users of VR apps are immersed in a completely digital universe with interaction options with parts of the synthetic environment, rather than keeping a view of the real-world landscape. Both AR and VR allow information to be mediated in a spatial relationship with the environment, but AR relies on data from a real-world scene that has been digitally improved, whilst VR apps rely exclusively on computer-generated content. The terms "resolution," "precision," and "frequency" are all incompatible [8]. These attributes can be customized for a specific HMI/operator to maximize material throughput. Using this probability paradigm, a separate low-throughput human–machine port can be described as a loud discrete information channel between a social purpose and a robot power system, as indicated in Figure 1.1.

A microprocessor or an FPGA chipset is used to create an HMI, and its software must include at least an application program and an operating system [9]. A typical HMI software system architecture is shown in Figure 1.2. Obviously, the software for the HMI must be compatible with the operating system and hardware of the system [10].

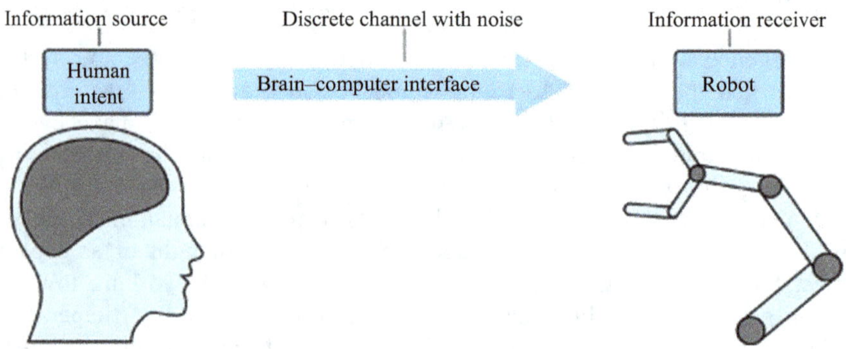

Figure 1.1 Modeling a low-throughput HMI as a noisy discrete information channel

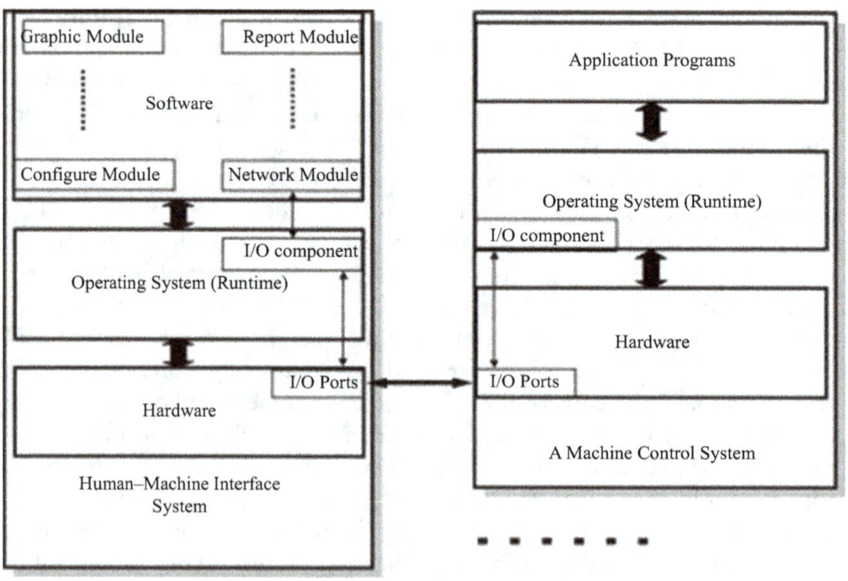

Figure 1.2 Architecture of HMI software system

1.3.1 Industry 4.0

Following the technological foundation for successful Industry 4.0 operations, the issue of what Business 4.0 truly implies arises. As a result, the following subsections clarify I4.0's meaning and scope by first discussing the term's beginnings and progression into a substitute for a quarter industrial rebellion, and then presenting paradigms that have been second hand in the literature toward highlighting the breadth and the notion of Industry 4.0 activities.

1.3.2 Cause and development of the term

Surprisingly, academics have not invented the word Industry 4.0, or its German equal, Industry 4.0, in scholarly papers. Instead, a portion of their high-tech strategy 2020, the German national government welcomed the Industrie 4.0 initiative, which had been begun a short time before by the German-Science Research Alliance. The phrase Industry 4.0 was originally offered to the general public and scientific group during the Fair in 2012. Industry 4.0 concept and concepts quickly originate their way into technical research, with an increasing number of studies devoted to various areas of the technology. Some of the initial contributions to Industry 4.0 were actually written in German rather than English. Only over time, as noted in the introduction, did Industry 4.0 become a stand-alone phrase suggesting a substitute for future IM and a fourth manufacturing revolution. Nevertheless, research from a variety of domains has used the name Industry 4.0 to describe breakthrough CPS- and IoT-integrated SM systems. Scientific articles regularly discuss the Manufacturing Internet, the Manufacturing Internet of Things, or simply IM, especially in the United States (Figure 1.3).

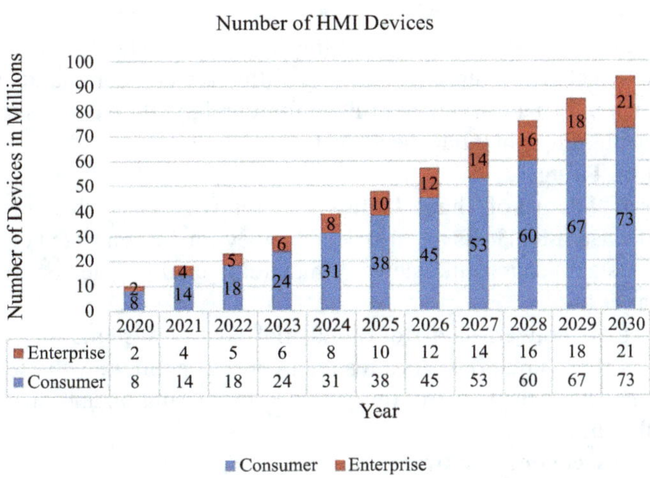

Figure 1.3 The growth of HMI software system

1.4 HMI features

HMIs can be used by operators to set up and break cycles, change put points, and operate other duties associated with controlling and engaging with a management process. Because the HMI is software-based, it can simply be altered and adjusted by replacing physical cables and controls with software parameters.

1.5 HMI benefits

Any software or equipment that allows a person or operator to interact with a machine, device, or system is referred to as an HMI. Machines have become more intelligent as the world has changed due to ongoing technological advancements and discoveries. HMIs have also been considerably developed in order to facilitate smoother interaction with the ever-changing technology.

HMIs can be found in a variety of places and applications, including automobiles, household appliances, portable devices (smart watches and smartphones), machinery, video games, and a wide range of industrial applications.

The benefits of HMI are as follows:

1. **Better productivity**
   An HMI improves the efficiency of a work in progress. Even if a person is capable of performing a task, this type of software/device allows for greater productivity. Using an HMI in the workplace allows for more work to be completed in less time.
2. **Fulfillment/hunt of happiness**
   Because of its ease of control, its application in entertainment and leisure systems has substantially enhanced people's lives by boosting comfortability. Gamers, for example, can play online games from various regions.
3. **Improve data saving/recording**
   They have exceptional data-recording capacities. The system to which an HMI is connected can automatically record data when commands are entered into it. Such information can be utilized for a variety of objectives in the future, such as debugging potential mechanistic issues.
4. **Internet of Things**
   The term "Internet of Things" refers to a group of gadgets that are all linked to the Internet. Because HMIs are also devices, they can be connected to the Internet. This enables applications like network monitoring and remote control access.
5. **Data conversion**
   Some HMIs convert data from industrial control systems into comprehensible visual representations. As a result, an operator can view the schematics of the systems and manage them, for example, by turning on and off pumps in an industrial plant.
6. **Decrease the cost of hardware**
   An HMI can help the industry save money on hardware such as consoles, panels, and wires. Hundreds of them can be replaced by an HMI, saving money.

1.6 Disadvantages of HMI

HMIs have almost no drawbacks. Their usefulness outweighs any shortcomings that may exist. They are really handy and well-liked nowadays. Any interface issue could arise as a result of the following.

1. **Security**
   HMI software systems had long been thought to be malware-free, but now that they are connected to the Internet, they are at risk of being hacked. A faulty HMI could be used by hackers to crash systems, resulting in the loss of sensitive data and security.
2. **Poor interface design**
   The majority of the machine and automotive accidents are blamed on human error, but when thoroughly investigated, it is frequently discovered that they were caused by poor HMI design, resulting in issues with machine-to-operator contact. A badly designed interface can lead to unfavorable operational outcomes and errors, which can have catastrophic ramifications for safety and performance.

1.7 Total global HMI dedicated AR/VR devices 2020–2030

The number of consumer HMI devices dedicated to AR and VR throughout the world is forecast to grow to more than 73 million by 2030, an increase of more than 65 million compared to 2020.

References

[1] Zhou J, Zhou Y, Wang B, and Zang J Human–cyber–physical systems (HCPSs) in the context of new-generation intelligent manufacturing. *Engineering* 2019; 5(4): 624–36.
[2] Wang L From intelligence science to intelligent manufacturing. *Engineering* 2019; 5(4): 615–8.
[3] Tao F, Qi Q, Wang L, and Nee AYC Digital twins and cyber–physical systems toward smart manufacturing and Industry 4.0: correlation and comparison. *Engineering* 2019; 5(4): 653–61.
[4] Zhou J, Li P, Zhou Y, Wang B, Zang J, and Meng L Toward new-generation intelligent manufacturing. *Engineering* 2018; 4(1): 11–20.
[5] Evans PC, and Annunziata M Industrial Internet: pushing the boundaries of minds and machines. *Report*. Boston, MA: General Electric Company; 2021.
[6] Tao F, Cheng Y, Xu DL, Zhang L, and Li BH CCIoT-CMfg: cloud computing and Internet of Things-based cloud manufacturing service system. *IEEE Trans Ind Inform* 2014; 10(2): 1435–42.

[7] Lee I, and Lee K The Internet of Things (IoT): applications, investments, and challenges for enterprises. *Bus Horiz* 2015; 58(4): 431–40.
[8] Pan Y Heading toward Artificial Intelligence 2.0. *Engineering* 2016; 2(4): 409–13.
[9] Thoben KD, Wiesner S, and Wuest T "Industrie 4.0" and smart manufacturing—a review of research issues and application examples. *Int J Autom Technol* 2017; 11(1): 4–16.
[10] Liu Y, and Xu X Industry 4.0 and cloud manufacturing: a comparative analysis. *J Manuf Sci Eng* 2017; 139(3): 034701.

Chapter 2

Human–machine interaction (HMI) technology—Malaysia National Technology Roadmap Industry4WRD leading the human intelligence transformation in smart manufacturing

Chee Fui, Wong[1]

Malaysia's manufacturing industry played a significant role in transforming the country into an industrialized nation. The manufacturing sector is constantly evolving from the conventional mass production lines using the intensive labor force to the use of smart manufacturing using human–machine interaction (HMI) technology to increase the productivity. This new phase of evolution of smart manufacturing using HMI technology has called for the need for a National Policy in Malaysia.

While smart manufacturing using HMI technologies has a potentially huge opportunity, there are concerns among many manufacturing companies as to how the HMI technologies would evolve the products and services that are produced and the business models or manufacturing processes that are used to produce them.

In 2018, the Malaysian government took the proactive measure and launched the "Industry4WRD: National Policy on Industry 4.0," with the objective to transform the Malaysian manufacturing sector to be resilient, smart, and systematic.

2.1 Smart manufacturing—a global overview

The "Fourth Industrial Revolution" is a revolutionary transformation of the industrial sector into smart manufacturing that includes the related economic activities as well as the way of living for each individual. The evolution of smart manufacturing will transform and change the traditional manufacturing process as well as the way the manufacturing business is conducted.

The "Fourth Industrial Revolution" or "Industry 4.0" is the era of "cyber physical systems (CPSs)" which comprises smart machines, storage systems, and

[1]Ir.Ts. Wong Chee Fui, Universiti Tunku Abdul Rahman, Selangor, Malaysia

production facilities capable of autonomously exchanging information, triggering actions, and controlling each other independently [1].

The transformation toward Smart Manufacturing to improve the manufacturing competitiveness and enhance the process efficiency is in line with the "United Nation Sustainable Development Goals" (SDGs) particularly Goal 9 on "Industry, Innovation and Infrastructure" and Goal 12 on "Responsible Consumption and Production."

Globally, most economies have recognized the importance of smart manufacturing in improving the process of making things better through the innovation in products and services, as well as the design, planning, and execution of the manufacturing processes. However, in order to encourage the transformation of traditional manufacturing toward the adoption of smart manufacturing, a clear and definitive policy and incentive from the government is necessary.

Internationally, many countries have introduced policies and roadmap on the adoption of new Industry 4.0 technology toward the transformation of smart manufacturing. Some examples of the national policy, roadmap, and initiatives adopted by the global economies are described as follows [2]:

1. Germany—"Industrie 4.0"
   The Industrie 4.0 policy focuses on the autonomy, interoperability, and sustainability with a focus on the industrial production sector in particular manufacturing, mining, and utilities sector.
2. Europe—"Roadmap for Industrial Process Automation"
   The European "Roadmap for Industrial Process Automation" promotes the core technologies such as human–machine interfaces and machine-to-machine communications, safety-human, machine, environment as well as the adoption of Industry 4.0 technologies.
3. United States—"Advanced Manufacturing Partnership 2.0"
   United States "Advanced Manufacturing Partnership 2.0 (AMP2.0)" focuses on innovation, talent pipeline as well as improving the business climate for the manufacturing sector.
4. China—"Made in China 2025"
   The Republic of China has introduced the "Made in China 2025" roadmap to develop the China manufacturing sector by focusing on research and development (R&D) and technological innovation.

This chapter focuses on the national policy that is being adopted in Malaysia toward smart manufacturing and transformation of the Malaysian manufacturing sector to be resilient, smart, and systematic.

2.2 Malaysia smart manufacturing using HMI technologies—the call for a national policy in Malaysia

Malaysia's manufacturing sector played a significant role in turning Malaysia into a global player in the international value chain. According to the "Eleventh Malaysia

Plan 2016–2020," the manufacturing sector is projected to grow at 5.1% per annum. This is projected to contribute 22.5% to Malaysia's GDP and 18.2% of total employment by 2020 [1, p. 8–21].

According to the Department of Statistics Malaysia (DOSM), "2018 Annual Economic Statistics—Manufacturing" report, the manufacturing sector's gross output in 2017 is RM 1,275.8 billion as compared to RM 1,120 billion in 2015, with an average increment of 5.7% per annum. The Malaysian manufacturing sector contributed to the employment of 2.2 million persons with a total salary and wages paid of RM 74.9 billion in the year 2017, as shown in Figure 2.1 [3].

The manufacturing sector in Malaysia is constantly evolving from the conventional mass production lines using intensive labor force to the utilization of smart manufacturing using HMI technology to increase productivity. This new phase of evolution of smart manufacturing using HMI technology has called for the need for a National Policy in Malaysia.

Smart manufacturing is transforming the future of manufacturing through the adoption of new technology with high efficiency, flexibility to sustain the sector competitiveness.

In Malaysia, there is an urgent need to drive the competitiveness of the manufacturing sector toward a higher value-added process through the utilization of the HMI technologies, digitalization, effective resources planning, and implementation of advanced manufacturing technologies. The smart manufacturing sector is paramount important to increase productivity and efficiency, create high skills jobs and talent, and enhance innovation to pave the way toward economic growth and social

Figure 2.1 Principal statistics of manufacturing sector, 2015 and 2017. Source: "2018 Annual Economic Statistics" Department of Statistics Malaysia (DOSM) [3, p. 14]

well-being. This would attract more high technology investments to Malaysia as the primary destination for smart manufacturing internationally.

While smart manufacturing using HMI technologies has a potentially huge opportunity, there are concerns among many manufacturing companies as to how the HMI technologies would change both the services and products that are produced and the business model or manufacturing process that are used to produce them.

The Malaysian government has realized this call for the transformation of conventional manufacturing to smart manufacturing using HMI technologies and is working toward how best to react to this need.

In 2018, the Malaysian government took the proactive measure and launched the "Industry4WRD: National Policy on Industry 4.0," with the aim to transform the Malaysian manufacturing sector to be resilient, smart, and systematic.

2.2.1 Industry4WRD: Malaysia national policy roadmap for HMI technologies

The rapidly changing international manufacturing environment calls for the manufacturing sector to review its current strategies and approaches to remain relevant and competitive.

In 2018, the Malaysian government launched the "Industry4WRD: National Policy on Industry 4.0" with the aim of transforming the Malaysian manufacturing sector to be resilient, smart, and systematic.

The Malaysia "Industry4WRD: National Policy on Industry 4.0" roadmap highlighted 13 broad strategies for the country to embark into a journey that will transform the manufacturing sector environment into smart manufacturing using HMI technologies.

2.2.2 Shift factors in Malaysia national policy roadmap for smart manufacturing using HMI technologies

The "Industry4WRD: National Policy on Industry 4.0" roadmap is anchored on three-shift factors: people, process, and technology. These shift factors need to be balanced and optimized in order to achieve the objectives of the national policy framework.

2.2.2.1 People shift factors for smart manufacturing using HMI technologies

The "Industry4WRD: National Policy on Industry 4.0" roadmap focuses on the human factor and the entire organization with an emphasis on the strategies toward having a suitable set of workforce. The focus on the people factor can be achieved through the development of human capital and sustainable transformation processes with regards to organizational strategies, collaboration, and governance.

The national policy roadmap puts people and the entire organization as a priority.

Malaysia's national policy has given a priority to develop, upgrade and retain the required talent and skills required as the "People Shift Factor."

2.2.2.2 Process shift factors for smart manufacturing using HMI technologies

"Process shift factors" for smart manufacturing using HMI process shift factors is an important factor to improve the manufacturing productivity and the whole business activities by promoting strategic and smart public–private partnerships for smart manufacturing using HMI technologies.

The national policy aims to provide a favorable business ecosystem and facilitate public–private partnerships by enhancing both business and manufacturing processes.

2.2.2.3 Technology shift factors for smart manufacturing using HMI technologies

The convergence of technologies and new technology advancement is increasing the competitiveness of the manufacturing industry. The manufacturing sector is beginning to implement and adopt smart manufacturing processes where hardware advancements are integrated with new software, data analytics as well as smart sensors. This technology shift will produce smarter processes, products, as well as increase the connectivity between customers and suppliers.

The implementation of HMI technologies is key drivers for conventional manufacturing to shift toward smart manufacturing. The rate of technological advancement and the convergence of technologies are led by applications of HMI technologies in the manufacturing sector, such as artificial intelligence (AI), big data analytic, advanced material, Internet of Things (IoT), and cloud computing.

2.3 Convergence of emerging technologies

Smart manufacturing with HMI technologies is rapidly evolving with the convergence of emerging new technologies. These new emerging technologies are transforming conventional manufacturing into smart manufacturing. The new enabling technologies introduce a new dimension to the industrial ecosystem and environment, which produce a dramatic increase in manufacturing productivity.

The digitalization of the production-based smart manufacturing is driven by the new technological drivers [4, p. 17].

1. Additive manufacturing

   Additive manufacturing is advancing with the use of new materials, opening completely new possibilities. It is revolutionizing traditional production, aided by a recent surge in metal additive printing.
2. Artificial intelligence

   AI is a concept that is made up of numerous subfields such as machine learning, which focuses on the development of programs that can teach themselves to learn, understand, reason, plan, and act when exposed to new data in the right quantities. AI technology will supplement the smart factory toward networked factories, in which data from supply chains, design teams, production lines, and quality control are linked to form highly integrated and intelligent engines.

3. Big data analytic

   Big data techniques are being applied in the manufacturing industry to improve customer experience and product quality, realize energy efficiency, and conduct predictive maintenance. It is now possible to collect masses of data from several different sources to direct decisions that anticipate product or equipment failure.

4. Advanced material

   New materials and nano-structures are being developed, allowing for beneficial material properties, for example, shape retention and thermo-electric efficiency. Together with additive manufacturing technologies, it will allow for massive customization and development of products that were previously not possible.

5. Cybersecurity

   Industrial communication is expanding and strongly connected, as such, digital security becomes a critical aspect that must not be overlooked in the industrial environment. It has now become more complex as it consists of connected devices and environments that cannot be protected by traditional cybersecurity approaches. Current cybersecurity has largely been developed for IT-centric devices and environments.

6. Simulation

   While many engineers are already familiar with simulations in the field of product design, the advent of augmented reality, AI, and big data are expected to take simulations to the next level. It will be possible to simulate manufacturing processes using different production settings to find the optimal way to manufacture a product. Simulations can also be used to test product usage under different operating environments using different types of materials.

7. Cloud computing

   Past industrial revolutions required significant capital as a ticket for entry. With cloud computing, many of the Industry 4.0 technologies can be made available to even smaller companies as a utility with minimal upfront capital investment. Companies can leverage cloud-based product design, simulation, AI, and big data solutions to improve their production processes and build products better suited for their customers.

8. Augmented reality

   While the initial augmented reality technologies are still in nascent stages, they are advancing at a rapid pace. Some of the first applications can be found in the delivery of information and training, for example, augmented reality can be used to deliver part replacement instructions to maintenance staff in the field.

9. Internet of Things (IoT)

   Industry 3.0 brought in an era of computing and interconnectedness but it often relied on humans to make even the most trivial decisions. Industry 4.0 technologies embody an unprecedented proliferation of sensors and connectedness among these sensors. Combined with other technologies such as AI and big data, it is now possible to envisage entirely autonomous systems that revolutionize manufacturing.
10. Autonomous robot

    Robots and machinery are rapidly transformed and enhanced in their technology and innovation. Robots can now work independently, including learning on the work scope and collaborating up with other robots and humans. This new technology advancement allows the robots to think, act, and react autonomously as well as to conduct remote decision making. The application of autonomous robots will contribute to a company's competitiveness, productivity, and profitability.
11. System integrations

    System integration is implemented in both vertical systems (within the value chain of the industry) and in horizontal systems (across multiple value chains). System integrations will achieve end-to-end digital integration and transformation across the whole value chain.

2.4 Malaysia readiness for Industry 4.0

The "Readiness for the Future of Production Report 2018," published by the World Economic Forum (WEF) and A.T. Kearney, is a global assessment of 100 countries and the report positions Malaysia in the "Leading" quadrant as shown in Figure 2.2. The "Leading" quadrant comprises countries with a "strong current production base" and who is "positioned well for the future." As shown in Table 2.1, Malaysia and China are the only two nonhigh-income countries in the "Leading" quadrant.

At international level, Malaysia has been acknowledged as fairly strong and competitive in the manufacturing industry. The "Global Manufacturing Competitiveness Index 2016" (by Deloitte) positioned Malaysia at 17th place among 40 countries. The report also projected Malaysia will progress four ranks to 13th by 2020 [5, p. 4].

Despite Malaysia's strong current position, the manufacturing industry in Malaysia needs to evolve fast toward smart manufacturing with HMI technologies as it faces challenges and competition from emerging countries with lower cost and rapidly changing technologies.

There is a new opportunity for Malaysia as international value chains and geographies of production are currently shifting and relocating their production industry. Malaysia will need to embark and transform its manufacturing industry to face the competition from other emerging countries in the region.

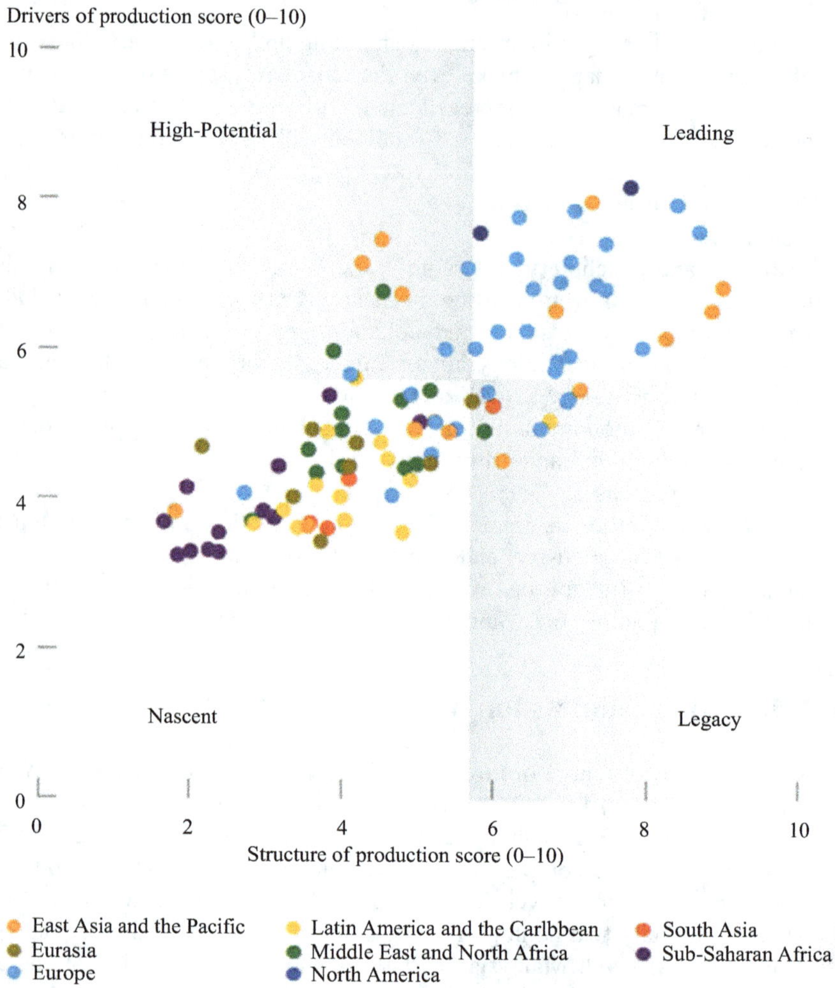

Note: Average performance of the Top 75 countries is at the intersection of the four quadrants.

Figure 2.2 Global map of readiness assessment results 2018. Source: Readiness for the Future of Production Report 2018 [6, p. 11]

Table 2.1 Readiness for the Future of Production Assessment Results 2018 (leading countries)

		Structure of Production		Drivers of Production	
Region	Country	Score	Rank	Score	Rank
Leading Countries					
●	Austria	7.46	9	6.79	18
●	Belgium	6.51	24	6.80	17
●	Canada	5.81	33	7.54	7
●	China	8.25	5	6.14	25
●	Czech Republic	7.94	6	6.01	26
●	Denmark	6.29	27	7.20	10
●	Estonia	5.75	34	6.00	27
●	Finland	7.00	14	7.16	11
●	France	6.87	18	6.89	14
●	Germany	8.68	3	7.56	6
●	Ireland	7.34	10	6.85	15
●	Israel	6.43	25	6.24	23
●	Italy	6.99	15	5.90	30
●	Japan	8.99	1	6.82	16
●	Korea, Rep.	8.85	2	6.51	21
●	Malaysia	6.81	20	6.51	22
●	Netherlands	6.32	26	7.75	5
●	Poland	6.83	19	5.83	31
●	Singapore	7.28	11	7.96	2
●	Slovenia	6.80	21	5.71	32
●	Spain	6.05	29	6.23	24
●	Sweden	7.46	8	7.40	9
●	Switzerland	8.39	4	7.92	3
●	United Kingdom	7.05	13	7.84	4
●	United States	7.78	7	8.16	1

[a]*Note*: Malaysia and China are the only two nonhigh-income countries in the "Leading" quadrant.
Source: Readiness for the Future of Production Report 2018 [6, p. 12].

2.5 Industry4WRD—framework

The Malaysia "Industry4WRD: National Policy on Industry 4.0" is a proactive initiative undertaken by the Ministry of International Trade and Industry (MITI) to transform the Malaysian manufacturing sector to be smarter, resilient, and more systematic.

The Industry4WRD Framework includes Malaysia's 10 years for the manufacturing sector, the specific national goals to be achieved, the shift factors, and the specific enablers that determine the strategies, policies, and action plans. This framework is illustrated in Figure 2.3.

2.5.1 Industry4WRD objectives

The objectives of the "Industry4WRD: National Policy on Industry 4.0" are as follows [4, p. 37]:

1. Attract stakeholders to Industry 4.0 technologies & processes, and further increase Malaysia's attractiveness as a preferred manufacturing location.

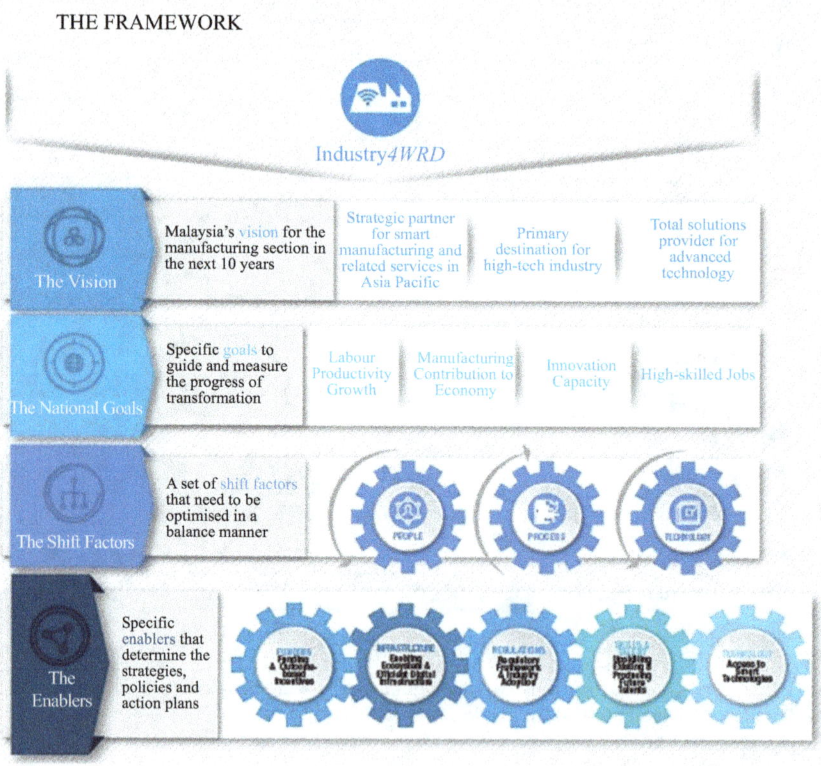

Figure 2.3 Industry4WRD framework. Source: Industry 4WRD: National Policy on Industry 4.0; MITI [4, p. 38]

2. Create the right ecosystem for Industry 4.0 to be adopted and align existing and future development initiatives.
3. Transform Malaysia's industry capabilities in both a holistic and an accelerated manner.

2.5.2 Industry4WRD strategic enabler

The strategic enablers that are used to determine the strategies, policies, and action plans in the Industry4WRD are as follows:

1. Funding—funding and outcome-based incentives
2. Infrastructure—enabling ecosystem and efficient digital infrastructure
3. Regulations—regulatory framework and industry adoption
4. Skills and Talent—upskilling existing and producing future talents
5. Technology—access to smart technologies and standards

2.5.3 Industry4WRD readiness assessment

The Government has allocated RM 210 million from 2019 to 2021 to support the migration and transition to Industry 4.0. The Malaysia Productivity Corporation (MPC) will carry out "IndustryWRD Readiness Assessments" to assist up to 500 small–medium enterprises (SMEs) to migrate to Industry 4.0 technologies.

"Industry4WRD Readiness Assessment (Industry4WRD-RA)" is a comprehensive program to assist companies in their capabilities and readiness assessment to adopt Industry 4.0 technologies and processes. The assessment evaluates the companies' capabilities and gaps using a predetermined set of indicators. The evaluation from the assessment will be utilized to enable the companies to prepare the feasible strategies and plans to adopt Industry 4.0.

The Industry4WRD-RA will assist the companies to:

1. Determine their state of readiness in the adoption of Industry 4.0 technologies.
2. Identify the gaps and areas of improvement for Industry 4.0 adoption as well as opportunities for productivity improvement and growth.
3. Develop feasible strategies and plans to perform outcome-based intervention projects.

"Industry4WRD Readiness Assessment" will produce an assessment report which includes the finding on the readiness profile and the recommendation for improvement plan.

2.6 Case study—Pentamaster—embracing Industry 4.0 automation

2.6.1 Background

Pentamaster Corporation Bhd is a manufacturing company in Malaysia that supplies equipment and provides automation manufacturing solutions to the

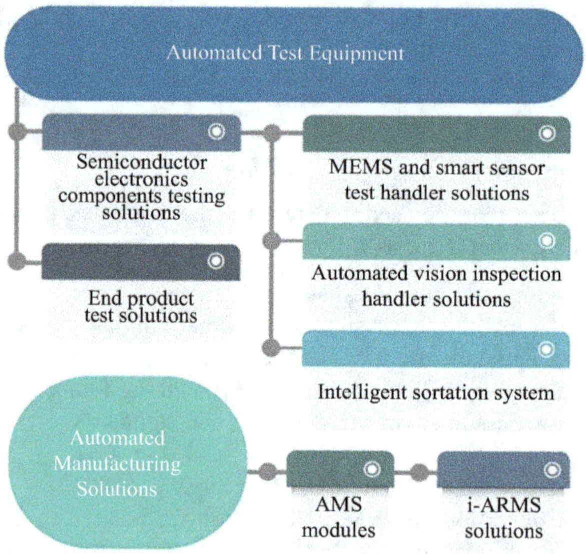

Figure 2.4 Digitized product by Pentamaster. Source: "The Race Towards Industry 4.0"; MPC [7, p. 27]

manufacturing and semiconductor industry. The company has a workforce of more than 400 staff.

The case study of Pentamaster achievement in embracing Industry 4.0 was published and reported by MPC [7]. MPC is an autonomous body under the MITI, where MPC roles are defined under the Malaysia Act of Parliament, known as MPC Act (Incorporated) 1966. (MPC website) (Figure 2.4).

2.6.2 Pentamaster implementation of Industry 4.0

Based on MPC [7], Pentamaster has implemented the Industry 4.0 technologies in the following areas:

1. Developed a software that can expedite the process to produce real-time data.
2. Optimized business planning through data collection and analysis.
3. Asset utilization based on the effective in-house solution.
4. Quality checking through autonomous robot.
5. Implemented real-time order and product quality status update for the customer.

The digitalized product by Pentamaster in the implementation of Industry 4.0 technology is shown in Figure 2.4.

2.6.3 Pentamaster implementation of Industry 4.0

The study by MPC [7] has reported that numerous positive attributes were achieved by Pentamaster especially on the efficiency and productivity after implementation of Industry 4.0 technology.

The result of cloud computing in Pentamaster has produced an increase in productivity by 5%, whereby the total machine downtime has reduced by 40%. The skills of the technical professional have recorded a rise by 35%. The inventory holding cost registered a deduction by 20% and maintenance costs and cost for quality also show reduction by 10% and 15% respectively [7].

The case study of the adoption of industry 4.0 technologies by Pentamaster has generated the following results:

1. Customized products produce higher margin compared to the mass manufacturer products.
2. Product innovation has generated a profit increment of 30% annually.

The case study of Pentamaster in embracing the industry 4.0 automation is based on the report published by MPC [7] entitled "The Race Towards Industry 4.0."

2.7 Conclusion—moving forward

The transformation of Malaysia's manufacturing industry is a long-term process, where many manufacturing firms have been facing significant challenges for decades. A clear national policy and supportive ecosystem are required to provide the manufacturing sector with stability and allow the companies to implement these long-term transformation strategies.

New HMI technology-based models of production are fostered from the new disruptive technologies. As the HMI technology and Industry 4.0 technologies become mature, the cost and barriers to adopt these technologies will decline. Malaysia manufacturing companies need to invest and innovate in the new HMI technologies to improve the productivity, quality, and effectiveness in order to remain competitive.

The Malaysian government has a comprehensive policy to transform the conventional manufacturing sector into a smart manufacturing sector that adopts the HMI technologies through Malaysia "Industry4WRD: National Policy on Industry 4.0."

This Industry4WRD policy framework aims to achieve Malaysia's 10 years vision for the manufacturing sector to be the smart manufacturing and related services strategic partner in the Asia Pacific; to be the primary high-tech industry destination and to be the total solutions provider for the advanced technology.

References

[1] Economic Planning Unit, Malaysia (EPU). (2015), Eleventh Malaysia Plan, 2016–2020 – Anchoring Growth on People. https://www.talentcorp.com.my/clients/TalentCorp_2016_7A6571AE-D9D0-4175-B35D-99EC514F2D24/contentms/img/publication/RMKe-11%20Book.pdf (Accessed 10 Nov 2021).

[2] CIDB. (2020), *Construction 4.0 Strategic Plan (2021–2025)*. Construction Industry Development Board (CIDB), Malaysia.

[3] Department of Statistics Malaysia (DOSM). (2019), 2018 Annual Economic Statistics Manufacturing Report. https://newss.statistics.gov.my/newss-portalx/ep/epDownloadContentSearch.seam?cid¼518903 (Accessed 10th Nov 2021).
[4] Ministry of International Trade and Industry, Malaysia (MITI). (2018), Industry 4WRD: National Policy On Industry 4.0. http://www.miti.gov.my/miti/resources/National%20Policy%20on%20Industry%204.0/Industry4WRD_Final.pdf (Accessed 10 Nov 2021).
[5] Deloitte. (2016), 2016 Global Manufacturing Competitiveness Index. https://www2.deloitte.com/content/dam/Deloitte/global/Documents/Manufacturing/gx-global-mfg-competitiveness-index-2016.pdf (Accessed 10 Nov 2021).
[6] World Economic Forum (WEF). (2018), Readiness for the Future of Production Report 2018, World EconomicForum's System Initiative on Shaping the Future of Production. http://www3.weforum.org/docs/FOP_Readiness_Report_2018.pdf (Accessed 10 Nov 2021)
[7] Malaysia Production Corporation (MPC). (2018), The Race Towards Industry 4.0. Selangor: Malaysia Production Corporation. http://www.mpc.gov.my/wp-content/uploads/2019/01/Ready-Set-Go-The-Race-Towards-Industry-4.0.pdf (Accessed 10 Nov 2021).

Chapter 3

Challenges and impact of human–machine interaction systems in smart manufacturing

Alex Looi Tink Huey[1,2]

The global digital economy today is quickly becoming the main driving force of economic growth. The accelerating pace of digital economy requires businesses and organizations to embrace digital transformation. Businesses and organizations that embrace digital solutions have greater resiliency in the face of economic uncertainty, a sudden drop-off in demand, and faster recovery [1].

Business sectors such as healthcare, manufacturing, real estate, retail, and financial services can leverage digital business models and digital technologies to enable a conducive and sustainable doing-business environment. Digital business models and digital technologies such as the Internet of Things (IoT), big data, artificial intelligence (AI) (robotics, automation, accurate diagnosis in healthcare systems, market and financial data analysis, self-driving cars, etc.), and additive manufacturing (aerospace and automotive components manufacturing, biocompatible materials for customized implants, life-saving devices in the medical sector, efficient and environmentally friendly energy and power sector components manufacturing, etc.) are driving the change of current business models and shifting the global economics and market structures.

The four main pillars of digital technologies are data, computational power, and connectivity (machine-to-machine and machine-to-product connectivity); analytics and intelligence (advanced analytics and AI to shift from detect to predict and prevent); human–machine interaction (HMI) (industrial automation such as collaborative robots, automated guided vehicles, and augmented reality); and advanced production methods (integrated and accelerated prototyping and manufacturing) [2].

According to McKinsey, Industry 4.0: Reinvigorating ASEAN Manufacturing for the Future, some of the key reasons companies are struggling and holding back the implementation of Industry 4.0 are due to the inability to define a business plan to meet digital strategic goals, siloed data (stored data that is not available to the entire organization but only some parts of it), no integration across business units (BUs), no digital talent or capacity building to execute the digital transformation

[1]Malim Consulting Engineers Sdn Bhd, Kuala Lumpur, Malaysia
[2]International Electrotechnical Commission (IEC) Young Professional 2018, Busan, Republic of Korea

Figure 3.1 What will manufacturing look like in the future?

road map, lack of awareness and fear of cybersecurity risk, and lack of coordination and transparency across BUs [2].

Implementing Industry 4.0 requires a clear understanding and a well-designed digital strategy roadmap, a shared and common vision of digital transformation and its potential, building an agile and competent digital talent pool, and a policy and education environment that encourages its implementation (Figure 3.1).

3.1 Smart manufacturing

Manufacturing is a process that converts raw materials into physical products or consumables through the use of labor, machinery, tools, and biological or chemical processes. In a factory or a plant, the manufacturing process involves a specified combination or sequence of steps by a particular operation on a particular material or component by managing fragmented communications protocols and manufacturing practices.

Today, enablers such as new and advanced digital technologies, sustainability, and process optimization have encouraged the digital transformation of the manufacturing industry, to become adaptive, fully connected with embedded sensors (IoT) in manufacturing equipment and product components, automation, ubiquitous cyber-physical systems (CPS), cloud computing, and analysis of all relevant data and information by creating more value add-ons and reducing costs [3]. CPS are integrations of computer systems and physical processes or mechanisms. Embedded computers or networks control or monitor these physical processes or mechanisms using computer-based algorithms. Applications of CPS include industrial robotics, autonomous vehicles, smart grid, and so on.

A report from the McKinsey Global Institute, The Internet of Things: Mapping the Value Beyond the Hype, estimated that the IoT applications could have a potential economic impact as much as US$ 3.9 trillion to 11.1 trillion annually in

2025 [4] in nine settings: factories (operations and machinery optimization), retails, worksites (safety and health, operations optimization), human (health and fitness), outside (logistics and navigation), cities (public health and transportation), vehicles (autonomous vehicles and condition-based maintenance), home (security and chore automation), and offices (security and resources).

Figure 3.2 illustrates the traditional manufacturing industry which is mostly operating manually on legacy systems and data silos. Take, for example, manual checking of the motor bearings and replacing them every 30 days regardless of the condition is time consuming and labor-intensive which in turn, increases the operations and maintenance cost. Without proper predictive measures in place, risk of equipment or machinery failure increases. Management will have limited to no data to guide decision-making as well as limited data for business budgeting and growth in the absence of data-driven manufacturing system. The manufacturing processes have no integration with one another and limited technology involvement. Manual recording can lead to inaccurate tracking processes and limited visibility on operations and productivity information.

The smart manufacturing industry on the other hand shown in Figure 3.3 is data-driven based on digitized processes and operations. Smart systems, connection and integrated operations, and centralized data provide complete real-time data for fast and efficient decision-making. Management will also have transparency and visibility on the factory's operations and production information to provide improved customer experience and plan for future factory or production line expansion and business budgeting. Digital technology enablers such as IoT and AI allow factories to track their assets accurately combined with improved resource

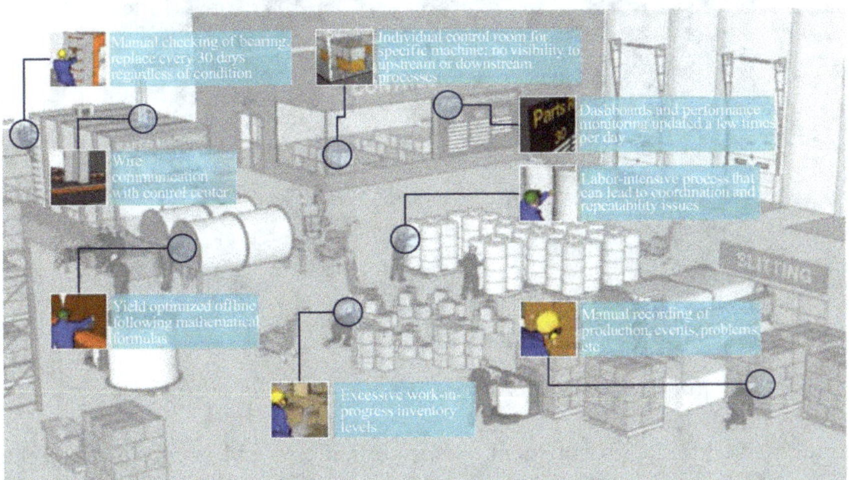

Figure 3.2 Traditional manufacturing processes—Industry 4.0 (McKinsey Digital—Industry 4.0: Reinvigorating ASEAN Manufacturing for the Future) [2]

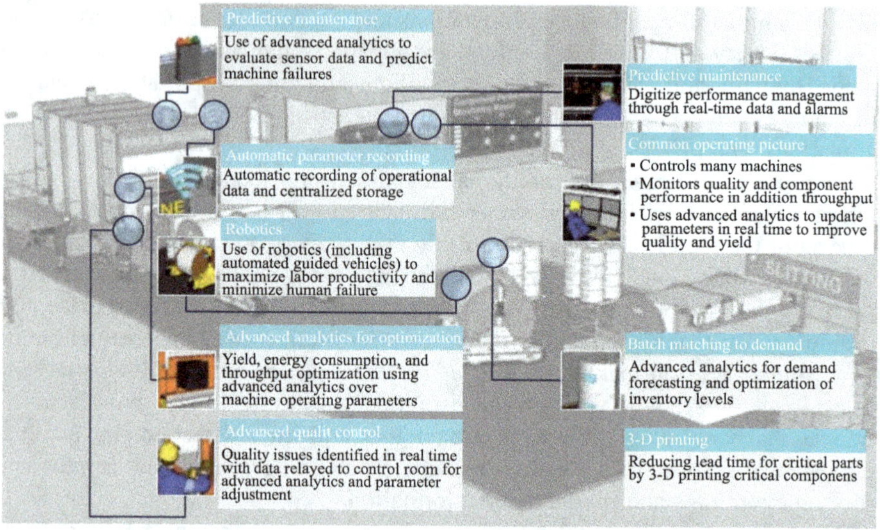

Figure 3.3 Industry 4.0, optimized world-class smart manufacturing system—Industry 4.0 (McKinsey Digital—Industry 4.0: Reinvigorating ASEAN Manufacturing for the Future) [2]

Figure 3.4 Robotics, automation, and artificial intelligence in Industry 4.0

utilization (workforce, procurement, energy efficiency, and energy demand management). The digital transformation in the manufacturing sector allows companies to take a more holistic approach to find responses to process innovation, greater customer-centric strategy, smart data-gathering, customer-specific adaptation, reducing manufacturing cost, shortened lead time, and efficient use of resources (Figure 3.4) [5].

Smart manufacturing encapsulates the whole manufacturing value chain and life cycle of a product, from concept (simulated in a virtual environment), research and development (R&D), design to production, marketing, distribution, and logistics [6]. It is a data and technology-driven approach that achieves efficiencies by integrating data from multiple technical systems between departments, factories, and even enterprises across domains, hierarchies, and geographic boundaries.

3.2 HMI

HMI refers to the communication and interaction between a human and a machine via a user interface which allows humans to control machines [7]. User services are a fundamental purpose for the existence of a smart system, and the user interface is essential to the access of those services [8]. The primary purpose of the user interface is to provide the facilities of the machine accessible to humans by providing the necessary set of commands and services to get work done conveniently and efficiently. The simplification of these applications reduces the user's learning curve for new interfaces. Today, natural user interfaces allow humans to control machines through natural and intuitive behaviors.

A well-designed user interface enhances user's experience further providing user friendly, powerful, and efficient control of the machine. A poorly designed user interface in contrast will be used reluctantly further diminishing its potential value to the user.

Industry 4.0 has brought a major change in human interaction with machines. The term "machine" in this case study refers to all kinds of dynamic technical system including automation (supervision and control system), decision support equipment, and software [9].

Figure 3.5 depicts the levels of manufacturing control system where machines used to occupy Level 1 (Sensing and Control Level), are now taking some functionalities from the upper levels such as Level 3 (Manufacturing Level, System and Management Level) and Level 4 (Enterprise Resource Planning) [9]. Industry 4.0 brings a paradigm shift in data-driven manufacturing system where machines now support users in business management (energy, procurement, predictive engineering, etc.) and enterprise resource planning (ERP) through digital technologies such as IoT, CPS, and cloud computing.

3.2.1 Socio-technical approach in HMI

Human context or human factor and its interaction with digital technologies are becoming increasingly more important to the point where digital technologies are becoming more human-aware [10]. Human-aware technology is a system that combines both human intelligence and AI algorithms. Therefore, Industry 4.0 initiative in smart manufacturing requires a socio-technical approach in the design and development of digital technologies to unlock new potential where human skills, reflective and conceptual learning capabilities are combined with the precision and speed of the computers [12] (Figure 3.6).

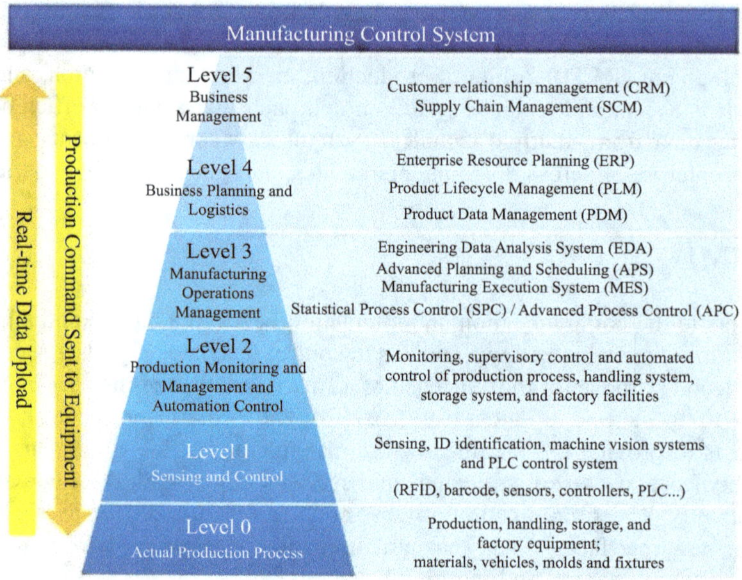

Figure 3.5 The levels of manufacturing control system [9]

Figure 3.6 Human beings, machines, and cyber-physical devices communicating with one another as naturally as in a social network

Having said that cyber-physical human systems (CPHSs) involve interconnected computers, cyber-physical devices, and people communicating and interacting with one another [11] as naturally as in a social network over small or vast distances. In short, CPS is a system that takes human responses such as intents, psychological states, emotions, gestures, voice, and cognitive science into consideration.

Mobile technology as a digital solution platform in manufacturing through the application of smartphones, tablets, and smart watches has enabled factory management and workers to easily connect and communicate via cloud technologies to create an agile, responsive, and truly collaborative environment. This also means that a worker is always connected to work and being contactable, observable, and controllable. The boundaries between private social life and work life are becoming increasingly blurred, tenuous, and indefinable. This leads to a sociological situation known as "a panopticon-type effect" whereby one behaves in a manner that allows for the possibility that one may be watched at any time or at all times [12]. A consequence of this effect is that the freedom for workers to be creative and innovative free from such observation is lost.

There are significant differences between product innovation and process innovation. Product innovation focuses on introducing new or significantly improved products or services, whereas process innovation focuses on the innovation of manufacturing facilities, skills, and technologies. Product and process innovation are known to have opposite employment effects. Product innovation, if developed in contexts of high demand growth, has positive effects, however, process innovation, often adopted to increase manufacturing productivity with the introduction of process automation and reducing labor costs, leads to job losses for lower-skilled workers [12]. The impact of automation on unemployment and mental health is one of the pressing problems in today's society globally, with growing pressures on workers to improve both the quantity and quality of their job performances.

According to a report by PricewaterhouseCoopers, PwC, the biggest challenges and barriers for businesses and organizations in adopting industrial digital technologies remain a lack of digital culture, digital talent and skills, and a clear digital operations vision [13]. In Germany, a recent study that explored the impact of digital change on skills and employment suggests that the "ability to plan and organize, to act autonomously," combined with business-specific and job-specific working experience, are crucial for the successful implementation of digital transformation.

The transition to digitalization and automation revolution has been accelerated by the COVID-19 pandemic, and businesses emerging from the crisis are facing major changes in customer behaviors and preferences, and workplace physical distancing practice as highlighted by McKinsey & Company (Figure 3.7) [14]. The manufacturing industry is reconfiguring its supply chains and their production lines into dynamic, interconnected systems; meanwhile, the service industry is adapting to prioritize digital-first customer interaction and engagement, as well as contactless operations. These digital transitions will have significant effects on the requirements for workforce skills and capabilities. Businesses will require people with the right set of skills to develop, manage, and maintain their automated equipment and machinery, digital processes and mechanisms, and to be able to do the jobs that computers or machines cannot. While computers or machines have become better at replicating human tasks (both predictable and repetitive nature of

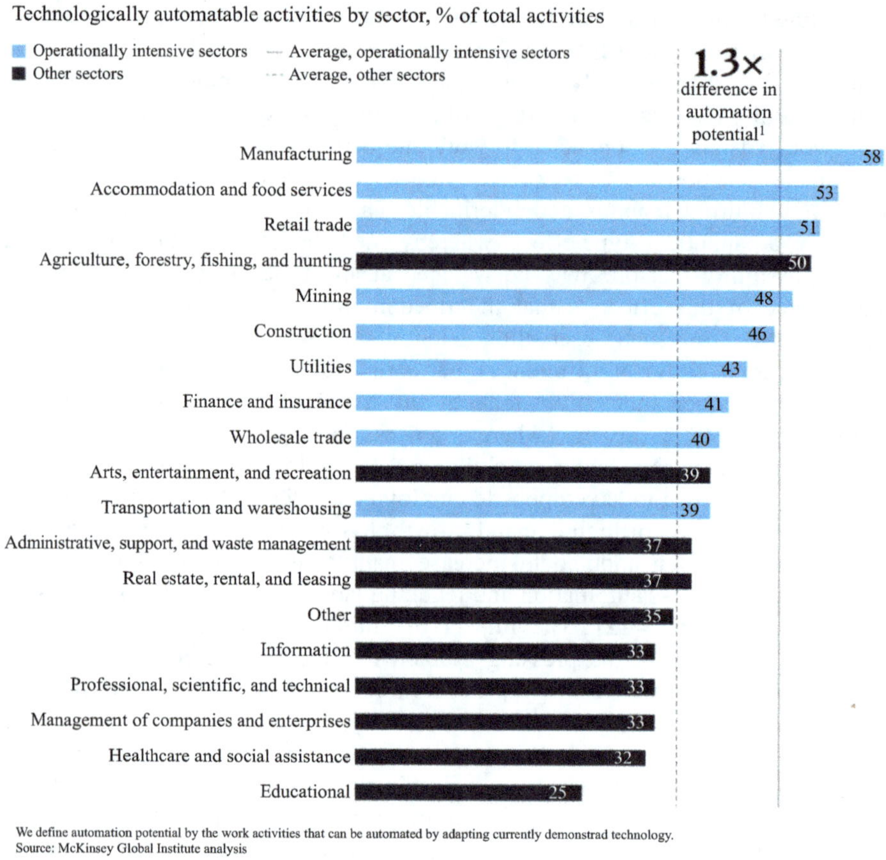

Figure 3.7 Operations-intensive sectors have 1.3 times the automation potential of other sectors (McKinsey & Company—Building the Vital Skills for the Future of Work in Operations) [14]

operational tasks), they have a hard time replicating logical thinking and problem solving, social and emotional capabilities, providing expertise, coaching and developing others, innovation, and creativity [15].

Education systems will need to evolve together with policy makers working to improve basic skills in the fields of science, technology, engineering, and mathematics (STEM), and putting emphasis on critical and systems thinking, complex communication, problem-based, hands-on learning, software development, big data collection and analysis, innovation, and creativity in line with digital transformation and technology trends. There is a growing need for multi-disciplinarity, where workers will need to have a more systemic view and, more importantly, soft skills.

It is also essential to understand that the adoption of digital transformation requires us to overcome human psychological barriers aside from education systems, government policy intervention, technological progress, and disruptive digital technologies. Digital transformation efforts must go hand-in-hand with people-centric culture and approach to ensure its successful implementation in the process of change.

3.2.2 Framework of HMI

It is said that as computation evolves, humans may no longer use computer devices, and instead, HMI will no longer require direct user attention and will become intuitive and behaves as if it is second nature [10]. Most CPSs involve control loops which keep humans as external elements to the control system, e.g., aircraft pilots deciding whether to engage the autopilot or when to take manual control of the plane, and cruise control systems for autonomous vehicles which simply maintain the desired speed without taking the driver's behavior into consideration.

Today, we are advancing toward CPHSs where human presence and behavior are no longer seen as external elements but have become a key part of the control system. By considering humans as an integral part of the system, the control loop's performance and accuracy can be greatly improved [10]. In this context, the cruise control systems for autonomous vehicles can now take the driver's psychological and physiological states into consideration (e.g., fatigue, attention levels, and irritation) and produce outputs such as alarms or suggesting cruise control activation, further improving driving safety.

Hence, it is crucial to develop and integrate reliable and accurate human behavior modeling techniques that allow computers or machines to learn and predict human behavior. The huge challenge to the development of CPHSs is the modeling of complex behavioral, psychological, and physiological aspects of human nature as human behavior is quite hard to predict (Figure 3.8).

To increase the technological readiness levels and improve human–machine collaboration of CPHSs, it is essential to research and study the key aspects of

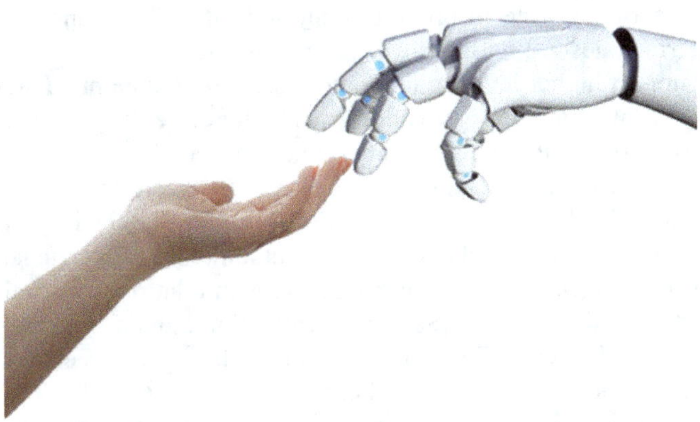

Figure 3.8 Envisioning the incorporation of human–machine collaboration into smart manufacturing

safety and situation awareness, programming and teaching, task dynamics, trust and acceptance, and sociology of work requirements [16].

Collaborative machines or robots in development for smart manufacturing applications are a new generation of robots designed to work, interact, and collaborate with humans. They are not bound to any type of cage or fencing and are not limited by the boundaries and workspace limitations of traditional industrial robot applications and automation. Collaborative robots also known as "cobots" have improved safety features for working in proximity with humans and programming which allows flexible application, enabling deployment and redeployment within the digital factory [16].

There is a distinction between cooperation and collaboration. In cooperation, humans and machines work on different subtasks that lead to a common end result or goal. In collaboration, all tasks are carried out by the human and the machine together and coordination between the two is required (sequence of shared actions toward a shared goal).

The context of safety and situation awareness explores how safe production operations can be carried out and risk prevention between cobots and human operators in a shared workspace, addressing the aspects of safety in relation to usability and performance. Situation awareness should come from a human operator-centered perspective on whether the human operator feels safe in a human–cobot collaboration. The study of ergonomics offers a great potential for realizing the full collaboration and participation of humans and robots in a complex value-added process, for example, the development of light-weight robots compared with the traditional heavy-weight counterpart and the development of sensible artificial skin for robots which allows the robot to notice an unscheduled contact and immediately stop or slow its own movement when in direct interaction with human operators [12].

Programming and teaching define simplified programming, intuitive implementation, and flexible and adaptive application. Simplified programming environments enable and support workers with a variety of skill levels (low to high) to have a more prominent role in programming and interacting (end-user programming and interface design) with robots in the digital factory.

Cobots support higher level collaboration, including dynamic communication, optimization, learning, and program adjustments; therefore, they have the ability to perform tedious, difficult, or dangerous operations. Task dynamics explore the transparency, control, and responsibility shared between humans and cobots. Studies also suggested that human operators should still feel in control of and responsible for their actions, and overall work outcome. It is also crucial to develop a socially responsible way of producing algorithms in relation to ethical principles.

Trustworthy designs in CPHSs are essential for human operators to accept intelligent systems. Factors influencing violation of trust include errors, reliability, and faulty robot behavior should be addressed.

Sociology of work is segregated into skills, training, and workload. This context discusses the focus on empowering and upskilling the workers to facilitate problem-solving competence with learning strategies and adequate competency

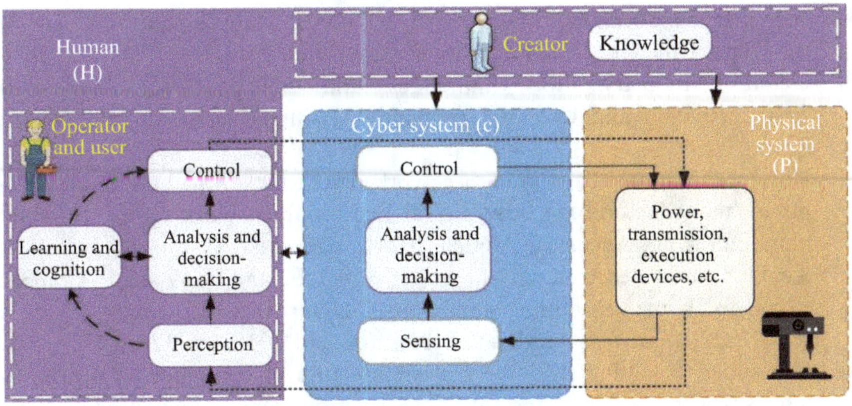

Figure 3.9 CPHS framework for smart manufacturing [17]

management. The introduction of cobots in a digital factory should reduce the workload of a human operator and not otherwise.

Figure 3.9 illustrates the framework of digital or smart manufacturing systems which are characterized by the integration of a cyber system between the human and the physical system. The cyber system consists of both hardware and software; its common goal is to complete the various tasks previously performed by the human operator. The cyber system's additional functions include sensing (IoT), analysis, decision-making, and control (big data collection and analysis, and AI). The human operator which is in control of the system provides the underlying analysis, computation and control models, methods, and rules drawn up by the knowledge and experience of the human operator [17].

Smart and intelligent manufacturing is realized by the integration of the strengths of humans, cyber systems, and physical systems which provide significant improvements in automation, quality, stability, problem-solving, and sustainability to businesses and organizations. In addition, aside from the reduction of manual labor, some of the mental work can be performed by the cyber systems through big data collection and analytics, further increasing the efficiency of knowledge dissemination and utilization within the organization.

References

[1] BDO International. COVID-19 is Accelerating the Rise of the Digital Economy. https://www.bdo.com/insights/business-financial-advisory/strategy,-technology-transformation/covid-19-is-accelerating-the-rise-of-the-digital-e, Accessed August 2021.

[2] McKinsey & Company. Industry 4.0: Reinvigorating ASEAN Manufacturing for the Future; 2018.

[3] The International Electrotechnical Commission (IEC). Factory of the Future. IEC; 2015.

[4] McKinsey & Company. The Internet of Things: Mapping the Value Beyond the Hype. Chicago, IL: McKinsey & Company; 2015.

[5] Deloitte. Industry 4.0: Challenges and Solutions for the Digital Transformation and Use of Exponential Technologies. London: Deloitte; 2015.

[6] How the Digital Factory Framework Facilitates Smart Manufacturing. https://www.iec.ch/blog/how-digital-factory-framework-facilitates-smart-manufacturing, Accessed August 2021.

[7] Ke, Q., Liu, J., Bennamoun, M., An, S., Sohel, F., and Boussaid, F. Computer vision for human-machine interaction (Chapter 5). In *Computer Vision for Assistive Healthcare*; Academic Press, London; 2018, pp. 127–145.

[8] Englander, I., and Wong, W. The architecture of computer hardware, systems software, and networking: an information technology approach. 6th ed. John Wiley & Sons, Inc., US; 2021.

[9] Nardo, M., Forino, D., & Murino, T. The evolution of man-machine interaction: the role of human in Industry 4.0 Paradigm. In *Production & Manufacturing Research*. London: Informa; 2020, pp. 20–34.

[10] Nunes, D., Silva, J.S., and Boavida, F. A Practical Introduction to Human-in-the-Loop Cyber Physical Systems; 2018.

[11] Sowe, S. K., Zettsu, K., Simmon, E., de Vaulx, F., and Bojanova, I. Cyber-physical human systems: putting people in the loop. IT Professional, 2016, pp. 10–13.

[12] Garibaldo, F. and Rebecchi E. Cyber-physical system. London: *AI & Society*; 2018, pp. 299–311.

[13] Upskilling is Crucial for the Successful Digital Transformation of Business. https://www.thepeoplespace.com/ideas/articles/upskilling-crucial-successful-digital-transformation-business, Accessed September 2021.

[14] Building the Vital Skills for the Future of Work in Operations. https://www.mckinsey.com/business-functions/operations/our-insights/building-the-vital-skills-for-the-future-of-work-in-operations, Accessed 10th September 2021.

[15] McKinsey & Company. A Future that Works: Automation, Employment, and Productivity. McKinsey & Company, 2017.

[16] Weiss A., Wortmeier A., & Kubicek B. Cobots in Industry 4.0: a roadmap for future practice studies on human-robot collaboration. *IEEE Transactions on Human-Machine Systems*, 2021;51:335–345.

[17] Zhou, J., Zhou, Y., Wang, B., and Zang, J. Human-Cyber-Physical Systems (HCPSs) in the context of new-generation intelligent manufacturing. Engineering, 2019, pp. 624–636.

Chapter 4
Robotics and autonomous systems in smart manufacturing

Abonab Nasser Salem Ali[1] *and Wai Yie Leong*[1]

4.1 Introduction

Robotics – involved in the design, construction and operation of robots – is an interdisciplinary research area. The aim of robotics is to design smart machines that can help people not only to run their daily lives, today, but also to keep everyone safe, in future. Robots are aimed at the advancement of information engineering, a discipline that studies how machines can be a substitute for human beings by emulating how human children repeat cognitive and behavioral skills by learning from parents and peers. Robots can be used in many practical cases for the purposes of many, but are today used in mainly dangerous environments (including inspection of radioactive materials, bomb detection and deactivation), manufacturing processes, and/or where humans will most likely perish (e.g., in space, underwater, in the heat and high cleaning and containment of hazardous materials and radiation) as shown in Figure 4.1 [1].

Robots can take any shape, but essentially they are like humans by theirs copying a particular human behavior. It is said that copying human behavior helps to use the robot in performing some repetitive behaviors usually performed by people given that robots can try to replicate a walk, a posture, a voice, a cognition or some other human operation. Information engineering is inspired by many of today's nature robots and contributes to the field of bio-inspired robotics. The idea of developing machines that can run independently dates back to ancient times, but the search mechanism and the use of robots did not expand significantly until the twentieth century. Throughout history, many scientists, inventors, engineers and technicians have always believed that today's robots are the result of closely observing human activity and imitating humans managing to fulfill tasks, such as human-fashioned tasks. Today, robotics is a rapidly growing field with continuous technological advances. The research, design and building of new robots serve a variety of practical purposes, such as at local, commercial and military levels. Many robots are built to perform functions that would put the life of people at risk to perform, such as defusing bombs, finding survivors in the rubble, exploring shipwrecks, and mining. More and more robots have also been designed to perform specific tasks in

[1]The Institution of Engineering and Technology, Malaysia

Figure 4.1 Robot photo [1]

the production process. For example, many robots have been built to operate assemblies, which may not be easily adaptable to other applications and are described as “assembling robots.” Welding seam is another application of robots, and some suppliers provide complete robot welding systems with any welding equipment. Robots also perform well at managing machinery and other items, such as spinning disks, integrated devices, etc. When it is an integrated automated device, it’s called a “welding robot” although it can be adapted to a separate manipulator for a variety of tasks. Some of the robots specifically built to tackle the heavy burden are defined as “big robots.”

We have evidence that a process, such as human figures emulate step-by-step in performing tasks, dates back to ancient times to Greece. The concept has been found in man-authored novels since the early nineteenth century. Despite these initial concepts and incipient representations of humanoids, it’s in the 1950s that the dawn of a technological revolution in Android started. George Devol invented the computer-operated robot and programmed it for the first time in 1954. In ancient China, it is written in the text of the third century BC, that King Mo Chu

Figure 4.2 History of robots [2]

was offered a life-size, mechanical figure on the human form of Yan Shi Road. The evolution and history of robot manufacturing is shown in Figure 4.2 [2].

Perhaps, you have never heard of FANUC, the largest company in the world for industrial robots. But chances are that you have used the products built by one of its organizations. With activities in 400,000, in 1956, the employees of this Japanese company automated cars for Ford and Tesla, and metal cases for Apple's iPhone. The company became differentiated from the crowd and laid the foundation at the end of the day for the manufacture of modern robots. About 270 BC, a Greek engineer named old Ctesibius made water clocks with Android or loose numbers. It is presumed that the Greek mathematician Archytas made birds based on a mechanical description of Taranto "dove," which was driven by steam. Heron of Alexandria was a mathematician, physicist and engineer who lived around 10–85 AD. His most important invention was the Aeolipile, the first steam turbine. Other inventions include automated machines for temples and theaters, surveying instruments, and military machines and weapons. This led to FANUC flourishing today and its shares jumped by 35% in the past six months, more than double the yield of 14% of Japan's Nikkei index. This booming market for robots shows little sign of slowing down. According to the International Federation of Robotics, sales of industrial robots units increased by 15% in 2015, while revenues rose 9% to $11 billion. In 2016, the turnover in North America rose by 14%, by $1.8 billion. ABI Research, a consulting firm, believes that the sales of the industry will triple by the year 2025, as shown in Figure 4.3 [3].

4.1.1 Development of robots

The development of robots (DevRob), sometimes referred to as genetic robots, is a scientific field that aims to study the developmental mechanisms, structures and limitations that allow for lifelong learning of new skills and the opening of new knowledge

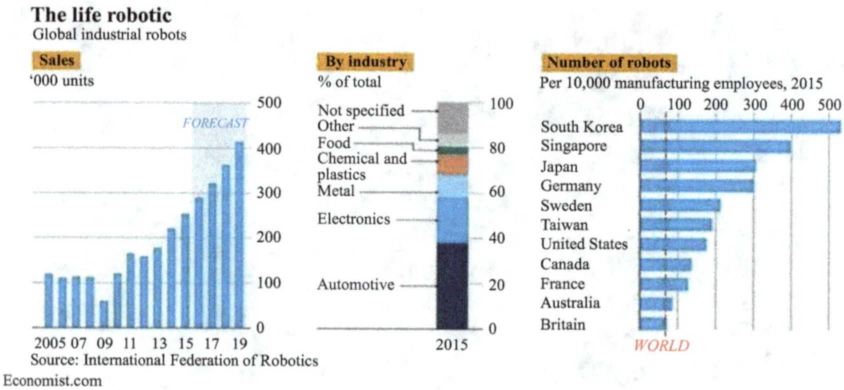

Figure 4.3 Statics of life robotics [3]

embodied in machines. As in children, it is expected to be cumulative, progressively more complex, and to have resulted from the exploration of the self in the world of social interaction, learning. A methodological approach typical of early human and animal development theories points to the importance of details in areas such as developmental psychology and neuroscience, science development and evolution biology, and linguistics, and to personalize the formal implementation of robots by often exploring variations or variants of robots. The exploration of these models of robotics helps researchers contrast the robots built with nature and, as a result, facilitate the creation of a robotics that also creates responses and new ideas on human and animal development theories. The development of robots is related to but separate from the evolution of robots. The evolution of robots uses the evolutionary robotics (ER) population of robots that develop over time, while DevRob is interested in how to structure the robot learning trajectory by tweaking one control system to develop through experience, over time. DevRob is also related to research in the field of robotics and artificial life.

4.1.2 Future of robotics

Many companies are working on consumer robots that can navigate their surroundings and identify common objects and perform simple household tasks without installing a specialized expert [4–12]. Perhaps, the process in 2020 that produced the "World Robotics" report for the first time is highly competent with a lizard-like mind that can be programmed for almost any routine chore. With projected increased computing power to make the current robots look simple, the second generation of robots with mouse-like mind that train automatically to pick up new knowledge may become possible by 2030. As well as software applications, these robots can host a range of programs called "air conditioning units" that generate positive and negative signal reinforcements under predefined conditions. Incidentally, the year 2040 is supposed to be year when the third-largest computing power generation robot with possible monkey minds becomes possible. These robots learn from psychological rehearsals that mimic physical, cultural and psychological factors. Physical properties include knowledge of shape, weight, strength, texture and appearance of things. Cultural considerations include the name and meaning of the object, the appropriate location and the intent of the object. Psychological considerations are related to humans and other robots, including goals, values, emotions and desires. Simulation to track external events, and models that are worth keeping in observable reality, would have to be valid. This should make it possible for the robot to learn through Consciousness. In the middle of the twenty-first century, the fourth generation of robots is slated to arrive with human minds having intellectual power capable of inference and generalization. Researchers hope that these machines will contribute to a good blending logic system for third-generation devices. Receiving his education properly, he is likely to become a big fourth generation of intelligent robots.

"Modern robots are a significant piece of the advancement of the assembling industry," says Junji Tsuda, President of the International Federation of Robotics. "Robots develop with many forefront advances. They are vision acknowledgment, aptitude learning, disappointment expectation using AI, new idea of man-machine-cooperation in addition to simple programming, etc. They will help improve efficiency of assembling and grow the field of robot application." The instrument flight rules (IFR) standpoint shows that in 2021 the yearly number of robots provided to manufacturing plants far and wide will reach around 630,000 units as shown in Graph 4.1 [1].

"IT, like mechanical autonomy, process mechanization is by a long shot the greatest donor in changing individuals' occupations," remarks Junji Tsuda, leader of the International Federation of Robotics. Organizations and governments must team up to guarantee laborers have the right stuff that the innovative change requires as shown in Graph 4.2 [1].

As indicated by the International Federation of Robotics, in excess of 3 million modern robots will be being used in plants far and wide by 2020. This implies the operational stock will dramatically increase within 7 years (2014–2020). Outside

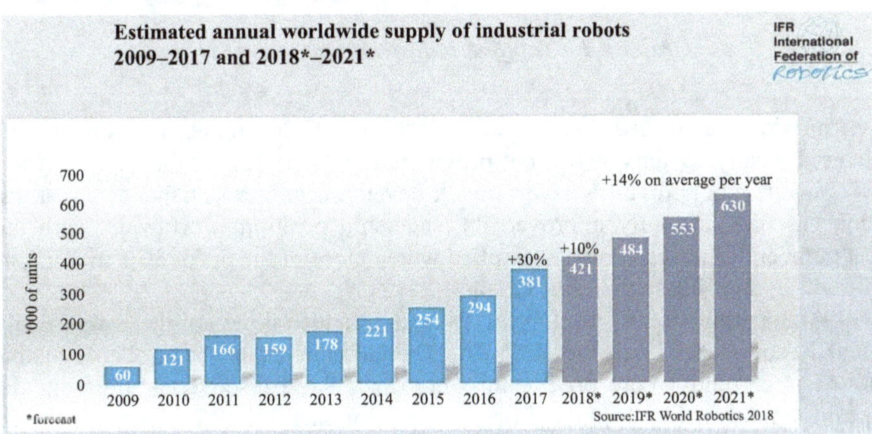

Graph 4.1 Global industrial robot sales doubled within 5 years [1]

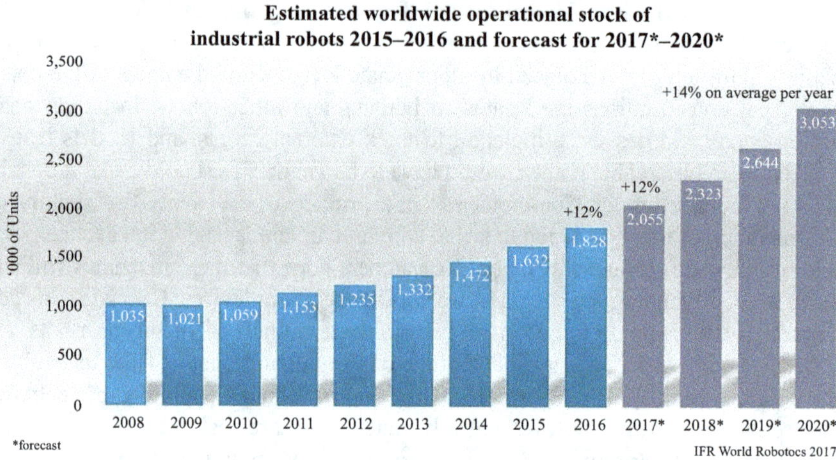

Graph 4.2 Factories around the world by 2020 [1]

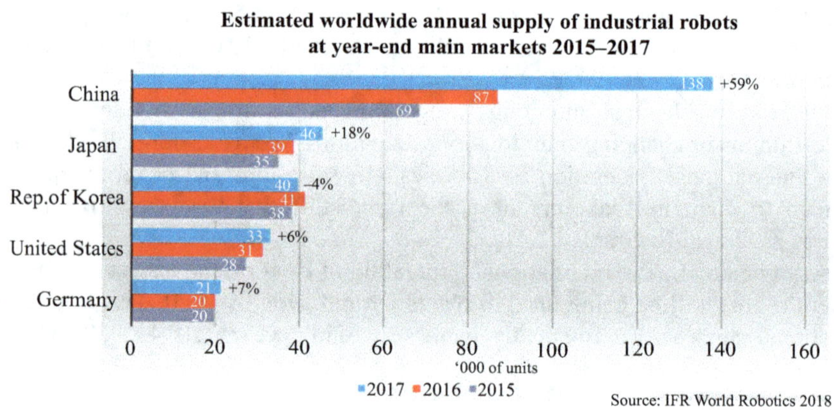

Graph 4.3 Top five industrial robotics [13]

robot providers expanded their deals by 72% to 103,200 units, including robots delivered locally by universal robot providers in China. This is the first occasion when outside robot providers have a higher development rate than the local makers. China has fundamentally improved its leadership position, worldwide, with the most number of industrial robots supplied with a piece of the pie of 36% in 2017, as indicated by the IFR statistics in Graph 4.3 [13].

AI intonation points for Object Detection undertaking of the Large-Scale Visual Recognition Challenge (LSVRC) Competition happened in 2014. In this test, AI is currently more precise than humans. These discoveries are from the opposition information from the pioneer sheets for each LSVRC rivalry facilitated on the Image Net site shown in Graph 4.4 [13].

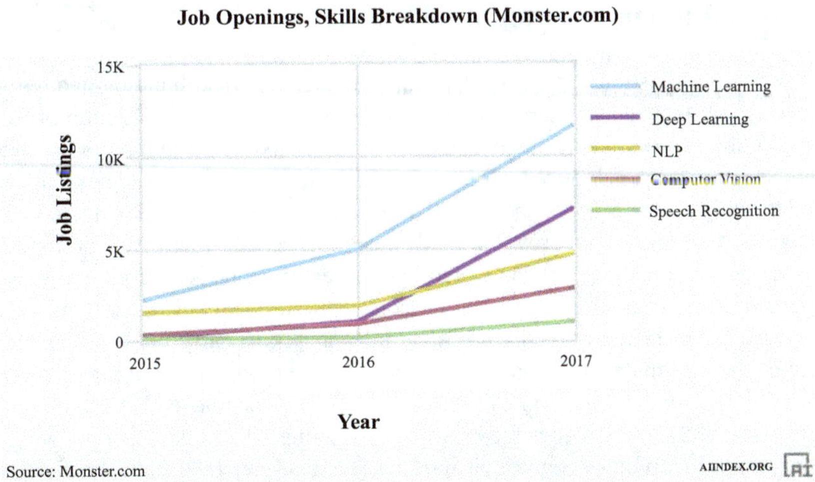

Graph 4.4 Artificial intelligence (AI) growth [13]

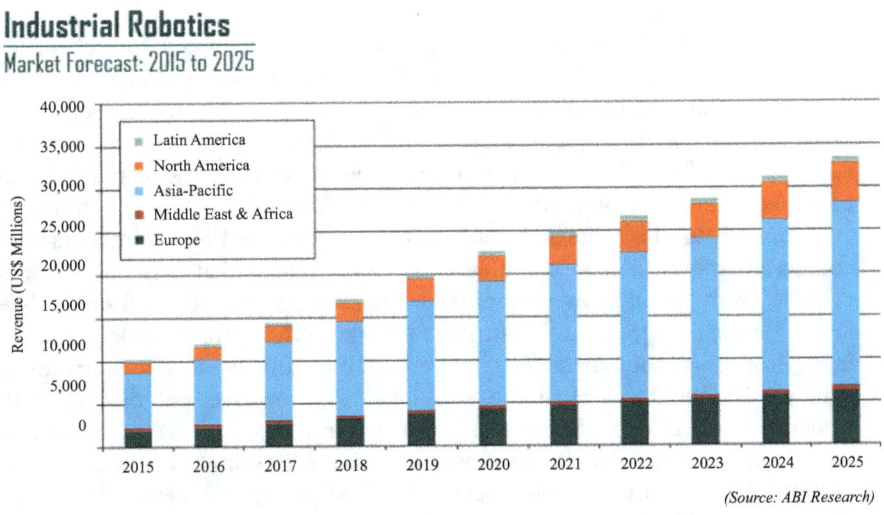

Graph 4.5 Army of Robots [2]

An Army of Robots Is Coming to the Factory Floor ABI Research figures of interest in plant floor robots are said to develop from not exactly a billion dollars in 2018 to $22 billion by 2027. In the course of recent years, we have seen startling advances in industrial facility robot innovation. Alongside, we have seen development in selection which is going to get a lot more extreme. With the plunge in price of both versatile and fixed-arm robots, this mechanization apparatus has become a simple decision-making to implement for small and big-sized producers to use industrial robots in the production process as shown in Graph 4.5 [2].

4.2 Introduction to autonomous systems

Robotics autonomy is the part of mechanical building, electrical building and software engineering that manages the structure, development, activity, and utilization of robots, just as a PC framework for the control, tactile criticism and data handling in computers. Researchers who can think on their feet are needed and innovative strategies that can be utilized to apply autonomy will show up. The mechanical field is perplexing and will be set aside as some effort to grow new advances is needed which will prompt solid development. Technologies are motivated by the need for precision commonly, where arrangements have been arrived at with a serious level of flawlessness taking place over many years. Science fiction media has a significant task to carry out in breaking reasonable boundaries that can rise as a method for venture topping, and in time fiction can turn out to be genuine innovation. A review of rising advancements that can be utilized in mechanical technology shows for example the cutting-edge robots that we will seek. Such advances incorporate counterfeit muscles, flexible electrical links, artificial skin, quick-moving robots, artificial noses or material finger gadgets and can carry fake gadgets to living life forms one bit nearer and a lot nearer to the human beings.

1. *Artificial strengths*

   Carbon is a biochemical component, denoted by the symbol C, which stands for one of the Universe's greatest plentiful elements. Carbon nanotubes are carbon allotropes and have a tubular nanostructure. Scholars use carbon nanotubes with the power of imagination to build artificial muscles and actuators. The development of muscles which can contract and twist from these materials is not easy, and in this case, in particular, while inspiration from nature is minimal.

   Examiners used a solitary walled carbon nanotube (SWNT) to manufacture counterfeit muscles. It is a troublesome development that includes twofold cathode layers isolated by an ionic fluid consisting of a chitosan electrolyte layer. Because of the various attributes of the SWNT framework, actuators made from this material have improved qualities all together to a greatness contrasted with past actuators with ionic electroactive polymer (I-EAP) actuators. Critical upgrades are made in reaction time of 19 ms, exceptionally wide recurrence run (handfuls to many Hz), incredibly high-pressure age rate (1,080 MPa/s), and ultra-high mechanical yield power thickness (244 W/kg). The robots will have reduced size and lighter weight later on, with there being exceptional contrasts in the manner in which the appendages are moved, and a way to deal with human expectation as shown in Figure 4.4 [14].
2. *Flexible electrical cable*

   A versatile link that can particularly be utilized in applying autonomy, and this is what we call a progressive innovation. Regardless of how well a robot may have been assembled, the unbending links that interface segments can turn out to be unattractive, the versatility can be confined, and their utilization has a high wear and tear after some time. The flexible links created by Asahi Kasei Fibers that are utilized in robots that follow shapes and help lessen impedance during movements help smoothen the development of a robot.

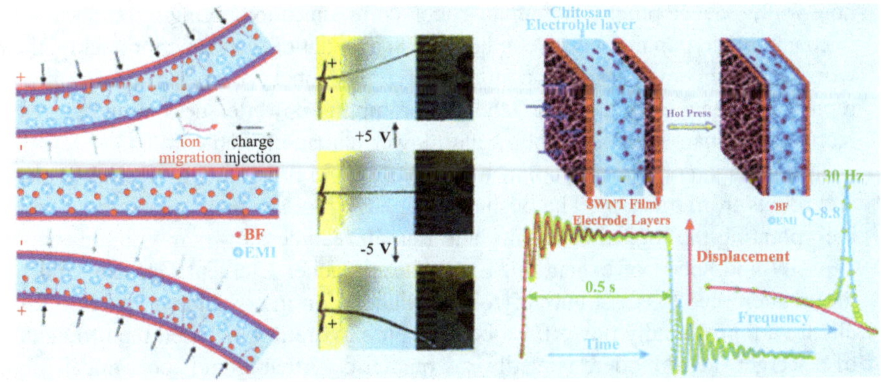

Figure 4.4 Reproduction strengths [14]

Figure 4.5 Elastic electric cable [15]

As far as their flexibility is concerned, the maker contrasts this substance and human skin. The link inside the sheath is spiraled to permit more prominent relative lengthening. The versatility empowering innovation can be utilized for force and sign links. This elastic cable is expected to become available to the market in the coming years as shown in Figure 4.5 [15].

3. *Tangible senses besides simulated skin*

Robots may, before long, sense material improvements and respond to them. This pivotal innovation was created at Laval University's Robotics Laboratory and is a noteworthy development in human–robot communication, bringing expectations from automation up in modern conditions. On account of their superior quality, their working arrangement and imperturbability, the robots from processing plants represent a genuine threat to the laborers, prompting powerlessness in humans who recognize human likeness in robots with lower human-created conflict.

In this way, complete isolation was actualized in the plants to defend workers and robots from mishaps. The business will have more to pick up, in any case, by incorporating the logical capability and human readiness of man with the robots' capacity and speed of execution. Exploration teacher Clement Gosselin and his group attempted to boost human–robot collaboration in a sheltered situation for a client with profoundly powerful robots. Their exploration included the production of a weight open minded, versatile, adaptable, deformable and economical robot skin. "The skins of robots we assemble are predominantly used to speak with them in our research facility," says Professor Gosselin. "At the point when you tenderly move such a skin on a robot, it is comprehended that there was a touch, and he derives the client's expectation. He deciphers the touch as a request to play out a development." Instead of being bound to unbendable frameworks, these model robots help the administrator react to his requests at the appointed time as though they were an augmentation of the specialist 's body as shown in Figure 4.6 [5].

4. *Imitating synapses in silicon*

Just as in sci-fi films, we see a robot in real life adopt the thought process of a person, in light of the fact that there is no such robot today. In this field,

Figure 4.6 Tangible simulated skin [5]

analysts have so far not gone excessively far, as of robots built not long ago. MIT researchers have built a silicon chip that can imitate one neurotransmitter conduct in the cerebrum. The chip is made out of 400 transistors and can reproduce the way toward adjusting to a neural connection's new information. Movements along these lines of adjustment is known as cerebrum pliancy and incorporates learning and retaining new abilities in data.

To recreate a full human cerebrum, scientists need in any event 100 billion neurons and a significantly more prominent number of neurotransmitters that are practically unthinkable for duplicating, especially regarding the exertion and the best approach to join and process with such technology. The MIT specialists structured their PC chip so the transistors could copy diverse particle channel conduct. Albeit most processors run in twofold (on/off) mode, the current streams in simple mode go through the transistors on the most recent cerebrum chip. As particles go through particle directions in a cell, a slope of electrical potential makes the flow go through the transistors as shown in Figure 4.7 [7].

Figure 4.7 Synapses in silicon [7]

5. *Bio Tack tangible radar*

   The human hand is an astonishing gadget which is utilized in pretty much every activity. The human body would not have been a finished instrument for gauging emotions if the hand could not detect what it is feeling. A group of analysts at the Viterbi School of Engineering at the University of California are using robots to evoke emotions in humans, or rather using robots to detect what they are coming in contact with. Till the point a robot cannot feel an egg or a cup as a snag, it would not feel the material surface as accomplished.

   The Bio tack radar comprises adaptable skin which covers a fluid filling. The vibrations that happen on an item's surface after finger development are perused and deciphered by the sensor. The sensor is amazing to such an extent that the temperature of the item can be distinguished, just as the course of the fiber because power contradicts movement. This current sensor's exactness is high-95%. Bio tack was tried with 177 normal materials and the greater part could be recognized as shown in Figure 4.8 [9].

The key goal behind the development of robots is to ease people's workload. The robots will perform tasks such as moving, choosing, packing, perceiving, manipulating objects, etc. by programming the robots to do these tasks. This makes robots capable to lessen manpower from performing boring and repetitive activities.

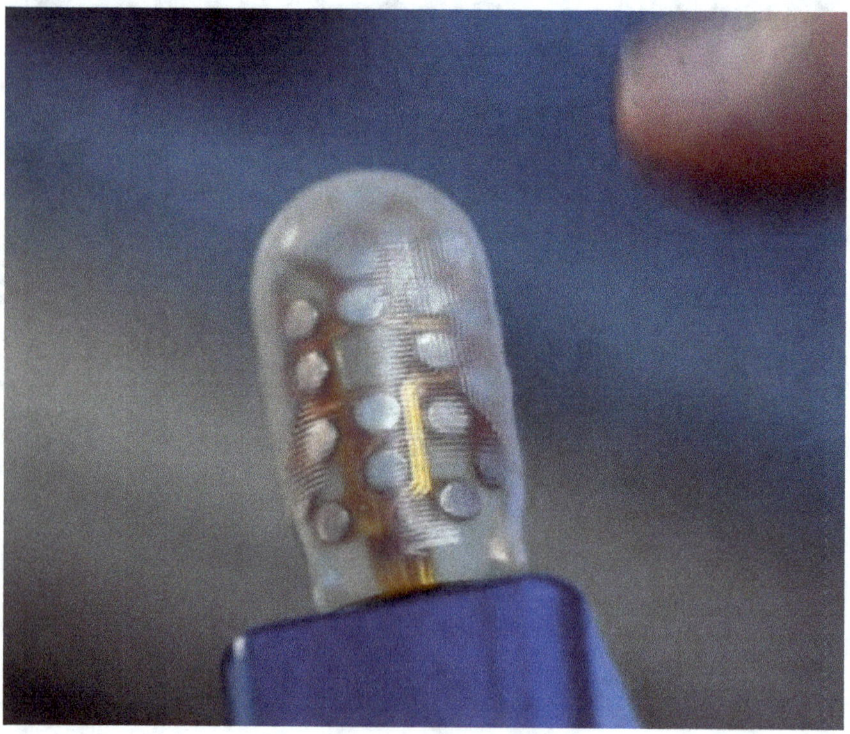

Figure 4.8 Bio tack radar [9]

Robots have mechanical architecture and structure to complete tasks, and they have electrical parts for driving and regulating various tasks precise to the need for which they are built. Robots have integrated minor computer programs which ensure proper functionality.

AI is a computer science sub-section that deals with the creation of programs that perform similar acts to human intelligence. AI is built with comprehensive psychological, mathematical studies, including psychology, physics, philosophy, sociology, neurology, computer science, and biology. The main goal behind the development of AI is to build expert systems that can serve as human help and allow machines to operate with human-like intelligence as shown in Figure 4.9 [11].

A gap analysis of what is needed for effective ground robotics is mandatory and this is why it has been carried out over the past 6 years with rescue robots in the field. It proposes a taxonomy of ground rescue robots as a function of vacuum and platform size, and then addresses the general robot system specifications for ground rescue robots. The operator control unit and the payloads are listed for the four missions to date (structural assessment, quest, identification of hazardous materials and medical assistance).

4.2.1 Concept of robotics laws

A robot may not injure a human being or allow a human being to come to harm through inaction. A robot must obey human orders, except where those orders

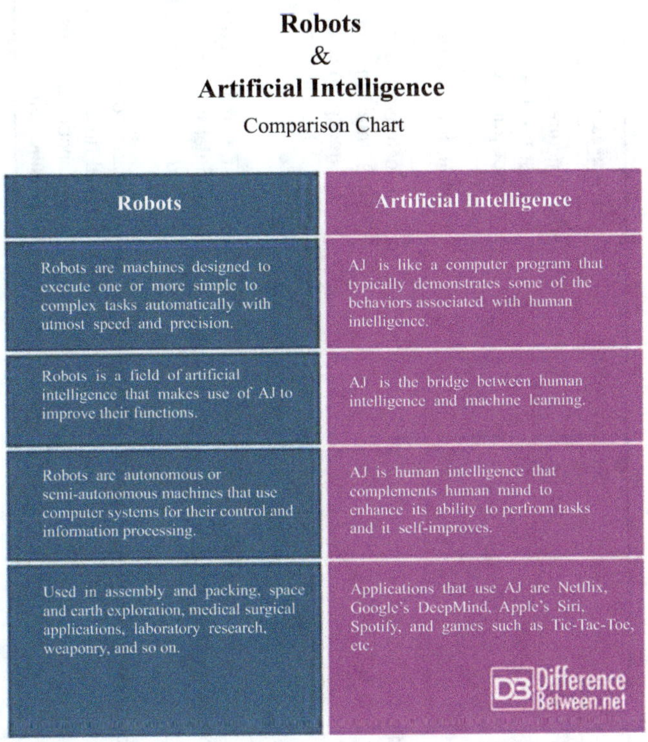

Figure 4.9 Comparison chart [11]

Table 4.1 Gap analysis and benchmarking for current product development [16,17]

Process	Relevant work	Key features	Technology/tools	Limitations
Pahl and Reitz	Pahl and Beitz 1988	Serial structured design process, function decomposition, core transformations (energy, material, information)	Stand-alone CAD/CAM/CAE	Lack of information sharing, social network, cooperation and collaboration, hard to implement in distributed and collaborative settings
Axiomatic design	Suh 1990	Serial structured design process, transformation matrix, independence axiom, information axiom	Stand-alone CAD/CAM/CAE	Lack of information sharing, social network, cooperation and collaboration, hard to implement in distributed and collaborative settings
Integrated product and process design	DoD 1996	Design for X. reduced overall time to deliver a product, reduced costs, improved quality	Trade-off study, uncertainty management, stand-alone CAD/CAM/CAE	Lack of social network, hard to implement in distributed and collaborative settings
Concurrent engineering	Prasad 1996	Parallelization of tasks, reduced time-to-market, cross-functional teams, information sharing	Collaborative decision-making, inter-connected CAD/CAM/CAE	Lack of social network, hard to implement in distributed and collaborative settings

Table 4.2 Gap analysis and benchmarking of business process [16,17]

S. no.	Business process	Needs	Fulfillment N	Fulfillment P	Fulfillment F	Existing	Odoo
I	The Division Of Sales/ Marketing	System Integration of data that can be monitored in real-time by sales/ marketing manager		√		Each sales have their own data that can not be monitored by sales/ marketing manager	Sales Module can integrate data between sales and can be monitored in real time by sales/marketing manager
2	Make a quotation	The quotation system with the data already entered into the system database and already passed through the verification of sales/ marketing manager. So in making the quotation, no longer need to open a lot of documents		√		Quotations are made in a Microsoft Excel document with a prepared quotation template. Data entered by copy and paste into available templates	In the Sales Management module there are quotations feature to make a quotation with data already entered and verified by the previous sales/marketing manager; so in the process of making the data quotation, stay selected without having to open many supporting documents again
3	Make a report	Reporting system for every sales activity that can be monitored by sales/marketing manager and director of operations in real to m". and there is a data analysis that supports the assessment of sales performance		√		Each salesman writes a report using Google Sheets about week-long activities and plans for the following week	In the Customer Relationship Management (CRM) module, there is My Pipeline feature that will record the sales activity automatically when following the Sales Flow that has been determined. There are two report features. namely Activities Analysis and Pipeline Analysis that can be used to view sales performance. On the Dashboard, using CRM sales module, one can see one's performance

would conflict with the First Law. As long as such protection does not conflict with the First or Second Laws, a robot must protect its own existence.

4.2.2 Communication system used in robotics

One of the major concerns in networked robotics is wireless communication, particularly in networked service robots. Wireless communication, however, is usually less reliable than wired communication, and more prone to interference. Thus, the design of the wireless connection and the wireless network is usually more demanding. A wireless communication link's design space requires many variables, such as selection of frequency and modulation, power and link-budget limitations, features of signal propagation, a wide range of wireless communication standards to choose from, and so on.

4.2.3 Advantages of robots

Robots can go deep down to unknown places where humans will perish, and they can give us information that humans cannot get by their own might. Robots can work 24/7 without any pay and food, and they cannot get bored. Robots can execute tasks quicker than humans and much more reliably and correctly, and they are more popular every day. Robotic pets can support anxious patients and keep them involved.

4.2.4 Disadvantages of robots

Robots have a power supply so that when people will lose jobs in factories, the robots will function provided maintenance to keep them working had been kept up to the point that robots are needed. It costs a lot of money producing or purchasing robots, and the software and the equipment needed for the robot are expensive. Robots are expensive in maintenance to be done and repair to be completed, and systems need to be revised to meet changing specifications, computers need to be made smarter, repair costs can be very high in case of a failure, procedures are needed to recover missing code or data which can be time-consuming not to mention that they are costly.

4.3 Fifth industrial revolution

Industry 5.0, or the fifth industrial revolution, will revolve around human-machine interactions given greater collaboration between the two is expected as cognitive computing will be better equipped to work alongside human intelligence. Staff would need to be experienced in assigning value-added output tasks, and much of this phenomenon is motivated by a broad market ready for greater customization of its activities and the customization of mass-produced products. The benefits of an Industry 5.0 model include greater customization potential which will allow for greater job creation unlike automated Industry 4.0 systems, where it is not just any job that creates value but high-value long-term careers that do the needful. Let the robots do the dull, dangerous and dirty things and let the people do the imaginative, fascinating stuff. The fifth industrial revolution would include industrial robots as a critical part although much of what Industry 5.0 describes includes a human being's ability to customize on a mass-scale and personalize a product, which is only possible with advanced robotic capabilities. Industrial robots help to close the loop on architecture by automating the entire production process end to end and efficiently, so that people are free to build and invent, without having to think about production constraints. Unlike in Industry 4.0, where robotic capabilities take center stage, industrial robots in Industry 5.0 are taking a back seat to human intelligence. However, they will remain a crucial component allowing for completely new production methods. While we are still at the beginning of understanding Industry 4.0's true vision, what's on the horizon is becoming apparent. Industry 5.0, though still out of reach for the time being, would bring major improvements for manufacturing. This delineation becomes hazier with Industry 5.0, as humans and robots start working side-by-side and even together. Such types of robots are known as collaborative robots, which will perform the requisite tasks of heavy lifting with accuracy while the skilled individual provides a craftsman's cognitive skills.

4.3.1 Robotics beyond 2030

By 2030, robots could replace up to 20 million production jobs worldwide, where people displaced from those jobs are likely to find that the automation has also squeezed comparable roles in the services sector. However, increasing automation will also boost jobs and economic growth. The firm called for action to prevent a harmful rise in inequality in income. The Revolution in Robotics quickly accelerating, as technical innovation quickly progresses in electronics, robotics, carbon recovery, man-made computer and human intelligence, there is going to be a convergence in learning. Consequently, the power of robots, and their capabilities to perform duties once carried out by men, will leap. Number of robots deployed in industrial usage multiplied worldwide in the past two to three decades to two and a half million. The use of robots in manufacturing would multiply faster in the next 20 years. By 2030, as many as 20 million robots will be used globally, estimated 14 million to be used in China. The consequences are immense, and the challenges ahead are equally overwhelming for governments and policy-makers in their capacity as leaders and think-tanks. Building robots can improve efficiency and growth in the economy. It will also contribute to the development of new jobs in still-existing industries, in a "creative destruction" phase. However, current business models in many sectors will be majoritarily disrupted. So tens of millions of current jobs will be lost, with human workers gradually being replaced by robots, as robots become much more sophisticated. The consequences of these job losses will vary widely for both individuals and companies across countries and regions, with a disproportionate toll on less-skilled workers and weaker local economies. In many places, the impact will exacerbate the social and economic stresses caused by unemployment and income inequality in times when political polarization is already a disturbing phenomenon.

4.3.2 Robots in architecture

The use of robotic manufacturing in architecture, sculpture and design ties the industry closely with the state-of-the-art research institutions. Robots were employed by industrial manufacturers for years, but they have been given serious consideration by architects not until recently. The relationship between architects in the age of digitization, virtualization and computerization, on one hand, and robots, on the other, is becoming deeper. Defining what a robot means is hard in architecture. The term is likely to refer to anything from robotic arms to 3D printers to CNC milling machines. In essence, robots are programmable automated mechanisms which aid in digital manufacturing processes. In fact, a robot is a more proficient process.

Grasshopper is a graphical algorithm editor, closely integrated with Rhino's 3-D modeling software, for designers who create new shapes using generative algorithms. Creating Grasshoppers does not involve programming or scripting skills, but also allows designers to create type generators from the basic to the awe-inspiring. A few years ago, a robot built a 72-foot brick building in Manhattan as shown in Figure 4.10 [3].

4.3.3 Five applications of robotics

1. *Industrial robots*

   The first robots were industrial robots that replaced human workers carrying out routine simple tasks. Factory assembly lines can work in a well-defined environment without the intervention of humans, where the robot will execute tasks in a specified order, acting on items precisely put in front of it. Flexibility is required when industrial robots interact with humans, creating strong safety requirements for both robotic and mobile robots. In particular, the robot's speed must be reduced, and the mechanical design must ensure that moving parts are not a source of user hazard as shown in Figure 4.11 [3].

2. *Autonomous mobile robots*

   Many mobile robots are operated remotely, performing tasks such as pipe inspection, aerial photography, and bomb disposal that rely on a system

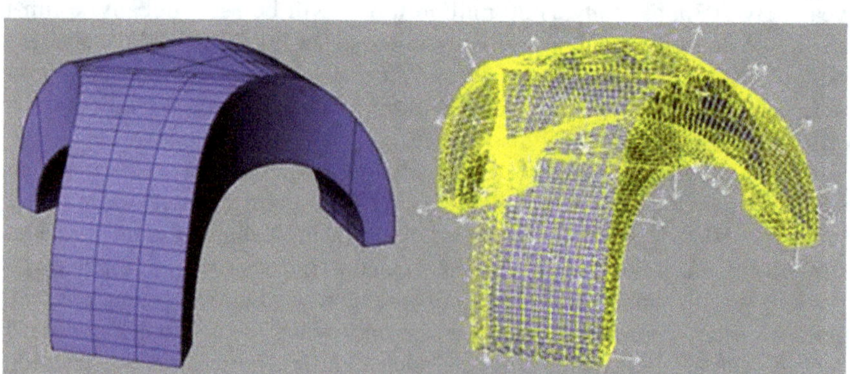

Figure 4.10 Fabrication of free-form structures [3]

Figure 4.11 Industrial robots [3]

controller. These robots are not autonomous; but they need a human operator to use their sensors to give remote access to dangerous, distant, or inaccessible locations to their operator. Any of them could be semi-autonomous, performing automated subtasks. A drone's autopilot stabilizes the flight as the human opts for the flight direction as shown in Figure 4.12 [3].

3. *Robotic arms*

   The robot needs an actuator—a component of a robot to act upon its environment—that affects the environment. Many robots, especially robotic weapons used in industry, have an impact on the environment through end-effectors, usually grippers or similar instruments. Mobile robots' actuators are the engines that cause the robot to move, as well as components such as a vacuum cleaner's vacuum pump. Educational robots are usually mobile robots whose motors and display devices such as lights, sounds or a screen are their only actuators. End-effectors can be built using robotic kits or using additional components with pre-assembled robots, although there are educational robotic arms available as shown in Figure 4.13 [3].

4. *Generic robot*

   A generic robot is one we use to present the algorithms in robotics. The generic robot's capabilities are similar to those found in educational robots, but the one you use may not have all of the capabilities presumed to exist in the presentations, so you will need to improvise. You do not yet understand all of the words in the definition below, But the specification needs to be formalized shown in Figure 4.14 [3].

Figure 4.12 Autonomous mobile robots [3]

Figure 4.13 Robot arms [3]

5. *Robotics kits*

   The LEGO Mindstorms robotics kits were introduced in 1998. A kit consists of standard LEGO bricks and other building components, along with motors and sensors, and a programmable brick that contains the computer which controls the robot components. The robotics kits have the advantage of flexibility: you can design and create a robot to perform a specific function, only as

Figure 4.14 Generic robot [3]

Figure 4.15 Robot kit [3]

far as your imagination sets limits. Also, a robotic kit can be used to teach mechanical design to students.

The disadvantages of robotics kits are in their being more expensive than simple pre-assembled robots, and that robotic algorithm exploration depends on one's ability to implement robust mechanical designs successfully as shown in Figure 4.15 [3].

4.4 Conclusion

Today, we find most of the robots in industries, factories, warehouses and laboratories working for people. There are many ways robots are helpful. It boosts the economy, for example, because businesses need to be efficient to keep up with competition from the industry. Having robots therefore helps business owners to be competitive, since robots can do jobs better and faster than humans can do in places where a human would be rendered redundant by his slower capabilities, like assembling a car. However, robots are incapable in jobs that are not repetitive, like where human beings must use their judgement and exercise choice, and so robots today support science and industry only. Finally, as the technology advances, new ways of using robots will arise which will offer new hopes and new opportunities.

References

[1] McKerrow PJ, McKerrow P. Introduction to Robotics. Sydney: Addison-Wesley, 1991, pp. I–XVI.

[2] Fu KS, Gonzalez R, George Lee CS. *Robotics: Control Sensing. Vis.* London: Tata McGraw-Hill Education, 1987.

[3] Arkin RC, Arkin RC. *Behavior-Based Robotics*. London: MIT Press, 1998.

[4] H Wenxing, Weng Y, Xie L, Maoqing L. Design and implementation of web-based DSS for online shopping mall. In 2009 IEEE International Conference on Control and Automation, 2009, pp. 1308–1313.

[5] Huang S, Zhang DH, Leong WY, Chan HL, Goh KM. Detecting tool breakage using accelerometer in ball nose end milling. In The 10th International Conference on Control, Automation, Robotics and Vision, 2008, pp. 927–933.

[6] Leong WY, Homer J, Mandic DP. An implementation of nonlinear multiuser detection in Rayleigh fading channel. *EURASIP Journal on Wireless Communications and Networking* 2006; 2006:1–9, Article ID 45647. DOI10.1155/WCN/2006/45647.

[7] Leong WY, Homer J. Blind multiuser receiver in Rayleigh Fading channel. In *2005 Australian Communications Theory Workshop*, 2005, pp. 155–161.

[8] Min HF, Yie LW. Investigating target detection and localization in wireless sensor network, *Procedia Engineering*, Vol. 41. New York, NY: Elsevier, 2012, pp. 75–81.

[9] Leong WY. Angle-of-arrival estimation: beamformer-based smart antennas. In The 3rd IEEE Conference on Industrial Electronics and Applications, 2008, pp. 1593–1598.

[10] Leong WY, Mandic DP. Towards adaptive blind extraction of post-nonlinearly mixed signals. In The 16th IEEE Signal Processing Society Workshop on Machine Learning for Signal Processing, 2006, pp. 91–96.

[11] Leong WY. Implementing blind source separation in signal processing and telecommunications. In The University of Queensland, Australia, 2005.

[12] Xiaofan W, Chong PHJ, Yie LW. Performance comparison of CSMA/CD, CSMA/CA, CSMA/RI, CSMA/PRI and CSMA/PR with BEB. In The 5th IEEE Conference on Industrial Electronics and Applications, 2010, pp. 1843–1848.

[13] Thrun S. "Probabilistic robotics." *Communications of the ACM* 2002; 45(3): 52–57.

[14] Murphy RR. *Introduction to AI robotics.* London: MIT Press, 2019.
[15] Kim S, Laschi C, Trimmer B. "Soft robotics: a bioinspired evolution in robotics." *Trends in Biotechnology* 2013;31(5): 287–294.
[16] Murphy, R.R. and Stover, S. (2006). *Gaps Analysis for Rescue Robots*. 2006. 333–339.
[17] Terminanto, A., Hidayat, R., and Hidayanto, A. Implementation of enterprise resource planning using Odoo module sales and CRM. Case study: PT Ecosains Hayati. *IOP Conference Series: Materials Science and Engineering*. 2017, 277. 012034. 10.1088/1757-899X/277/1/012034.

Chapter 5
Artificial Intelligence Implementations in HMI for smart manufacturing

Rian Abduallah Ba Sunbul[1] *and Wai Yie Leong*[1]

5.1 Introduction

Artificial intelligence (AI) is an arm of modern science that has spread its use in our daily lives and academia rapidly in recent years, reaching many industry domains and fields of study, driven by robots and intelligent services for governments and businesses. That is, by the automated behavioral characteristics we interact with we mean those faculties of human beings on which various computer programs depend, and that is consistent with human mental capabilities in different works. The machine's ability to teach itself new knowledge and adjust to new situations by making the right decision is among the most important of those capabilities. This kind of modern system has the capabilities to accurately differentiate between the many cases to choose from and do accordingly.

AI can interpret and, thus, make correct decisions based on analysis taking into account all the possibilities and by understanding their outcomes, and only then choosing the right options that lead to the desired outcomes. It can also find out mistakes from human errors and correct those fast and, thus, enable future developments to take place. AI aims at designing a machine capable of achieving a specific goal in a manner similar to (or exceeding) human capacity. They are often algorithms (specific software) that run on a computer or a group of computers (like the one in the picture). It is not an alternative electrical device (a misconception by some) for the human being.

5.2 Applications

AI can be applied to different ends:

- The classification of medical image: especially, in biomedical image recognition for diagnostic sciences, AI is very useful.
- The use of AI in accounting databases: is being investigated as a basis for alleviating the problems of accounting databases, like the need for accounting information classifications are not met by decision-makers.

[1]The Institution of Engineering and Technology

Figure 5.1 AI in traffic light [1]

- The use in computer games: for computer games have come a long way from small, three-dimensional text-based games to graphics-based games featuring complex and vast universes, which are structured as having graphic interfaces, audio play, and incorporating user feedback.
- Artificial neural networks in transportation (Figure 5.1): the use of neural systems in a few road haulage areas is depicted. The IT age today forces us to manage more and more details and AI is invaluable in this area.

5.3 Advantages and disadvantages of AI

Advantages:

- Indefatigable worker: which is due to machines' ability to act constantly, without feeling tired or distracted;
- Immense storage capacity: the storage and processing of a massive quantity of data;
- Meticulous worker: which is due to a precision programmed into it and reduced margin of error, and so human use of AI systems helps to minimize the percentage of error that can occur during the time of performance;
- Expert medical attention: done using surgical simulators or applications that help diagnose neurological disorders or that allow the patient to know the side-effects of medication;
- Imperishable: given AI systems can do things that humans cannot do, such as excavations (emergency or fire) and exploring places that are hard to reach, like the ocean floor.

Disadvantages: The biggest cost of using AI systems is from modification and maintenance, besides these systems not being capable of applying human consciousness of ethics to judgment so that the decisions they make are not applicable to situations demanding human capability to discern between right and wrong and then react.

- The inability of these systems to adjust to same data received anytime with changed time–space matrices and improve their method of work;
- Inability to adapt to circumstances and changes that can arise in the workplace by inventing ways to react and develop as humans;
- Inability to forge ties with workplace so that they replace workers and employees by hordes without repentance.

5.3.1 The components

1. *Data representation*: How to represent data or a computer problem so that the machine can process it and generate the correct output (or rather: how to position the problem in a machine-appropriate image so that the system understands it and can solve it). It should be noted that in the data representation process, languages are used like OWL and RDF, which are used to represent the data.
2. *Research*: This is what humans take care of most of the time, like understanding where the machine's checks and tests of the options available to it are according to the parameters set for it, or where it has created these itself and then it decides on the best solution.
3. *Future*: a system will be able to handle all the house chores, from welcoming and receiving guests, to warning about a danger, such as a fire or a small child's falling down a height or a house being robbed. The AI system can be designed to prepare coffee for the homeowner and his guests, and to fulfil the needs of monitoring the house from the store using AI technology. Driving our vehicle, recognizing while communicating with our mental and mood states, and selecting items that match our preferences and desires, and checking out where we can find them, all need an organizational and ethical framework regulating the work of AI and protecting its fruitful employment. The cycle of smart automation, formulating rights, will impact all these while AI will in every field either in medical care or in our daily life. This also makes it difficult to think about the nature of work in which humans will function, as advances in AI systems have reached the point that makes programmers in their advanced stages unable to grasp the dimensions of the capabilities of these devices they design, create and program. Here are several potential possibilities that AI systems are likely to cater to in different areas such as: entertainment, data and information security, and transportation.

5.3.2 Working principles

AI uses think-tanks to mimic people's intellect to construct something which is called a propensity model. The machine has to learn how to respond to certain actions and, so it uses historical data and algorithms. AI may do a lot more than this, but these are common marketing applications and features. So while the robots seem ready to rise to take over from humans, it is humans who are still required to do the majority portion of the job. Mainly, we are using AI to save us time—from substituting for people to email automation thus enabling AI to do a lot of the work while humans focus on other more important tasks.

History: A computer's intelligence is paramount to classify jobs it intelligently does in its ability to mimic humans. The root of the word AI dates back to the 1950s, and specifically to the year 1950 AD when scientist Alan Turing created what is known as the Turing Test, which meant calculating the intellect of a machine and intelligently classifying it in its ability to imitate the human mind. The speed of development of AI began in the early twenty-first century when interactive robots were a source of excitement in stores, but the matter went beyond that when a robot communicated with humans by sensations and by facial expressions, and with other robots [2].

Statistics: Figure 5.2 shows the revenues from the AI from 2016 to 2025 (in million US dollars) [3].

Eighty-four percent of businesses believe investing in AI would offer more strategic advantages. Seventy-five percent think AI will open up a new business though also offering new ways for competitors to access their markets. Sixty-three percent assume that the cost reduction burden would require the use of AI.

As we can see in Figure 5.3, from the growing tide of adoption of AI over the next decade, AI can drive huge growth in the global economy through growing industries. The market intelligence firm [3] predicts a rise in global annual revenues for AI software from \$3.2 billion in 2016 to \$89.8 billion by 2025, a 28-fold increase. This prediction is substantially higher than their previous business revenue estimates for AI. Owing to an improved outlook for a range of different use

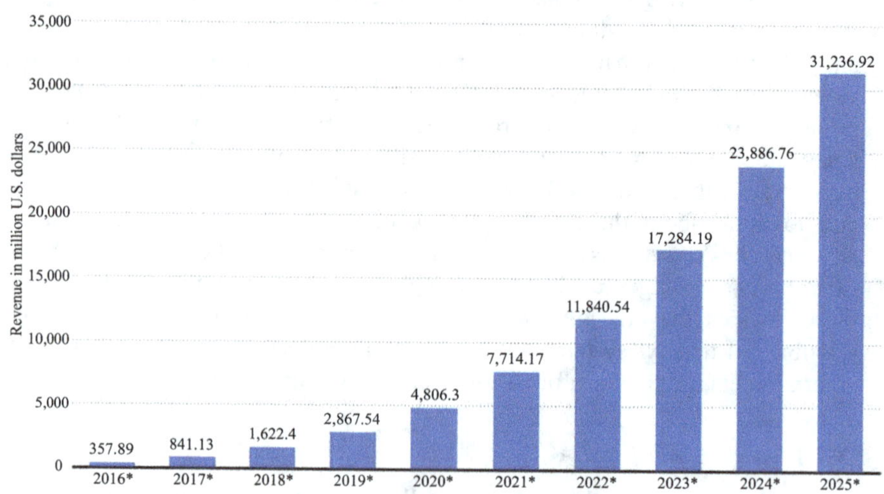

Figure 5.2 Investment in AI [3]

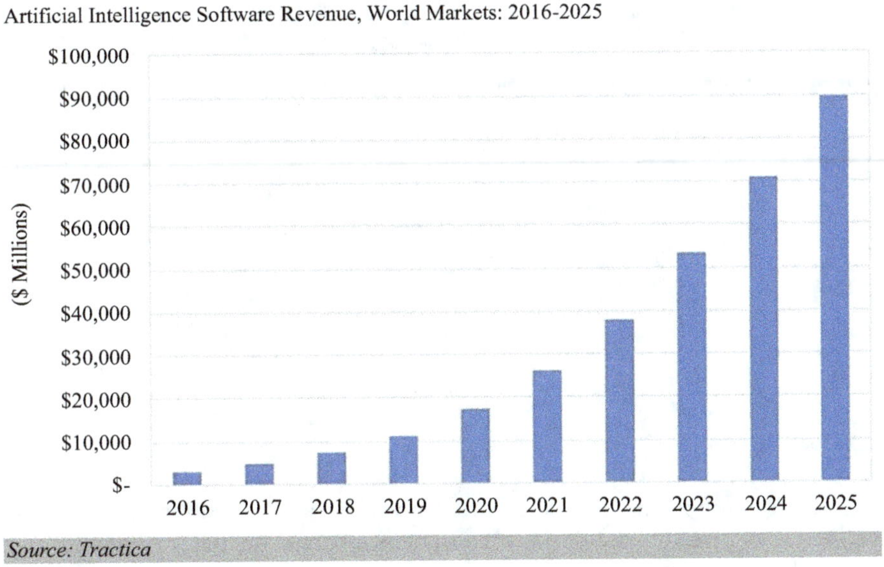

Figure 5.3 Growth of AI [4]

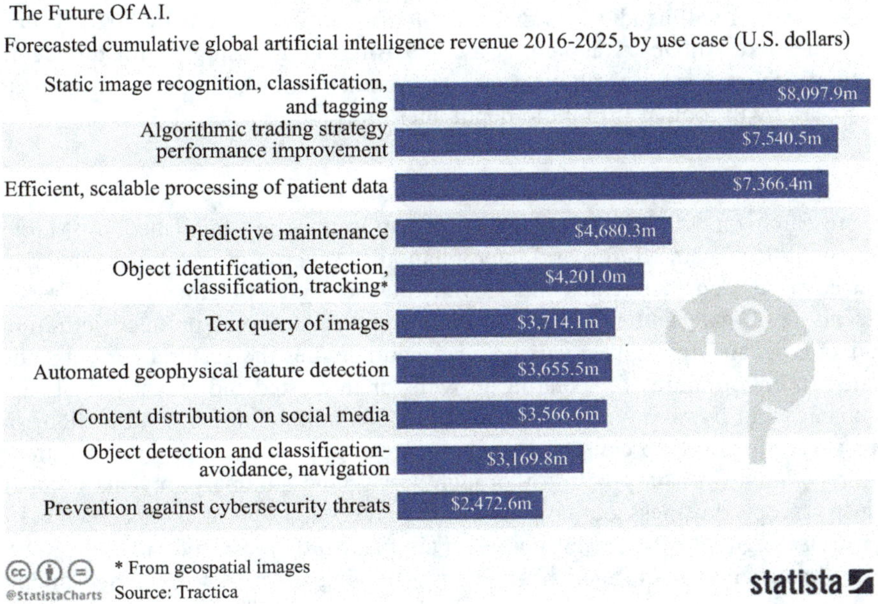

Figure 5.4 Areas of AI technology [5]

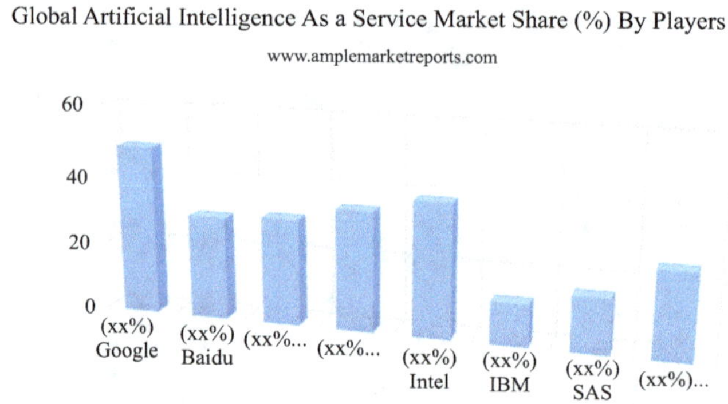

Figure 5.5 AI in the global platform [6]

cases across multiple industries, reported revenues in Q2 2017 rose. "Artificial intelligence is already key to how consumers operate on the Internet today—enabling them to deploy hyper-personalized services through an 'AI First' strategy," explains Aditya Kaul, Tractica's research director. The rest of the enterprise and government markets are still catching up on AI adoption and have yet to fully understand its value, including the Building Research Establishment (BRE).

As we can see, AI is used to an extensive degree in e-commerce! And Gartner predicts that AI will handle over 80% of all customer interactions by 2020. Today, companies like Alibaba, Rakuten, eBay, and Amazon are using AI for the identification of fake reviews, chatbots, product recommendations, big data management, etc. and we will look at it in more depth later on (Figures 5.4 and 5.5).

5.3.3 Conclusion

The research on "Artificial Intelligence" Platforms Study published by Market Survey AMR is a straightforward interpretation of specific industry-classified data focused on the characteristics that govern business growth globally. The report presents up-to-date and valuable market insights that reveal product definition, product type, and application variety. The report studies the sectoral current status related to opportunities for creating growth for interested individuals, rivals and companies and by taking advantage of suitable conditions for growth to flourish. This study aims to help consumers achieve green growth in their particular areas. The report states a study with a thorough survey and summary, reveals product/industry scope, business outlook, and status by 2026. This market research [6] report also highlights the opportunities, main growth drivers, study of top competitors, threats and risks to market growth by studying the market by understanding the competitive advantage of key players by using SWOT analysis. The market research insights give a global market value of millions, where the key players have included the company description, product photos, financial information, and marketing survey.

5.4 AI technology

The AI sector denotes the mechanization of human intelligence and the study of its intellectual abilities, which is one of the most important reasons for studying AI. This means we are led to understand our human mind, and the mental processes of every decent reader in a way that departs from the science of philosophy, psychology and anatomy, where all three study the human mind, in turn. The study of AI is trying to construct knowledge of the human mind to the degree of simplicity where this knowledge requires comprehension or it can be further simplified. Another reason to research this science is that our smart software is useful in many areas of our lives, which has become digital in and of itself! While no one can predict the details of the future, it is obvious that computers with human intelligence will have an enormous and clear impact on our daily lives and the manufacturing goals of our civilization. AI is a curious mystery: whether biological or electronic, how can this small brain, understand, perceive, predict, and interact with a world that is larger and more complex than the brain? How can we take a path that makes such a small brain complete in all its complex characteristics? That is a difficult issue. But unlike the quest for a quicker means of transport than the speed of light, the AI scientist and his student find that this work is based on stable and plausible foundations. All he has to do is look at the mirror to find an example of a vibrant smart system. AI currently has many applications, whether they are general purpose applications, such as perception and reasoning, or special purpose tasks, such as playing chess or doing medical diagnoses! Subject matter experts and scientists also turn to AI to retain their memories and perceptions of having lived their lives through experiences by programming AI systems to regurgitate outcomes of experiences lived before. The AI systems save the experiences of scientists and subject matter experts and retell the experiences that embody someone's life. The field of AI fits all types of experiential orientations.

The AI acronym stands for AI (Figure 5.6), a term employed to the science of the new computer systems, where this science belongs to the current generation of

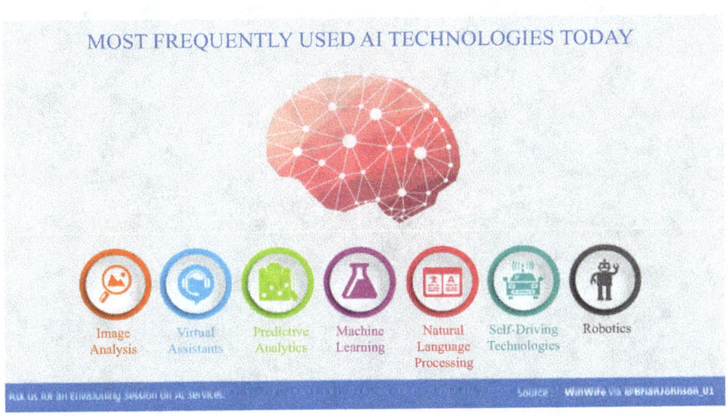

Figure 5.6 The most frequent technology used in AI [7]

computers and seeks to replicate the processes of thought that take place within the human mind, so that the machine is capable of solving problems and making decisions in a rational, structured and similar way of thinking as a human brain faced with a problem will react. AI is an exciting science and we do not have an explanation for this excitement, just as we do not have a precise definition of it because we do not originally have a precise definition of intelligence. Intelligence is being studied as a basis for relieving accounting database problems. Some of the issues with current accounting database systems are as follows.

Technology (Table 5.2):

1. Robotic process automation: refers to and automates the operation of organizational systems by imitating human activities. It is important to bear in mind, in this particular sphere, that AI is not intended to replace human beings but to assist and complement their abilities and talents (Figure 5.7). This method is important to companies such as Pega systems, Automation Anywhere.
2. Machine learning (ML) platform: is an informatics sub-discipline known to all as important for AI (Figure 5.8). It aims to develop new techniques that enable computers to learn, and thus become smarter. ML platforms are becoming more popular using algorithms, API (application programming interface), development tools, tools for research, big information, and applications. Applications range far, from filtering emails to computer vision, where traditional algorithms are difficult or unfeasible to perform the necessary tasks.
3. Speech recognition: is used for converting and transforming human ability for talking into a useful, robust format for computer application processing (Figure 5.9). To transcribe and turn human language into useful formats is frequently witnessed nowadays in companies such as Great, Complexity Communications, OpenText, and Verint.

Figure 5.7 Using robotics in AI [8]

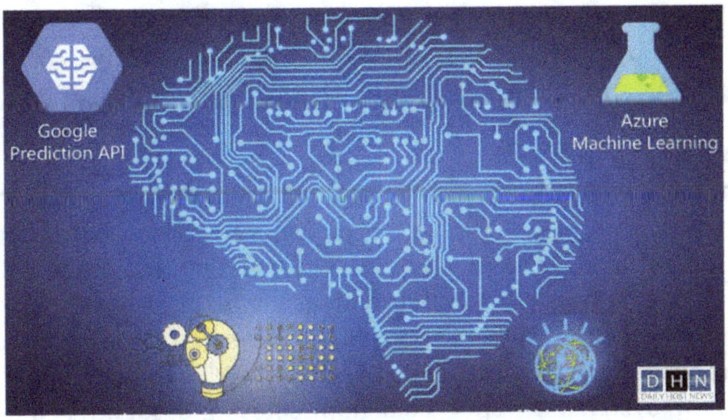

Figure 5.8 ML in AI [9]

Figure 5.9 Speech recognition in AI [10]

Systems provide the services of speech recognition.

The concept of Speech recognition enables hands-free control of various devices and equipment (a particular boon to many disabled persons), provides input to automatic translation, and creates print-ready dictation. Wordsmith is a popular Automated Insights tool, which applies NLP to generate news stories. Applications such as (Siri) are just one of those programs that can help you understand computer functioning better.

More and more technologies are being developed each day that can memorize the human language and touch hundreds of thousands of lives via interactive voice-response and mobile systems apps.

NICE, Nuance Communications, OpenText and Verint Systems are among the organizations providing speech recognition services.

4. Deep learning: platforms use a type of ML involving artificial neural circuits with different layers of abstraction which can imitate the human brain, process

data and create decision-making patterns (Figure 5.10). It is now mainly used for pattern recognition and classification of applications that are only compatible. Technologies all have profound learning possibilities worth exploring.

All of the innovations have tremendous learning opportunities worth pursuing.

5. Image recognition: This is the method of recognizing contents in images where the identification of an object or feature is progressively stacked in a digital image or video and AI is used to great effect on the technology (Figure 5.11). To identify, AI will search for images on social media sites and compare them to a wide variety of data sets to determine which ones are the most important for a search. Clarifai provides customers with image recognition systems to spot near-duplicates and identify uncategorized images similar to these.

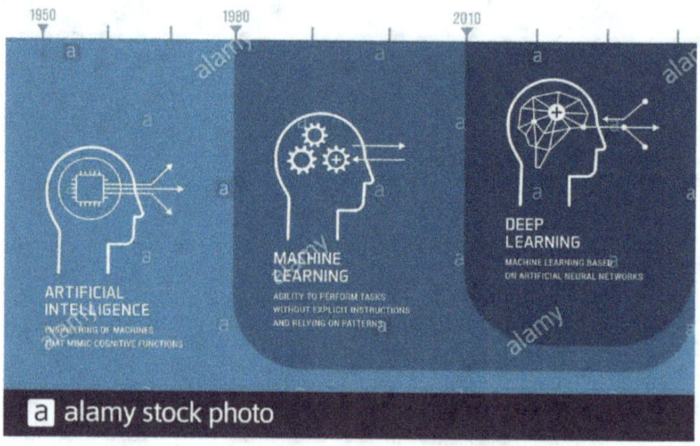

Figure 5.10 Deep learning in AI [11]

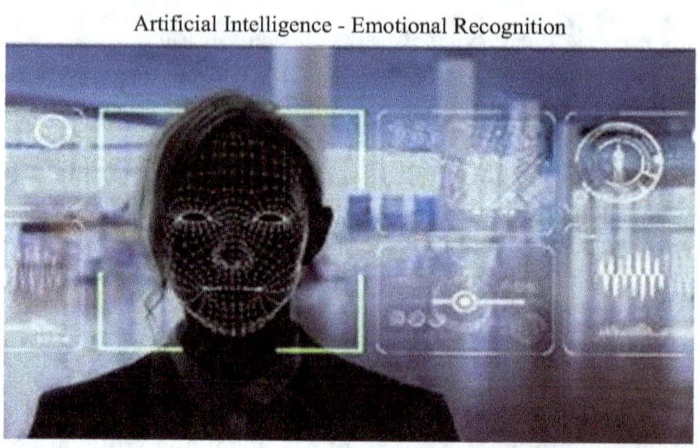

Figure 5.11 Image recognition in AI [12]

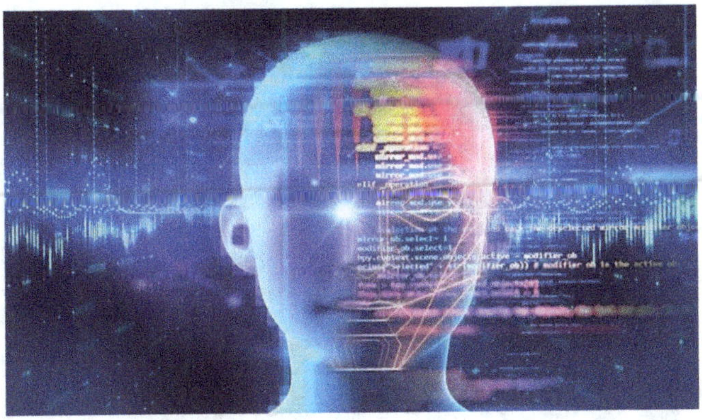

Figure 5.12 Biometrics in AI [13]

This is one of the pioneers in that field and improves face recognition technology that can be applied to bank card authentication and other applications for payment and picture processing. The mission of GumGum is to unlock the value of AI-based images and videos created online.

6. Biometrics: This technology will define, calculate and analyze the structure and form of the human body's behavior and its physical aspects (Figure 5.12). It allows for more natural human–machine interactions, including touch, image, voice and body language–recognition interactions, and is widely applied in the field of market research. Biometrics can be used to create better and more open records of the work of the employee's hours, such as entry and exit time. All biometrics companies are working hard to develop this area

   Applications are based on:

- Universality: everyone should have that characteristic.
- Distinctiveness: based on that attribute, any two individuals should be sufficiently different to distinguish one another.
- Acceptability: citizens should be able to consider the biometric program and not feel intruded, invaded, and so on.
- Behavioral Biometrics: is based on the measurements of human activity and related user action data; and, thus, it indirectly measures certain characteristics of the human body. This group includes signature, gait, gesture and finger-stroke as the key to the recognition of human identity.

5.5 The ethics of AI

The ethics of AI and robotics is often focused on "concerns" of civilization's loss of humanness and human beings being denigrated to the status of automatons,

Table 5.1 GAP analysis and benchmarking

Author (year)	Objective of the study	Methodology/approaches/ tools/techniques used	Key findings	Remark
Ezgi Ertu Rk *et al.* (2014) [14]	Computer fault forecasting	SVM, ANN and ANFS10-fold cross-validation test modeWEKA tool used	Performance measures using SVM, ANN and ANFS were 0.7795, 0.8685, and 0.8573	They used NASA, MDP, and maximum accuracy was 0.857 for data using ANFS method
Malkit Singh *et al.* (2013) [15]	Business Technology Fault prediction at an early stage of the life cycle of device development	Levenberg-Marquardt (LM) algorithm based ANN Polynomial function (PF) based on ANNMATLAB R2011a	Accuracy with LM-based ANN was 88.1%Accuracy with PF-based ANN was 78.8%	They used data collection Ant 1.7 and the average accuracy they achieved using ANN based on LM was 88.7%
Saiqa Aleem *et al.* (2015) [16]	Comparative ML approaches with program prediction mode for publicly accessible data	10-cross validation test mod-eMLPSVMBaggingRandom forest	The mean precision of the software fault prediction model for the 15 data givenDates DataSVM (89.29%)Bagging (89.38% Random forest (89.08%)	They used NASA, MDP data set and maximum accuracy was 99.52 for PC2 data using MLP method
Venkata *et al.* (2005) [17]	To predict software fault	70% training and 30% testingDecision treeLogistic RegressionNeural NetworkWEKA tool usedNASA MDP (CM1, JM1, KC1, and PC1)	IBL and 1-R have better consistency accuracy when compared with other methodsIBL-MAE (0.0543)1-R-MAE (0.0351)	We used the NASA MDP data collection and the closest accuracy was 0.0543 using MAE performance on CM1 data
Martin Shepperd *et al.* (2014) [18]	Factors for predicting performance of device faults	Software fault prediction meta-analysis for previous studies, such as all relevant and quality primary studies	They conclude that research into defect prediction should be carried out on blind analysis, improving reporting protocols and intergroup protocols	It was a factor survey that affects predictive accuracy of the software faults

Xi Tan *et al.* (2011) [19]	Using recall and precision to improve the performance of software fault prediction model	Eclipse 3.0 data90–10% data split	Cluster-based prediction of defects in both recall and precision better than class-based modelReminder (31.6–99.2%) precision-Precision (73.8–91.6%)	They used Eclipse 3.0 data and the maximum accuracy of 99.2%(Recall)
Qinbao Song *et al.* (2011) [20]	They proposed and evaluated a general framework study to test the concept of software fault prediction	NASA MDP and AR Data sets90–10% data splitTwo data preprocessorsTwo attribute selectionsNaïve BayesJ48One R	Using the maximum precision scored on PCI data set Naïve Bayesian Method (89.7%)	They used NASA MDP and 1.7; the maximum accuracy was 89.7 for PCI data using Naïve Bayesian
Okutan *et al.* [21]	Promise data repository	Nine data sets the application repository NOD for developer number and the source code and consistency LOCQ	Bayesian network	LOC, LOCQ, and RFC are the most effective metrics
Luc, Véronique and Pascal [22]	An open-source Eclipse Foundation and NASA IV and IV and V facility dataset	Halsteed line of code and Mc cape complexity metrics	Fifteen Bayesian Network (BN) classifiers	The Augmented Naïve Bayes classification shows better performance compared to rising Naïve Bayes

Table 5.2 The characteristics of data science, ML, deep learning and AI

Data science	ML	DEEP learning	AI
• Need to encompass entire analytics universe • Branch that deals with data	• Combination of machine and data science • Machines utilize science techniques to learn about the data; hence, called ML	• Specific branch of ML that deals with different flavors of neural network Examples: • Simple neural network • Convolutional neural network • Recurrent neural network • Long short Memory	• Empowering machines to decide on their own • Thus machines can understand and react rapidly according to the situation
• Data operations related to date i.e. • Data cleaning • Data gathering • Data subsetting • Data Manipulation	• Building model evaluation model and validation	Mainly utilized in – Object detection in image and video – Speech recognition – Neural language processing and understanding	• As the name suggests, it imparts humans' natural intelligence to machines

which response is not new to the arrival of technologies. Some such concerns turn out to be rather quaint (trains are too fast for souls); some are predictably incorrect in suggesting that technology will radically alter human beings (telephones will disrupt personal correspondence, writing will kill memory, video cassettes will make it redundant); some are usually right but fairly relevant; (digital technology will destroy industries that make photographic film, cassette tapes, or vinyl records); but some are broadly correct and deeply relevant (cars will kill children and fundamentally change the landscape). Innovations, such as nuclear power, cars or plastics, have given rise to ethical and political controversy and substantial policy attempts to regulate the progress of such innovations, typically only after some harm has been done. In addition to these "ethical issues," emerging innovations question existing norms and conceptual frameworks which are of particular philosophical interest. Finally, once we have understood technology in its context, we need to shape our societal response, including regulation and law [23]

- The characteristics of AI:

The five AI characteristics identified here relate to complexity, localization, uncertainty, capital intensiveness and awareness. Some of these features are in principle common with other high-profile technologies, but still bear some aspects specific to AI, albeit only the intensity with which they characterize the technology. AI technology's complexity stems, in part, from its multidisciplinary nature. Some AI sub-areas are locally bound, and thus local technological skills are required to build on the basic concepts. The ambiguity in AI projects stems primarily from the two factors of high change rates and the difficulties in quantifying the benefits from the transition of cognitive load from human to machine. Expensive human skills and enabling technologies are required to develop, maintain and use AI systems efficiently, making such projects highly capital intensive [24].

5.6 Stage of intelligence

We need to define intelligence and define what we mean by human intelligence. The stage of intelligence applies to any situation seeking to solve complex problems. According to the Oxford dictionary definition, intelligence is the ability to understand, learn and think. Several capabilities can be considered indicators of intelligence, namely: learning or understanding from experience, extracting meaning from a matter that is likely to mean extracting meaning from conflicting or conflicting messages. From this definition it can be concluded that if I have knowledge on the machine vested with some intelligence, then we must have created it with three abilities: understanding, learning and thinking. So the new way of trying to simulate human intelligence has created an emergency of new science called AI. There are different definitions of AI, as one of its meanings (in the words of John McCarthy who launched it at the Dartmouth Conference) is to make the machine behave in a way that can be called intelligence, if humans behave the way the machine behaves. We have found AI to give the computer the ability to solve problems and make the right decisions in a logical way. The goal of AI is to build machines that carry out tasks that require human intelligence in normal situations. However, all of these definitions describe one meaning, which is simulating human intelligence. The field of AI has opened up new technological applications such as robotics, expert systems, games, language processing (NLP) and image and shape recognition given the subject of AI has evolved with many sciences. In addition, innovations related to new computer systems inspired by biological sciences clearly demonstrate that this science is multifaceted knowledge. It is the application of expert systems, which makes the computer see humans as the person sees it, so it analyzes the image as it is analyzed by the common man and distinguishes them in terms of differences in the face, for example, or in the handprint or the eye. Among its military, medical, agricultural and space applications, in the military, an expert system has been developed that helps to know the enemy's strength, number, and how many weapons and other weapons they need to confront the enemy and where he is. Specifically, in monitoring, if a person enters a place where entry is forbidden, AI can tell everything about the prohibited entrant. In the medical field, there are many expert systems used, including "datage", which works on AI. In medical corridors, the requirement from AI is to distribute the patients and their files according to what is required and according to the departments designated to the patients. As for agriculture, it helps by giving suggestions and information to improve agriculture and hybridize it for the better. AI and expert systems were used in space science where robots were sent to the moon and Mars, taking samples from the heavenly bodies and analyzing them for results.

Even now, in this crisis of Covid-19, some of the countries are using AI, including the United Arab Emirates, which has widely used a system of thermal cameras to measure the temperature of objects, and used robots and "Drones" aircraft for sterilization.

5.7 AI in 2030

Intelligence is a rare and valuable commodity (Figure 5.13). From the mysterious brain of the octopus and the swarm intelligence of ants, to Go-playing deep learning machines and driverless vehicles—intelligence is the most powerful and precious resource in existence. Despite recent advancements in AI that allow it to win games and drive vehicles, there are endless untapped opportunities for advanced technology to have major and beneficial worldwide impacts. Especially so when AI and robotics intersect. Driven by three major trends, we are now in the midst of a powerful new wave of AI. Today we employ both robots and AI, but the two are not yet combined in any profound way. We are in the early days of the evolutionary saga of intelligent machines. Factories, for example, are filled with not-very-smart robots, fulfilling their destiny by fastening bolt after bolt, weld after weld. Various tech companies, meanwhile, are starting to deploy "embryonic" AI systems that do nothing more than flap their metaphorical machine-learning limbs right now. They have no sense of the situation. Such devices, in line with the metaphor, neither know what flying is, nor why one may want to fly from point A to point B, but they can attain flight someday. ML systems are the equivalent of the old pre-internet joke of computing: if you had. One business is enthusiastically embracing robotics coupled with AI, realizing the potential for lower costs, increasing efficiency and health, allowing new product innovations and providing operational insights into the business. It is interesting to

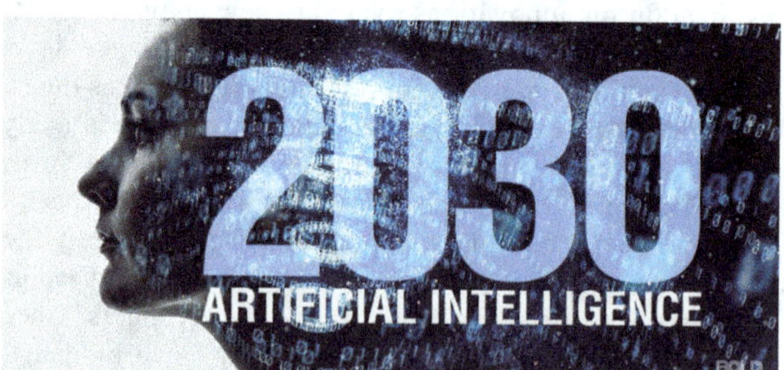

Figure 5.13 AI in 2030 [25]

note that only a few business leaders are inspired by possible labor force cuts or risky "moonshots" to use AI technology to transform existing operations.

These technologies are highly tailored to specific tasks though impressive. Typically, any application requires years of specialized research and careful, unique construction. Substantial improvements in potential uses of AI technology in similarly focused applications, including more self-driving vehicles, healthcare diagnostics and personalized therapies, and physical assistance for the elderly can be achieved. The study panel found no reason for alarm that AI is an immediate danger to civilization, contrary to the more sensational forecasts for AI in the mainstream press (Figure 5.14). No machine has been developed with self-sustaining long-term goals and intent, nor are they likely to be developed soon [26].

1. Healthcare
   The entry of the technology giants like Microsoft, Google, Apple and IBM in the healthcare sector holds significant importance for the industry. AI is currently being used for many healthcare needs, including data mining for pattern recognition and the more precise medical condition detection and treatment, medical imaging, medication management, drug discovery and even robotic surgery (Figure 5.15). A case in point, in this regard, is the application of IBM Watson (an AI tool) to extract the meaning and context of a collection of structured and unstructured data which may be critical to the selection of a treatment plan.
2. Retail and e-commerce
   Retail and e-commerce (Figure 5.16) are perhaps the only space where the application of AI is the most discernible to the majority of end-users [29]. As a competitive space, organizations always look for techniques to find patterns in consumer behavior and thus align their strategy to outsmart their competitors.

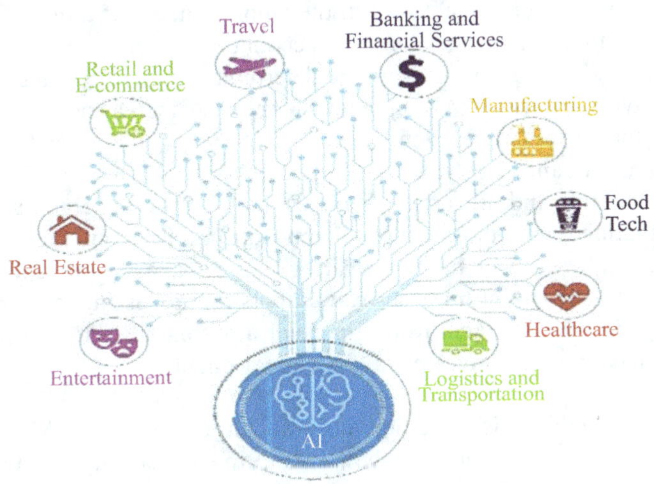

Figure 5.14 The significant AI applications [27]

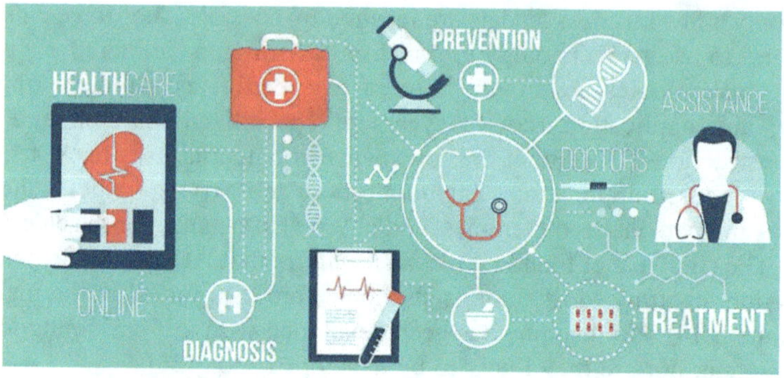

Figure 5.15 Healthcare in AI [28]

Figure 5.16 Retail and e-commerce [30]

Surely AI has found a sweet spot in the whole scheme of things. The product recommendations on your Amazon account are merely an application of complex AI algorithms in real-time to determine which categories and types of products you are using. AI applications are also increasingly being used to enhance the customer experience.

3. Banking and financial services

   The Banking and Financial Services industry is undergoing a massive transformation due to the onset of AI applications. There are plenty of AI uses cases in this sector. In many scenarios, human agents are replaced within fractions of a second by smart software robots for processing loan applications. Similarly, in split seconds, Robo-financial advisors sift through various data rates to make the best investment choices for clients.

4. Travel

   The travel industry is set to derive significant benefits from AI riding on the widespread use of chatbots. Chatbots are a proven means to enhance customer service and interaction mainly due to their 24 × 7 presence and instant question resolution.

Given advanced AI algorithms power chatbots with improved efficiencies, enabling them to respond more accurately to customer queries, therefore it follows that many major travel agencies are turning to AI companies to develop their AI-powered customer service assistants.

5. Entertainment and gaming

   Having entered the entertainment industry, AI is helping program producers and broadcasters to identify which shows or programs to recommend to individual users based on their activity on their websites. It helps Netflix and Amazon give users a more personalized experience, and ML algorithms are widely used to study user behavior where such algorithms are becoming more intelligent over time, to the extent that they can determine whether a user wants to buy a product for themselves or gifting purposes, or whether family members have different watching preferences.

6. Marketing

   Marketing is multifaceted activity and deals with many operations that could benefit from total or partial automation (Figure 5.17). For this reason, marketing technology, or martech, is crucial to success in business. Martech is the software used to improve marketing campaigns and strategies, and AI technology is currently transforming it [31].

7. Transportation

   AI and ML algorithms are used for prediction, monitoring, and traffic management (Figure 5.18). Let us take one self-driving car as an example. Still in the research phase, self-driving vehicles seem to have a promising future in the transport sector. Experts are working hard to construct algorithms for such a complex product where the new system can provide new features, such as analyzing and optimizing data from many sources, routing, mapping, and navigation by situations.

   - How AI will impact communication in the future

     AI is not restricted to any generation, but AI can affect communication in the future which can be used in many fields such as self-driving cars, healthcare and many other applications. AI technology is neither good nor bad *per se*. It is, in fact, a reflection of the person's heart using it or unleashing it by automation

Figure 5.17 Marketing of the industries [32]

Figure 5.18 Transportation in AI [33]

- There are ways in which AI will help communication [34–41].
- AI analyzes the global environment of what I call global respiration, living networks and reports on concrete insights, trend analysis and real-time alerts (up to the minute). Let imagination run wild. How would you do if every minute of the day you had the breadth of detail at your fingertips?
- AI will deliver news to target audiences (even micro-target audiences) in new and innovative ways using virtual and augmented reality applications. Editors, analysts and staff would be able to digitally "play" from the comfort of their offices into a meeting room (or White House briefing room), and attend a news conference, briefing, or information session.
- AI will enable faster responses to crises, following preset parameters as part of human-centric contingency plans. AI bots will be programmed to assist crisis communication leaders—and they would not be swayed by emotions in heated crises.
- AI is set to deliver better metrics for corporate communications. Knowing what I know about ML and deep learning algorithms, I predict that the metrics from marketing automation services that corporate communications teams will get a decade from now will be better than what today's marcom teams are getting.
- AI will be the method through which a new concept called identity-based corporate communications is expected to appear. The accuracy will be of incredible performance, with communications targeted to individual needs.

5.8 Conclusions

AI is useful to the community which helps us to dispense extraneous employees. From the above, we arrive at the conclusion that AI is a name given to a set of new methods where the methods in programming systems can be used to develop systems that simulate some element of human intelligence and allow them to make inferential processes about facts and laws that are represented in computer memory. Many theories of this new science are still under research and development, but there are some techniques dependent on it that started to emerge from the research in the scientific field, and they have proved their effectiveness by completing works that were almost impossible to do using traditional programming. It also shows us that the saying The saying AI is computer software that simulates the way people think is proved wrong since scientists in this field have not been able to reach that aim as yet and, instead, most computers still show a clear degree of industrial stupidity.

References

[1] Asilayi. (2018), Pic of Using Artificial Intelligence, https://medium.com/civic-analytics/smart-traffic-lights-or-intelligent-traffic-lights-are-a-vehicle-traffic-control-system-that-ab5429ca0c40 (Accessed 10 November 2021).

[2] J Scott. (2017), The History of Artificial Intelligence, http://sitn.hms.harvard.edu/flash/2017/history-artificial-intelligence/ (Accessed 10 November 2021).

[3] L Columbus. (2018), 10 Charts That Will Change Your Perspective On Artificial Intelligence's Growth, https:// forbes.com/sites/louiscolumbus/2018/01/12/10-charts-that-will-change-yourperspective-on-artificial-intelligences-growth/#4ab26c554758 (Accessed 10 November 2021).

[4] M Feldman. (2016), Market for Artificial Intelligence Projected to Hit $36 Billion by 2025, https://www. top500.org/news/market-for-artificial-intelligence-projected-to-hit-36-billionby-2025/ (Accessed 10 November 2021).

[5] M Armstrong. (2016), The Future of AI, https://www.statista. com/chart/6810/the-future-of-ai/ (Accessed 10 November 2021).

[6] R Tuffcon. (2020), Artificial intelligence in Global Platforms, https://thecantonsentinel.com/technology/61962/artificial-intelligence-as-a-service-market-prescribes-strong-growth-business-boosting-strategies-growth-opportunities-by-forecast-2025/ (Accessed 10 November 2021).

[7] B Jhonson. (2018), Pic of Used Technology, https://medium.com/@Brian.johnson_62680/artificial-intelligence-ai-top-use-cases-and-technologies-used-today-3c22e1a63e78 (Accessed 10 November 2021).

[8] M Parsons. (2017), Robotics Technology Picture, https://www.supplychaindigital.com/technology/automation-and-robotics-supply-chain-future (Accessed 10 November 2021).

[9] S Beloussov. (2017), Machine Learning Picture, https://www.dailyhostnews.com/quick-comparison-machine-learning-platforms-amazon-microsoft-google-ibm (Accessed 10 November 2021).

[10] C Neolle. (2019), Speech Recognition, https://www.processmaker.com/resources/customer-success/blog/how-artificial-intelligence-will-disrupt-speech-recognition/ (Accessed 10 November 2021).

[11] S Bagachi. (2020), Pic of Deep Learning, https://www.cxotoday.com/news-analysis/deep-learning-shows-promising-growth-amid-challenges/ (Accessed 10 November 2021).

[12] V Bogdanov. (2019), https://becominghuman.ai/smart-face-control-how-facial-recognition-chan ges-different-industries-f03204496eaf

[13] T Qu, (2019), Biometrics Pic, https://www.scmp.com/tech/science-research/article/3023540/china-rapidly-catching-us-battle-ai-supremacy-adoption-and (Accessed 10 November 2021).

[14] E Erturk and E Sezer (2014), A comparison of some soft computing methods for software fault prediction. Expert Systems with Applications. 42. 10.1016/j.eswa.2014.10.025.

[15] M Singh and D Singh Salaria (2013), Article: software defect prediction tool based on neural network. *International Journal of Computer Applications* 70 (22):22–28.

[16] S Aleem, L F Capretz, and F Ahmed, Benchmarking Machine Learning Technologies for Software Defect Detection", e-prints, 2015.

[17] V S Vongala (2005), Knowledge-based fault detection using time-frequency analysis. LSU Master's Theses. 3716. https://digitalcommons.lsu.edu/gradschool_theses/3716

[18] J S Martin, B "https://dblp.org/pid/39/7630.html" David, and H "https://dblp.org/pid/69/3214.html" Tracy (2014), Researcher bias: the use of machine learning in software defect prediction. "https://dblp.org/db/journals/tse/tse40.html#ShepperdBH14" *IEEE Transactions in Software Engineering* 40(6): 603–616.

[19] X Tan, X Peng, S Pan, and W Zhao (2011), Assessing Software Quality by Program Clustering and Defect Prediction. 244-248. 10.1109/WCRE.2011.37.

[20] Q Song, Z Jia, M Shepperd, S Ying, and J Liu (2011), A general software defect-proneness prediction framework. *IEEE Trans. Software Eng* 37:356–370.

[21] A Okutan (2018), Use of Source Code Similarity Metrics in Software Defect Prediction. NASA's IV&V Program, https://www.nasa.gov/centers/ivv/about/index.html.

[22] M Luc, P Véronique and P Pascal. (2010), Orekit: An open source library for operational flight dynamics applications.

[23] M Anderson and SL Anderson. (2007), Machine AI Ethics. Stanford Encyclopedia of Philosophy, https://plato.stanford.edu/entries/ethics-ai/ (Accessed 10 November 2021).

[24] AS Tabandeh. (1994), The Characteristics of Artificial Intelligence Springer, https://link.springer.com/article/10.1007/BF01195055 (Accessed 10 November 2021).

[25] A Sears. (2018), Figure 1, https://www.boldbusiness.com/digital/ai-will-change-four-major-industries-by-2030/ (Accessed 10 November 2021).

[26] M James. (2019), Artificial Intelligence in 2030, https://qeprize.org/news/future-of-ai (Accessed 10 November 2021).

[27] A Taykar, (2018), Figure 14, https://www.leewayhertz.com/ai-applications-across-major-industries/ (Accessed 10 November 2021).

[28] A Choudhary. (2020), Figure 15, https://datafloq.com/read/artificial-intelligence-smart-systems-healthcare/7421 (Accessed 10 November 2021).

[29] H Wenxing, Y Weng, L Xie, L Maoqing, Design and implementation of web-based DSS for online shopping mall, 2009 IEEE International Conference on Control and Automation, pp. 1308–1313, 2009.

[30] C Olivera. (2019), Figure 16, https://medium.com/@Countants/how-artificial-intelligence-is-transforming-the-e-commerce-industry-countants-scalable-custom-73ae06836d35 (Accessed 10 November 2021).

[31] A Takyar. (2018), Artificial Intelligence in Industries, https://www.leewayhertz.com/ai-applications-across-major-industries/ (Accessed 10 November 2021).

[32] R Rynoso. (2017), Marketing in AI, https://learn.g2.com/industries-using-ai (Accessed 10 November 2021).

[33] V Chadouray. (2019), Figure 18, https://hackernoon.com/examples-of-ai-implementation-in-transport-c1a451b90fb7 (Accessed 10 November 2021).

[34] S Huang, DH Zhang, WY Leong, HL Chan, KM Goh, Detecting tool breakage using accelerometer in ball nose end milling, The 10th International Conference on Control, Automation, Robotics and Vision, pp. 927–933, 2008.

[35] WY Leong, J Homer, DP Mandic, An implementation of nonlinear multiuser detection in Rayleigh fading channel, *EURASIP Journal on Wireless Communications and Networking* 2006;2006:1–9, Article ID 45647. DOI10.1155/WCN/2006/45647.2006

[36] WY Leong, J Homer, Blind multiuser receiver in Rayleigh Fading channel, *2005 Australian Communications Theory Workshop*, 2005, pp.155–161.

[37] HF Min, LW Yie, Investigating target detection and localization in wireless sensor network, *Procedia Engineering*, vol. 41, Elsevier, New York, NY, pp. 75–81, 2012.

[38] WY Leong, Angle-of-arrival estimation: beamformer-based smart antennas, The 3rd IEEE Conference on Industrial Electronics and Applications, 2008, pp. 1593–1598.

[39] WY Leong, DP Mandic, Towards adaptive blind extraction of post-nonlinearly mixed signals, *The 16th IEEE Signal Processing Society Workshop on Machine Learning for Signal Processing*, 2006, pp. 91–96.

[40] WY Leong, *Implementing Blind Source Separation in Signal Processing and Telecommunications*, The University of Queensland, Australia, 2005.

[41] W Xiaofan, PHJ Chong, LW Yie, Performance comparison of CSMA/CD, CSMA/CA, CSMA/RI, CSMA/PRI and CSMA/PR with BEB, The 5th IEEE Conference on Industrial Electronics and Applications, 2010, pp. 1843–1848.

Chapter 6

5G and beyond environment for smart manufacturing

Sultan Salah Sultan Melhi[1] *and Wai Yie Leong*[1]

6.1 The current communication system

This applies to the fourth era of cell remote networking and is the continuation of 3G and 2G standards while 4G is a broader concept. It may also include requirements external to IMT. 4G will update existing communications networks and aims to have a robust and stable IP solution such as voice services.

Consumers are supported with mobile and video access "anytime and anywhere," at significantly higher data speeds relative to previous generations.

Following are the most important goals of the fourth generation communication networks:

1. High network capacity-more users per cell at the same time.
2. An average data rate, between any two points in the world, of at least 100 Mbps.
3. Transformation over heterogeneous networks smoothly.
4. High network availability over several networks in regional roaming.
5. High-quality multimedia service for the next generation, for example, real-time voice and high-speed data.

6.2 Introduction

The new mobile technology occurs almost every 10 years after the first version of the first mobile telephone contact network in 1979, and the first 2G system service began in 1991. Also, for the third-generation first operation in 2001 and fourth- generation first operation in 2009. That indicates to the world that in every 10 years, during each period the generations establish touch. There are also variations between 4G users and 5G technologies, for example, which are anything but optimizing productivity:

1. Low battery uptake.
2. Reduced probability of discontinuation, in the context that expansion coverage is suggested.

[1]The Institution of Engineering and Technology, Malaysia

3. Competitive rates in larger parts of the service area.
4. Cheaper or no traffic charges due to lower installation of the network cost.
5. Higher total capacity at one time for several users.

In several papers and research projects, the 5th Generation Cellular Networks (5G) is a concept that is used to describe a potential big stage of mobile connectivity standards beyond the fourth generation of current standards. 5G does not define any requirements in any official document issued by any agency responsible for telecommunications standards.

While the revised specifications that identify capabilities that surpass those defined in the current 4G are being considered, those new capabilities are still being compiled according to the current 4G specifications and the initial fifth-generation specification was released at the end of 2017 with the identification of the candidate's most relevant requirements and technologies.

6.3 Differences between 4G and 5G

Table 6.1 shows the differences between 4G and 5G.

6.4 Why is 5G a big deal?

Over the past few years, 5G has been picking up, and a lot of it is justified. From a technology viewpoint, 5G can offer peak speeds of up to 10 Gbps compared to LTE-Advanced Pro, which can enable peak speeds of 3 Gbps. So on average, 5G is arguably ten times better than 4G LTE. It is important to note that it all depends on your mobile operator as to how much capacity they allocate, which configurations they use [1].

6.5 What makes 5G faster?

The shorter the frequency, the greater the bandwidth, according to communication standards.

Table 6.1 Differences between 4G and 5G

4G	5G
IPV4 address is 32 binary cells	IPV6 address is a 128-digit binary
The data speed for mobile devices is 100 Mbps	The data speed for mobile devices is 1 Gbps
The data speed for stationary users is 1 Gbps	The data speed for stationary users is 10 Gbps
The delay time in the fourth generation is up to 10 ms	The delay time in the fourth generation is up to 1 ms

In 5G networks, the use of shorter frequencies (millimeter waves between 30 GHz and 300 GHz) is the reason why 5G would be quicker. This high-band 5G spectrum provides the anticipated boost not only in speed but also in power, low latency, and quality.

The 5G download speed, however, can vary widely by region.

According to Fortune Magazine's February 2020 issue, 5G speed steps are taken on average.

6.6 Difference between 1G, 2G, 3G, 4G, and 5G

Every day we learn about the difference between faster communication and improved output of new technologies. Nonetheless, communications networks were beginning to develop innovations to produce the best performance.

6.6.1 First-generation 1G

First communication generation, which works with the Frequency Division Multiplexing technology known as FDM, was only concerned with calls and operated on analog waves. There are no SMS or Internet connections.

6.6.2 Second-generation 2G

This generation is the most famous in the communication world and has been recognized as GSM or Global Mobility Network. This generation relied on modern digital signal technologies as the first generations focused on digital signs–used technologies such as TDMA and CDMA, which opened the door to new services like SMS and E-mails. This connectivity generation was used to create new innovations in it before the 2.5G or GRPS technology began to hit 144 Kbps of data speed.

6.6.3 Third-generation 3G

This generation relied on UMTS technology, which allowed more data processing and a speed up to or just over 2 Mbps. Other new features including video calls and GPS features have been introduced. Despite the great benefits that occurred with this century, it has created some minor faults, which are higher costs and higher energy usage the network developed into 3.5G with the introduction of HSDPA and HSUPA technologies which increased download speeds to 14.4 Mbps and uploads to 5.8 MB.

UMTS provides a clear evolutionary path to high-speed packet access (HSPA). HSPA refers to the combination of high-speed downlink packet access (HSDPA) and high-speed uplink packet access (HSUPA). HSDPA allows data rates up to 14.4 Mbit/s in the downlink and 5.8 Mbps uplink whilst HSPA+ data rates are in the order of 84 Mbps and 21 Mbps respectively.

6.6.4 Fourth-generation 4G

This version is based on the LTE and WiMAX specifications, providing super speeds of up to 173 Mbps, which were built up to 25 Mbps using the latest LTE-A specifications and enabling you to download a film in less than half minutes within 800 MB.

6.6.5 Fifth-generation 5G

In terms of speed and performance, the picture alone is enough to express what fifth-generation network changes can deliver. The technology is still under trial, with just a speed of 1 Gbps anticipated to be issued commercially by 2020. We don't really have enough details about the system because it's still being checked. I think that we need to wait for the news.

6.7 The evolution of the 5G

The fifth era's key trademark is the web for everybody or the web of things (Figure 6.1). This is, each individual and everything will be associated with the web, such a significant number of devices or gadgets will be associated with the web in the home, in the city, or any work environment, and that is the thing that drives us to the term shrewd urban communities, in light of the fact that the information is created anyplace by everybody. Any PC that can be assessed as effectively as conceivable to derive important data in an opportune way, for example, following the wellbeing status of patients and the old, testing gadgets and gear at home, and evaluating whether a substance is faulty or not, just as surveying the traffic circumstance in the lanes and supporting and educating drivers concerning the risks. This likewise makes it ready for moving into self-driving vehicles. Versatile contact regularly assumes a key job in permitting this data to be transmitted proficiently and securely from machine to machine and without human impedance preparing it and making fundamental moves at the max throttle and with

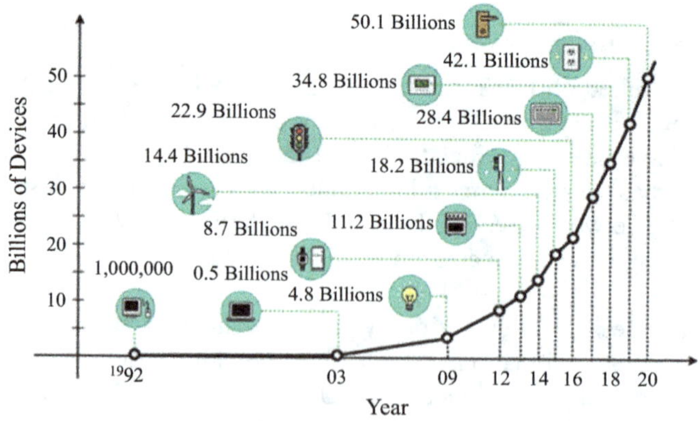

Figure 6.1 Interpretations of the rise for the fifth generation in smart devices [2]

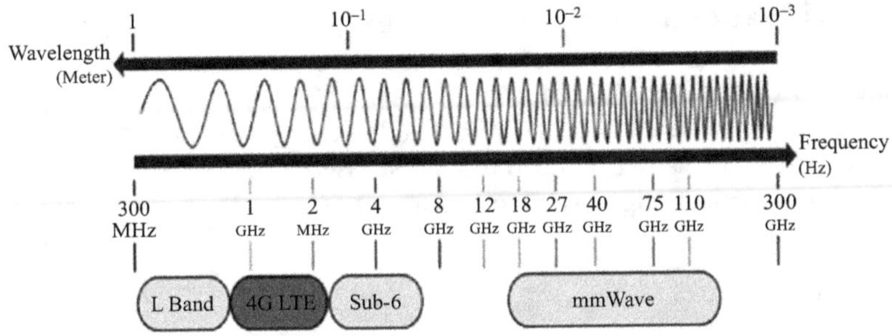

Figure 6.2 Networking spectrum bands [3]

less postponement. It also shows issues dealing with PDA information. The immense number of one systems and sensors that are introduced wherever to be an enormous scope connected with full limit and dependability. Another serious issue is to make sure on delivery and update on information to administrators.

6.8 How does 5G work?

The 5G norm splits the work of the new generation into two sections (Figure 6.2), the first section relies on the technologies and frequencies used in the preceding generations and the second section is independent reliance on new waves. They need to explain how communication networks operate in general, to fully understand how they operate.

The 5G revolution uses a great deal of electricity and affects the environment. The fifth-generation 5G is more powerful than any other technology. For example, the 5G network needs less power than the 4G network to transmit the same amount of data.

The 5G system also makes it possible to move sensors and controls more easily, thus improving connectivity between linked equipment and improving the management of energy consumed during these connections. The fifth-generation 5G will help achieve environmental and climate targets in this way.

6.9 Features and advantages of 5G technology

1. It has a very high velocity, high power, and low cost per piece.
2. Video, speech, and Internet support.
3. It also provides global access and portability of service.
4. It has a very high speed of upload and download.
5. It provides a high resolution and broad bandwidth bi-directional for heavy smartphone users.
6. 1 Gbps or greater network bandwidth.
7. Complex access to information.

6.10 Disadvantages of 5G technology

1. It's a science under development.
2. High speed in some parts of the world is difficult to obtain.
3. Protection and privacy also need to be discussed.
4. Most old computers are not going to support 5G.

6.11 Applications

1. It allows you to know the temperature and the location and monitor the OC's with your handsets.
2. To make school learning even faster.
3. This has applications in the scientific sector at the same time.
4. Natural disasters can be detected; the universe, stars, and galaxies can be visualized.
5. Smart systems with features of artificial intelligence.

The new generation of wireless Internet networking is 5G or 5th generation. 5G should remedy the latency issues. Yet it must be noted that these ideas are all theoretical and fictional and it will take a lot of effort from government and mobile network operators to make this research come true. There is still a need to work out on the security part of 5G as 5G opens the door to a new level of hazard, with a larger number of users and better services.

China, South Korea, Japan, and the United States are the main countries that have already launched their 5G networks. The United Kingdom has already launched, and several more countries are making fast steps toward the launch of their 5G network. That's the worldwide example of objective 5G technology.

5G is expected to hit the market by 2020, as shown in Figure 6.4. By 2021, the quantity of 5G associations is expected to reach a figure of between 20 million and 100 million. A few appraisals put the figure at 200 million. Likewise, spending on 5G portable framework for that equivalent year is estimated to be at around 2.3 billion US dollars.

The diagram in Figure 6.3 will perceive in what manner 5G will reform administration. The quicker downloads would not just empower better quality and increasingly solid portable video, yet in addition move some escalated handling to the cloud, opening the route for progressively enlarged and augmented reality, multichannel video, and propelled games to multiply on versatile.

From Figure 6.6, 5G is supposed to be at least 10 times faster than 4G, but the unfortunate thing is that we are going to have to wait until 2020. Several nations are sadly stuck with plain old 2G.

From Figure 6.5, the worldwide 5G chipset market report covers verifiable market patterns, current market elements, division-by-locale showcase appraisal, nation-level investigation for each fragment, piece of the pie examination of key players, serious scene, and gracefully chain investigation.

From Figure 6.7, the market size of the little cell 5G arrange was evaluated at \$344 million in 2018 and is relied upon to hit \$6,876 million by 2026, ascending at

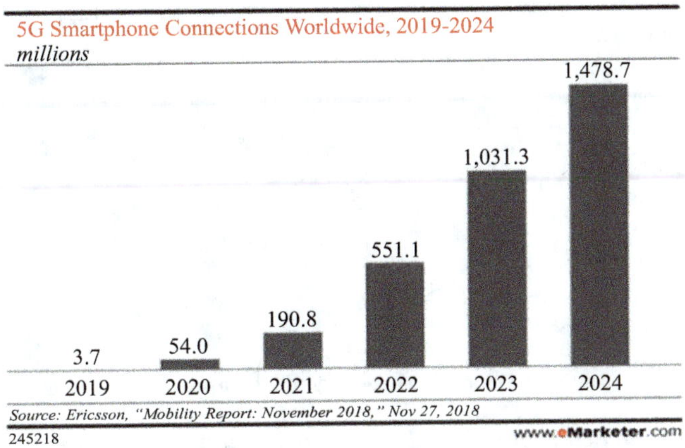

Figure 6.3 5G smartphone connections worldwide, 2019–2024 [1]

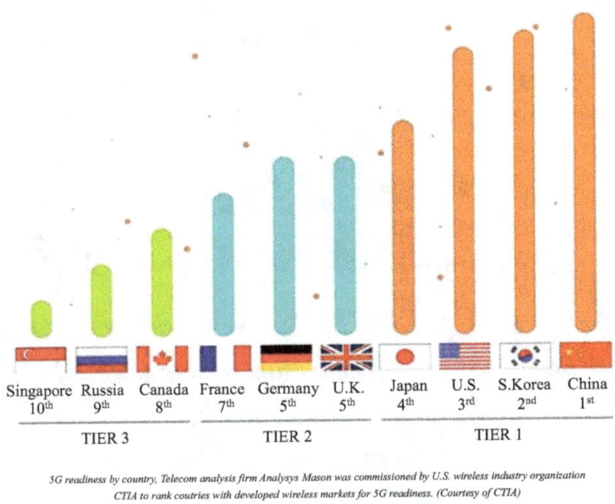

Figure 6.4 Statistic 5G network [4]

a CAGR of 45.3% from 2019 to 2026, North America commanded the little cell 5G organize showcase in 2018 and is anticipated to stay prevailing over the estimate time frame.

6.12 5G innovation

6.12.1 Introduction to 5G innovation

The new 5G arrangement innovation should be accessible by 2020, and understanding the way of the specialized advancement toward 5G is significant. This

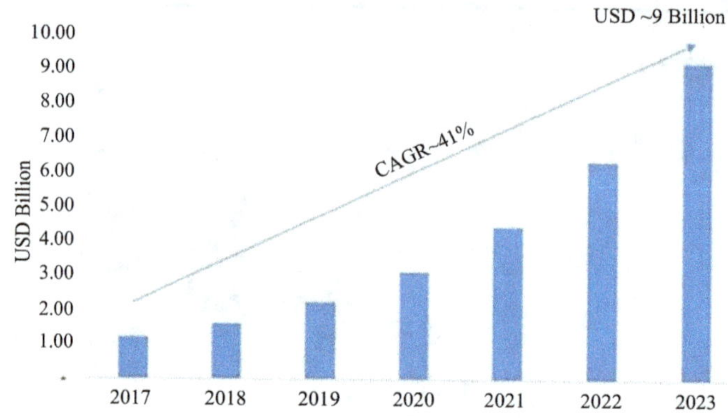

Figure 6.5 5G chipset market business opportunities, target audience [6]

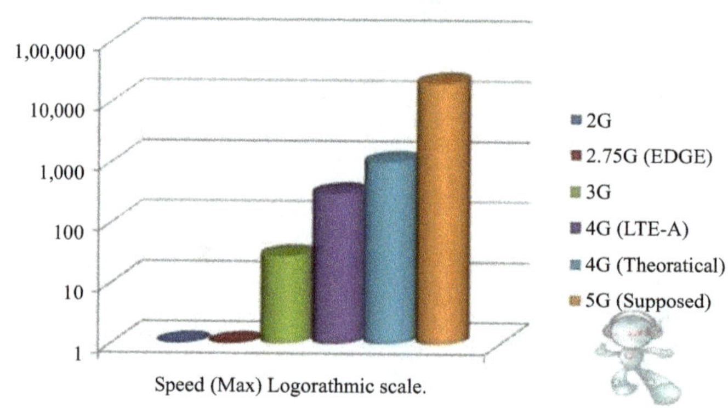

Figure 6.6 Speed (max) logarithmic scale [5]

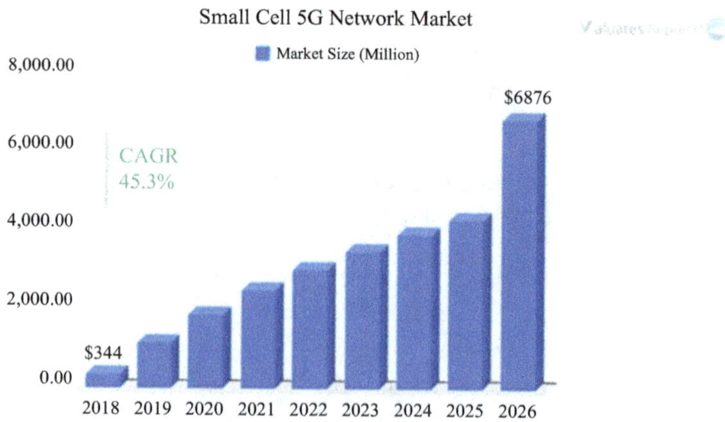

Figure 6.7 Small cell 5G network market [7]

chapter gives a broad and thorough outline of late exertion toward 5G development. It features basic highlights, for example, flexibility, network, and cloud-based administrations, as these must be guaranteed.

The world has seen an advancement of cell correspondence advances from 2G framework to 4G long-haul development. The fundamental inspiration was the more data transfer capacity and less dormancy necessities.

As the genuine pace of transmission of information is throughput, it could not be exactly the usable data transmission. The other significant dormancy boundary is the time delay between sending the message to the sender and disentangling the beneficiary in the wake of getting the equivalent. Dormancy fundamentally relies upon the speed at which every hub is handled through which the information stream navigates. We experienced higher system speed and quicker download speed making constant video calls as we ventured from the 2G GSM framework into the 3G UMTS framework. LTE and resulting LTE-An offered us expanded system transfer speed, diminished post-ponement in getting to application servers, making it conceivable to remotely get to significantly increase play traffic (information, voice and video) anyplace. Be that as it may, 4G is truly horrendous. Along these lines, 4G is extremely portable broadband. Although 3G was the principal portable broadband norm, it was initially planned for voice with some thought of sight and sound and information, where 2G was situated as the first computerized portable voice correspondence standard with more prominent inclusion than the conventional voice administrations simple to 1G.

5G should improve not just the information preparing velocity of the versatile system yet additionally the system's adaptability, unwavering quality, and vitality utilization. It is expected that by 2020, 50 billion gadgets will interface with the apparently changing worldwide IP arrangement.

6.12.2 Past technologies

Many things have changed in the past years because we become more modern day by day. As a simple example before this, we used letters to communicate with our friends and we had to wait like a week to get a reply from them. Now many technologies have made our lives so much more comfortable. This is the past technology we used to live in. First, WIFI is the most popular technology and it will connect with an electronic device to exchange data or connect with the internet wirelessly by using radio waves. In addition, with this technology, we can connect any digital device to a wireless network.

Moreover, before this, we used mobile phones and the communication devices are connected to wireless networks through radio waves and satellite transmission. Most mobile phones provide short message service (SMS), multimedia message service (MMS), and voice communication. In the 2000s era, mobile phones also provided a variety of other services such as email, Internet access, and Bluetooth. The purpose of using a mobile phone is for keeping in touch with family members or for business.

Another past technology is Bluetooth. Bluetooth is the wireless standard technology that is used for exchanging data over short distances. Bluetooth is used to exchange data using short-wavelength UHF radio waves in the industrial and medical radio bands from 2.402 GHz to 2.480 GHz and it also built their personal area network. Bluetooth were existing in products such as laptops, gaming equipment, headsets, telephones, media players, and watches. It is useful for transferring information between two or more devices that are near to each other.

More than that, there is one popular VoIP application in the entire world. It is a free premium that you can download and use for free. This is the most popular application like wildfire from the time of its release. Skype is used by people of any age to communicate chat easily using voice, video calls, or text with anyone on the account list. Small businesses have made the most use of this application that can make them interact with their customers and colleagues.

Last but not least, broadband is used to connect and to get information, and share it to each other. Broadband has such a high access speed Internet. It also offers four different forms like DSL, fiber-optic, cable, and satellite. It also refers to high-speed data transmission in which a single cable can carry a large amount of data at once.

6.12.3 Gap analysis and benchmarking analysis

The gap analysis and benchmarking analysis on the various communication techniques for 5G adaptation are shown in Table 6.2. Worldwide 5G selection is experiencing an intense time at this moment, Channels (Worldwide Innovation Market Examiner Firm) is gauging a splendid future for cutting-edge cell phones.

Figure 6.8 engaged with the overall estimate for 5G handsets by Waterways are very great at under 1.9 billion units dispatched somewhere in the range of 2019 and 2023, with an estimation of just about 800 million units in 2023 alone. If the

Table 6.2 Gap analysis and benchmarking analysis

Year	Author(s)	Analysis (objective with used techniques)
2018	Maji *et al.* [8]	Enhance the secrecy capacity in a dual-hop energy harvesting cognitive network using decode—and—forward relay employing a time switching relay scheme
2018	Chen *et al.* [9]	A novel wireless energy harvesting cooperative jammer aided transmission scheme for the maximization of achievable secrecy rates in presence of multiple eavesdroppers
2018	Qin *et al.* [10]	Two secondary user selection schemes to enhance the primary user security as a minimal interference-based scheme and maximal jamming rate-based scheme
2018	Nandan *et al.* [11]	To secure communication in a two-cell MIMO NOMA-based CRN using case ceded zero-forcing beamforming tech unique
2018	Baubella *et al.* [12]	Secondary user relay with multiple and tennis gives better secrecy performance than having one antenna. Secrecy per romance when transmitting from sec notary user to secondary receiver

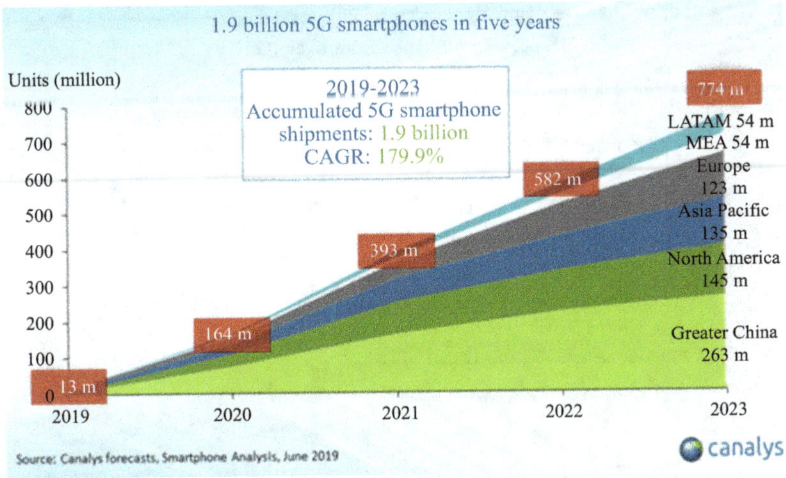

Figure 6.8 Smartphone market sales forecast June 2019 [13]

expectation seems, by all accounts, to be right, 51.4% of the worldwide versatile market in 2023 will mean this mind-boggling number of 5G telephones.

An increasing point-by-point Canals map contains an aggregate number of 774 million 5G telephones which will be conveyed in 2023. China would immerse no under 263 million 5G telephones followed by North America 145 million and the Asia Pacific 135 million, as per this table, Europe is relied upon to be the last district past the 100 million imprints with 123 million units.

Trenches examiner Mo Jia uncovers that this year Chinese administrators will toss not exactly US$ 5 billion for neighborhood 5G arrange development. Somewhere in the range of 70,000 and 90,000 5G base stations will be worked all through China this year. In the interim, it looks as if numerous different nations are occupied with restricting and unbanning Huawei and ZTE hindering the reception of 5G.

6.12.4 Benchmarking analysis

As this remote innovation of the cutting edge is multiplying over the globe and as a pioneer in 5G space, we wanted to test certifiable 5G advantages and bring a more profound plunge into its effectiveness. We in this manner charged a thorough worldwide investigation from Signs Exploration Gathering (SRG), one of the most regarded remote investigator organizations with profound cell execution skills.

With industrially accessible 5G gadgets worked by Qualcomm Snapdragon Portable Stages, SRG got a great many gigabytes of the test information from versatile systems in South Korea, the UK, Switzerland, and different US urban areas. It took a long time to dissect that included several boundaries.

Figure 6.9 shows the results of data levels reported on SKT, EE, Swisscom, and Verizon networks. Compared to existing LTE networks, the average capacity gains observed were at least 2×. Gains in peak performance topped 10×.

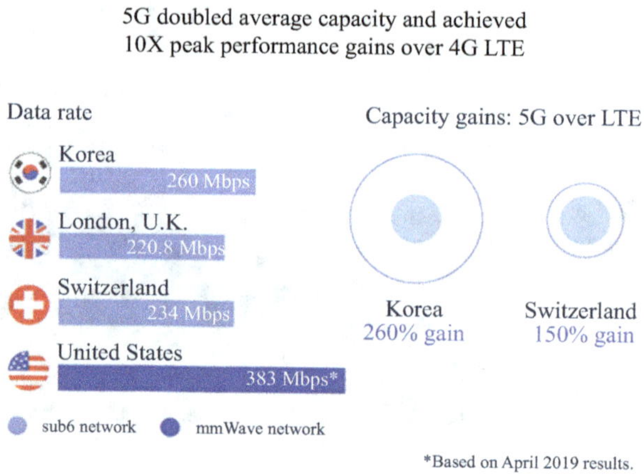

Figure 6.9 Testing multiple sub-6 GHz and millimeter wave [14]

6.12.5 Suitable concept(S) law (S) for 5G

The organization of 5G raises intriguing lawful and administrative issues because of the one-of-a-kind highlights of the innovation (Figure 6.10). 5G can be conveyed across frequencies, reasonable for various 5G applications—low groups of under 1 GHz are appropriate for longer arrangements (e.g., portable, broadband, and enormous IoT).

Mid groups between 1 and 6 GHz have a more extensive range for improved broadband ("eMBB") and strategic applications, whereas groups over 6 GHz are fitting for various applications requiring high information rates.

Some industry players have suggested that the 5G range be authorized broadly on a select premise to give speculation affirmation and range effectiveness, given the enormous scope of interest in new national systems required for 5G.

6.12.6 Mathematical model(s)

6.12.6.1 New dimension

The new dimension includes the demand, managing traffic and mobility, and using new methods for effective control of interferences in complex heterogeneous deployments.

6.12.6.2 Spectrum usage

Spectrum usage includes extended consideration of the operating spectrum band as well as operating in new spectrum regimes for the complete concept of a system spectrum that carefully addresses the needs in any scenario use.

6.12.7 The communication system and modulation system for 5G

Methods examined for 5G will in general utilize various bearers to accomplish ghostly adequacy. As of now, 4G LTE is utilizing QAM (quadrature playfulness

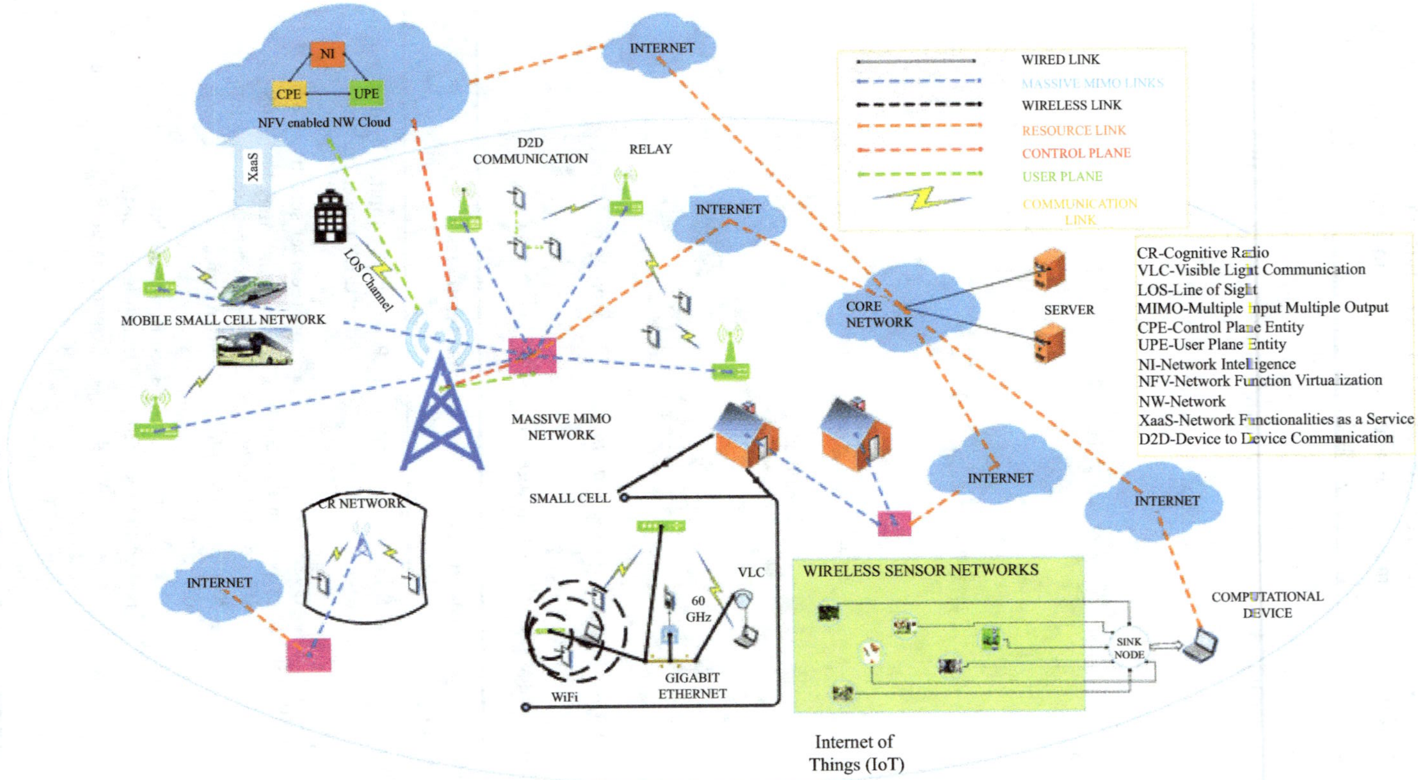

Figure 6.10 A general 5G cellular network architecture [15]

tweak) as balance with OFDM (symmetrical recurrence division multiplexing) and OFDMA (OFDM numerous entrance) as access plot. 5G can offer a high piece rate and the transfer speed should be utilized, adequately [9–12,16–18]. A few proposed recommendations.

Quadrature abundancy balance transmits two simple message signs or two advanced piece streams, by evolving (regulating) the amplitudes of two transporter

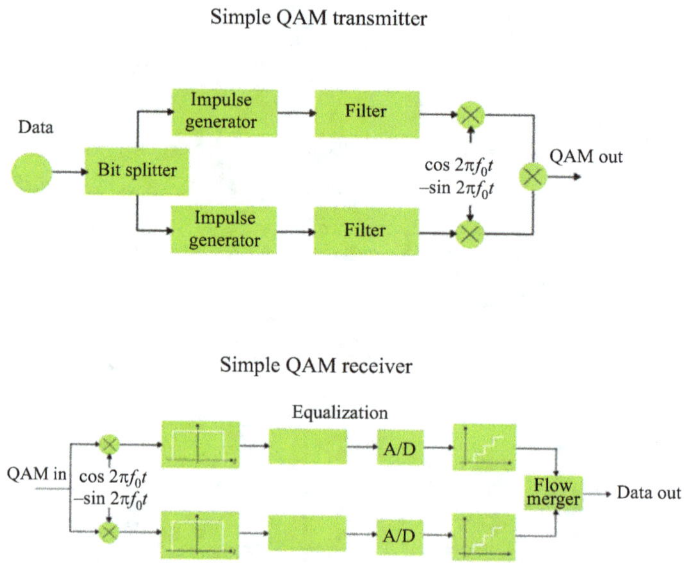

Figure 6.11 Explain simple QAM transmitter and Simple QAM receiver [15]

Table 6.3 Advantages and disadvantages 5G

Optical millimeter wave modulation schemes	Advantages	Disadvantages
Intensity modulation (double sideband with carrier format)	• Simple to generate	• Performance limited by dispersion-induced RF power penalty
Optical single sideband with carrier (OSSB+C) format	• Overcomes dispersion-induced RF power fading • Minimal bit walk-off impact	• Low receiver sensitivity • Performance may be limited by phase decorrelation in a long-reach environment
Optical carrier suppression technique	• Overcomes dispersion-induced RF power fading • Uses half the required LO frequency	• Suffers from bit walk-off for longer transmission distance • Requires large RF drive power

waves, utilizing the computerized tweak plot for adequacy move keying (ASK), or the simple adjustment conspires for playfulness balances (AM).

The two bearer rushes of a similar recurrence are out of the stage by 90° and are in this way called quadrature transporters. The regulated waves are summarized, and the last waveform is a blend of both stage move keying (PSK) and sufficiency move keying (ASK) or stage adjustment (PM) and adequacy balance (ASK) in the simple case (Figure 6.11). The advantages and advantages 5G are given in Table 6.3.

6.13 Beyond 5G

6.13.1 Introduction

Remote interchanges are the transmission of data between at least two focuses not genuinely identified with one another. Separations can be short, which is utilized for remote control of TV and even far separation utilized for profound space radio correspondences. The paper manages innovative turns of events and their preferences and near investigation on 3G, 4G, 5G, and 6G innovation review.

6G innovation alludes to an innovation of the sixth century. Coordinating 5G innovation to a local reach is proposed. Sight and sound video and fast web availability and the Earth symbolism satellite systems are utilized for asset observing and climate data. Coordinating each of the three satellite sorts, for example, media communications, route, sight and sound systems that give worldwide areas, rapid web access, and climate data administrations for portable clients are three primary objectives for 6G innovation. Various advances were executed for the progress from 4G to 5G that decreased dormancy (the deferral among sender and collector) and sped up.

Now the choice for any technology included in 6G remains uncertain but engineers are already working on different technologies. For example, work on more advanced modulators in Germany is being done to connect future mobile receivers to a fiber optic back-end. An improved MIMO transmission technique will also be needed to simultaneously process the new mobile streams in the base stations researched for years by various academic centers.

In this month's news, China is already working on developing the new future 6G standard. What that means is that the Chinese government is investing in finding communication techniques that allow mobile technology to be lifted to a higher level. The European Union has funded similar research into 5G applications at a time when 4G still had to be rolled out. It is farfetched to say that while there is no race, China will lead the race. The 6G technology will be developed with the help of international providers, network companies, academics, governments and other organizations, and an international institute with representatives from Europe, Asia, and North America will set the final standard for this.

It is hard to predict what 6G will look like now. Some say the way we use our mobile phones and mobile data will revolutionize. Others claim that holding the greater volumes of data that we can need or supporting the greater and more advanced applications and the quality of the screen would just be a quicker version.

6.13.2 One advanced communication technology beyond 2030

6G is one of cutting-edge correspondence past 2030. A cutting-edge age of versatile systems is commonly described by a scope of cutting-edge correspondence advances that offer remarkable presentation, for example, information rate, idleness, and abilities. Monstrous various info, numerous yield (MIMO) and mm Wave correspondences, for instance, are likewise essential empowering agents of 5G systems. So as to meet the prerequisites recently portrayed, 6G systems are required to depend on customary range yet in addition on the recurrence groups that have not been thought of yet for cell gauges, specifically the terahertz band which is it tends to be abused to apportion a monstrous measure of data transmission to short proximity and greatly pillar framed correspondence and Obvious Light Correspondence (VLC) that give organize administrators a savvy and down to earth approach to improve indoor inclusion.

Furthermore, to the new range, 6G will give remote systems by fusing a scope of advancements that have as of late been permitted by improvements in research on physical layer and circuits yet are not part of 5G. We expect 6G to coordinate full-duplex abilities in the correspondence stack to empower persistent downlink transmission with uplink to recognize or control message all the while and it likewise increments the multiplexing capacities and the general framework without utilizing extra data transfer capacity, for example, misuse novel channel estimation method, for instance, out-of-band estimation and packed detecting to build the effectiveness of the 6G control arrangement and consolidate correspondence with detecting. It is additionally improving control tasks and giving Novell client administrations.

Finally, the complexity of communication technologies and network deployments will probably prevent closed-form and manual optimizations. While the applications of intelligent techniques in cellular networks are already being discussed in the 5G domain, we expect 6G deployments to be much denser, heterogeneous, and with stricter requirements in terms of performance with respect to 5G. Therefore, intelligence will play a more prominent role in the network, going beyond classification and prediction tasks that are being considered for 5G systems. Notice that standards may not specify the techniques and learning strategies to be deployed in networks, but data-driven approaches are tools that network vendors and operators can use to meet the 6G requirements.

6.13.3 Features in 6G

6G will beat 5G in the categories of network coverage and network reliability. 5G supports the highest speed of mobile devices at 500 km/h. But 6G will increase the speed of mobile devices to 10 devices/square kilometer. 6G networks will be able to use higher frequencies than 5G networks and provide substantially higher capacity and much lower latency. One of the goals of 6G internet will be to support one microsecond-latency communication. This is 1,000 times faster – or 1/1000th the latency – than one millisecond throughput.

Table 6.4 shows the features in 6G.

Table 6.4 Features in 6G

Features	6G
Frequency bands	Sub-6 GHz mm Wave for mobile access exploration of THz bands (above 140 GHz) Non RF bands
Data rate	1 Tbsp
Latency	<1 MS (radio: 0.1 music)
Architecture	Cell-free smart surfaces at high frequencies (mm Wave tiny cells are used for fixed and mobile access) Temporary hotspots served by drone-mounted BSs or tethered balloons Trials of tiny THz cells
Devices types	Sensors and DLT devices CRAS XR and BCI equipment Smart implants
Spectral and energy efficiency gain	1000 $\times$ in bps/Hz/m^3
Traffic capacity	1–10 Gbps/m^2
Reliability	10
Localization precision	1 cm on 3D
User experience	10 Gbps 3D everywhere
Application types	MBRLLC mulch HCS MPS

6.13.4 Characteristics of 6G

Operating at terahertz frequency bands, 6G will deliver a peak data rate of 1,000 gigabits/s having air latency less than 100 microseconds. When we talk about 5G vs 6G network speed, 6G speed is expected to be 100 times faster than 5G with enhanced reliability and wider network coverage.

Table 6.5 shows the characteristics between 5G and 6G.

6.13.5 Future system for 6G

The NTT DoCoMo (Japan's largest mobile operator) whitepaper has a list of six technologies for 6G to achieve:

1. Extremely high-speed and high-capacity communications for example peak data rate go >100 Gbps.
2. Extreme coverage extension, including coverage in high altitude, under sea, and in space.
3. Extremely low power consumption and cost reeducation, including alternative charging technologies.
4. Extremely low latency for example 1 ms end-to-end latency.

Table 6.5 Characteristics between 5G and 6G Source: [19–21]

Characteristics	5G	6G
Performance	Gbps speeds Low latency (1 ms) Developing in-door coverage Improved device autonomy Improved performance	1–10 Gbps everywhere Up to Tbsp speeds Very low latency (0.1 MS) Super long-range and more indoor coverage High-precision 3D positioning Device autonomy: almost free of battery charge Guaranteed performance
Spectrum	6 gigahertz (GHz) and above 24.25 GHz Development of use in unlicensed spectrum Sharing between operator and incumbent (such as government users) High bands: limited number of national licenses	95 GHz to 3 terahertz (THz) Operation in shared spectrum bands Inter-operator sharing High bands: large number of local licenses
Network	Limited number of mobile network operators Mainly outdoor macro cell, small cells mainly in hotspots Infrastructure owned by the operator Massive machine-type communications	Growing role, a large number of local network operator More indoor small cell deployments Slices of network offered on a leasing basis Machine support: broadband for everything
Other aspects	Limited AI use	Massive AI use

5. Extremely high reliability, for example, availability to improve from “five-nine” to “seven-nine” (99.99999%).
6. Extremely massive connectivity and sensing, for example, handling 10 times as many connections as 5G does in comparable space.

6.13.6 Architecture of 6G

The disturbance brought by these correspondence innovations will empower new 6G arrange models, yet in addition, conceivably require auxiliary updates with the regard to current versatile system plans. For instance, the thickness and the high access information pace of terahertz correspondences will make obligations on the fundamental vehicle arrange, which give both more purposes of access to fiber and a higher limit than today backhaul systems. This is the accompanying of engineering 6G.

- Cell-less engineering, with tight reconciliation of various frequencies correspondence innovations, to ensure a consistent portability backing and adventure the correlative qualities of various recurrence groups, for example, the sub-6 GHz layer for control, and terahertz interface for the information plane.

- Disaggregation and virtualization of the systems administration hardware, including the lower layers of the convention stack (physical and Macintosh layers), to diminish the expenses of the systems administration gear and make hugely thick sending monetarily possible.
- Advanced get to backhaul joining, to adapt to the enormous increment in the thickness of passage.
- Energy-reaping methodologies for low-power utilization arrange activities, to make 6G gadgets increasingly productive and less vitality expending as for current ones.

6.13.7 Five real applications on the 6G technology

These applications help to manufacture a psychological sensory system that incorporates man-made reasoning (man-made intelligence) and remote observation, with the objective of giving keen reactions, upheld by a speed expected to arrive at 1 terabyte for each second, a phenomenal degree of speed, which will build the presentation of 5G applications to beside growing capacities to help progressively new and creative applications over the fields of remote recognition, remote sensor and imaging.

Compared to 5G innovation, 6G advances will get lower inactivity, higher speed, and more data transmission. This trend-setting innovation will help interface this present reality to the virtual computerized world, as it will make item plan and innovative work progressively proficient and diminish its expenses fundamentally while giving computerized items creation in the physical world through cutting edge innovation including 3D printing.

In terms of monetary turn of events, 3G advances helped web-based business, while 4G supported online business and portable installments. The development and use of the 5G foundation denote the start of the brilliant assembling of Chinese endeavors and give the premise to the fast advancement of the segment.

The 6G computerized economy will turn into the definitive factor for the seriousness of nations later on, and 6G innovation will turn into the primary driver of the advanced economy.

Sixth-age innovation will turn into the motor of another round of monetary turn of events. Governments, organizations, and exploration foundations should improve collaboration among them to make progress in rivalry for the advancement of the 6th era.

6.14 Conclusion

As we know, every day the world gets further into technology. We're trying to improve our lives and make them easier and better. As we know, every day the world gets further into technology. We are trying to improve our lives and make them easier and better. One of the best ways to improve our communication with the world is via 5G as the next wireless network. It is safe and secure for the public, who needs the solutions. Wired society today goes wireless and if it has difficulty, 5G is surely the solution. 5G technology will offer computers and laptops competition though. With more reliability than previous mobiles, it will be available in the

2020 market at an affordable cost. In conclusion, 5G will lead the greatest technological transition of our lives, with infinite possibilities.

References

[1] Bretnall, A. (2019, February 26). *eMarketer*. Retrieved from eMarketer.com: https://www.emarketer.com/content/getting-ready-for-5g (Accessed 10th NOV 2021).

[2] Ahmad, S. (2013, December 8). *wccftech*. Retrieved from wccftech.com: https://wccftech.com/the-next-generation-of-mobile-networking-5g/ (Accessed 10th NOV 2021).

[3] OkPortal technology. (2019, May 13). *Quora*. Retrieved from Quora.com: https://www.quora.com/5G-Communications-Which-countries-are-using-5G-technology (Accessed 10th NOV 2021).

[4] Beal, V. (2020, May 22). *Webopedia*. Retrieved from webopedia.com: https://www.webopedia.com/author/Vangie-Beal (Accessed 10th NOV 2021).

[5] Eva. (2019, September 17). *Cool Blue*. Retrieved from coolblue.nl: https://www.coolblue.nl/en/advice/difference-smartphone-mobile-phone.html (Accessed 10th NOV 2021).

[6] IT,NEW MEDIA AND SOFTWARE. (2020, February 6). *Open PR*. Retrieved from Open PR.com: https://www.openpr.com/news/1921844/5g-chipset-market-business-opportunities-target-audience (Accessed 10th NOV 2021).

[7] Media and Telecommunication. (2020, April 8). *Open PR*. Retrieved from Open PR.com: https://www.openpr.com/news/2003392/small-cell-5g-network-market-size-share-trends-growth (Accessed 10th NOV 2021).

[8] The 34GG Blog. (2020, January 26). *The 34GG Blog*. Retrieved from blog.34Gg.co.uk: https://blog.34Gg.co.uk/search/label/6G (Accessed 10th NOV 2021).

[9] Leong W.Y., Homer J., and Mandic D.P. (2006). An implementation of nonlinear multiuser detection in Rayleigh fading channel, *EURASIP Journal on Wireless Communications and Networking*. 2006: pp. 1–9. DOI 10.1155/WCN/2006/45647

[10] Leong W.Y., and Homer J. (2005). Blind multiuser receiver in Rayleigh Fading channel, *2005 Australian Communications Theory Workshop*, pp. 155–161

[11] Min H.F., and Yie L.W. (2012). Investigating target detection and localization in wireless sensor network, *Procedia Engineering, Elsevier*, 41: pp. 75–81.

[12] Leong W.Y. (2008). Angle-of-arrival estimation: beamformer-based smart antennas, *The 3rd IEEE Conference on Industrial Electronics and Applications*, pp. 1593–8.

[13] Nebula Solution Corporation. (2013, December 31). *10 Communication Innovation in The Last 10 Years*. Retrieved from nebulasolutionsinc.com: https://www.nebulasolutionsinc.com/blog/10-communication-innovations-last-10-years/ (Accessed 10th NOV 2021).

[14] Quallcomm Corporation. (2019, October 23). *Qualcomm*. Retrieved from qualcomm.com: https://www.qualcomm.com/news/onq/2019/10/23/benchmark-global-study-proves-5g-performance-coverage-and-power (Accessed 10th NOV 2021).

[15] Rajah & Tan Asia. (2019, April 12). *Lawyer who knows Asia*. Retrieved from eoasis.rajahtann.com: https://eoasis.rajahtann.com/eoasis/lu/pdf/2019-04_5G-Legal-and-Regulatory-Implications.pdf (Accessed 10th NOV 2021).

[16] Leong W.Y., and Mandic D.P. (2006). Towards adaptive blind extraction of post-nonlinearly mixed signals, *The 16th IEEE Signal Processing Society Workshop on Machine Learning for Signal Processing*, pp. 91–6.

[17] Leong W.Y. (2005). *Implementing Blind Source Separation in Signal Processing and Telecommunications*, The University of Queensland, Australia.

[18] Xiaofan W., Chong P.H.J., and Yie L.W. (2010). Performance comparison of CSMA/CD, CSMA/CA, CSMA/RI, CSMA/PRI and CSMA/PR with BEB, *The 5th IEEE Conference on Industrial Electronics and Applications*, pp. 1843–8.

[19] Agrawal, D. (2016, January 26). *Research Gate*. Retrieved from researchgate.net: https://www.researchgate.net/publication/291421086_5G_mobile_technology_A_survey (Accessed 10th NOV 2021).

[20] Gupta, A. (2015, August 15). *Research Gate*. Retrieved from researchgate.net: https://www.researchgate.net/figure/A-general-5G-cellular-network-architecture_fi1G9_280873356 (Accessed 10th NOV 2021).

[21] Nistor, C. (2019, July 1). *Note Book Check*. Retrieved from notebookcheck.net: https://www.notebookcheck.net/Canalys-5G-phone-shipments-will-overtake-4G-in-2023.426822.0.html (Accessed 10th NOV 2021).

Chapter 7

Drone supports applications in smart manufacturing

Ba Kowaina Aseel Salem[1] *and Wai Yie Leong*[1]

7.1 Introduction

Having the greatest use for military purposes such as surveillance and defence or in civil works such as fire-fighting and monitoring of pipelines, systems like a plane remotely piloted, or pre-programed for a route, or often bearing a payload to accomplish their tasks as camera systems or even missiles, are being used increasingly. Used in difficult and dangerous missions vis-à-vis their conventional use constitutes a qualitative revolution, as companies, organizations and governments have become forerunners in using the products of an industry that provides them with effective tools that help them perform their duties effectively, professionally and with high degree of safety with the opportunity to save billions of money and human lives. The greatest use of drones is for military purposes, such as surveillance and attack, but their use in civilian tasks such as fire prevention and pipeline monitoring has increased significantly, where they are being used alongside the traditional aircraft in meeting difficult and dangerous tasks by providing many of the pilot's requirements in such situations, such as cabin and aircraft controls, and environmental requirements. Unmanned aircraft or the so-called "unmanned vehicles" have become the third revolution in wars in this century being pre-programed or programed aircrafts as unmanned aerial vehicles that usually vary in size according to their different uses, such as photographing, carrying shells, or for the purposes of surveillance and attack, in addition to their uses in fire-fighting and pipeline control, and equipped with the most deadly weapons used in modern wars.

Historically, we find that the first practical experiments of this plane were in England in 1917, while this aircraft was developed without a drone in 1924. Its first practical use was in the Vietnam War, and then drones were used in October 1973 war but the desired result was not achieved because of the weak capabilities of the system at the time and because of the presence of the Egyptian "missile wall" having its first effective use in the Battle of the Bekaa Valley. Between Syria and Israel, it resulted in the shooting down of 28 Syrian aircrafts without losing any Israeli aircraft in 1973.

[1]The Institution of Engineering and Technology, Malaysia

7.2 Applications

Drone applications are not limited to the military world only; rather they are serving the needs of a big part of the economy with advanced mechanisms and impressive capabilities. With the growing interest in drone technology, presently, drones are being used in many other areas like the agriculture industry, even as the use of drones grows with continuous advancements in technologies like the internet. Here we are going to talk about the most wonderful applications of drones and their capabilities [1].

7.2.1 Amazon Air Service for drone transportation

Amazon Air Service began several years ago using motorized planes to deliver light customer orders on trips of half an hour or less. These aircrafts are equipped with sensors that enable them to avoid obstacles on the way to their destination and return to base after landing the shipment in front of the customer's home.

7.2.2 Automated aircraft used in agriculture

Robots were used in the agricultural industry in Japan thirty years ago to spray pesticides and perform other tasks, and helped increase the produce of the rice grown in Japan by a third, given automated planes shortened ten days of manual work to just 2 days.

7.2.3 Construction aircraft

There are those who believe that, in the near future, automated aircrafts will work side-by-side with humans to lay the foundations of buildings and perform similar tasks, and there are others who have begun experiments in this field, such as the Swiss Federal Institute of Technology, Equinox's Drones Pvt Ltd. The Institute has entered into a partnership called the aerial construction project and its goal is the "investigat(ion) and the development of methods and techniques for automated aerial construction," where the team has already built a bridge with ropes between two buildings 24 feet tall using quad-helicopters (which were modified for the project). The bridge successfully carried the weight of an adult who walked from building to building. Another project of the same university saw the construction of a 20-foot tower made of cork with the use of four quadruple planes.

7.2.4 Advantages of drones

1. Rescue and extinguishing fires
2. Law enforcement
3. Preserving natural diversity
4. Defence
5. Saving on operating cost

7.2.5 *Disadvantages of drones*

1. Suicidal operations
2. Security threats
3. Spying

7.2.6 *The component of drone*

Drone is about learning a bit about the dynamics of the air, the methods of calculating the plane's mass and its weight in the air and other details.

1. Design the plane
   One of the most important parts of a drone is the tire because it supports the motor and other electronics. Ready-made tires are procurable from any location easily but, for assembling by hand, the use of sturdy and lightweight materials like carbon fiber, wood or plastic is known. Carbon fiber is the best choice for making a tire because it is a relatively light, strong and stable material to carry the weight of all parts of the quadcopter.
2. Choosing engines
   There is a common type of drone aircraft engine called (Brushless DC motor) characterized by durability and superior rotational speed, and when choosing an engine these are two main factors that must be taken into account.
3. Electronic speed controllers
4. Batteries
5. Choosing a GPS module
6. Measure the balance, height, and acceleration rate
7. Install the engines
8. The electronic circuit

7.3 Future of drone technology

Drone technology is constantly evolving, so the future of drone technology is currently undergoing groundbreaking progressive improvement. According to airdronecraze, an Amazon Services LLC affiliate advertising program website, drone technology has seven potential generations, and the majority of current technology sits in the fifth generation (5G) and sixth generation (6G).

Generation 7: Complete commercial suitability, fully compliant safety and regulatory standards-based design, platform and payload interchangeability, automated safety modes, enhanced intelligent piloting models and full autonomy, full airspace awareness, auto action (takeoff, land, and mission execution).

The next generation of drones, Generation 7, is already underway, as 3D Robotics announced the world's first all-in-one Smart Drone called Solo. Smart drones with built-in safeguards and compliance technology, smart accurate sensors, and self-monitoring are the next big revolution in drone technology that would provide new opportunities in transport, military, logistics, and commercial sectors. As these technologies continue to evolve and grow, drones will become safer and more dependable. This would allow for their subsequent mass adoption, provided the strict US Federal Aviation Administration (USFAA) legislation surrounding drone technology and usage is loosened to some degree [iii].

In the future, men will use drones in the water deep around 60 m and it can stay hidden there for months. Also, during wars, it can take off from the home easily like playing a game. Here is an image of the size of a commercial drone fleet (Figure 7.1).

The improvement in drone technology is taking place so fast through the years, and it's expected that it will keep continuing in this fashion because nowadays people and companies have put drone technology as their first choice while investing in groundbreaking hardware for process improvement.

Based on Figure 7.2, the investment will keep increasing because drones will be used a lot in future, together with IoT and AI to change our lives.

Drones have become an integral part of the operation of numerous companies and governmental agencies over the past few years (Figure 7.3), and have continued to penetrate new domains; where, other industries have stagnated or lagged behind. From fast deliveries at rush hours to the scanning of an unattainable military base, drones prove to be extremely beneficial in places where people cannot be reached.

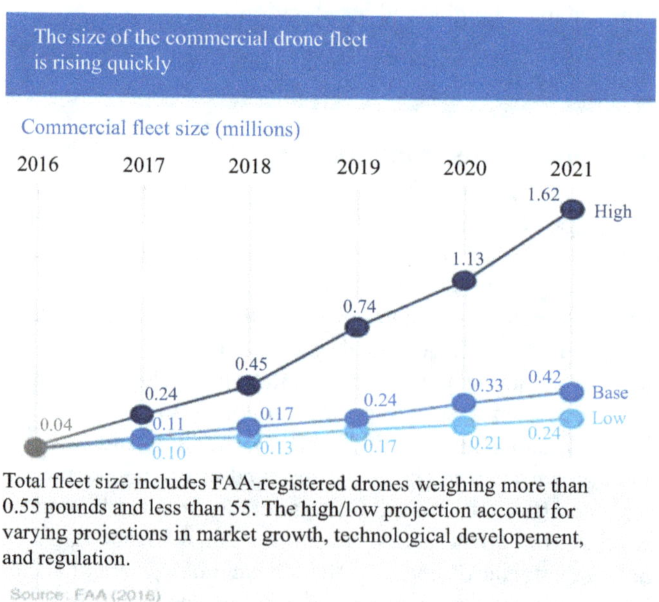

Figure 7.1 The future of drone technique [2]

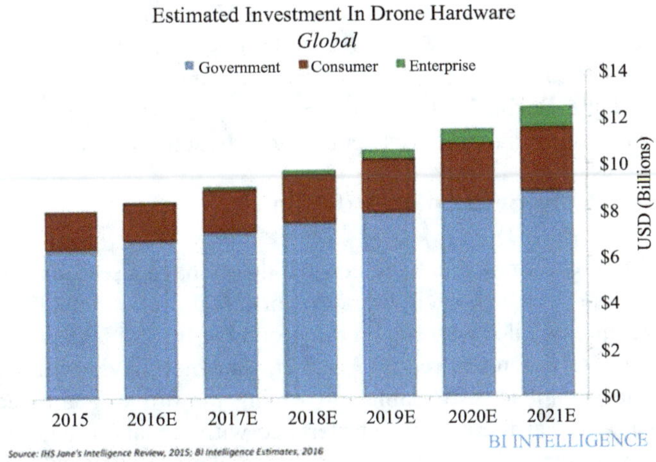

Figure 7.2 Estimated investment on drone hardware [3]

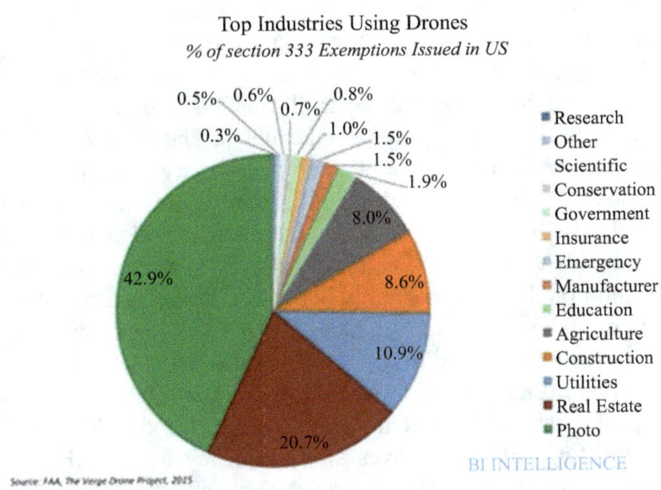

Figure 7.3 Industry revolution during past 10 years and how it can affect other sector [2]

7.4 The history of drone

In 1917, first wireless or rocket was invented. The technology for this would later be adapted by the Germans for their V1 rocket program in WWII. It was the beginning of its concept as a search and rescue equipment when the 1960 American fighter jet (U-2) broke down and plunged in Russia. The 1962 Cuban missile problem also saw drone technology being used.

Its first practical application was during the Vietnam War. In October 1973, drones were used. The weaknesses in the technology meant the desired outcome was not accomplished. Egyptian “missile wall” saw its first effective participation in the Battle of the Bekaa Valley between Syria and Israel, and resulted in the shooting down of 82 Syrian aircrafts against zero Israeli aircraft.

7.5 The working principle of drone

Drones are complex technological devices that do not have a seat for a pilot sit. But the aircraft’s name says how such aircrafts work. It does not actually direct itself fully, it also requires a pilot who sits on the ground at the guide depot and controls it wirelessly remotely. Drones require, in fact, that they have a pilot at the base station. This floor pilot is responsible for driving the drone, guaranteeing that it does not crash, and that there is no interference with signals in emergency cases, but he is doing all that with a joystick like a toy plane joystick. The pilot has to determine the path points of the plane, and using which the plane is directed by those coordinates. It is the automatic flying system by which the drone is guided because this system determines the location of the plane in the air through an enormous amount of data, which it collects with available technologies: a compass, an altimeter, a speedometer, and a GPS navigation device.

Drones, hundreds of kilometers away from the ground guidance station, are no longer controlled by radio waves but by satellites. This is because the wireless connection by satellite is almost always never interrupted. The drone can detect obstacles it faces with sensors installed on board, such as ordinary optical cameras, infrared cameras and radar, and the automatic flight system sends this information to the human pilot at the earth station. Drones also have traffic collision avoidance system (TCAS) shock avoidance devices. But the process of avoiding collisions is not performed by the pilot on the ground station, but by the drone’s autopilot system which sets into action on its own because the radio communication between the plane and the earth station cannot be relied upon in such cases. The main element at work in drones with simpler designs is good user control, as the movements of the plane are often controlled by remote control devices, when these devices issue movement commands in the form of radio waves at a frequency of 2.4 GHz, though some designs depend on Wi-Fi connections to reach these signal waves to sensors in the plane which, in turn, translates them into motion orders issued to the plane’s engines. The principle of equilibrium in helicopters depends on the opposite action of their engines, meaning that any imbalance in one of the helicopters from loss of control in any of the engines is automatically reversed by the other engines. Conversely, winged aircraft relies on the principle of equilibrium air pressure at their wings in order to secure the necessary balance and stability. These planes usually have batteries with a capacity that lasts between 10 and 12 min. The reason for not being able to provide a longer flying time is in the design trade-off between increasing the capacity of the battery that supplies energy vis-à-vis an increase in its size and additional weight, which in turn will affect the dynamic plane’s ability, in addition to the engines having to bear an additional load and that requires more energy to take off.

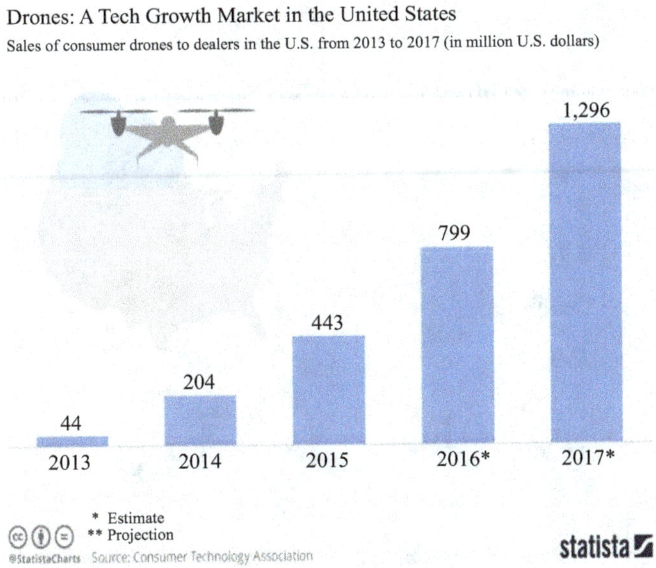

Figure 7.4 The growth of drones market in US [4]

How do you get the drone into this position? You could increase the rotation rate of rotors 3 and 4 (the rear ones) and decrease the rate of rotors 1 and 2. The total thrust force will remain equal to the weight, so the drone will stay at the same vertical level. Also, since one of the rear rotors is spinning counterclockwise and the other clockwise, the increased rotation of those rotors will still produce zero angular momentum. The same holds true for the front rotors, and so the drone does not rotate. However, the greater force in the back of the drone means it will tilt forward. Now a slight increase in thrust for all rotors will produce a net thrust force that has a component to balance the weight along with a forward motion component.

Here is the statistic of the sales of drones from 2013 to 2017 (Figure 7.4). One can see how much the sales have grown, how the sales performance will be in the future, and how the drone has a big future.

7.6 Conclusion

The name of drones does not fully indicate the way these aircrafts operate, as they do not fully direct themselves, but also need a pilot sitting at a guidance station on the ground whose hand holds the automated system that defines the path of flight of the drone and controls it via satellite. In fact, this technique is a double-edged sword, where you can use it for many purposes, both positive and negative. An example of the negative use of drones in wars, espionage, etc. is known. However,

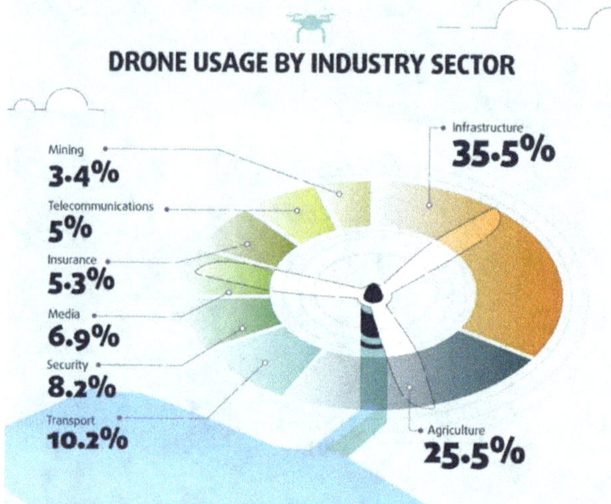

Figure 7.5 Drone usage by industry sector [5]

in contrast, there are many positive sides to drone use where it can be used in photography, shipment delivery, and industrial and commercial works (Figure 7.5). It is necessary to develop this technique and make it better, and factories are already doing it at these moment. The most important thing in this technique is that it was made for different ages to handle, where the small and the big can operate a drone.

7.7 Introduction to drone use

Drones were not used initially as we know them today, but rather as a dangerous and majestic air weapon used for the first time in the beginning of the last century. Used militarily by military groups to train in hitting targets, the engineers developed an aircraft carrying ammunition that dropped bombs on the desired target. The drone is real as we know it today, it is used only once. During the Cold War period, the United States researched on this technology and began to use drones for espionage and for dangerous missions with the need to avert serious harm to its people. These techniques have been used in warfare for a long time, but the time has come today to use them in completely different directions; directions, that allow us to take advantage of the eyes in the sky in ways that save humans and facilitate their lives instead of ending them. Little by little, these planes began to move away from their military origin and approach an elevated purpose, and this wave started with cameras that gave us stunning views of landscape from the top. Its uses evolved from films to real life, and some of these modern drones like the Black Hornet, a tiny aircraft weighing no more than 18g, carrying three cameras and flying over a distance of more than a kilometer and a half, have been known to exist. One of the companies made a drone for ambulance use case in crowded places. Once you raise a call, the plane is sent to your place by the GPS and the paramedic gives you the necessary instructions as he monitors your actions through the cameras and speaks to the person via a microphone on the plane itself. One of the other uses is in spraying crops with pesticides and fertilizers, as in Japan, where it is also used in drawing accurate maps of the towns as drones accomplish what took the traditional methods a year in 3 days. Recently, Amazon obtained a license to try its Air Prime service in the United Kingdom. In the United States laws are being made, for example, freedom was allowed to Google to use its new wing project, but this legislation is still very limited so far. But the idea is that all aspire to become Google and Amazon in their initiatives to carry out the lesser-delay-days delivery of goods to the consumer, meaning that this plane will deliver the product to you within half an hour of your placing an order.

The wonderful use of drones does not stop here. Rwanda has partnered with the drone company to produce aircrafts that specialize in transporting medicines to remote locations, thereby facilitating the delivery of aid to those who need it as quickly as possible. The United States of America is studying a similar package as well.

This industry is still in its infancy, and every month there is a new development that promises more good outcomes and progress for humanity, and with the support of countries such as the United Arab Emirates and others, we will be able to emphasize the peaceful and effective use of drones, and we may see them flying over our heads and feel safe for their presence and forget their past. We can look forward to a safer future combining luxury and happiness with the services they will provide us.

There is a panoply of technology to be used in the drone use for the drone to work well.

Past techniques:

1. Radar positioning and return home

   Drones could work in either GNSS or non-satellite style. DJI drones, for instance, could fly in either P-Mode (GPS & GLONASS) or ATTI mode, which does not use GPS. Super reliable drone navigation is very essential while flying, specifically in aviation technology such as 3D mapping, landscape scanning, and SAR (Search & Rescue) operations.
2. Obstacle detection and collision avoidance technology (Figure 7.6)

   The new high-tech drones now feature automatic braking technologies. They use obstacle avoidance sensors to scan the environment, where the software algorithms and simultaneous localization and mapping (SLAM) technology turn the photos into 3D maps that enable the drone to recognize and escape. To sense and minimize these processes, one or more of the proceeding sensors are fused

   - Sensing vision
   - Ultrasound

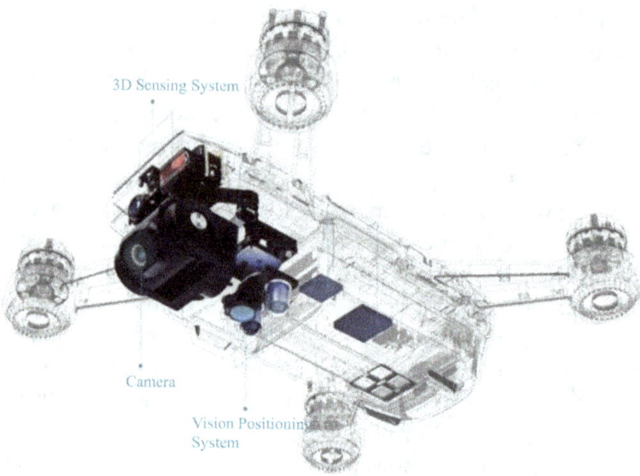

Figure 7.6 Drone cameras [4]

- Infrarot
- Lidar: Lidar
- Day to fly
- Sensor view

The quadcopter Skydio 2 uses 6 × 4k cameras to create a 3D image of its environment that would include branches, people, animals, vehicles, buildings and much more, Sovereignty technology Skydio 2 captures and measures what's going on in the drone. It can then smartly foresee what might happen soon, and produce realistic decisions several times a second.

3. UAV drone propulsion technology (Figure 7.7)

   Its propulsion system (engines, electronically controlled controllers and hydraulics) is the drone technology that pushes the UAV into the sky and flies in any position or in any way. The motors and hydraulics operate on a quadrotor in pairs with two motors/propellers rotating in clockwise direction (CW propellers) and two reverse rotational speed (CCW (counter clockwise) propellers) rotating engines. They collect data from the radar altimeter and the electronically controlled controllers (ESC) for either flying or hovering in the position of the drone engine.

   Extremely advanced UAV drone engines and power devices, including the following components, are known to exist:

   - Engine Stator
   - Bell Motor (rotor)
   - Curves
   - Lagers
   - Refreshment program
   - Electronic velocity sensors
   - Updater to ESC

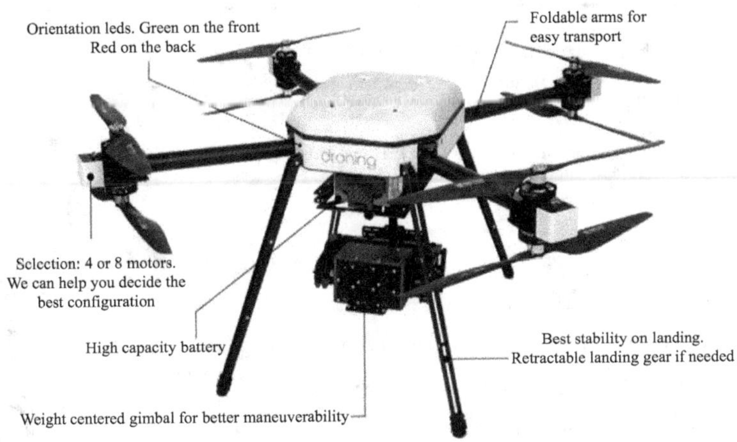

Figure 7.7 The drone UAV [6]

- Pushers
- Webbing AND Arms

4. GBS mode drone technology

Upon calibration of the compass, the drone then searches for the position of the GPS satellites. It enables the drone to move in "Safe To Fly" mode, if more than six are identified.

5. FPV live video transmission drone technology

FPV is "view of the first man." On the unmanned aerial systems, a video camera is installed, but this webcam transmits the live stream to the ground pilot. The floor pilot flies the aircraft as if they were crew on board the aircraft, rather than with a view of flying the aircraft from their actual ground position of the driver.

FPV enables the unpiloted aircraft to fly over a larger area, much larger than one could ever cover from the ground viewing the aircraft. First hand view makes flying more accurately particularly around barriers.

FPV makes it very easy for unmanned aircraft to fly indoors, either through forests and around residences, Such a live feed is linked to the gain of the amplifier between both the drone's control tower.

This same DJI territory 2 does have a 5-mile (8 km) FPV live footage range with 1,080 p image quality transfer.

Certainly, similarly aged UAV drones like DJI dji phantom and demon 4 Pro, can broadcast video feed up to 7 km (4.3 miles) away. The dji mavic Pro and Inspire 2 just use newest DJI electricity grid, Lightbridge 2.

6. LED light indicators (Figure 7.8)

These are located on the drone's front and rear. The drone LEDs will usually be green, yellow, or red. The LED signs at the front light up to show the UAV nose. The flight indicators for the rear LEDs light up to indicate the drone's different status when powering on, upgrading firmware, and flying.

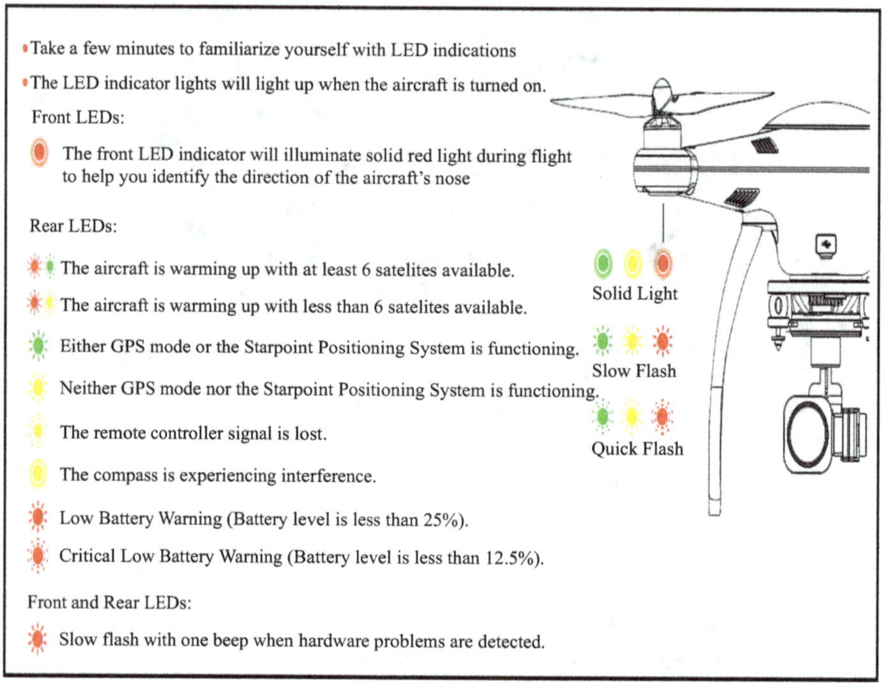

Figure 7.8 The drone light system [4]

Table 7.1 Comparison of satellite and aircraft and UAV

Operating	Payload capacity	Operation mode	Degree of availability	Resolution	System
Low	Limited	Autonomous or remote control	High	cm to meter	UAV
Med	Limited	Human pilot	High	100 m	Helicopters
High	Much more than helicopter	Human pilot	Moderate	Up to 50 m	Airborne
Too high	Limited	Autonomous	10–1 km	10 m–1 km	Satellite

Source: Singhal, Gaurav, Bansod, Babankumar, and Mathew, Lini. (2018). Unmanned Aerial Vehicle Classification, Applications and Challenges: A Review. 10.20944/preprints201811.0601.v1.

It is a good idea to know what your multicopter's flashing LEDs indicate.

7. FPV over 4G/LTE networks

A new live stream that emits an infinite range of low-latency content over the 4G/LTE network was confirmed in 1996. This aircraft technology has been introduced on Sky Drone FPV 2 and includes a display panel, a data subsystem and a 4G/LTE modem (Table 7.1).

7.8 Data collection and analysis

The rising popularity of drones with infinite opportunities for its use pushes us to work out the approaches for collecting and analyzing data for use at home and work and efficiency therefrom. The processing and analysis of UAV data is a lengthy process, involving data gathering, image pre-processing, classification techniques for feature extraction, numerical equations computation based on Reflections such as Indexes, and then the development of an appropriate model for visualizing and analyzing data. This section explains each step and contrasts the UAV data collection and management tools in Tables 7.2 and 7.3.

Table 7.2 Commercial drones available in global market

Camera	Weight kg	Speed	Flight time	Flight mode	Model
12 mp single	1.6	35 mph	Up to 25	x-star GBS ATTI	Autel
Shot			Min	IOC	Premium
EFC-721	95 g		05-Oct	Bind-n-fly	Blade BLH740A

Source: Commercial drone market size, share and trends analysis report by product (fixed-wing, rotary blade, hybrid), by application, by end-use, by region, and segment forecasts, 2021 – 2028, Apr, 2021, Report ID: 978-1-68038-482-6.

Table 7.3 Gap analysis and benchmarking

Literature review			
S. no.	**Title of paper**	**Author**	**Summary**
1	Structural analysis of a composite Target-drone	Yong-Bin Park *et al.*	Structural static and dynamic analysis of a wing and landing gear
2	Design and structural analysis for an autonomous UAV system consisting of slave MAVs with obstacle detection capability guided by a master UAV using swarm control	Lakshmi Narashiman Aswin *et al.*	Master and slave frame structural analysis
3	Low-velocity impact analysis of a composite mini unmanned air vehicle during belly landing	Serhan Yuksel	Studied impact stresses induced during belly landing
4	Design and aerodynamic analysis of a flapping-wing micro aerial vehicle	Bor-fang Tsai *et al.*	Aerodynamic study during different angle of attack (AOA) and K
5	Design and analysis of engine mounting frame of an UAV	Santhosh *et al.*	Frequency analysis of frame

Source: Chowdhury, Mostafa, Shahjalal, Md, Ahmed, Shakil, and Jang, Yeong Min. (2020). 6G wireless communication systems: applications, requirements, technologies, challenges, and research directions. *IEEE Open Journal of the Communications Society*, 1-1. 10.1109/OJCOMS.2020.3010270.

The most common use of advanced camera carrying drone units is to capture media coverage from inaccessible locations. Earlier, only high-profile industries were able to capture critical shots as they were able to afford helicopters. But today, almost all small-scale media houses and local journalists are able to capture crystal clear air footage for impressive news coverage. Drones can also move easily to tighter areas due to their small size; so, they are also utilized for recording face-to-face interviews.

7.9 Applications

1. Aerial photography

   Drones are now being used to capture footage that would otherwise require expensive helicopters and cranes. Fast-paced action and sci-fi scenes are filmed by aerial drones, thus making cinematography easier. These autonomous flying devices are also used in real estate and sports photography. Furthermore, journalists are considering the use of drones for collecting footage and information [7] in live broadcasts.
2. Humanitarian aid and disaster relief

In addition to emergency response, drones have proved useful during times of natural disaster. In the aftermath of hurricanes and earthquakes, UAVs have been used to assess damage, locate victims, and deliver aid. And in certain circumstances, they are helping to prevent disasters altogether.

To help monitor and combat forest fires, surveillance drones outfitted with thermal imaging cameras are being deployed to detect abnormal forest temperatures. In doing so, teams are able to identify areas most prone to forest fires or identify fires just minutes after they begin [8].

7.10 Conclusion

The future to come relies on drones given these will build a new industry. Nearly all industrial organizations are developing and embracing drones, covering precision production, transportation and transport infrastructure. Future engineering focused on higher strength, payload, enhanced contact between humans and UAV, and consistent rules and regulations for safe operation and stable UAV service. In addition, the integration of information technology into drone technology would enable the drone to make choices and be independent of human operators.

7.11 Introduction to drone and telecommunications

The new technology – the 5G of mobile technologies, or 5G – is currently being introduced around the world at select locations. And that poses a problem which is evident. What factors will fuel the growth of the sixth mobile communications generation? How will 6G be different from 5G, and what kinds of interactions and operations would make what is possible with 5G now not be possible with 5G later?

Nowadays, thanks to the efforts of Razvan-Andrei Stoica and Giuseppe Abreu at Jacobs college in Bremen, Germany, we have an idea of some sort. These men have also mapped out 5G constraints and will start driving 6G development by the influences they think will 6G will have on people's life and work. Our inference is that artificial intelligence (AI) is the primary engine of digital technology and also that 6G is the enabler behind an entirely new generation of applications for machine intelligence (Figure 7.9).

Some backstory first. 5G is a major move over the preceding 4G standards, by any means. The first 5G networks now deliver up to 600 mb per second of data speeds and have the ability to go even higher. By contrast, 4G normal response at up to 28 Mbits/s – and perhaps most cell phone users from time-to-time will have experienced this rate of crushing to zero for reasons that are not always clear.

In this way, 5G is clearly better, and might even replace several landline phone connections.

But perhaps, the most important benefits go further than those estimates in the headline. For instance, 5G base stations are built to handle up to one million connections versus 4,000 which 4G access points can handle. This should make a significant difference in contact at big events like sporting events, demonstrations, and so on, and it could enable all kinds of applications for the Internet of Things (IoT).

And there's latency – the time it would take to pass across the network – for the signals. 5G is equipped with a delay of just one millisecond, compared to 50 ms or more on 4G. Any player can tell you how vital this is, because it makes the playing characters using the TV remote more real-time. Yet, different telecommunications providers have shown how the same benefit makes it easy to monitor drones more reliably and even use a mobile link to conduct telesurgery.

All this would be accessible with lower power consumption for booting, and existing reports say that 5G networks should have 10 times the 4G battery capacity.

So how could 6G do even better? Of course, 6G will offer even faster upload speeds – the current belief is that they could reach 1 terabit per second [9–12].

Figure 7.9 The 6G over the world [14]

But, really, what revolutionary changes could it offer? According to Stoica and Abreu, the response is that it will require rapidly evolving large-scale partnerships between smart agents to solve complex flying challenges and negotiate solutions to complex issues. SoftBank [13] notes that in addition to enhancing the high-speed connections, low latency and "massive device connectivity" capabilities expected during the 5G era, 6G needs to enable high reliability and much-improved energy efficiency, and so has been exploring some of the architectural options for 6G. SoftBank is "taking up the challenge of resolving numerous issues in different areas," with the operator, which plans to invest 2.2 trillion yen (US$20.13 billion) during the next 10 years on its 5G and 6G networks, pinpointing a dozen main challenges. Those challenges are architectural, technological, and social, as the SoftBank team points out in Figure 7.10.

Take the question of self-driving cars being driven around a big city. That's a big challenge, given that every day around 2.7 million cars and trucks enter a city like NYC.

It's an interesting futuristic vision. There is a lot of negotiation and horse-trading to be completed even before a set of 6G guidelines can be outlined, much less completed. But if Stoica and Abreu are right, machine intelligence will be the main driver shaping future communication systems.

Even when advertising mobile phone suppliers and retailers of materials are diligently working to construct 5G, and as the community eagerly awaits its rollout, researchers across the globe are starting to work to produce the next generation of the wireless world: 6G.

What is 6G? None ever understands it fully yet, and the internationally recognized bodies, hundreds or thousands of organizations around the world, will work it out in the next decade. But New York University's research shows that by 2035, 6G would contribute to the ability to transmit wireless signals at human computational speeds.

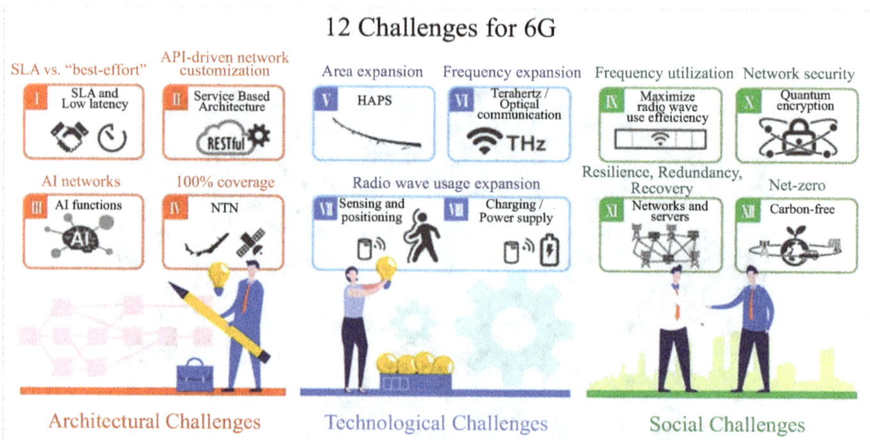

Figure 7.10 The 12 Challenges for 6G

Dubbed by researchers as "wireless cognition," 6G could mean that human intelligence could eventually be transmitted spontaneously over air over 6G mobile devices, TV remote, robotics and AI applications for human understanding could be allowed.

Precise placement and location finding, combined with outstanding sensing and image analysis, will become achievable with 6G just as people use their eyes to clearly see things both near and far.

The new wavelengths in the sub-Terahertz levels will transmit superfast computations over a broad range of wavelengths, allowing for much higher capabilities for future cell phones.

For instance, the future 6G phone will be decided by examining the air over you for allergens, explosive materials or dangerous substances, and determine whether your food is safe to eat. Using night vision, it can help you see in the dark, and render pictures much clearer than the human brain can see. Even as the mobile phone has substituted the standalone camera and digital watch for most and is starting to substitute the pocketwatch, 6G phones can sometime replace eyeglasses with goggles that could double as headphones with incredible image analysis and quadraphonic sound.

7.12 What is 6G technology?

1. Increase performance and maximize your data throughput and IOPS
2. Protect your system and secure your data
3. Ease your service and build efforts and expand your data center configuration options

7.13 6G concept

6G technology will help develop a computational nervous system that combines AI and wireless awareness that provides intelligent responses, backed by a speed estimated to exceed 1 terabyte per second, an unparalleled level of speed that will lead to an improvement in 5G applications. And 6G technologies can get lower latency, higher speed and more bandwidth compared with 5G technology. This advanced technology will help to connect the real world to the digital world, as product design and research will increase efficiency and reduce costs significantly while providing digital-enabled production processes in the physical world through advanced technology, including 3D printing.

7.14 What do we expect from the 6G?

Anything that depends on Internet connection will improve greatly when using the 6G networks, while today we have strong virtual reality and augmented reality systems. Thanks to the 5G networks, smart cities and farms that are fully connected to the Internet, and smart robotic systems that work in factories, the

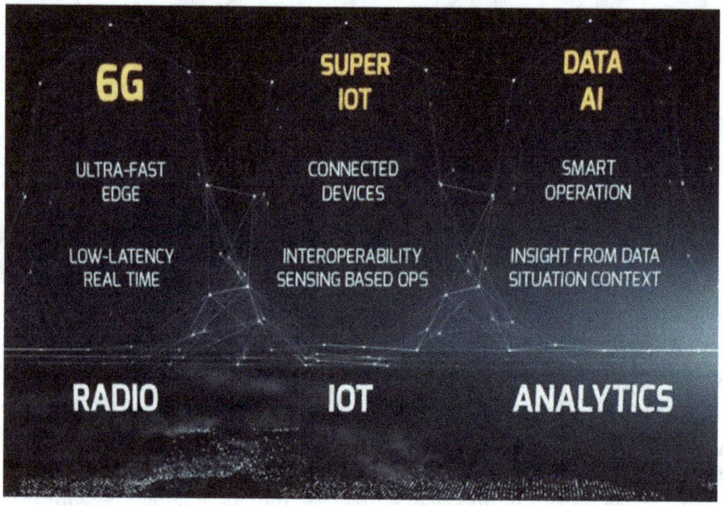

Figure 7.11 showing the way and types [1]

next generation will continue to support these sectors. In the best possible way, we will probably discover new fields of application that we could not reach today. The 6G networks will also play an important role in the revolution to take effect in AI and the IoT devices we are witnessing now (Figure 7.11). For example, self-driving cars must be aware of their location, the surrounding environment and the changes of this environment. Also, road users such as pedestrians, riders and other cars must be taken account of. To communicate with them, so that accidents and road congestion are avoided, drone flight time must be reduced as much as possible.

7.15 Service requirements

The 6G wireless system will have the following key factors:

- Enhanced mobile broadband (eMBB)
- Ultrareliable low-latency communications (URLLC)
- Massive machine-type communication (mMTC)
- AI integrated communication
- Tactile Internet
- High throughput
- High network capacity
- High energy efficiency
- Low backhaul and access network congestion
- Enhanced data security

The 6G network is expected to also have continuous cellular communication 1,000 times better than the 5G system. URLLC, a key 5G feature, will again be a key driver in 6G communications by offering an end-to-end (E2E) delay of under 1 ms.

The comparison of 6G and 4G and 5G communication systems are shown side-by-side in Table 7.4.

Quite obviously, 5G will not be in a position to satisfy consumer demands after 2030. Then, 6G will fill the void between 5G and requests in the industry. Based on past trends and predictions of future needs, the key goals for 6G networks are to increase data levels per user, (ii) a very huge proportion of wireless networks, (iii) global connectivity, (iv) very low-latency, (v) reduction of energy usage with charger-free IoT devices, (vi) ultrahigh stable interconnection, and (vii) machine-learning powered intelligence.

Every new cellular system generation is driven by innovative applications (Figure 7.12). 6G is no exception. It will be borne out of an unparalleled emergence of exciting new applications and technological trends that will shape its performance targets while radically redefining standard 5G services. In this section, we first introduce the main applications that motivate 6G deployment and, then, discuss ensuing technological trends, target performance metrics, and new service requirements.

Table 7.4 The comparison of 6G and 4G and 5G communication systems

6G	5G	4G	Issue
1 Tbps	10 Gbps	1 Gbps	Per device peak data rate
100 bps/Hz	30 bps/Hz	15 bps/Hz	Maximum spectral efficiency
Fully	No	No	Satellite integration
Fully	Partial	No	AI
Fully	Partial	No	Autonomous vehicle
Fully	Partial	No	XR
Fully	Partial	No	Haptic communication

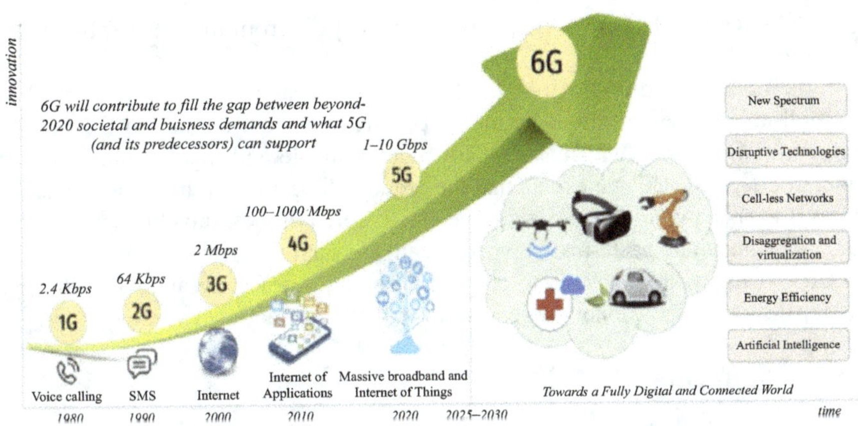

Figure 7.12 The growing of generations [15]

7.16 Applications of 6G

1. Multisensory extended reality (XR) applications

   XR will yield many killer applications for 6G across the AR/MR/VR spectrum. Upcoming 5G systems still fall short of providing a full immersive XR experience capturing all sensory inputs due to the irksome inability to deliver very low latencies for data-rate intensive XR applications. A truly immersive augmented reality (AR), virtual reality (VR), mixed reality (MR) experience requires a joint design integrating not only engineering (wireless, computing, storage) requirements but also perceptual requirements stemming from human senses, cognition, and physiology [16].
2. Connected robotics and autonomous systems(CRAS)

   A primary driver behind 6G systems is the imminent deployment of CRAS including drone-delivery systems, autonomous cars, autonomous drone swarms, vehicle platoons, and autonomous robotics. The introduction of CRAS over the cellular domain is not a simple case of "yet another short packet uplink IoE service." Instead, CRAS mandate control system-driven latency requirements as well as the potential need for eMBB transmissions of high definition (HD) maps are in the offing. The notion of Query Optimization in a Protective Environment (QoPE) applies once again for CRAS; however, the physical environment is now a control system, potentially augmented with AI. CRAS are perhaps a prime use case that requires stringent requirements across the rate-reliability [17].
3. Block chain and distributed ledger technologies (DLT):

   Blockchains and DLT will be one of the most disruptive IoE technologies. Blockchain and DLT applications can be viewed as the next-generation of distributed sensing services whose need for connectivity will require a synergistic mix of URLLC and massive machine-type communications (mMTC) to guarantee low-latency, reliable connectivity, and scalability [18].
4. Wireless brain-computer interactions (BCI)

Beyond XR, tailoring wireless systems to the human user is mandatory to support services with direct BCI [19–27].

The drones in 2030 are taking form as we expect from them the following in the future:

In the future, the drone will take a technological leap and a new concept of the drone will take shape, as it will be more accurate and more rapid, and the 6G telecommunications service will make its use more valuable and more common among people. In the year 2030, we expect that the drone will be adopted in terms of research and travel to space with ease. So, with the 6G-powered drones, travel to space will be made more accurate in terms of speed, cameras and information. We must not forget also that it will be a force to contend with and a deadly weapon against opponents for the army that cannot be dispensed with all accuracy that you own.

7.17 Conclusion

This report presented a bold new 6G network vision outlining developments, threats and related studies on drones. And while other concerns will emerge as part of a normal 5G operation, new study avenues such as Large Intelligent Surface (LIS) communications, 3CLS, holographic radio and the others will create a powerful research program for the next decade. The drones have a special place in the repertoire of research to be taken up in future, and the 6G promises a dazzling and impressive future. Scientists are experimenting to bring out something supernatural and it will definitely change the world. On the other hand, the drone technology will change the whole world and, in the future, we can hope to see what will be a new idea of how we use technology.

References

[1] L Brown (2019). Drone application. Filmora. https://filmora.wondershare.com/drones/drone-applications-and-uses-in-future.html (Accessed 10 November 2021).

[2] A Meola (2016). Drone usage expected to triple by 2020. https://www.businessinsider.com/faa-says-drone-usage-will-triple-by-2020-2016-3 (Accessed 10 November 2021).

[3] Business Insider Team (2016). How drones will change the world in the next 5 years. https://www.businessinsider.com/the-drones-report-research-use-cases-regulations-and-problems-2016-4-1 (Accessed 10 November 2021).

[4] R Allian (2017). How do drones fly. Wired. https://www.wired.com/2017/05/the-physics-of-drones (Accessed 10 November 2021).

[5] D Joshi (2019). Future of drones. Business insider. https://www.businessinsider.com/drone-technology-uses-applications (Accessed 10 November 2021).

[6] D Loesche. (2017). Drones: a tech growth market in the United States. https://www.statista.com/chart/9525/sales-of-consumer-drones-to-dealers-in-the-us/ (Accessed 10 November 2021).

[7] CK Writer (2019). Medicines delivered by a drone!. https://youngzine.org/news/technology/medicines-delivered-drone (Accessed 10 November 2021).

[8] E Comberg (2018). How construction workers are using drones on building sites. https://www.archdaily.com/898598/how-construction-workers-are-using-drones-to-enhance-building-sites (Accessed 10 November 2021).

[9] World Time Zone. Available online: https://www.worldtimezone.com/5g.html (Accessed on 28 March 2020).

[10] W Saad (2019). Vision of 6G. https://www.researchgate.net/publication/331396903_A_Vision_of_6G_Wireless_Systems_Applications_Trends_Technologies_and_Open_Research_Problems (Accessed 10 November 2021).

[11] M Chan (2019). 6G wireless system. https://www.researchgate.net/publication/331396903_A_Vision_of_6G_Wireless_Systems_Applications_Trends_Technologies_and_Open_Research_Problems (Accessed 10 November 2021).

[12] M Bennis (2019). Vision of 6G. https://www.researchgate.net/publication/331396903_A_Vision_of_6G_Wireless_Systems_Applications_Trends_

Technologies_and_Open_Research_Problems (Accessed 10 November 2021).

[13] Ray Le Maistre. (2021). SoftBank outlines 12 challenges for 6G, 6G Research and Innovation Channel, Aug 2.

[14] Sphere Drones Team (2014). Drone technology and business. https://spheredrones.com.au/ (Accessed 10 November 2021).

[15] N Joshi (2017). Stunning application. Allerin. https://www.allerin.com/blog/10-stunning-applications-of-drone-technology (Accessed 10 November 2021).

[16] Carlos (2020). Ways drone will impact society. cbinsights. https://www.cbinsights.com/research/drone-impact-society-uav/ (Accessed 10 November 2021).

[17] A Loftus (2019). The evolution and state of 5G technology as we transition towards 6G. http://www.tmcnet.com/voip/features/articles/444011-evolution-state-5g-technology-we-transition-towards-6g.htm (Accessed 10 November 2021).

[18] KB Letaief,, W Chen, Y Shi, J Zhang, Y-JA Zhang (2019). The roadmap to 6G: AI empowered wireless networks. *IEEE Communications Magazine* 57, 84–90.

[19] H Wenxing, Y Weng, L Xie, L Maoqing (2009). Design and implementation of web-based DSS for online shopping mall, 2009 IEEE International Conference on Control and Automation, pp. 1308–1313.

[20] S Huang, DH Zhang, WY Leong, HL Chan, KM Goh (2008). Detecting tool breakage using accelerometer in ball nose end milling, The 10th International Conference on Control, Automation, Robotics and Vision, pp. 927–933.

[21] WY Leong, J Homer, DP Mandic (2006). An implementation of nonlinear multiuser detection in Rayleigh fading channel, *EURASIP Journal on Wireless Communications and Networking* 2006, 1–9, Article ID 45647. DOI 10.1155/WCN/2006/45647.

[22] WY Leong, J Homer (2005). Blind multiuser receiver in Rayleigh Fading channel, *2005 Australian Communications Theory Workshop*, pp. 155–161.

[23] HF Min, LW Yie (2012). Investigating target detection and localization in wireless sensor network, *Procedia Engineering*, Elsevier, pp.75–81, Vol .41

[24] WY Leong (2008). Angle-of-arrival estimation: beamformer-based smart antennas, The 3rd IEEE Conference on Industrial Electronics and Applications, pp. 1593–1598.

[25] WY Leong, DP Mandic (2006). Towards adaptive blind extraction of post-nonlinearly mixed signals, *The 16th IEEE Signal Processing Society Workshop on Machine Learning for Signal Processing*, pp. 91–96

[26] WY Leong (2005). *Implementing Blind Source Separation in Signal Processing and Telecommunications*, The University of Queensland, Australia,

[27] W Xiaofan, PHJ Chong, LW Yie (2010). Performance comparison of CSMA/CD, CSMA/CA, CSMA/RI, CSMA/PRI and CSMA/PR with BEB, The 5th IEEE Conference on Industrial Electronics and Applications, pp. 1843–1848.

Chapter 8
VoIP technology in manufacturing

Mohammed Mohammed Moqbel Ali[1] *and Wai Yie Leong*[1]

8.1 Introduction

8.1.1 What does VoIP means?

It is an acronym for Voice over Internet Protocol. It is a method for connecting voice conversations via the Internet or any network that uses the Internet Protocol. Thus, any number of people connected to a single network that uses the Internet Protocol (IP), such as the Internet, can speak by phone using this technology.

In 1995, a few clients started to understand that sound can be transmitted over the system (the Internet) rather than transmitting it by the method of phone lines. This can be utilized through a system connecting all pieces of the world to give huge aggregates that he pays to worldwide calls where the principal programming program was built up. Its clients can speak with one another and just require a sound card, radio, and web association. These projects were not around then getting a charge out of the immaculateness of sound and the ideal quality, yet they were the sign of the way toward transmitting voice over the system conceivable and promising [1].

Telephone communications over the network are no longer limited to computers only, especially after it is now possible to use services (voice transmission via the Internet Protocol) without even the need to operate the computer via phones that support this technology. It is worth noting that the technology of communication via the network protocol (Internet) is a technology built on software, it takes advantage of VoIP technology to transmit voice and data over a computer network. This innovative technology includes several superior advantages, for example, call transfer, remote conference (remote) conferences, and voice messages, to strengthen communication networks and reduce the total cost. To use VoIP services, we only need a broadband Internet connection such as a DSL service, an Internet router, and phone connections (Figure 8.1).

8.2 History of VoIP technology

VoIP innovation is changing the conventional telephone world. Conventional telephone lines are gradually easing back with the advantages and favorable circumstances of VoIP innovation around the world.

[1]The Institution of Engineering and Technology, Malaysia

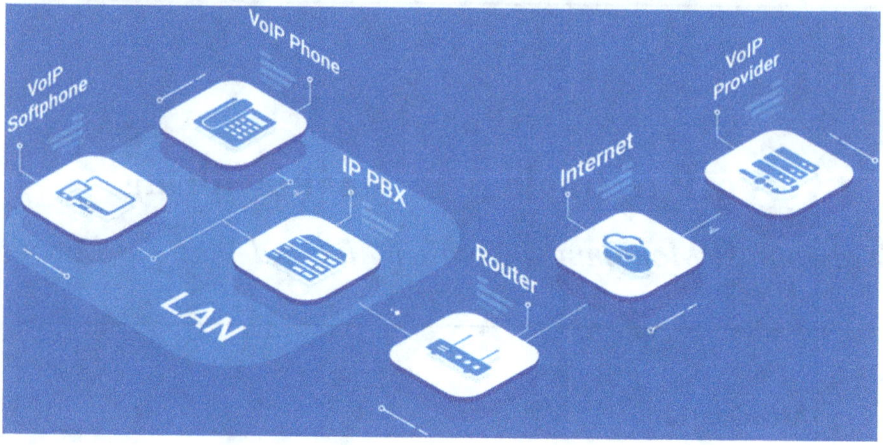

Figure 8.1 VoIP technology [2]

The historical backdrop of VoIP innovation shows that this innovation began in 1995 when it began a little organization called Vocaltec, and it was accepted to be the primary Internet telephone program. This program was intended to run on a home PC and fundamentally the same as the telephones utilized today, and it utilizes sound cards, amplifiers, and earphones. The program was classified "Web Phone" and uses the H.323 convention rather than the SIP convention, despite the fact that it has become progressively famous today.

Vocaltec began its prosperity with the Internet telephone, and its prosperity was in 1996. The principal disservice that showed up in 1995 is the absence of broadband limit, and on this premise, this product was utilized in modems that prompted poor sound quality when contrasted with a standard call. It is likewise valuable to take a note that one of the principles to vacate representatives is likewise the organizer of whichvoip.com.

By 1998, VoIP traffic had expanded the rate to roughly 1% of all voice messages in the United States. Consideration was centered on setting up and designing gadgets that empowered a PC's association with a PC-to-telephone and a telephone to a telephone-to-telephone. Systems administration organizations, for example, Cisco and Lucent, have given the hardware that can oversee VoIP Traffic and subsequently by 2000 VoIP represented over 3% of all voice messages [3].

Among the notable and spearheading organizations in this field is Skype, which showed up in the mid-nineties, and there are a large group of different projects, for example, Net2Phone, PC2Call, and Zero Phone. In 2005, the issue of guaranteeing voice transmission quality turned into a need over information transmission, so this innovation turned out to be progressively solid for intelligible voice transmission without intruding on calls. With the exponential development rate with VoIP innovation and with the utilization of remote advancements, the eventual fate of this innovation has gotten astonishing.

8.3 Technology working principle

This innovation changes over simple signs from the telephone to advanced signs. This sign is separated into packets and utilized the Internet Protocol to send these advanced packets in a few ways over similar information network. At the point when these bundles arrive at the predetermined goal (recipient), they reassemble the parcels sent so as to be heard unmistakably dissimilar to the standard associations, they utilize one explicit way, and if the other party (the collector) is an ordinary telephone, the sign is changed over back to sound signs to be comprehended from the beneficiary.

8.4 Specialized activity steps

Changing over Analog Signals into Digital Signals. Great bundle pressure (parcel width is excessively little). There are a few (conventions) that you can decide to pack the parcel in a propelled manner altogether not to defer sound.

Coordination of sound parcels into information bundles utilizing the RTP—Real-Time Protocol.

We need signs to contact the client (ringer) ITU-T H323.

- Upon getting, the bundle is breaking down, information is removed, the advanced signs are changed over to sound again and sent to the telephone.
- It must occur progressively all together not to have a sound cut.

8.5 Requirements for the technology to work

- PBX ought to be utilized to characterize the course of the telephone discussion.
- Convert sound signs, simple signs to advanced signs, computerized signals.
- Use Internet Protocol (IP).
- Digital signs must be separated into little parts called advanced pillars for transmission in a few ways through the information organize itself.

Bundles must be compacted well (parcel width is excessively little). There are a few conventions that you can decide to pack bundle in a propelled manner all together not to get a deferral in the sound.

- There ought to be a cradle space for cushion gathering all together not to defer sound.
- It must occur progressively all together not to have a sound cut [4].

8.6 Some of the benefits of the VoIP technology

A modern approach to voice communication, VoIP stands for Voice over Internet Protocol and refers to a system that allows users to make voice calls over an internet connection, rather than through a phone line. As internet connectivity becomes more omnipresent, more businesses are switching over to VoIP for its many advantages over conventional phone lines.

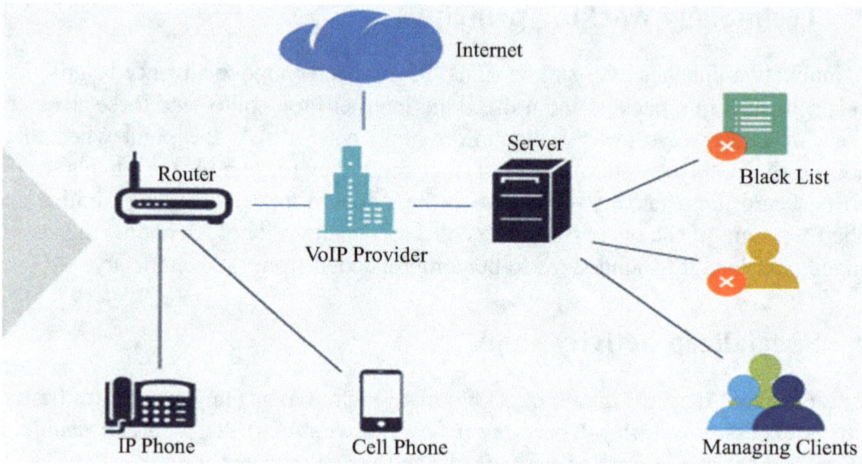

Figure 8.2 VoIP server helps in the functioning of the Internet-based telephone system or VoIP phone system [5]

VoIP technology works by converting sound into digital voice communication and then transferring it through an internet connection. Let's dive in and discuss some of the key features of VoIP that have pushed so many Connecticut businesses to make the switch.

- Lower costs
- Increased accessibility
- Complete portability
- Higher scalability
- Advanced features for small and large teams
- Clearer voice quality
- Supports multitasking
- More flexibility with softphones

Figure 8.2 shows the benefits of the VoIP technology.

8.7 Minimize cost

The association by means of this innovation spares a ton of cost since what is paid for this administration is fixed as a month-to-month membership and the call is open nonstop. The expense of worldwide calls through this innovation is less expensive than some other methods for correspondence. Cost reserve funds to specialist co-ops.

Since this innovation relies upon the product more than it relies upon the gear, which encourages the procedure of support notwithstanding the issues that are frequently known and the hardware is easy, the expense is less (the expense of establishment might be high, yet then the cost will diminish).

8.8 Mobility

The client of this administration should not be in one spot in particular, yet it tends to be utilized anyplace even on account of the movement to another nation and inner calls keep on being gotten as though it was in a similar nation and with a similar tax since it relies upon the system (the Internet).

It permits clients to introduce their telephones anyplace in the workplace by associating their telephones to the nearest organize space while keeping their present number. Calls can be moved anyplace on the planet because of the SIP convention highlights.

8.9 Scalability

For the capacity to develop with the expanding needs, it is conceivable to extend and increment newlines effectively by buying the vital hardware (VoIP Phone) and associating it to the system without the requirement for isolated associations or by introducing the program on the gadget.

8.10 Features

It contains standard telephone utilities notwithstanding different administrations, for example, the capacity to amass talks—and voice message, the capacity to modify ringtones, the capacity to keep and include the subtleties of the individuals calling them, and the capacity to control numbers can forestall a call or get a specific number and others.

8.11 Easy to use

The VoIP telephone framework permits you to handily fix and alter your telephone framework due to the GUI-put together interface and with respect to the Internet, which encourages the procedure of establishment, framework control, and announcing, while other private telephone gadgets regularly experience issues being used because of their generally structured interfaces for the master to introduce and utilize these gadgets.

The chance of joining voice, picture, and information advancements and simultaneously the utilization of voice over the Internet innovation permit clients to utilize all applications that occur on the Internet, for example, dealing with any program or utilizing email notwithstanding correspondences where a similar system is utilized for voice, picture, and information. Along these lines, it permits the client to set aside exertion and cash simultaneously [6].

8.12 VoIP technology standards

It is a set of rules and conditions that govern the process of telephone communications and is divided into the following (Figure 8.3).

Figure 8.3 VoIP technology standards [7]

8.12.1 Closed systems

It depends on shut (not free source) principles, for example, Skype and the well-known Skinny Client Control Protocol SCCP, a shut terminal control convention initially created by Selsius Corporation, which is claimed and created by Cisco System Inc. One of the most mainstream gears working under this convention is the Cisco 7900 arrangement.

8.12.2 Open systems

Open gauges that depend on open-source conventions include the following.

8.12.2.1 H.323 convention

It is a lot of standard conventions radiating from a framework created by the ITU-T to move sound and picture documents over a PC organize utilizing Packet-Based bundles. This convention is utilized in most well-known applications, for example, NetMeeting.

8.12.2.2 Taste convention

It is an abbreviation for the Phase Initiation Protocol, which is a convention for phone signals related with Internet conventions, which are utilized to start, alter and end VoIP calls. The Internet Engineering Task Force (IETF) built up this convention, and it was at first distributed as RFC 3261. Taste can depict the correspondence important to begin a call and the stage start convention has become a forward leap in the realm of VoIP telephones. This convention is fundamentally the same as the HTTP convention, as it is a book convention, straightforward and adaptable to utilize, so the SIP convention has supplanted the standard H323 convention in across-the-board use.

8.12.2.3 IAX2 convention

It is a correspondence convention between Asterisk programming and is an open-source phone trade program that permits voice over Internet Protocol between Asterisk servers and IAX2 customers and packs the bundle well. To get a thought of the measure of waste created during VoIP, recall that a compacted voice that possesses a region of 5.6 Kbps will require 18 Kbps from the transmission capacity, the distinction between 5.6 Kbps and 18 Kbps comprises of the headers of the bundles that will transmit these information. These headers contain all the vital data

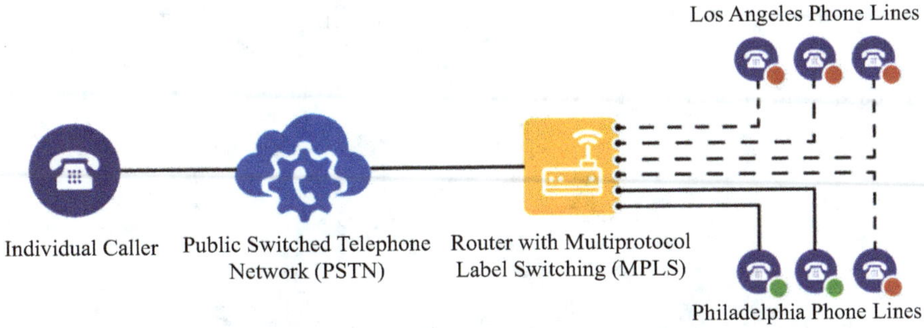

Figure 8.4 Technology working principle [8]

(e.g., IP address) to convey voice parcels to the beneficiary. The IAX2 convention radically diminished this loss by constraining the measure of extra bits permitted to use for every bundle, and furthermore abused the rule of collection discussions sent to a similar goal and remembering them for similar parcels (Figure 8.4).

Taste trucking looks like something to do in the midst of a get-away in warm atmospheres with at any rate one fruity beverage available. Indeed, it looks nearer to these air terminal staff waving huge light sticks on planes (the activity title is really Marshall or Signaler). Before we leave, the guide should help shed some light on what is secured beneath. Peruse on for answers to questions like: What are SIP and SIP text styles? How is the SIP association not quite the same as the PSTN association? What are the benefits of the SIP network?

What is SIP? Before examining what a SIP association is, it is important to clarify what the SIP is. Taste represents Session Initiation Protocol. Basically, SIP is the flagging and the correspondence of the board strategy.

Each gathering of companions has that individual who arranges everything and is a specialist in uniting everybody. Taste is like that individual; however, as opposed to interfacing with individuals, it manages IP-based applications, most quiet voice, video, visit, and other sight and sound associations.

Taste works by sending parcels of data between gadgets that help SIP. There are two kinds of bundles—sign and media. Signal bundles set up and end an association between two gadgets, permitting trade of media parcels.

Obviously, interchanges rely upon the two gatherings knowing the other and having the option to comprehend what is being said. Taste directs these two things by following the gadget's IP locations, and setting decides that codecs fit for various meeting types with the goal that all the gadgets included can deal with the media accurately [9].

8.13 How VoIP is transferred?

VoIP works especially like some other information you send over the Internet like messages in that it changes over voice correspondence into information parcels and

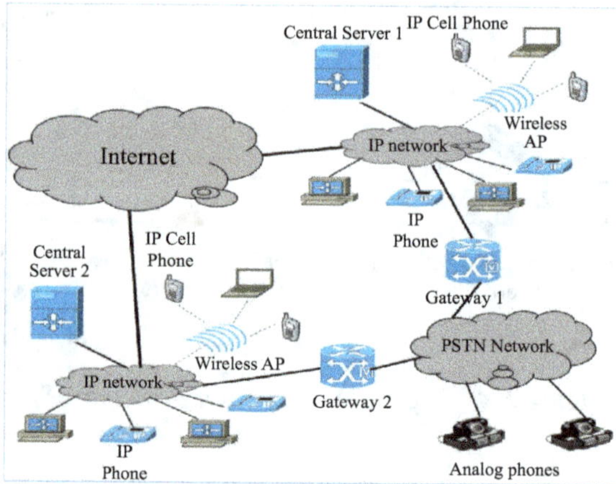

Figure 8.5 How VoIP is transferred [10]?

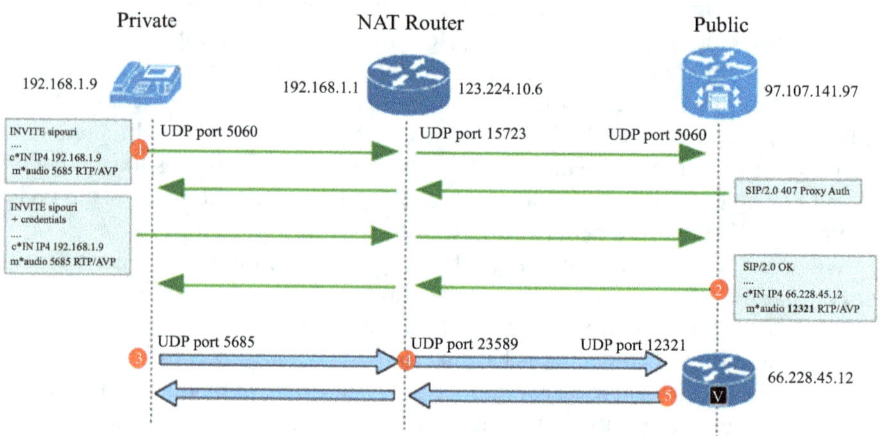

Figure 8.6 NAT diagram [12]

sends it over the IP organize. With VoIP, you do not have to have a PTSN in your business or home, as calls just rely upon your web association (Figure 8.5).

VoIP came to change the conventional telephone world. Customary telephone lines are bit by bit easing back as VoIP innovation gives preferences and advantages in trendsetting innovation. It is useful to stop and take a gander at the historical backdrop of Phoebe and locate an additionally energizing future [11].

8.13.1 NAT diagram

In actuality, NAT is more unpredictable than this, changing port numbers too, however. We have held the idea to the uncovered fundamentals to plot why NAT harms VoIP (Figure 8.6). NAT accommodates numerous advantages, including address reuse and fundamental security. This security exists since parcels that show up from the open Internet without being mentioned from inside the LAN will be disposed of. Be that as it may, it is this security component that makes VoIP troublesome when utilizing NAT. The idea of putting a VoIP call to a gadget behind a NAT necessitates that the NAT conveys spontaneous bundles from the Internet to the VoIP gadget. This is an intricate theme, and as we will see later on, NAT traversal can cause all sorts of difficulties for VoIP.

8.13.2 Advantages of VoIP

1. Integrate multiple locations through one phone system.
2. Dial by extension to any user in the network with no toll charges.
3. Save on long distance.
4. Competitive international calling rates.
5. Line number portability.
6. Forward voicemails to your email address for access anywhere.
7. Easily and affordably scale from 1 to 5000 users.
8. Add features as you grow and as your needs change.
9. Web-based management tools for service and feature control.

8.13.3 Disadvantages of VoIP

While VoIP is frequently less expensive and more adaptable than the customary telephone frameworks, it merits remembering potential issues. For instance:

1. Some VoIP administrations do not work during power blackouts or when the web association is down. It is a smart thought to introduce a reinforcement power gracefully or have a cell phone as a reinforcement.
2. The 9-1-1 help given by VoIP administration is additionally not the same as the customary telephone line. Customised telephone administration by a specific telephone number with a fixed location however VoIP administrations can be utilized from any web networks. In this way, it is imperative to give a precise location to the VoIP supplier and update this data promptly if this data is changed.
3. The quality of the broadband connection and the efficiency of your computer influence the calling rate. So, make sure you have a secure link to the internet, and both your hardware and software are running correctly.

8.13.4 What to look for in a VoIP provider?

The world is getting increasingly more interconnected with every year as we continue walking forward into this web age. Thus, worldwide calling is going to keep getting progressively significant for a wide range of organizations. Each VoIP supplier will have shifted plans depending on various needs, for example, nations permitted, the measure of minutes allowed in a month, and whether the client is eager to sign a drawn-out agreement or not. Finding the privileged VoIP supplier for your particular needs can be troublesome, as you will have to take a gander at a wide range of measurements. We should see explicitly what to take a gander at to decide whether a VoIP supplier is beneficial for you.

8.14 Literature review

8.14.1 Introduction

VoIP transmits voice over the Internet instead of over the traditional landline network. In addition to making international calls, a lot of professional online communications such as video chats, chatting, and file sharing that become more flexible when using such technology.

VoIP innovation is changing the customary telephone world. Conventional telephone lines are gradually easing back with the advantages and preferences of VoIP innovation around the world. It is valuable to stop and take a gander at the historical backdrop of VoIP and you will locate an all the more energizing future.

The historical backdrop of VoIP innovation shows that this innovation began in 1995 when it began a little organization called Vocaltec, and it was accepted to be the principal Internet telephone program. This program was intended to run on a home PC and fundamentally the same as the telephones utilized today, and it utilizes sound cards, receivers, and earphones. The program was designated "Web Phone" and uses the H.323 convention rather than the SIP convention, despite the fact that it has become increasingly mainstream today.

As far as VoIP technology is a good choice for many individuals, it reflects a fundamental advantage in the world of business and companies, because it improves the productivity of companies by providing an excellent internal communication network between workers in leading companies with large human resources, in addition to making many communications from domestic and international for marketing purposes, customer support, sales increase, and service provision (Figure 8.7). Thus, VoIP

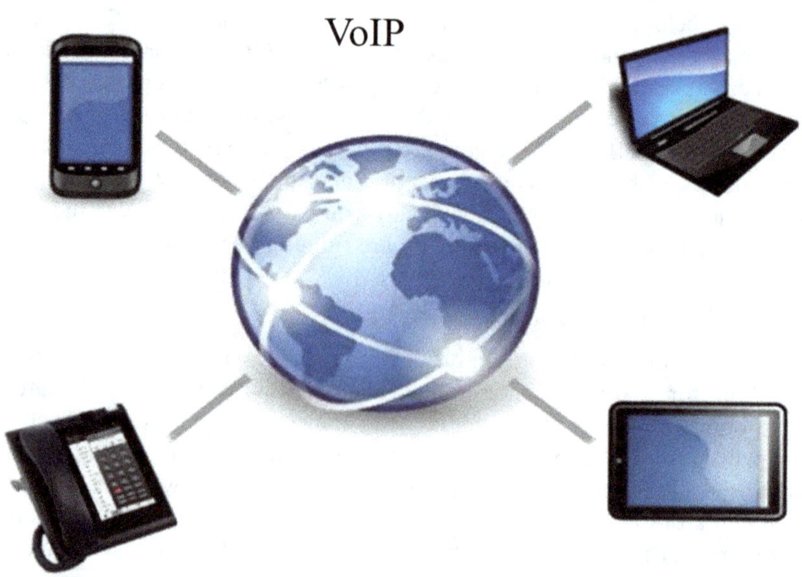

Figure 8.7 VoIP innovation [14]

technology provides unprecedented flexibility in organizing your contacts while taking advantage of many other advantages [13].

Steps to working VoIP technology:

1. Convert analog signals to digital signals.
2. The sound packets are compressed well so that there is no delay in sound.
3. Transfer voice packet over VoIP
4. Upon receiving, the package is analyzed, data is extracted, the digital signals are converted to the audio again and sent to the phone.

Digital communications make employees communicate with each other on the one hand and with customers on the other, the easier task. It enables them to communicate through their computer while forwarding incoming calls to their mobile devices. Especially since using your VoIP smartphone, sharing your information and ideas in record time is smooth. You can also hand out projects with your colleagues' collaboration with instant communication tools, group chats, and video chats.

The era of the communications revolution. This period was full of many inventions in the field of communications, where many devices were invented, including:

1. Phone: The first appearance of the phone was in 1876 CE, when Alexander Graham Bell was patented when he made his first phone call to Thomas Watson, from Ranford to Paris, and in 1915 Watson received the first intercontinental phone call.

   VoIP began with an organization called VocalTec in 1995. They spearheaded the principle of broadly accessible Internet telephone. It was called, essentially, Internet Phone. It permitted one Internet client to call another, and it was associated with the speakers and an amplifier. No video was accessible at that point, and the arrangement necessitated that the two clients be on a similar programming.

   In 1996, Internet voice message applications woke up. Clients could send voice messages over the Internet to the goal telephone. This was additionally full of intricacies, for example, poor sound quality, times of quietness, and loss of association. Around the same time, VocalTec declared their Internet telephone programming related to Microsoft NetMeeting.

   With the trouble of characterizing a particular idea for telephones with numerous highlights, purchasers are quick to incorporate savvy gadgets in their lives and quest for a solitary gadget that consolidates various highlights.

   Independent gadgets known as the PDA showed up and these gadgets spread generally among clients in the nineties, at that point, technologists endeavored to coordinate the highlights of the individual right hand with the wireless so the vast majority does not convey two gadgets together.

   Whats more, the main organization to work in the creation and conveyance of the individual partner on the planet around then is Palm, "an American organization spend significant time in the assembling of individual computerized associate (PDA) and different gadgets established in 1992."

2. Radio: In 1896, radio was patented for Marconi, who succeeded in transmitting the first radio transmission of sound, and the experiment had succeeded in transmitting the signal at a distance of 2 km, after many failed attempts.

   The origins of broadcasting in the world are due to the use of short-wave because of its ability to reach long distances, which is a great scientific revolution in the field of communications, the discovery of the short wave is due to a number of scientists, including "Max Weil" who proved the existence of electromagnetism, or radio waves in 1867 AD, "Henry Jackson" and the Italian "Galileo Marconi" and Thomas Edison, who preferred to produce the necessary wavelengths for transmission through a short wave, but Heinrich Hertz, the German naturalist, was the first to conduct radio research, and the first to conduct short-wave experiments, where he verified that the electric current. The variable creates waves that can be transmitted through space without using wires, at the speed of light, in the year 1888 CE.

   In the conversation about the emergence of broadcasting in the world, it is worth noting that wireless communication has multiple uses in the military during the First World War, and commercially in the navigation process between ships, then experiments began to broadcast radio programs in America and other countries until the first radio stations appeared in the 1920s in America, which was broadcasting regular programs that last all day and were followed by other stations in Europe and the world, and most of the radio stations belonged to the owners of laboratories specialized in the manufacture of receivers [15].
3. Television: In September 1927, the inventor of television, Philip Farnsworth, demonstrated the first television set working by transmitting images, where he discovered the possibility of encoding radio waves with an image and then re-displayed it on the screen.

   Television is one of the most powerful media outlets to attract the attention and influence of the audience, and it carries with it the advantages of "cinema" in terms of image and color and the advantages of "radio" and "theater" sound broadcasting in terms of movement and vitality that it enjoys and in talking about the origins and development of television (Figure 8.8). It is worth noting that the word television consists of two sections, Tele, which means remote, and vision, which means vision, and means remote vision, as this system allows the transmission and reception of the moving visual image and the accompanying sound by electromagnetic waves.

   Scientists' experiences contributed to the emergence and development of television since the 1920s in America, and they have benefited from various experiments and studies carried out by previous scientists in the field of photography, electricity, and telecommunications, among them was the Russian scientist Vladimir Zorikin who immigrated to the United States of America, where he was able, through a coin in the Westinghouse Company, to invent the first electronic scoop device named the electronic television camera.

   Philo Farnsworth developed the electronic television camera, and Allen Dumont invented the first home TV receiver and also developed the reception screens. John Logip Bird from United Kingdom researched on the field of television, he has conducted a lot of research to better display the image on television.

Figure 8.8 *Television is one of the most powerful media outlets to attract the attention and influence of the audience*

In 1927, one of the researchers succeeded in sending a closed-circuit television picture, which was from Washington to New York, and the following year, General Electric began experiments in television broadcasting, and in 1935, NBC started regularly broadcasting its programs through the New York station that it established, and in 1940, TV became increasingly popular and RCA photographed New York City by air with a mobile device and then transferred on TV, and research and experiments are continuing to this day to create a better way in TV broadcasting with better technologies and cheaper even to the era of the high-definition picture. And what was mentioned was a detailed brief talk about the origin and development of television [16].

4. The Internet emerged because of the endeavors of numerous individuals who added to its turn of events, it cannot be said that there is one individual who created the Internet in the entirety of its subtleties, the principal Internet thought was found in the year 1961 AD after the distribution of the main research in the field of data moves through systems by Leonard Kleinrock, the idea of the Internet started to create in a genuine manner in the late sixties of the only remaining century through a gathering of conversations and gatherings that addressed this idea, and the initial segment of the system was developed in August in 1969 AD, which is known as the system connector.

   In 1969, Charley Kline (University of California student) sent the first message over the Internet, which is about 483 km away, and this was considered the first attempt in the history of the Internet.

   The ARPANET network promises the emergence of the Internet, as it was developed by the US Department of Defense in cooperation with several universities, and researchers in universities and employees of the Ministry of Defense were allowed to share information with colleagues in other cities across it, by logging in from the University of California to the University of Stanford, and by 1971, approximately 30 universities were participating in Arbant.

By 1986, the National Science Foundation (NSFNET) had become the center for supercomputers connected to the Internet at a speed of 56,000 bps. The number of hosts on the Internet exceeded nearly 20,000 by 1987. The European Nuclear Research Organization (CERN) has been able to develop hypertext markup language (HTML) that is still used in web pages.

5. The invention of the World Wide Web

   The World Wide Web was invented by Sir Tim Berners-Lee in 1989 – originally he was trying to find a new way for scientists to easily share the data from their experiments. Hypertext (text displayed on a computer display that links to other text the reader can immediately access) and the internet already existed, but no one had thought of a way to use the internet to link one document directly to another. to easily share the data from their experiments. Hypertext (text displayed on a computer display that links to other text the reader can immediately access) and the internet already existed, but no one had thought of a way to use the internet to link one document directly to another.

   Berners-Lee created the world wide web while he was working at CERN, the European Organization for Nuclear Research in Switzerland. His vision soon went beyond a network for scientists to share information, in that he wanted it to be a universal and free 'information space' to share knowledge, to communicate, and to collaborate. You can find out more about how his work on the world wide web at CERN.

   There are three main ingredients that make up the world wide web. URL (uniform resource locator), which is the addressing scheme to find a document; HTTP (hypertext transfer protocol), which connects computers together; and HTML (hypertext markup language), which formats pages containing [17].

Gap analysis and benchmarking of VoIP:

Authors	Traffic and tools	IP transition mechanisms	Year
T. Hoeher, M. Petraschek, S. Tomic, and M. Hirschbichier	VoIP traffic with SIP tool	IPv4, IPv6to4 tunnel, and Teredo tunnel	2007
C. Lee, Y. Yu, and P. Chang	Viseo traffic with NS-2 tool	Dual-Stack Transition Mechanism	2009
R. Yasinovskyy, A.L.Wijesinha, and R. Karne	VoIP traffic with Open-swan & MGEN tools	IPV6-to-4 tunnel with IPSec &NAT	2009
T. Sanguankotchakom and M. Somrobru	VPIP, Video, FTP traffic with NS-2 tool	Dual-Stack Transition Mechanism	2005
J. Lee and K. Chon	Video traffic with video camera	Pure IPv4 and IPv6 across the globe	2006

8.14.2 Conclusion

The governmentally subsidized Lincoln Lab at MIT creates, as it keeps on doing today, trendsetting innovation in light of a legitimate concern for national security (Figure 8.9). Weave McAuley and Charlie Radar build up the main voice bundle

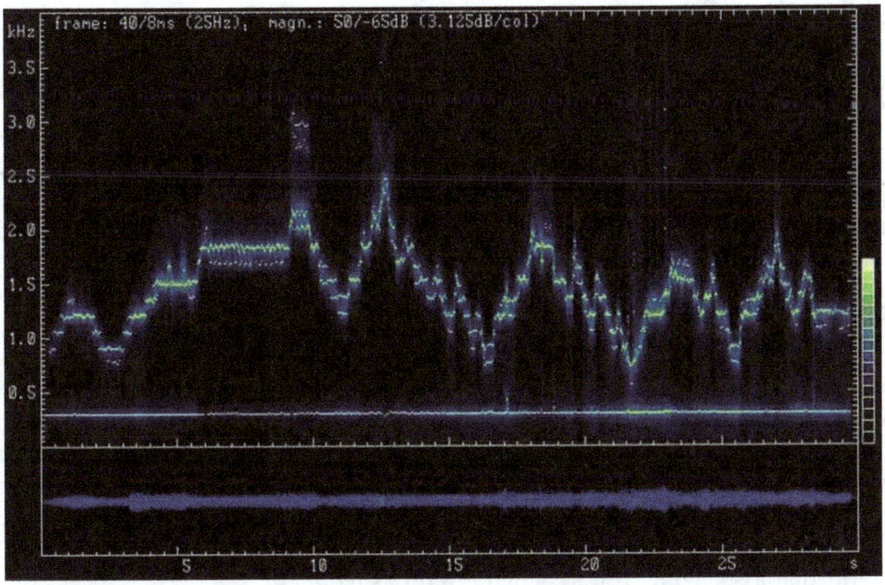

Figure 8.9 Trendsetting innovation in light of a legitimate concern for national security

over the early ARPANET. It utilizes something many refer to as an LPC, or direct prescient coder, to transmit voice. In 1974, the lab effectively ran a test among them and Culler Harrison, Inc., which was in California. By 1976, Culler Harrison and Lincoln Labs had a phone call over LPC. In 1982, they accomplished a significant achievement by utilizing LPC to associate over a nearby link organize, a portable parcel radio net, and an interface with the PSTN.

The number of VoIP clients increase from 150,000 in year 2003 to 1.2 million in year 2004.

Cost of Enterprise VoIP Phone Systems. In a similar way, clients will not be charged to send and receive messages. For clients of VoIP, calls over the Internet will not be charged per call. Facilitated VoIP telephone frameworks frequently charge per telephone line or expansion, giving an adaptable, operational cost model. There are different highlights and plans offered relying upon the degree of brought together correspondences and versatility highlights wanted by the organization. Facilitated VoIP telephone administrations will in general be progressively moderate over the long haul as capital costs are decreased with the specialist co-op assuming liability for equipment and programming updates. Facilitated VoIP is likewise entirely adaptable, permitting organizations to use on-premises hardware whenever wanted, in a half and half arrangement. To bring down your business' TCO for its undertaking telephone framework, get in touch with us today to perceive what the intensity of facilitated VoIP can accomplish for your association.

8.15 Beyond 2030

8.15.1 Introduction

The development of the cell phone and other cell phones has occurred unmistakably more quickly than for some different advancements, and changes proceed apace as we move into 2020. The times of utilizing agonizingly moderate and constrained 3G and 4G information associations while we fretfully trust that the 5G overhaul will come are currently, fortunately, ancient history. A great many people assume that they can download the most recent application in a brief moment or stream video in a hurry. The appearance of 5G LTE gave us speeds about multiple times quicker than 4G and carried with it definitely a larger number of advantages than quicker web speeds on our cell phones. The speed of 5G additionally implied that we could get to VoIP approaching every one of our gadgets finally, with portable applications accessible on our telephones as well as on all versatile working frameworks. It is ordinary for telephone bearers to incorporate VoIP into the dialer application alongside typical calls. Additionally, advancements are accessible from numerous administrators that can deal with a bring over 5G information association, cell system, or WIFI as required.

To make a VoIP call we need access to two things: a VoIP account and a web association. We can setup a new login account in a short time. However, we might not be able to gain access to a web network quick enough for VoIP using our cell phone. This was everything except unthinkable with 3G however explained somewhat with 4G. For buyer VoIP calls, 4G is surely satisfactory; however, it isn't in every case adequately powerful for organizations. Fortunately, administrations, for example, IDT Express offer worldwide VoIP availability and market-driving rates, permitting organizations to completely misuse the capacities of the current innovation. So, in what capacity will the following stage into the domain of the 5G and 6G information association help to improve things further for business?

It is anticipated that 6G will surrender paces to multiple times quicker than as of now accessible. Yet, once more, speed isn't the main advantage. With the utilization of 6G innovation, there is a principal change in how data transmission is utilized. With 5G, a set measure of data transfer capacity is allotted toward every path by a system supplier and this is fixed notwithstanding concerning the traffic being dealt with. With 5G and 6G innovation, this can be balanced so if there is blockage one way, data transmission can be utilized from the other bearing utilizing a mix of programming and equipment. Users always request for higher speeds compared to what is available. As more gadgets come online consistently, this capacity to modify transfer speed allotment "on the fly" and speed up will turn out to be progressively significant (Figure 8.10).

5G will open up expected open doors for associations moving to VoIP as web velocities and data transfer capacity will never again be a limitation. The manner by which the two workers and outside contacts can impart will be smoothed out and greater sound will be accessible for some calls. With the improved productivity of data transmission, the board may likewise receive rewards for organizations that

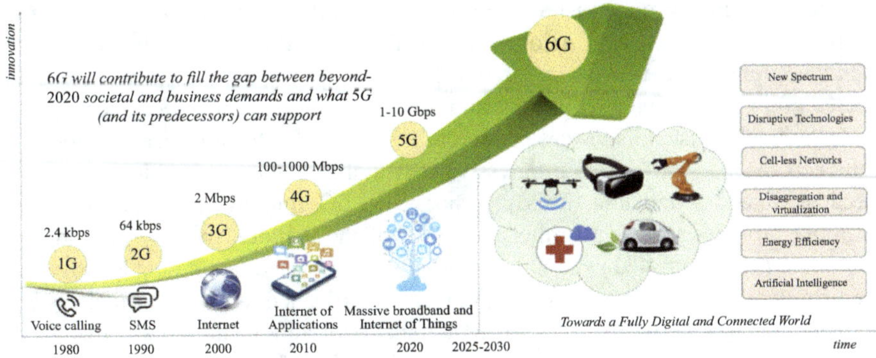

Figure 8.10 6G will surrender paces to multiple times quicker than as of now accessible [18]

use video conferencing and bunch sound calling. All things are considered, the number of potential members will be expanded, the nature of the video will be improved, and buffering will be diminished. Information regarding the substantial applications will likewise observe upgrades, with sped-up and diminished slack. Similarly, as the move from 3G to 4G LTE saw new administrations being made, the hop to 6G will see associations finding better approaches to incorporate video conferencing or calling into the everyday lives of representatives and accomplices. Energizing and inventing new advancements with the utilization of expanded or computer-generated reality will turn out to be progressively ordinary.

8.15.2 Beyond 2030 in 6G technology and the improve for VoIP

Similarly, as 5G came to fruition following quite a long while of R&D concentrated on past 4G (B4G) advances, the ICT business is in a comparative circumstance with 6th era (6G) remote as driving affiliations center around B5G advertise openings. There will be a requirement for considerable designing endeavors extending from the advancement in materials science, electrical building, power the executives, and numerous different zones as a feature of a multidisciplinary methodology.

Driving organizations and NGOs are centered around both the advancement of 5G just as totally new advances, capacities, and arrangements that will be one of a kind to the market past 5G. Research and development contemplations are driven by ICT environment financial aspects, for example, new B5G showcase driven plans of action and openings, just as cultural variables, for example, the United Nations Sustainable Development Goals. Furthermore, the 6G market will likewise concentrate on filling 5G ability holes similarly as B4G planned for LTE.

Alongside newly opened doors will come with new difficulties as the 6G market will require considerable designing and imaginative methodologies in numerous regions including the need to manage essentially higher radio frequencies than 5G. This is on the grounds that B5G is probably going to use terahertz recurrence groups just as the lower levels utilized in 5G are dependent on millimeter waveforms. As a major aspect of this specialized test, advancement in materials science and the electrical building will be required to devise brilliant surfaces important to help with radio sign engendering and the executives.

As far as the past 5G showcase sway on business tasks, 6G advances are foreseen to quicken the pattern toward progressively virtualized, programmable, and appropriated organized foundation. This incorporates a 6G framework inside customary transporter arranges just as a further extension of the venture, mechanical and government-claimed/controlled gear and programming. Unbiased facilitating and oversaw administrations will keep on developing in significance as will virtual-systems administration and the miniaturized scale administrator administration model.

From 1G to 5G, going through Universal Mobile Telecommunication Systems (UMTS) and Long-Term Evolution (LTE) advancements, every age of portable innovation has been intended to address the issues of system administrators and last purchasers. 5G, specifically, is making a huge advance toward building up a low idleness material access organize, by giving new extra remote nerve tracts, i.e., information pipes. However, these days social orders are turning out to be perpetually information-driven, information needy, and robotized and will present progressively increasingly tough prerequisites (as far as ultra-high unwavering quality, limit, vitality productivity, and low dormancy) which may immerse the limit of customary advances for remote frameworks. We present the defense that 6th era (6G) frameworks will add to filling this hole specifically (Figure 8.11).

- 6G will encourage the Industry 4.0 unrest, i.e., the computerized change of modern assembling, however, advancement headways in the field of semiconductor and coordinated circuit, and will drive radical robotization of efficiency.
- 6G will reform the social insurance part through developments like versatile edge figuring, virtualization, and man-made brainpower and will dispose of

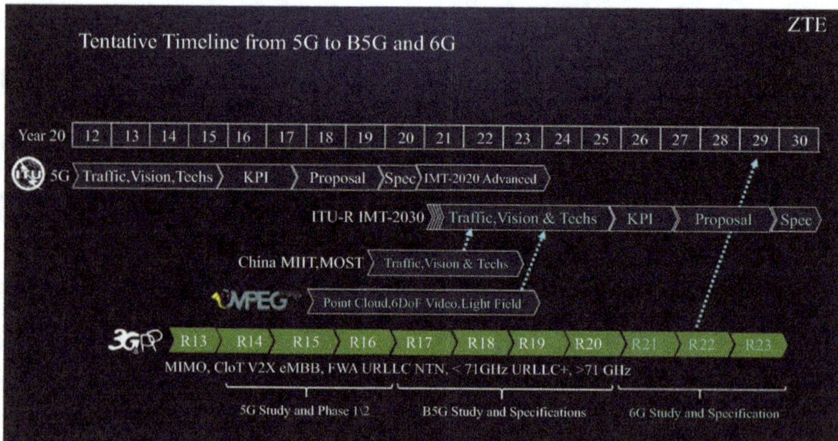

Figure 8.11 The 6G market will likewise concentrate on filling 5G ability holes similarly as B4G planned for LTE [19]

existing boundaries through remote medical procedure and health-care work process advancements.

- 6G will make ready for the coming time of associated and self-governing vehicles and flying vehicles, for example, drones, offering the capability of more secure voyaging, improved traffic the board, and backing for infotainment applications through advances in equipment and programming just as spearheading network arrangements.
- 6G will quicken the selection of answers for brilliant urban communities, focusing on life quality upgrades, ecological checking, and city the board computerization through help for client-driven machine-to-machine correspondence and vitality gathering. 6G will likewise advance mechanical progressions in the field of virtual/enlarged reality and holographic telepresence.
- 6G advances will envelop limit development methodologies to offer a huge scope network to the clients, in any event, when common correspondence foundations might be undermined (e.g., if there should arise an occurrence of crisis or fiasco circumstances). The wide reason for this paper is to see how future 6G frameworks can be created to fulfill the needs for a completely associated, canny computerized world. We outline rising advancements and improvements at all layers of the convention stack, from center physical specialized strategies to systems administration structure and organization, that have a critical guarantee for future 6G frameworks.

Another age of versatile systems is for the most part described by a lot of novel correspondence advancements that give extraordinary execution (e.g., regarding accessible information rate, dormancy) and capacities. For instance, enormous Multiple Input, Multiple Output (MIMO) and mm Wave interchanges are both key empowering influences of 5G systems. To meet the prerequisites recently depicted, 6G systems are required to depend on the regular range (i.e., sub-6 GHz and mm Waves) yet additionally on recurrence groups that have not been thought of yet for cell gauges, to be specific the terahertz band (which can be abused to dispense a huge measure of data transfer capacity to short proximity and greatly beamformed correspondences) and Visible Light Communications (VLC) (which could give arrange administrators a savvy and viable approach to improve indoor inclusion). Other than the new range, 6G will likewise change remote systems by utilizing a lot of advances that have been as of late empowered by headway in physical layer and circuits research yet are not part of 5G. We expect 6G as well (i) coordinate full-duplex abilities in the correspondence stack, to empower constant downlink transmission with synchronous uplink affirmations or control messages (or the other way around), and increment the multiplexing capacities and the general framework throughput without utilizing extra data transfer capacity; (ii) abuse novel channel estimation methods, for example, out-of-band estimation and packed detecting, to expand the effectiveness of the 6G control plane; and (iii) consolidate correspondence with detecting and system-based confinement capacities, to improve control tasks and give novel client administrations.

With 6G ready to work on a lot higher recurrence than 5G (for instance, the FCC conceded >95 GHz for test use), the shorter frequencies will take into consideration higher limitation exactness, conceivably down to centimeter-level situating. One result of such accuracy will be full advanced portrayals of the physical world, or "computerized twins", by likewise combining information from different sources including system information. System administrators will likewise have the option to produce interconnected and cooperative advanced twins and computerized portrayal of bigger articles and their condition. Nokia Bell Lab exhibited an advanced twin of a New Jersey road with drone-caught high-goals information for remote system streamlining, for instance, exact sign engendering forecast.

These utilization cases should be bolstered by new, progressed fundamental advancements that will give rules to investigate in the order in the coming years. New range advances are featured by all speakers as such an area. This remembers both radio innovation for the purported D-Band (140–180 GHz) or more and progress in material sciences. Ringer Lab's Viswanathan called attention to that handset plan for such radio frequencies will be progressively modern and may need to utilize glass interposers rather than silicon. ZTE likewise observes "Past Silicon" as one of the main 6G challenges.

System design is another key innovation prerequisite that requirements to progress in the approach 6G. One such headway is the thing that Nokia Bell Lab's Viswanathan finds in the pattern of RAN-Core assembly. This is principally determined by the need to draw the center nearer to RAN for low inertness administration just as to make the RAN progressively incorporated toward the cloud. A related pattern featured by Viswanathan is the interest for hyper-specific cutting. He accepts that the system cutting should move from asset reservation in 5G to giving separate programming stacks and capacities by utilizing diverse miniaturized scale administrations (Figure 8.12).

Both ZTE's Fang and InterDigital's Alain Abdel-Majid Mourad, Director of Engineering R&D, focused on the significance and interest for development to

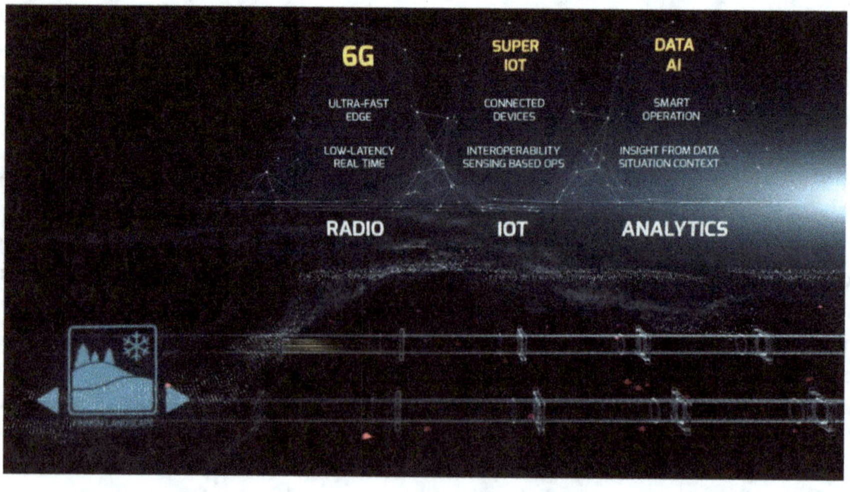

Figure 8.12 The features of 6G [20]

meet 6G's new KPIs. System security in 6G is additionally featured. While Nokia Bell Lab's Viswanathan saw in 6G an "intuition," for instance utilizing constant investigation of sensor information by AI, Ericsson's Prytz accepted that the all-encompassing arrangement of equipment-based security confided in figuring and made sure about enclave will frame the base of things to come processing systems.

With regard to the planning, the speakers had an agreement that it would be around 2030 when 6G will begin commercialization. ZTE trusted 3GPP will begin progressively concrete 6G particular work in R22, which the organization hopes to see in 2029. See the outline underneath for ZTE's point-by-point expectation for the timetable from 5G to Beyond 5G (B5G) and 6G.

Specifically, the speakers at the Summit hope to share all intents and purposes with their perspectives on what they expect 6G to resemble in a year. The comprehensive sharing on what the key assessment and evolution aspects will be conducted in the years to come. While there is no assurance these forecasts will end up being right, Nokia Bell Lab's Viswanathan put it best when he stated, "We have 10 years to be refuted, and now can have a fabulous time foreseeing what's to come."

8.15.3 The disturbance brought by these correspondence advances

The disturbance brought by these correspondence advances will empower new 6G arrange models, yet additionally, conceivably require auxiliary updates as for current versatile system plans. For instance, the thickness and the high access information pace of terahertz correspondences will make imperatives on the hidden vehicle arrange, which needs to give both more purposes of access to fiber and a higher limit than the present backhaul systems. In this specific situation, we imagine the presentation or potentially sending of the accompanying compositional ideal models:

- cell-less design, with tight incorporation of various frequencies and correspondence advances, to ensure a consistent portability backing and endeavor the integral attributes of various recurrence groups, for example, the sub-6 GHz layer for control, and terahertz interface for the information plane.
- disaggregation and virtualization of the systems administration gear, including the lower layers of the conventions stack (e.g., physical and MAC layers), to diminish the expenses of systems administration hardware and make greatly thick sending financially achievable.
- propelled get to backhaul reconciliation, to adapt to the gigantic increment in the thickness of passages.
- vitality collecting techniques for low-power utilization organize activities, to make 6G gadgets progressively proficient and less vitality expending as for current ones.

At last, the multifaceted nature of correspondence advancements and system arrangements will presumably forestall shut structure and/manual improvements. While the use of insightful strategies in cell systems is as of now being examined in the 5G space, we anticipate that 6G arrangements should be a lot denser (i.e., as far as various passages and clients), heterogeneous (as far as coordination of various innovations) and with stricter prerequisites as far as execution as for 5G. Along these lines, insight will assume an increasingly conspicuous job in the system, going past grouping and forecast undertakings that are being considered for 5G frameworks. Notice that the standard may not determine the methods and learning techniques to be sent in systems, however information-driven methodologies can be viewed as instruments that organize sellers and administrators can use to meet the 6G prerequisites [21].

8.15.4 Conclusion

The historical backdrop of Internet telephones and VoIP goes back farther than you most likely know, and moves quicker than basically any correspondence innovation, aside from possibly cell phones. VoIP has a rich history, with acclaimed names, government offices, and instructive foundations all creating their imprint on this brilliant innovation. After over too many years of hypothetical research, logical papers, and so forth, the administration organization Advanced Research Project Agency (ARPA) constructs the primary bundle exchanged system. Parcel exchanging works by gathering information into datagrams and sending them freely, as opposed to having a devoted circuit that must consistently be associated start to finish. The ARPANET worked by having little PCs interconnected by modems. Vint Cerf and Bob Kahn are credited with making the conventions that would make the Internet. From 1G to 5G, going through Universal Mobile Telecommunication Systems (UMTS) and Long Term Evolution (LTE) advancements, every age of portable innovation has been intended to address the issues of system administrators and last purchasers. 5G, specifically, is making a huge advance toward building up a low idleness material access organize, by giving new extra remote nerve tracts, i.e., information pipes. However, these days social orders are turning out to be perpetually information-driven, information needy, and robotized and will present progressively increasingly tough prerequisites (as far as ultra-high unwavering quality, limit, vitality productivity, and low dormancy) which may immerse the limit of customary advances for remote frameworks. We present the defense that 6th era (6G) frameworks will add to filling this hole specifically.

References

[1] R. Valdes, "How Stuff Works," 9th May 2001. [Online]. Available: https://computer.howstuffworks.com/ip-telephony1.htm. [Accessed 15 May 2020].
[2] A. Chanda, "inaani," 4 Oct 2017. [Online]. Available: https://www.inaani.com/blog/things-must-know-voip-server/. [Accessed 12 May 2020].
[3] Y. Masjedi, "nexteva blog," 09 Jun 2019. [Online]. Available: https://www.nextiva.com/blog/what-is-voip.html. [Accessed 14 May 2020].
[4] R. Yonatan, "Get voip," 08 Jun 2012. [Online]. Available: https://getvoip.com/library/how-voip-works/. [Accessed 14 May 2020].
[5] A. Chanda, "inaani," 4 Oct 2017. [Online]. Available: https://www.inaani.com/blog/things-must-know-voip-server/. [Accessed 12 May 2020].

[6] E. Hippo, 04 November 2017. [Online]. Available: https://www.atlantech.net/blog/the-top-6-benefits-of-moving-to-a-voip-telephone-system. [Accessed 15 May 2020].

[7] B. M. Varghese, "medium," 9 Dec 2019. [Online]. Available: https://medium.com/@2014eeb1046/introduction-to-voice-over-internet-protocol-voip-10c00f790213. [Accessed 14 May 2020].

[8] C. Johnson, "nextiva," 6 Jun 2019. [Online]. Available: https://www.nextiva.com/blog/how-does-voip-work.html. [Accessed 14 May 2020].

[9] J. Myers, "plum voice," 19 Sept 2010. [Online]. Available: https://www.plumvoice.com/resources/blog/what-is-voip-how-it-works/. [Accessed 15 May 2020].

[10] I. Malik, "networks training," 6 April 2010. [Online]. Available: https://www.networkstraining.com/ip-telephony-and-voip-tutorial-part-1/. [Accessed 15 May 2020].

[11] H. Lamin, "Cisco," 03 June 2011. [Online]. Available: https://www.cisco.com/c/en/us/solutions/small-business/resource-center/collaboration/voip-how.html. [Accessed 15 May 2020].

[12] Categories, "voipstudio," 1 Sept 2014. [Online]. Available: https://voipstudio.com/blog/sip-nat-traversal/. [Accessed 15 May 2020].

[13] S. O. Brien, "AZ Big MEDIA," 9 Dec 2019. [Online]. Available: https://azbigmedia.com/business/technology/how-voip-is-changing-the-telecom-sector/. [Accessed 27 May 2020].

[14] Z. Kida, "randercom," 9 Aug 2018. [Online]. Available: https://randercom.com/news/what-is-voip/. [Accessed 22 May 2020].

[15] C. Johnson, "NEXTIVA BLOG," 11 Oct 2019. [Online]. Available: https://lynn-library.libguides.com/c.php?g=591810&p=4092057. [Accessed 27 May 2020].

[16] B. Waters, "FALLON," 8 Feb 2011. [Online]. Available: https://www.fallonsolutions.com.au/Handy_Hints/a-brief-history-of-television. [Accessed 28 May 2020].

[17] K. A. Zimmermann and J. Emspak, "Live sci nce," 27 Jun 2017. [Online]. Available: https://www.livescience.com/20727-internet-history.html. [Accessed 28 May 2020].

[18] Boston, "mmwave," 12 June 2019. [Online]. Available: http://mmwave.dei.unipd.it/research/6g/. [Accessed 10 June 2020].

[19] W. Shi, "telecoms," 17 March 2020. [Online]. Available: https://telecoms.com/503142/consensus-on-6g-is-gradually-forming/. [Accessed 12 June 2020].

[20] M. DeCarlo, "Slashgear," 21 March 2019. [Online]. Available: https://www.slashgear.com/6g-wireless-network-planning-terahertz-applications-security-iot-post-5g-21570705/. [Accessed 13 June 2020].

[21] V. Simkiene, "SK Telecom," 20 June 2019. [Online]. Available: https://voip.review/2019/06/20/sk-telecom-samsung-nokia-ericsson-partner-6g-rd/. [Accessed 13 June 2020].

Chapter 9

Industrial Internet of Things solutions in smart manufacturing

Naqid Marzoq Abdulmalek[1] and Wai Yie Leong[1]

9.1 Introduction

It seems that the global community is becoming a place of imaginative craziness given the World Wide Web has successfully turned the world into a small village where everything – from reaching people by email to meeting people by Google Meet – is easy with the result that we can move between imagination and reality with the speed and ease of moving between moments. This networking technique of coming close to people that the Internet unleashed on the ordinary was unknown to the majority. But after the growing conversation around IoT, known as the Internet of Things (IoT), which is stated to become the main factor affecting our lives in the future, this technique that can make our life easy has aroused curiosity, is confusing to many, and remains shrouded in mystery as many people have no idea how the technique works, its characteristics, and its domains of functioning. What is the definition of this technique? What is the current communication system? And how does it work? And how will this technology affect our future life? The Definition: a modern technological approach that seeks to connect objects represented by devices and sensors and link them with Wi-Fi with the aim to transfer data [1]. One common requirement for the devices to work is the presence of Internet. IoT is defined as a developing principle of the Web such that all objects in our lives have the potential to link to the Internet and/or to transmit and receive data and execute different tasks across the network connecting them.

Key components of the IoT:

(A) Tangible things: such as appliances, coffee machines, or buildings.

(B) Sensors (sensitive) [2]: devices connected to the surrounding environment by a local network and to the Internet, with capability for intelligence so that they can work without the need for human intervention, that makes them smart sensors.

Types of sensors:

1. Temperature and temperature sensors
2. Pressure sensors

[1]The Institution of Engineering and Technology, Malaysia

3. Humidity level sensors
4. Light intensity detectors
5. Humidity sensors
6. Detection close-up
7. Radio frequency identification (RFID)

(C) *The Cloud*: The IoT collects massive data from devices, applications, and users that must be managed in an effective way in the cloud, where the cloud system integrates billions of devices, sensors, and gates, protocols, data storage, and provides predictive analyses. Companies use these analytical data to improve products and services, install proactive measures for executing specific steps in the production process, and accurately build their new business model. The (IoT gateway), that we find in some Internet systems, acts as the middle layer between devices and the cloud and keeps data for interim processing or just in between transferring to the cloud. This provides a certain level of security to the network to protect the system from harmful attacks and unauthorized access.

(D) *Network*: Things Internet Components use a wired or wireless network to communicate with each other in an Internet system.

(E) *User interface*: User interfaces are the visual and tangible part of the IoT that users can access to. Designers should design a user-friendly interface that is easy to use and contains options Interactivity and compatible with common wireless standards

9.2 Application area

The IoT network can include everything around us in our daily lives. Anything that can connect with the Internet, for example, IoT baby's products to monitor your baby all the time. Also, in the farms, animals can be linked – with sensors on collars on animal bodies – to the network to increase the production of farm products, as well as take care of your animal's health while controlling the temperature of the house by the pressing of just one button. All this data you'll collect from your devices will help you to avoid any mistake or prevent overload in the system. Imagine if all of this can be done with 4G, then what revolution can be brought in IoT during the 5G era. There are various useful applications of this concept, like:

(A) *Smart and wearable devices*: this technology can be found in a lot of semiconductor industries. It is one of the first industries to deploy loT at its service almost anywhere. Everyday we can see smart watches or any other device that monitors our health and/or family health. This device can be planted on the body, or even tattooed on the skin. It consists of microprocessors that function to send and receive data via the Internet and track specific systems in the body.

(B) *Smart home applications*: if we wish to have (Iron Man) a home with automated systems controlling everything inside it, for example, switching the light on or off at your voice command, IoT is the solution. Also, it includes a smart program that recognizes your face so that there is no need any longer for a key or access card to enter your home. This kind of house allows you to monitor everything around you and take care of your family. This house also contains sensors, external alarm devices, heat detectors, fire detector sensors and cameras all over the house. It will forewarn the house owner when food, say, is wasted in the fridge.

(C) Detect fumes rising nearby, and siren accordingly.
(D) Automatic arrangement of our everyday meetings.
(E) Give warning in case there is a hazard in an environment. Fusing the sensor to measure the amount of fuel in the heat station and alert if it is close to being empty.
(F) Provide guidance in finding parking spaces for your car parked in a parking lot if you forget the place. The possibility to notice certain natural conditions, like soil quality and its quality, is also possible by IoT networks.

IoT future application: In the coming years with scientific progress, several fields of applications on the planet are opening up to have IoT technology:

(A) *Communications*: This industry interacts the most with IoT. Because this technique of IoT is to store data of any equipment or device and monitor them to make sure this equipment and devices performance is not faltering, it must be connected to the cellular network by using GSM/3G /4G.
(B) *Healthcare*: This sector is also a principal field where the IoT can be implemented because doctors can monitor their patients with ease, create patients in the healthcare systems with electronic documents, then track patients' health by using wearable devices on patients that can enable patients to exercise and/or help them sleep, and carry out other healthcare routines, with analysis of different data, as well as help in maintaining wellness. These also give a warning, if there is any danger on the life of the patient.
(C) *Retail*: By using this industry (IoT), if we use its technology as customers, we can get the best in the shop and fresh goods. As an owner of the shop, I can easily monitor the foodstuff and control the buying and selling, as well as the accounting operation.
(D) *Factories and companies sector*: This field has much capacity to combine with the IoT as any other. After getting connected, we can easily do different tasks that can save money and time like collecting and analyzing data from all machines, material, tools, and staff. Classification all material will be more flexible and efficient.

The advantages:

- You can insert any information and monitor your equipment from far.
- By this technique, the pairing of devices is more efficient and simpler.
- Transferring data packets over a network reduces time and saves money.
- This technique is the most important at this time; lessening dependence on human intervention so that the result is fewer mistakes and more effectiveness in the completion of tasks.

The disadvantages:

- There is a lot of danger of spreading personal information, due to using the web.
- Because of its complicated system, a loophole in the system can cause a lot of problem for every device.
- With automation, unemployment will increase and many people will lose their job.
- Excessive use of technology leads to addiction and social isolation.

Challenges facing the applications of IoT:

1. Internet security things: According to a developer report for the 2017 Internet Explorer Forum, safety is the biggest concern of developers. They find that three security challenges should be addressed: Computer attacks: Attacks target cloud service providers, user privacy, or modification of components of electronic devices. That can cause economic losses.
    - Software vulnerabilities: Another major security challenge lies in the vulnerabilities of applications and internet programs of things as well as the presence of some defects and gaps in their programming,
    - Data interception during transmission: Data hijackings are considered during communications between devices [3].
2. Privacy of IoT: Private information sometimes leaks that is intercepted on consumer electronics as information is transmitted mostly to cloud service that often contains third parties for storage [4].
3. Understanding IoT: According to Kavanaugh, he predicted in 2018 that the rate of change has exceeded [5].
4. Lose the trust: according to a cyber-security company that showed the effect of the security on Gemalto, 90% of its customers do not have trust in the IoT system. Over 2/3 of customers as well as approximately 80% of companies help the police by giving them the user's data to take part in security beefing. Also, according to Said Jason Hart, Director of Gemalto, the customers and businessmen faced big issue because of IoT safety issues and less trust in ISPs. They hope that the App developers will be able to protect their data from hacking while using their device to send and receive or store the personal information on it [6,7].
5. The ecosystem and IoT: combine the IoT into the ecosystem. An IoT ecosystem is a vast collection of interdependent and connected technologies and devices that work together to form a cohesive system that meets a specific goal. The IoT ecosystem is the vehicle that makes "things" (sensors or devices) "smart" by providing connectivity with other "things." This interconnectivity is what allows products to interact with us and each other within the IoT ecosystem.

The IoT service providers:

1. Amazon Web Services (AWS): Amazon Web Services, Inc. is a subsidiary of Amazon providing on-demand cloud computing platforms and APIs to individuals, companies, and governments, on a metered pay-as-you-go basis.

These cloud computing web services provide a variety of basic abstract technical infrastructure and distributed computing building blocks and tools. One of these services is Amazon Elastic Compute Cloud, which allows users to have at their disposal a virtual cluster of computers, available all the time, through the Internet. AWS's virtual computers emulate most of the attributes of a real computer, including hardware central processing units and graphics processing units for processing; local/RAM memory; hard-disk/SSD storage; a choice of operating systems; networking; and pre-loaded application software such as web servers, databases, and customer relationship management.

2. Microsoft Azure: includes cloud to save data of the device, computing, and the processing summary.
3. Bosch IoT: this company offers the wearable tech cloud, as well as open source to store data.
4. Salesforce Cloud: offers services to easily link equipment with this service; then, stores the data of all equipment; and, then, analyze and use the good data of the customer.
5. Cisco: it is among the first organizations offering to settle into the web business field. It has useful reports about this technique.

The IoT future: By 2025, this service will be able to get 79.4 Zettabytes of information. IDC has predicted that this technology is estimated to rise by 28.7% from 2020 to 2025 as well as a complex long-term increase in the growth trajectory ending 2025. Estimated 44 billion equipment will link to the web with this technology according to a projection of the Statistics Research Department. The next huge step towards linking the whole world is by using this technology.

Artificial intelligence (AI) growth and this technique: This technique allows data flow between the device and AI to help manage this data without any human mistakes because many organizations are indifferent to how this amount of data should be administered. AI is also regarded as the key driver of this technique of revolutionary development.

Future communication (5G) of this technique: 5G is fundamental to billions of these technique apps. Between 2020 and 2030, this technique's applications might well increase from 75 to 100 billions, and the increase from 4G to 5G is most important as regards the growth of IoT. On one cell, up to 5,500 going up to 6,000 NB-IoT devices can be supported by the current 4G network. With a 5G network, a single cell can handle up to one million devices.

Protect the Internet to block unwanted access: The IoT makes your home smarter and more efficient when it is secured against cybercrimes. Your home devices like smart TV, security cameras, door locks, and more can add a level of luxury and comfort to your life, but they can also make your home unsafe. The router will play an important role here by securing the Internet's entry point into your home. While it is impossible to defend your smart IoT devices, the router will protect the Internet entry point by creating passwords, firewalls, and configuring them to allow only the trusted devices on your network.

Statistic and graphs:

1. Data rates: In the last 30 years, data rates have increased from 2 kbps to 1 Gbps, facilitating seamless transfer (Roopali Sood and Atul Garg) of heavy

data files (see Figure 9.1) for various cellular technology generations). The transition from the first generation of cellular to the second generation changed the way communications messages were sent [8] from analog to digital. The transition from the second generation to the third; and then the 4G that is much faster than 3G has seen the unraveling of limitless speed in how we communicate and are entertained. But, in the future, a revolution in data transfer rate which will increase the efficiency of the IoT is on the cards.

2. Number of connected devices globally: The fifth-generation cellular wireless promises greater speed and the ability to connect more smart devices at the same time. Faster networks mean the data accumulated by your smart devices will be gathered, analyzed, and managed to a higher degree. That will fuel innovation at companies that make IoT devices and boost consumer demand for new products, as shown in Figure 9.2 [10].
3. Spending on IoT worldwide by vertical in 2015 and 2020: As you can see, Smart Cities (23%), Connected Industry (17%), and Connected Buildings (12%) are the top three IoT projects in progress that have increased at this time shown in Figure 9.3. IoT analytics found that nearly half of the Smart City projects (45%) are in Europe, while the Americas lead in the area of Connected Health, with 55% of global projects today. The Americas are leading the world in Connected Car IoT projects as well, with 54% of them worldwide.

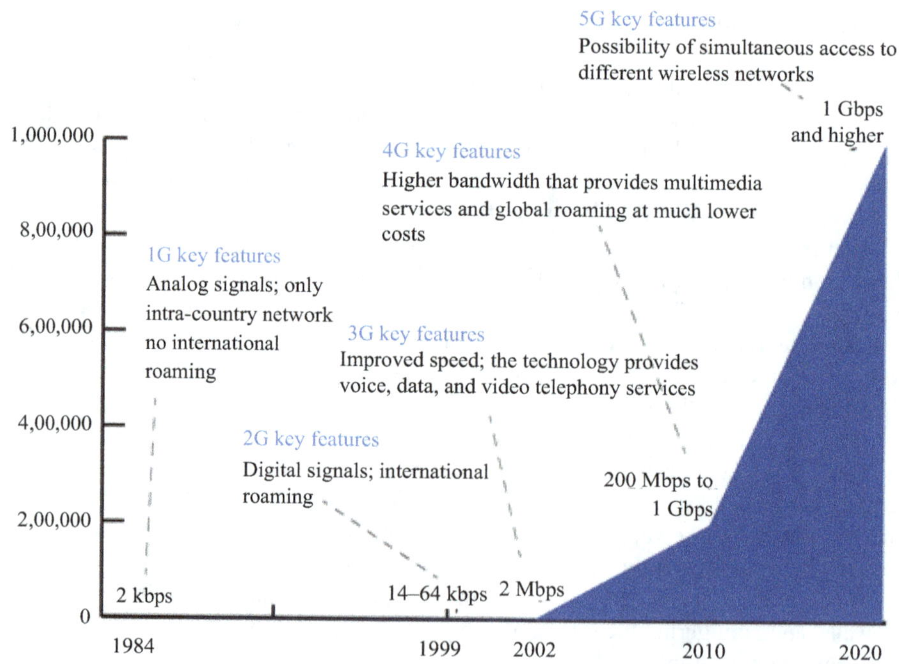

Figure 9.1 Data rates based on various cellular technology generations and the years representing the start of deployment [9]

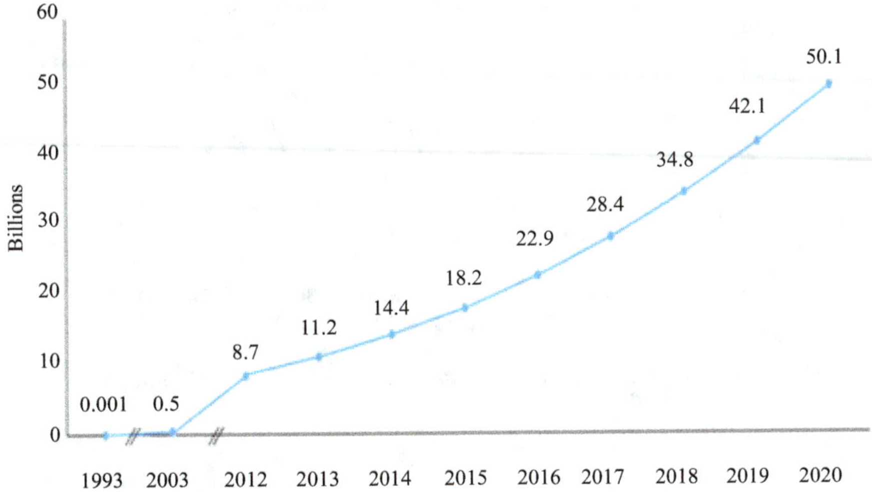

Figure 9.2 Number of connected devices globally [11]

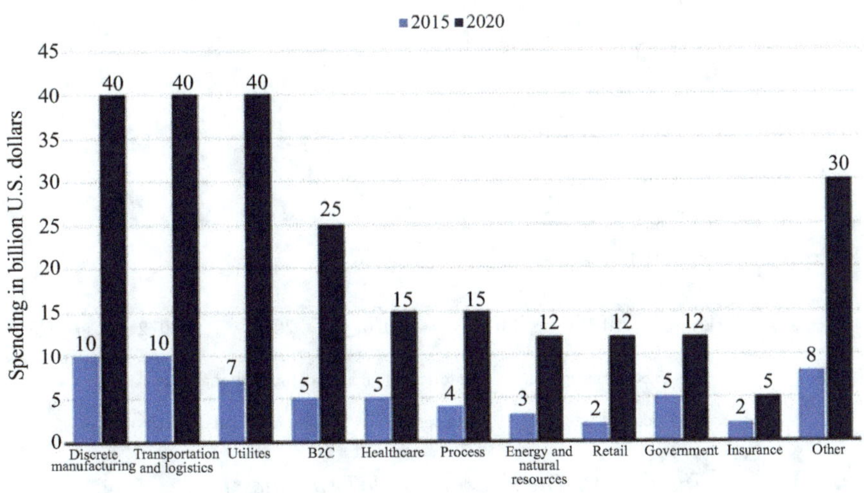

Figure 9.3 Spending on IoT worldwide by vertical in 2015 and 2020 [12]

According to Gartner, 20.4 billions IoT devices will be in use by the year 2020. The image given below will show you the spending of IoT worldwide and thereby depict the growth of IoT.

4. Global IoT malware attacks (using sonic wall data, Figure 9.4):

IoT malware attacks Skyrocketed in 2018, and the trend continues to 1H in 2019. SonicWall reports that IoT malware attacks in 2018 (rise from

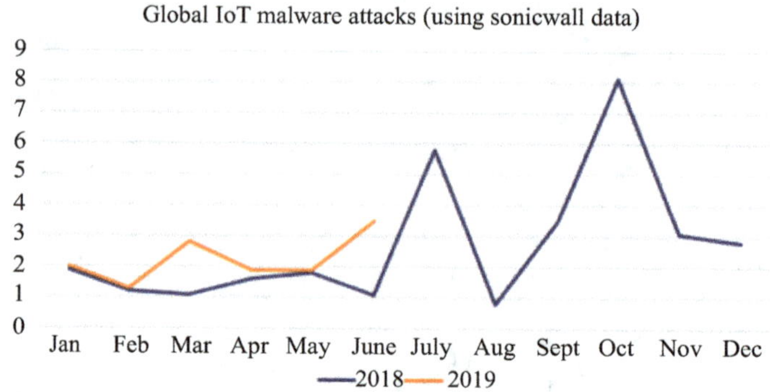

Figure 9.4 Global IoT malware attacks (using sonic wall data) [14]

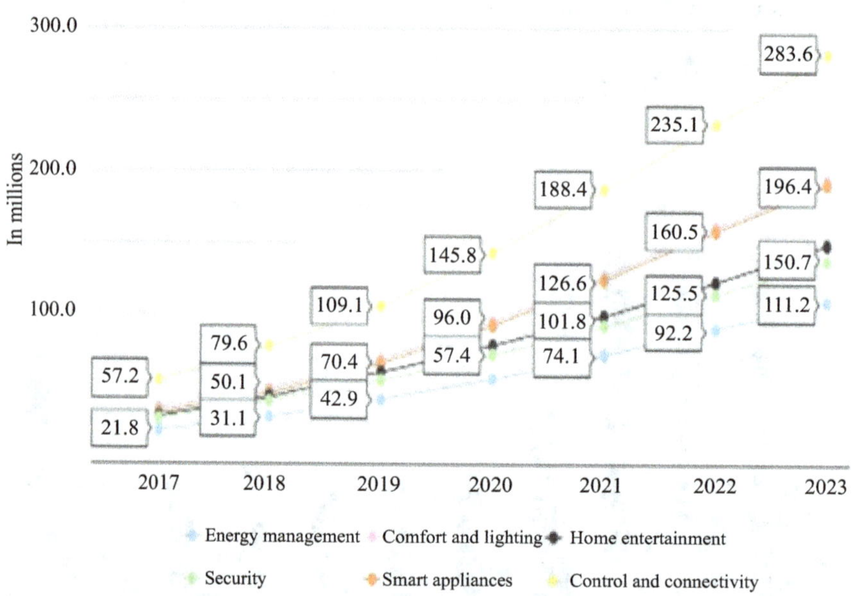

Figure 9.5 The smart home IoT statistics [15]

10.3 millions in 2017) jumped 215.7 % to 32.7 million. The first two quarters of 2019 already outperformed 55% in the first two quarters of 2018. If this rate continues it will be another record-setting year for attacks on IoT malware [13].

5. Smart Home IoT Statistics (Figure 9.5):

Currently, the most popular use of IoT involves smart home devices, such as smart thermostats, smart fridges, and smart locks. People use devices in their homes for all kinds of reasons. Consumers are willing to access devices that offer advanced security and convenience. According to Cisco, only 0.06% of the things could be connected to the Internet. IoT connected devices are everywhere, but there is still so much potential for innovation.

IOT history: the coining of the term IOT dates back to 1999 when for the first time it was mentioned in a presentation made by the co-founder of Auto-ID center at MIT, Kevin Ashton, in front of Procter & Gamble. However, it was not explicitly referred to after that. The term began to evolve gradually in conjunction with the development of wireless communication technologies. As accurate electromechanical systems and information technology developed with data analysis, it revealed the future visions of an institution. Coca-Cola was aiming to use machines to sell its products at Carnegie Mellon University, and the web began to be used to replenish the machines with cold drinks after checking to see whether the machines were supplied with cold drinks or not. The development of this Internet went through several stages, the first of which is machine to machine communication (M2M): this depends on the communication between machines by relying on the system [11]. The best evidence for its existence for some time was the Coca-Cola Company which aimed to place its own devices on the machines on campus to sell its products to the Carnegie Mellon University and has relied on the web to check whether a cold drink is available or not at the machines. This technique has developed over years, but it's so much easier to connect (machine with another device by Wi-Fi). The IoT is a sensor network of billions of smart devices, systems, and other connected applications. The M2M idea also needs an up and working IoT link. The IoT is a natural extension of Supervisory Control and Data Acquisition Systems (SCADA) and is one of the "software" categories of software applications related to handling control. Data is collected remotely to control equipment and use cases related to this. This is the SCADA for both programs, as these devices work to feed machines with and collect data from the computer on which the program application was installed after processing and presenting it appropriately. The concept of IoT can be considered an ecosystem, yet it was not realized until mid-2010 when the Chinese government announced that it would make IoT a strategic priority within its 5-year plan.

9.3 IoT principle

The creation of the IoT system depends on the abundance of the major components: smart devices that are supported by the web to harness their capabilities, mainly, to collect data from the environment, to process data, and to transfer data to the beneficiary, all connected by communications devices of all kinds. The secret of the action is that all IoT devices communicate with special sensors to capture and analyze the necessary data. Then, they return critical information without any human intervention, supported by special protocols [11] for network communication and web-supported devices. For example, standard Wi-Fi, Bluetooth, or satellite communications [45–48].

9.4 Conclusions

The IoT provides service opportunities that we can't do without and that have data storage needs that are limitless almost being based on big data. Collected at the same time, these detailed data raise profound security and privacy concerns, as expected. The coming years will witness the widespread use of the IoT in every field for business improvement, environmental control, and management. It will not only make life easier for people but also improve efficiency of companies which will have a lot to bear with in trying to attract users to these products, creating innovative and useful business models, and continuing to address challenges and enhancing user confidence by increasing security coverage on products, addressing privacy concerns, and spreading the concept of the IoT to improve the culture of using IoT optimally.

9.5 IoT vs Artificial Intelligence, RFID, and wireless communication

In the nomenclature of modern technology, one of the most famous terms is IoT. Its presence gives us the freedom to control our surroundings by controlling equipment. To me, I cannot imagine a world without the development of this technology and how it will make our life easy and we can control everything around us just by one voice command. The goal of the IoT is to unify all that is in our environment under a single framework, which then will keep us updated about the status changes of these things. For automated equipment, the future that we see would make things better in human life by using the products run by Artificial Intelligence, RFID, and wireless communication. All equipment will be ready to communicate and permit numerous functions in the environment everywhere [16].

9.6 Introduction

Web technology is developing everyday with the use of the growing repository of modern technological innovations. It reflects Internet Creativity. It is the extension of the new generation of Internet techniques and it offers a system that coordinates the sharing of the data among the devices rather than control these pieces of equipment. Often actually it enables the communication of 1,000,000,000 of devices every day via 5G telecommunication lines powering the Internet that has a huge ability to revolutionize our lives. This technique has multiple uses in all areas, including the technology of monitoring sick people, our health, supermarket store, etc. This technology utilizes some commands that specify an operation framework, with advanced technology such as RFID that is linked to the wireless sensor networks. The architecture of this system was propelled into place largely by the interests of multinational companies, who will have significantly profited first from providing enhanced levels of service and using customer data. IoT technology enables you to program and monitor items that has enabled companies to put into place more effective, rapid operations, to minimize delays, to deter fraud and to integrate complex and versatile organizational structures by IoT so it is a technical innovation that reflects the development of high computational networks and its advancement, which depends on data [17]. The IoT describes a world where all objects are linked to a network to share information. But it's about people as much as about inanimate objects. We already have millions in Yemen that have smartphones. But it's not enough. Smartphones are just the starting point of this technique, and in a few years, we would be having sensors that monitor our health in the best ways and provide a suitable environment. This helps us socialize and navigate the world. There are many more linked instruments in the world than individuals. The resulting networks and data will support a wide range of products. This technology is the present and future for companies. However, the potential drawbacks of major challenges are also present, as with any new technology. For this technology to flourish, security and privacy violations are the most frequently cited and the most crucial concerns to take into stride and bring to a halt [18].

9.7 Discussion

IoT was initially inspired by Radio-frequency identification (RFID) community members who referred to it as a chance to find data for the objects of daily use through the Internet or build a dataset using the RFID or NFC address. During 1999, according to the RFID Journal, when computers become enabled with sensors on objects around us, we would become capable of tracking and counting everything we come in contact with the data they collected about us, minimize data loss, prevent system failures, and avoid unnecessary expenses. You would know if things needed to be replaced, repaired, or bought anew and were fresh or stale. We need to make computers empowered to collect information so that they can see, hear, and smell the world for themselves. It allows devices to be observed, identified, and understood. IoT is an RFID technique, sensor technology, nanotechnology, and embedded intelligence.

Building blocks of IoT: An IoT system consists of a functional block for its smart applications. These are sensing systems, and the identification and management blocks for communication. The communication standards and protocols followed on the cloud server make it possible for objects to communicate with each other, and with the server, that requires network connectivity such as Internet access, Wi-Fi, Bluetooth, Zigbee, VPN, and/or mobile phones 2G/3G/4G.

1. Device: IoT devices are the IoT building blocks for sensing the data and for controlling the device activity with embedded instructions. These devices can be connected via wireless or wired channels to several other devices.
2. Communication: Communication building blocks are used to create communication between server and the user interfaces on the devices. The communication link is established within the data link layer, network layer, transport layer, and application layer.

3. Services: IoT offers different services such as device modelling and device handling, device discovery and data analysis, data publishing.
4. Management: Millions of IoT devices on various network topologies, with protocols, must be efficiently managed. The software for device management should make it possible for devices to interface with other devices and the server, and manage security during device handling.
5. Security: IoT network requires device and network-level security. The safety of IoT devices from the initial design to the final operating environment of the network shall be taken into account.

IoT technology: by using data that sensors collect, the IoT server controls different data sources like a fridge in a smart home, where a communication link, like Bluetooth, powers its IoT network making it possible for everyday objects to talk enabling the identification and status change of physical objects to be communicated to the server through the Internet. This technology uses different equipment elements like embedded RF, sensor system, Internet, computers, communications networks, and global roaming to be applied. It includes not only the sophisticated computer and communications technology that include many new assistant technologies but also data collection, remote access technique, control data transfer, data monitor analysis of sea measures, and telecontrol.

1. Radio frequency identification (RFID) system (Figure 9.6) sends an item or individual's status change signals wirelessly to the server on the cloud, and from the cloud using Internet to the data centers of the organization. The first one to utilize this technology was in Britain during World War II during 1948. The RFID system has a significant role in tackling the issue of entity recognition and its surroundings (Figure 9.7). Divided into three classes depending on the RFID tags' technique of supplying data, the RFID 's principal tools are name, editor, antenna, connection monitor, script, and file. It is highly effective, more productive, more stable, cheaper, more specific than conventional barcode tags. This technique has a broad variety of Wi-Fi usage including delivery, mapping, patient tracking, military uses, etc. [19].

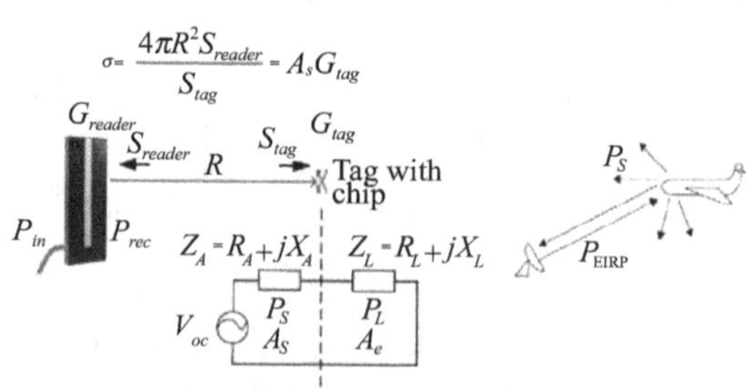

Figure 9.6 Label formula showed RFID system attached to an issue of radar cross-section that's symbols indicate A: efficient scattering [20]

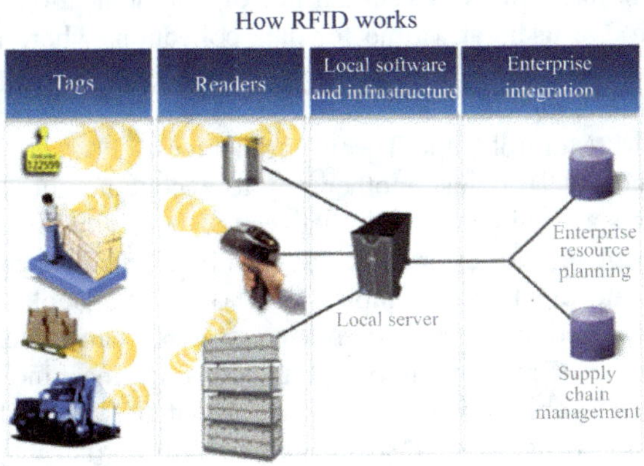

Figure 9.7 How RFID works [21]

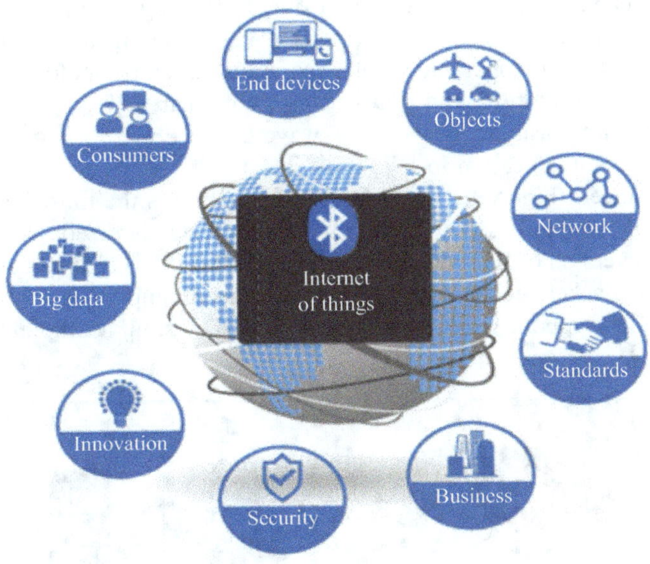

Figure 9.8 The relationship between IoT and Bluetooth [21]

2. Bluetooth (Figure 9.8): is a low-cost, radio wave technology that connects without cables the equipment such as laptop PCs, phones, PDA devices, and other tools with a 10- to100-m efficiency range. To communicate, generally, in a range of less than 1Mbps, the IEEE 802.15.1 standard is used by Bluetooth. In 1994, the company Ericsson started a new project called "Bluetooth." At the time, it was being used for networking in the Personal Area (PAN). Piconet is a

set Bluetooth system that has a similar line of communication. This Piconet can be used with 2–8 users at a time for data collection, where information sent could be text, picture, video as well as Four tones. The Bluetooth Specific Interest Community includes over 1000 Intel, Cisco, HP, Aruba, Intel, Ericsson, IBM, Motorola, and Toshiba companies.

3. Barcode (Figure 9.9): is just another way to encode numbers and letters with the use of bars and wide spaces in between. The original intention of Behind Bars is descriptive but not critical. Palmer (1995) recognized that alternative methods of data input are available in the Bar Code Book. Quick Response (QR), first designed for the automotive industry in Japan, codes the registration mark for a type of matrix barcode. Bar codes are machine-readable optical labels attached to objects which record item-related details. Thanks to its easy readability and greater storage space compared to normal ways of identifying objects, the QR code framework has recently become common outside of the automotive industry. There are three types of barcodes: alphanumeric, numeric, and 2-dimensional barcodes. The barcodes are meant to be accessible by computer and are designed to be machine-readable. Usually, they are read by laser scanners, though these can also be read using a camera.
4. Z-Wave (Figure 9.10): The Z-wave is a key enabling technology for IoT devices. It is an interoperable wireless communication technology designed for controlling, monitoring, and status reading for smart IoT devices. It can support a data rate of up to 100 kbps. Z-wave uses AES-128 Encryption to safeguard the communication in IoT.
5. LTE advanced (Figure 9.11): Long-term evolution is the most important and successful mobile-based wireless technology standardized by 3GPP. An advanced version of LTE that is based on 3GPP UMT is also a 4G technology. LTE is a packet technology-based core telecommunication technique with good VoIP support. Orthogonal Frequency Division Multiplexing Access

Figure 9.9 An IoT-based smart shopping cart using the contemporary Barcode scanner [22]

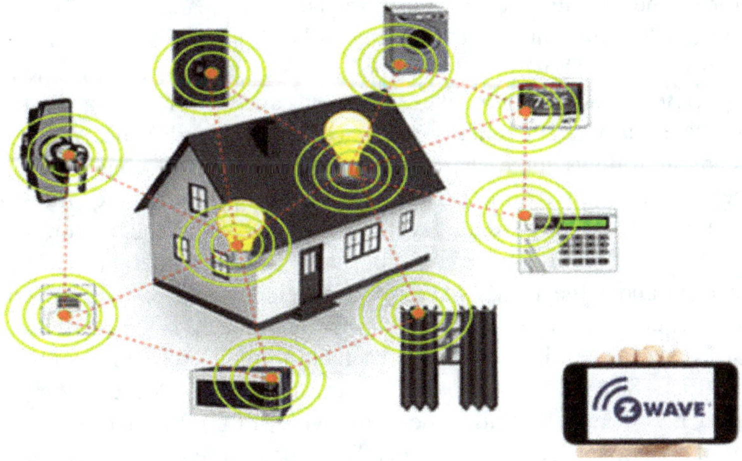

Figure 9.10 An IoT by using z-wave [23]

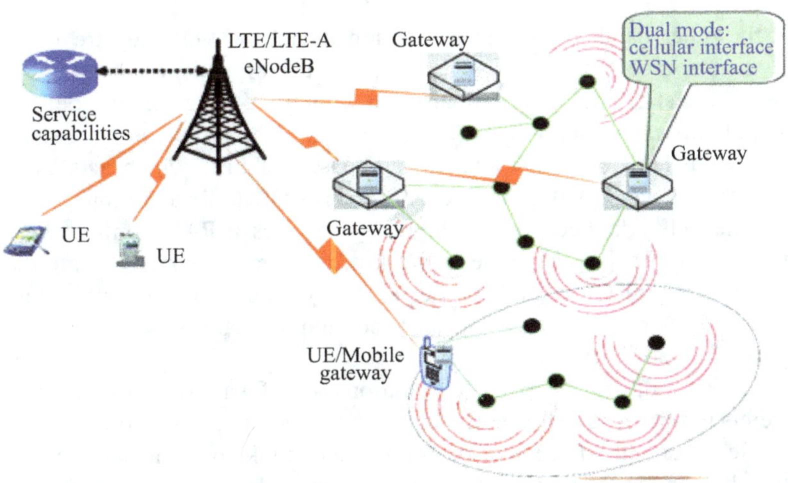

Figure 9.11 An IoT by using LTE also shows the range [24]

(OFDMA) technology providing high spectral efficiency is used in Downlink, and SC-FDMA is used in the uplink process. LTE can operate in various radio channels from 1.4 to 20 MHz. LTE operation is possible in both TDD and FDD modes, and it can reduce latency and is self-optimizing.

6. IEEE 802.15.4: this protocol has been developed to establish a sub-layer for (MAC), (PHY) for low-scale personal networks (LR-WPAN). It is often used for IoT, M2M, and WSNs due to its features such as low-power usage, low data rate, low cost, and fast message latency. It offers secure connectivity, on-

platform functionality, and can accommodate significant count of nodes (about 65k). Its offers minimum security standards [25], confidentiality and authentication facilities. However, it does not provide QoS assurances. The basis for the other technical protocol like Zigbee, it can start providing the minimum data rate assistance on energy-restricted devices and build WSN network protocol. It follows three-channel frequency bands and implements a direct sequence spread spectrum (DSSS) system. The physical layer transmits and consumes data at three data levels, depending on the common frequency channels: 250 kbps at 2.4 GHz, 40 kbps at 915 MHz, and 20 kbps at 868 MHz. Higher (f) and wide bands have good efficiency and low latency, whereas lower frequencies have better flexibility and reach larger distances. IEEE 802.15.4 MAC uses the CSMA/CA protocol to reduce possible collisions. The basic IEEE 802.15.4 allows two forms of Network Nodes: complete and reduced networks. The Full Function Device (FFD) will act as the administrator of a personal area network (PAN) or just as a regular node. A supervisor shall be responsible for establishing, monitoring, and managing the FFDs which will store a routing table within their memory and have a complete MAC installed.

On the other hand, the reduced function devices (RFD) [26] are very basic nodes with resource constraints. They can interact only with a handler and are restricted to the star topology. Normal protocol networks are the star, peer-to-peer (mesh), and cluster-tree topologies (see Figure 20). The protocol of the star topology includes at least one FFD and other RFDs. The FFD which acts as a PAN controller should be at the topology core and aims to handle and monitor all other nodes in the network. Peer-to-peer topology requires a PAN administrator, and other nodes connected in the same network or to other networks via intermediary nodes. A cluster-tree topology is a special case of the peer-to-peer topology and consists of a PAN coordinator, a cluster head, and normal nodes.

Cycle of Hype [27]:

The *X*-axis indicates people's expectations of IoT variations to arrive and the *Y*-axis refers to time factor. IoT is an emerging technology, as stated in Gartner's IT Hype Cycle, where the IoT revolution is projected to take place in about 5–10 years to market. It is an approach to represent emergence, adoption, maturity, and impact in the application of specific technologies. See the data image in Figure 9.12.

Research Project: European FP7 Research Project (Figure 9.13):

This would work as a scheme, for the structure of IoT tangible layout; style: modern architecture physical layer (ARM); developed by European FP7 Research Project Partners IoT-A; derived from business considerations, application -based requirements, and current technologies.

Analysis: numerous researchers have studied the IoT field, and more effort is required to mature it. As more countries and their economy sectors focus on this field, and try to develop it by carrying out some project, it would avoid future bad consequences. First of all, the security of our data will be enhanced, where many companies try harder than others to be reliable to the customer and increase the

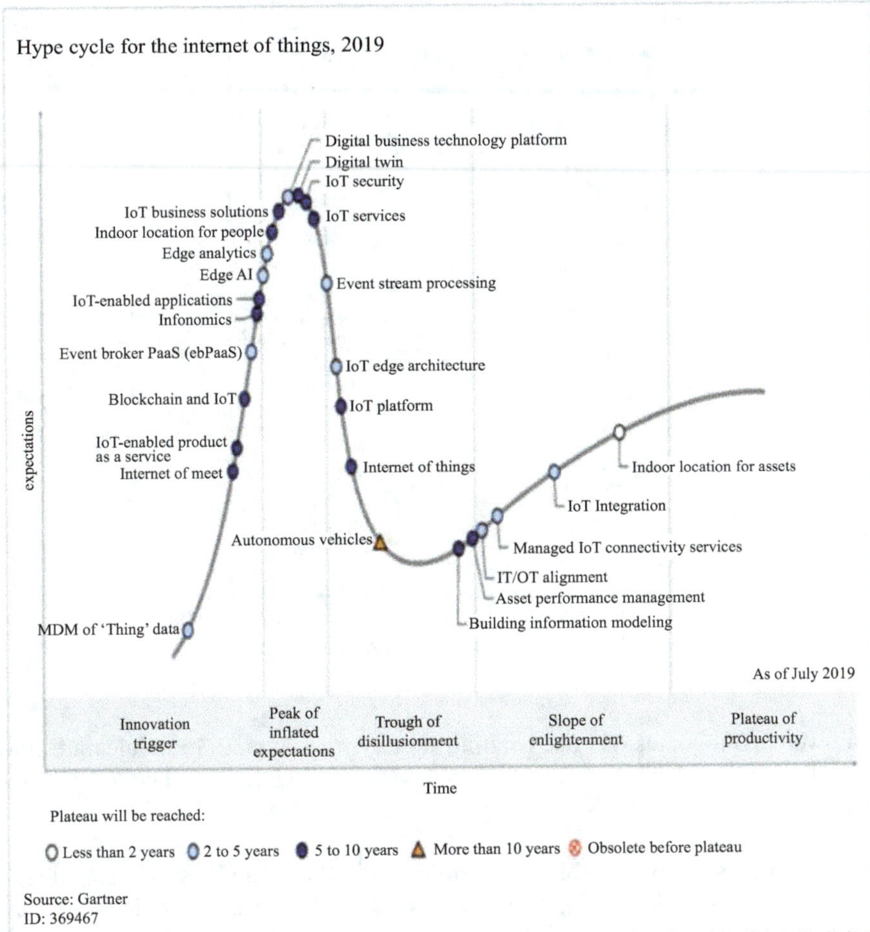

Figure 9.12 Cycle of Hype of IOT [28]

performance. Several experiments apply at improving the facilities; whereas, others want more study and experiment. It is normal because these kinds of technology want highly faithful Wi-Fi everywhere. As you know, some countries are still using 3G or 4G, so that this kind of technology still need time and in about 20–30 years will be applied everywhere.

Remark:

In the early of 2017, a great deal of attention was paid to both the 4G-based LPWAN wireless technology, LTE-M, and NB-IoT modules. However, even when over 50 mobile providers had announced LTE-M, NB-IoT, or both network deployments by the end of 2018, both technologies have been slow to become available to the market. Despite the sluggish adoption of LTE-M and NB-IoT, two major recent developments show positive turn of events. First, pricing for

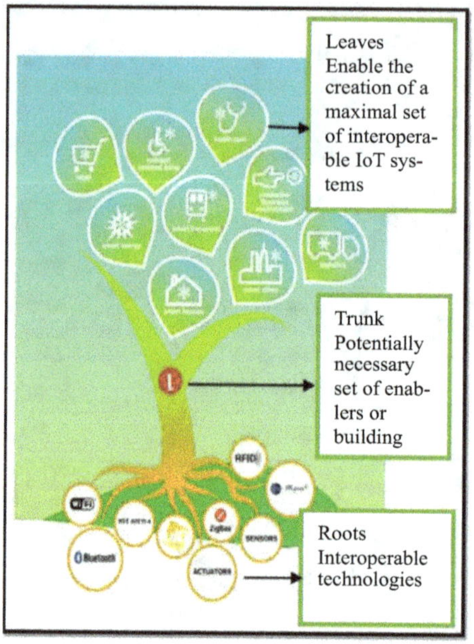

Figure 9.13 European FP7 Research Project [29]

low-power IoT modules has been falling rapidly. In China, 2G modules are now selling for under $2 while NB-IoT modules are selling for $3. The 2G module pricing is lower than NB-IoT.

Remark:

An express system goes through various stages that include issues related with (FCAPS), and its parts are included in this equipment. So, it provides the new updates to improve smart protocol to avoid any future problem. This technique will help the company provide the IoT with less faults as much as this technology will be dependable in future. This technology will change our lives, as Table 9.3 shows, these service providers include protocols, SKD/language and, hardware.

9.8 Introduction

Compared with the existing 4G communication system, the newest communication systems (5G as well as the 6G technology) are assumed to deal with many different evolutions. Storage efficiency, wide-coverage, low-power consumption, elevated performance, improved flexibility, high reliability are some common and effective benefits involved in the service performance of 5G and 6G communication systems. The future communication will include network connection efficiency and effectiveness that are faster than the present network, and with fewer connection difficulties. Only 5G, 6G and beyond will operate IoT with high performance.

Table 9.1 Gap analysis and benchmarking analysis of different IoT issue

S. no.	Author	Issue address	Key points	GAP analysis
1	Andreas *et al.* (2016) [30]	A survey is conducted on Mobile phone computing in various areas with IoT	Mobile Computing and Io T/ Web of Things	Open issues and research challenges have been verified in different applications
2	Razzaque *et al.* (2015) [31]	A survey is conducted on IoT and its characteristics and middleware in IoT with its requirements	Middleware requirements, functional requirements, service-oriented approach, agent based, tuple based, application-based approach	Many proposals have been focused on the hardware development requirement. A survey has been conducted and compared with these existing techniques
3	Rimavici u (2015) [32]	A survey is conducted on business approaches with respect to security, privacy, legal matter and technology advances	Business consideration-based approaches, legal matter and technology changes	Paper describes the various movements, technical approaches, and strategic planning for the IoT
4	Madakam *et al.* (2015) [33]	This paper provides "An Overview of IoT and various technologies, architectures used in daily life	This article focuses on IoT concepts through systematic review of scholarly research papers, case studies.	This paper provides an overview of IoT and various technologies, architectures used in daily life
5	Luigi Atzori *et al.* (2010) [34]	In this paper, a survey is conducted for communication solutions, identification and tracking the new technologies for the promise of several communications solutions	Various IoT enabling technologies for architecture. Applications, research and security issues are discussed.	We can see the future of IoT as an Internet of the future for better communication and network solutions
6	Ray (2016) [35]	A survey is conducted on IoT and its	Architecture, application and net-	Background definition of IoT de-

(Continues)

Table 9.1 (Continued)

S. no.	Author	Issue address	Key points	GAP analysis
		tools, architectures, utilities, technology and methodology; the future research challenges are also identified.	work protocols	fined and detailed analysis of research challenges are also mentioned
7	Ding *et al.* (2016) [36]	A survey on demands of security and mobility and distributed networks has been conducted to design the future of the Internet.	Data networking, content aware searching, retrieval and streaming make up the Internet architecture for content centric, mobility centric, or service centric network.	Future Internet architecture projects have been verified. These projects are NDN, COAST, Mobility First, XIA and SCION.
8	Anne Ngu *et al.* (2016) [37]	In this paper, the need for IoT Middleware and IoT applications are designed for real time prediction of blood alcohol containing smartwatch sensor data.	Applications and Architecture of IoT Middleware, service based, cloud based algorithm, lightweight-actor basedArchitecture and summary, key issues are also discussed.	IoT middleware: and IP addresses. Elaborated on two non-ontological solutions for addressing key challenges in IoT service discovery.

Table 9.2 IoT with different network generation

4G RELEASE 13, RELEASE 14 **2016, 2017, 2018**	**5G (phase 1) RELEASE 15** **2018, 2019**	**5G (phase 2) RELEASE 16** **2020**	
LPWA Standards (release13): EMTC(Cat-M1) NB-lot (Cat-NB1) EC-GSM-lot	Enhancements (release 14): EMTC (Cat-M2) – Positioning (OTDOA) – Single-cell multicast – Inter-frequency – Measurements – Higher data rate – VOLTE NB-lot (Cat-NB2) – Positioning (OTDOA) – Single-cell multicast – Power reduction – Latency reduction – PRB – Mobility and service – Higher data rate – New power class	Further Enhancements (release 15): EMTC – Extended coverage mode – Faster system acquisition, – Early data transmission – Wake up radio – Higher density NB-IOT – Faster system acquisition – Early data transmission – Wake up radio – Higher density – Cell-SIZE extension. – TDD support	5G Migration (release 16): EMTC/NB-loT – In band eMTC/NB-loT – 5G NR deployment options (SA/NSA) – Mesh networking – Non-orthogonal multiple access (NOMA) – Grant-free uplink

Table 9.3 Comparison between different platforms

	Microsoft azure	AWS	IBM-Waston IOT	The Things IO
Protocols	HTTP, AMOP, MQTT.	HTTP, MQTT. WEBSOCKETS	HTTP, MQTT	HTTP, MQTT, COAP, WEBSOCKETS
Certified hardware	Intel, Raspberry PI2, Texas. Freescale, Instruments	Broadcom, Marvell, renesa, Texas instrument	ARM mbed, Texas instrument, raspberry pi Arduino uno	Hardware agenostic
SDK/language	Net and UWP, Java, C, NodeJS	Java, C, NodeJS	C#, c python, javq, node is	Paython, node.jsMQTT, node.js HTTP, NODE, JSCOAP, JAILEB NODE
Connection	Easy configuration	Easy configuration	Difficult configuration	Easy configuration
Analysis	Medium configuration	Easy configuration	Difficult configuration	Easy configuration
Pricing	Paying for loT Hub unit related to number of devices and messages per days	Paying millions of messages traffic	Paying related to number of data traffic and data storages	Paying related to number of data traffic and the platform maintenance

Furthermore, 5G communication systems may have many challenges to be countered like assisting enormous heterogeneous traffic and achieving service levels for service quality mentioned above. The new technique, the optical wireless communication, is a suitable choice to satisfy the needs of 5G wireless communication. 5G communication will be ready to be used in 2020. The next-generation 6G will appear between 2027 and 2035. It is challenging to achieve the 5G/6G and IoT targets on the basis of tactile Internet. Supporting great capacity, high bandwidth, high speed, high safety, reduced use of power, high QoE, and extremely reliable interconnection for this type of network are needed, but all of this will include a lot of other big issues. The better working conditions for people achievable by the future 5G/6G and IoT networks will not be via RF-based systems. OWC techniques seem to be the best complementary RF networking solution. With the cooperation of radio frequency, OWC can help in the implementation of such systems. The future system will improve productivity and improve multiple folds over 5G, and client Quality of Service by using features, like its helping to protect user information. The 6G wireless network will be an international medium of communication (Figure 9.14). In several situations, it is assumed a per-user bit rate in future generations of network which could go upto about 1 terabyte per second. The next-generation system's simultaneous wireless connectivity will be 1,000 times better than the present generation's. [38]. In addition, the interesting situation is the incorporation of AI for driving autonomous structures. In the 6G system, video traffic is likely to feature among various data traffic systems. Terahertz band, AI,

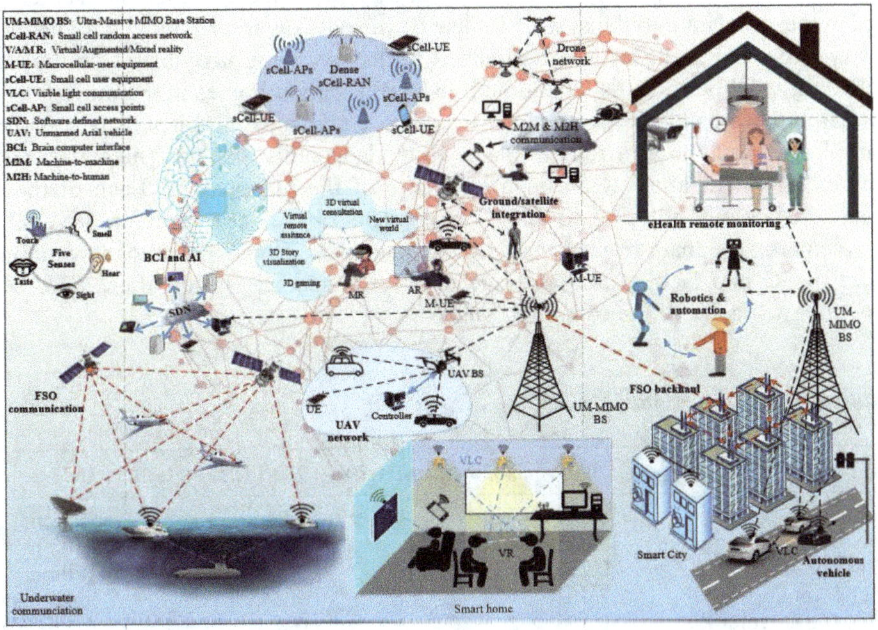

Figure 9.14 Possible 6G communication architecture scenario [39]

OWC [40], 3D networking, unmanned aerial vehicles (UAV), and wireless power transfer are the most important technologies that will be supported by 6G.

Discussion: since, in the future, 6G network is needed for humans' demand for applications that will only work with futuristic networks making customers happy. Scientists all over the world are now researching how 6G wireless communications systems will look like in 2030 and, therefore, talking about possible drivers for effective 6G communication systems is pertinent. Some of the important influential trends why the evolution of 6G network is slated to happen are extremely high data rate, extremely stringent safety, lower latency, low energy consumption, spectral efficiency, wide wavelength range, green network system, intelligent networks, network quality, and connection integration, localization, computer science, control, and detecting; 6G would be a completely digital and interconnected structure.

Expectations, if AI is incorporated with the 6G network, are those of applying AI to control all system equipment, physical layer signal processing, and strategic planning. This would promote the Industry 4.0 revolution, which also symbolizes a digitalization of manufacturing output. A few other main outcomes, as well as 6G projects, are explained in brief below.

Super intelligent community: 6G superior features will help to design the intelligent community. Human life will improve. The environmental monitoring, and automation using M2M communication and energy harvesting will be based on AI. This technology will make our lives so smart that our needs will be understood before we have to voice it or by just raising a finger. We can use smartphones, autonomous vehicles, etc. provided there is an Internet connection to send our needs to the data center where raw data becomes knowledge, and there is a server to feed processed data into the system for use by the data owner. Moreover, numerous towns across the globe will be designated to deploy flying taxis based on the 6G technology. All our dreams will come true because smart homes in the movies will be real and we can give instructions by voice commands, and it will be easy to control any machine at a remote place. Extended reality systems (hereafter XR) such as AR, MR, and VR are also key parts of the 6G systems. Each of these characteristics will be using AI and 3-dimensional objects as their main drivers.

Remark :The main reasons that led scientists to develop 6G: due to the winding down of 5G and lack of 5G coverage to meet consumers needs after 2030, 6G networks will achieve the following goals:

Table 9.4 Global trends of wireless connectivity

Issue	2010	2020 (Predicted)	2030 (Predicted)	Unit
Mobile subscriptions	5.32	10.7	17.1	Billion
Smartphone subscriptions	0.645	1.3	5.0	Billion
M2M subscriptions	0.213	7.0	97	Billion
Traffic volume	7.462	62	5016	EB/month
M2M traffic volume	0.256	5	622	EB/month
Traffic per subscriber	1.35	10.3	257.1	GB/month

(i) increased data rates for every piece of hardware, (ii) a more reliably linked consumer devices, (iii) international interconnection, (iv) quite low latency, (v) low battery-free energy level, free IoT, (vi) super-fast verifiable and more trusted interconnection, and (vii) linked intelligence with machine learning abilities.

6G system and needs:

- eMBB• another demands is URLLC: a main driver in 6G that offers end-to-end delay of below 1 ms
- massive machine type communications• AI interconnected system • Tactile World Wide Web • high data rates • increased web capacity • good power performance • low backhaul and network infrastructure • improved information safety.

It is also assumed that the 6G platform can have 1k times higher simultaneous Internet capabilities than the 5G spectral volume efficiency, compared to the frequently used spectrum. The 6G system will provide brain-computer interface (BCI) that handles all apps that will be part of our intelligence community, like home equipment and health care equipment. 6G can also facilitate the human brain to communicate with external devices. The brain–computer interface requires electrical signals which transfer to a digital device, collecting and analyzing the commands into instructions or, directly, the activities.

Communication technique of 6G:

(A) THz – The radio frequency band is close to full and, therefore, it cannot accommodate the growing demands of Wi-Fi technology. The Terahertz communication band can play a significant role in the next generation to provide extra broadband, excellent ability to connect, ranging from 0.1 to 10 Terahertz. The Terahertz communication band may enable the growth of small nanometer-to-micrometer cells supplying extremely speedy interactions for in-network coverage of up to 10 m and promoting the Nano-things web. Techniques that use radio frequencies below 0.1 THz cannot sustain Tbps connections, so 6G would have been the first system that provides high-speed Tbps.

(B) Cell-free communication: UAV is being recommended for use in other generations in locations where no facilities for human existence exist. The above technique may, after all, be properly utilized through 6G, enabling cell-free interaction. The participant's call must be converted to some other cell area only when a user equipment (UE) jumps from one cell coverage area to the next. However, during transmission between cells, the whole transmission could be rendered ineffective as well as the recipient's call being revoked in some situations, and the QoS in the module decreased. 6G will take care of the network coverage issue, because user equipment (UE) is linked to the entire system, not with a specific cell.

Using UAV, the various concepts in 6G transmission can be integrated so that the UE could use the modern technologies with ultimate convenience but with no manual operation on the device.

(C) AI:

This technique is not included in 4G or any of the older network. It is supported in 5G and can cause a change in the way we communicate, where information technology becomes the watchword for remarkable applications. AI in the future would be totally assisted by 6G for automation purposes, ranging from handover and web selection to financial collection, especially in important apps. The much more significant developments in 6G are AI and machine learning. The 6G supports web integrity where it makes sure there is error-free data transmission among transmitting and receiving entities with no loss of data.

5 sensory data transmit: people are using their five senses of hearing, sight, taste, smell and touch to explore the world around us. The 6G transmits information collected mostly from five senses wirelessly. The technology enables sensory integration also to neurological processes. Five sensory data transmit: people are using their five senses of hearing, sight, taste, smell and touch to explore the world around us. The 6G transmits information collected mostly from five senses wirelessly. The technology enables sensory integration also to neurological processes. It identifies the senses first from organisms and the environment and makes good use of all and contribute to the society and natural conditions. A BCI technique may increase a certain production technique. Automation investigators try to focus on automation, robotics, and autonomous systems. 6G will promote these technologies which will provide direct interaction and better contact among endpoints and the server, i.e.: robot-to-robot communication and robot-to-server communication.

9.9 Future application

Internet of Everything (IoE) is the seamless integration and independent cooperation among a huge variety of software components and detectors, items or equipment, people, processes, and information using the World Wide Web. This technique will be applied in intelligent communities, such as intelligent vehicles, intelligent health care, and intelligent industries. For example,

For example, the used of virtual reality and augmented reality to connect cars autonomously (Figure 9.15).

1. VR, AR technique: With connected vehicles as well as Artificial Intelligence-powered drones, it is expected that massive shopping experiences will be unveiled, where detectors will surround us and without the need for mobile devices to connect us to the outside world: IoT such as this will appear after 2030. This new method will allow users to seek your aspirations, like projecting your beloved short story onto your retina via plunging VR. I can imagine how exciting life will be. The whole technology is made up of goggles and a couch which can be managed to turn in all directions. Gamers may contact massive 3D worlds and explore new journey styles.
2. Intelligent town (Figure 9.16): linked vehicles using the IoT systems: the world will be rendered safe so that almost no car crashes will occur in the future.

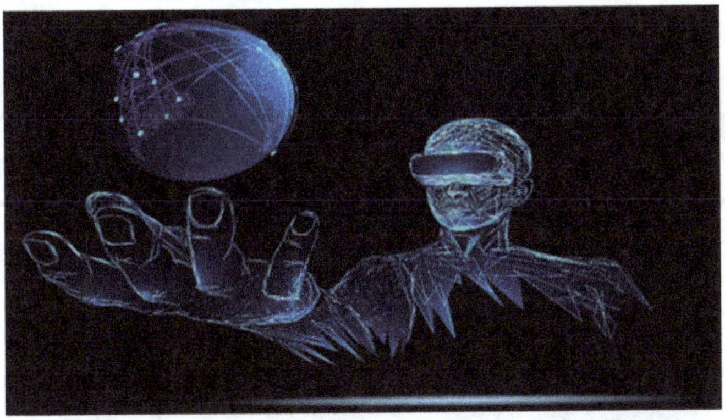

Figure 9.15 Virtual reality and augmented reality [40]

Figure 9.16 Smart city in future [41]

Detectors can detect all obstacles across a car and it will react immediately. By 2030, approximately every car should become part of the IoT. Licenses will become obsolete as well as driving schools will vanish, and there will be no need for petrol stations and automobile repair shops even. In the future, everyone's cars will interact while driving avoiding traffic jams and without the need for GPS for better location tracking.

3. Intelligent healthcare (Figure 9.17): because of 6G and other technologies such as augmented reality and virtual reality, holographic telepresence, and AI, all these techniques will assist in making a great system. It will monitor everything as well as carrying out a complicated surgery. In the years ahead, the

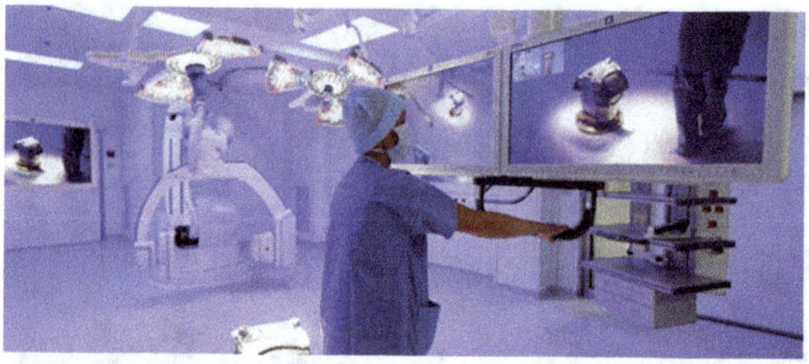

Figure 9.17 Smart health care [42]

Figure 9.18 Future communication [43]

body parts that no longer work will be replaced by other parts created by 3-dimensional printers' artificial ones. In future, IoE will be attempting to make prostheses.

4. Smartphones (Figure 9.18): Microsoft announced the end of the smartphone and Trend one innovation researchers predicted that we would not have to produce any more devices for digital communication by 2027 at the latest. Both were right. For example, HoloLens models are more practical. If I want to know whether or not I still have broccoli at home, I ask my HoloLens via voice control. I have the option to use it as glasses or as a smart contact lens. Either way, she tells me the result – and if I agree, the vegetable goes right onto my virtual shopping list, which directly reports to the grocery store.

Device 5-Shopping of the Future (Figure 9.19): Do you want me to show you how to shop in the future? Look, on the shopping window, which we are about to

Figure 9.19 Smart shopping in near future [44]

pass, and you see an offer flashing which is tailored to my needs. Today, this is much easier and more seamless: sensors, RFID technology and NFC chips make it possible. Even paying is easier. I just have to stop at the shopping exit where the system scans the chips or RFID on the products automatically. Vendors no longer have to be physically present to see the payment transactions through at the checkout. I, the buyer, pays via NFC chip built into my bracelet. I simply hold it up against a terminal and confirm my purchase via personal code – in a few seconds, I will be done.

9.10 Conclusion

This chapter discussed on the IoT. We conclude that the IoT will be the future for the next-generation technology. All cities will be smart cities with smart parking systems, better waste management and traffic management systems, even as our rural countryside will have accurate weather reports on their smartphones and will be able to monitor their health problems by using smart devices connected to the best health consulting agency or hospitals. IoT will help to shape the future and make cities and villages a better place to live. There will be billions of devices and IoT developers will make applications to connect people worldwide using these applications. The IoT industry will look completely different in future than it is today. However, one of the major challenges in the future of IoT is 'Hacking'. As a whole, IoT is a future for the smart world and smart devices with improvement in privacy and security.

References

[1] Cambridge Editors (September, 2019). "'What is the Internet of Things", https://cambridgealert.com/what-is-the-internet-of-things/.

[2] Dataflair Team (September 15, 2018). "How IoT works – 4 main components of IoT system". dataflair.training, https://data-flair.training/blogs/how-iot-works/

[3] J. Roettgers (May 13, 2014). "Stealth startup alert: skype co-founder ventures into health and wearables with Project Florida", GigaOM, https://gigaom.com/2014/05/13/stealth-startup-alert-skype-co-founder-ventures-into-health-and-wearables-with-project-florida/>.

[4] D. Roe (February 7, 2018). "7 Big problems with the Internet of Things", CSM, https://www.cmswire.com /cms/internet-of-things/7-big-problems-with-the-internet-of-things-024571.php

[5] D. Roe (January 12, 2018). "7 Big problem with The Internet of Thing", CSM, https://www.cmswire.com/cms/internet-of-things/7-big-problems-with-the-internet-of-things-024571.php

[6] T.H. Davenport and J. Lucker (2015). "Running on data: activity trackers and the Internet of Things", BICORNER.COM;chrome.extension://oemmndcbldboiebfnladdacbdfmadadm/https://www2.deloitte.com/content/dam/insights/us/articles/internetofthingswearabletechnology/DR16_running_on_data.pdf

[7] A. Davis and J. Hart (May 5, 2016). " $(ISC)^2$ security briefings", https://www.brighttalk.com/webcast/6319/198941/data-the-new-oil-find-out-why-and-what-it-means-for-your-organization

[8] R. Sood and A. Garg (2014). "Digital society from 1G to 5G: a comparative study", *International Journal of Application or Innovation in Engineering & Management,* 3(2):191.1G.

[9] R. Sood and A. Garg (2014). "Digital society from 1G to 5G: a comparative study", *International Journal of Application or Innovation in Engineering & Management,* 3(2):191.

[10] National Cable & Telecommunications Association (April 22, 2015). "Broadband by the numbers", https://www.ncta.com/broadband-by-the-numbers.

[11] National Cable & Telecommunications Association. "Broadband by the numbers", https://www.ncta.com/broad-band-by-thenumbers (Accessed April 22, 2015).

[12] An American Worldwide Consulting Agency McKinsey (November 4, 2018). "10 Charts that will challenge your perspective of IoT's growth", https://discoperi.com/iot_devices_market/

[13] Mehavarunan (October 7, 2019). "The future of IOT: 4 predictions about the Internet of Things", Thrive global, https://thriveglobal.com/stories/the-future-of-iot-4-predictions-about-the-internet-of-things/

[14] Thesslstore (September 4, 2019). "20 Surprising IoT statistics you don't already know", https://www.thesslstore.com/blog/20-surprising-iot-statistics-you-dont-already-know/

[15] iPropertyManagement.com (2020). "Smart-home-statistics", https://ipropertymanagement.com/research/iot-statistics

[16] D. Evans (2011). "The internet of things: how the next evolution of the internet is changing everything", CISCO White Paper.

[17] J. Gubbi, R. Buyya, S. Marusic and M. Palaniswami (2013). "Internet of Things (IoT): a vision, architectural elements, and future directions", *Future Generation Computer Systems*, 29:1645–1660.

[18] Anusha (January 24, 2019). "IoT application areas". quora.com/Do-you-know-where-IoT-is-applicable

[19] D. Mocinfar, H. Shamsi and F. Nafar (2012). Design and implementation of a low-power active RFID for container tracking @ 2.4 GHz frequency: Scientific Research.

[20] S. Kahng (January, 2009). "Design fundamentals and advanced techniques of RFID antennas", https://www.researchgate.net/figure/Tag-equation-of-an-RFID-system-analogous-to-a-radar-cross-section-problem-where-the_fig221787601

[21] Open Synergy GmbH (2018). "Open synergy – the world's Bluetooth® provider", https://www.opensynergy.com/bluetooth-for-iot/

[22] J. Z. Gao, L. Prakash and R. Jagatesan (2007). Understanding 2D-barcode technology and applications in m-commerce—design and implementation of a 2D barcode processing solution. 31st Annual international of computer software and applications conference, 2007. COMPSAC 2007. IEEE. ISBN: 978-0-7695-2870-8, ISSN: 0730-3157.

[23] Cambridge Innovation Institute (2020). "Z-Wave launches framework to standardise IoT security", https://internetofbusiness.com/z-wave-launches-framework-to-standardise-iot-security/

[24] J. Gozalvez (2016). "New 3GPP standard for IoT". IEEE Vehicular Technology Magazine, 11(1):14–20.

[25] X.-Y. Chen and Z.-G. Jin (2012). "Research on key technology and applications for the internet of things". *Physics Procedia*, 33:561–566, http://dx.doi.org/10.1016/j.phpro.2012.05.104

[26] S. Chen, H. Xu, D. Liu, B. Hu and H. Wang (2014). "A vision of IoT: applications, challenges, and opportunities with China perspective", *IEEE Internet of Things Journal*, 1(4), 349–359.

[27] Gartner's Hype Cycle Special Report for 2015, Gartner Inc. (2015). http://www.gartner.com/technolo-gy/research/hype-cycles/

[28] H. Wang (December, 2015). "Technology hype cycle, IoT marked. Original image courtesy of Gartner Inc", https://www.researchgate.net/figure/Technology-hype-cycle-IoT-marked-Original-image-courtesy-of-Gartner-Inc_fig282703898

[29] EuropeanFP7 Research Project. URL: http://www.iot-a.eu/internal/resources/pr-material/illustrations/IoT-A%20Tree.png/view?searchterm=tree (Accessed December 2016).

[30] I. Andrea, C. Chrysostomou, and G. Hadjichristofi, G. (2015, July). "Internet of Things: security vulnerabilities and challenges", In *2015 IEEE Symposium on Computers and Communication (ISCC)* (pp. 180–187). IEEE.

[31] M. A. Razzaque, M. Milojevic, A. Palade, and S. Clarke. (2015). "Middleware for Internet of Things: a survey". IEEE Internet of Things *Journal*, 3, 1–1.

[32] L. Al-Rimawi (2020). "Securities disclosure regime under Shari'a when contrasted to UK general law -with reference to the principle of caveat emptor (Part One)".

[33] M. Somayya, R. Ramaswamy, and T. Siddharth (2015). "Internet of Things (IoT): a literature review", *Journal of Computer and Communications*, 3(5).

[34] A. Luigi, I. Antonio, and M. Giacomo (2010). "The Internet of Things: a survey", *Computer Networks*, 54(15), 2787–2805, ISSN 1389-1286.

[35] R. Partha Pratim (2016). "A survey of IoT cloud platforms", *Future Computing and Informatics Journal*, 1(1–2), 35–46, ISSN 2314-7288.

[36] Wenxiu Ding, Xuyang Jing, Zheng Yan, and Laurence T. Yang (2019). "A survey on data fusion in internet of things: towards secure and privacy-preserving fusion", *Information Fusion*, 51, 129–144, ISSN 1566-2535.

[37] A. Ngu, M. Gutierrez, V. Metsis, S. Nepal and Q. Sheng (2016). "IoT middleware: a survey on issues and enabling technologies", *IEEE Internet of Things Journal*, 1–1. 10.1109/JIOT.2016.2615180.

[38] S.J. Nawaz, S.K. Sharma, S. Wyne, M.N. Patwary and M. Asaduzzaman (2019). "Quantum machine learning for 6G communication networks: state-of-the-art and vision for the future", *IEEE Access*, 7:46317–46350.

[39] S. Junaid Nawaz (April, 2019). "Quantum machine learning for 6G communication networks: state-of-the-art and vision for the future", https://www.researchgate.net/figure/Illustration-of-different-types-layers-sides-and-levels-of-B5G-communication-networks_fig332219159

[40] S. Knight (May 9, 2020). "Is virtual reality missing its moment", https://www.techspot.com/article/2025-is-vr-failing/

[41] Infineon Team (January, 2019). "The car of the future – mobility in 2030", https://www.infineon.com/cms/en/discoveries/car-future/

[42] M. Walker (November 11, 2016). "Healthcare in 2030: goodbye hospital, hello home-spital". https://www.weforum.org/agenda/2016/11/healthcare-in-2030-goodbye-hospital-hello-home-spital/

[43] J. Ashley (October 31, 2015). "How Hololens sensors work". https://www.imaginativeuniversal.com/blog/2015/10/31/how-hololens-sensors-work-2/

[44] J. Sims (June 21, 2015). "The future of shopping is both bricks and clicks", https://www.raconteur.net/retail/the-future-of-shopping-is-both-bricks-and-clicks

[45] W. Y. Leong (2005). Implementing Blind Source Separation in Signal Processing and Telecommunications. Diss. University of Queensland.

[46] U. Mehrotra and W. Y. Leong (2009). "NSEEAR: a energy adaptive routing protocol for heterogeneous wireless sensor networks", *2009 35th Annual Conference of IEEE Industrial Electronics*, pp. 2647-2652, doi: 10.1109/IECON.2009.5415255.

[47] W. Liu, E. Lupito, Y. L. Sum, B. Tay and W. Y. Leong, "Ultra wideband antenna for real time location system application," 2009 35th Annual Conference of IEEE Industrial Electronics, pp. 2738-2742, doi: 10.1109/IECON.2009.5415424.

[48] F. M. Han and W. Y. Leong (2012). "Investigating target detection and localization in wireless sensor network," *Procedia Engineering*, 41, 75-81, ISSN 1877-7058.

Chapter 10

Metal powder bed fusion: an overview on processes, materials, and challenges

Tan Koon Tatt[1], Khairur Rijal Jamaludin[2] and Sivakumar Paramasivam[3]

Additive manufacturing (AM) is one of the key pillars of the fourth industrial revolution. AM technology is developing rapidly in the industry for manufacturing custom industry, functional or finished machined parts. Printed parts, either semi-finished or finished, are widely used in the medical, automotive, and aerospace industries. AM technology has more competitive advantages than subtractive processes. The advantages include the capability of producing small and complex parts. Among the available AM technologies, powder bed fusion is promising for metal parts production. This chapter presents an overview of the four metal powder bed fusion processes, namely, direct metal laser sintering, selective laser sintering, selective laser melting, and electron beam melting. Metal materials commonly used and the key challenges of these processes are presented with the size of the global market.

10.1 Introduction

Additive manufacturing (AM) which started in the 1980s with stereolithography is now showing rapid development in the industry [1]. AM was previously known as rapid prototyping (RP) to gain a competitive manufacturing advantage in producing small and complex parts. Compared to traditional subtractive manufacturing (SM), AM reduces the manufacturing cost through the layer-upon-layer deposition process without wasting any material. Furthermore, the capability of the AM process of producing near-net-shape parts eliminates the post-manufacturing cost. AM may be deemed as a "designed freedom" process where the parts can be produced directly from computer-aided design (CAD) data.

[1]School of Science and Technology, Wawasan Open University, Penang, Malaysia
[2]Utm Razak School of Engineering and Advance Technology, Universiti Teknologi Malaysia, Kuala Lumpur, Malaysia
[3]Department of Mechanical Industrial Engineering, University of Technology and Applied Sciences, Muscat, Oman

The AM process is suitable for polymer, ceramic, and metal [2]. Polymer, specifically thermoplastics, was the earliest material used for additive processes. To date, many additive technologies have become available. The most common additive technique is the fused deposition modeling (FDM) technique. FDM was introduced in the 1990s to quickly fabricate low-budget thermoplastics [3]. In addition to industry, FDM is now very popular in domestic and household interest by an individual. Meanwhile, metal additive materials and processes have received considerable interest due to digital transformation needs.

Metal AM can be classified using the fabrication technologies or state of the raw material used during the metal AM process. Binder jetting, power bead fusion, sheet lamination, and direct energy deposition are the common technologies used for metal AM. The state of the raw material could be liquid, powder, or wire. The application of liquid and wire-based raw materials is relatively new in metal AM. Meanwhile, powder-based fusion is deemed the most popular metal AM technology for its capability of producing high-quality and high-precision metal components. Furthermore, powder-based fusion can fabricate a strong metal structure to support practical applications. Powder-based processes include direct metal laser sintering (DMLS), selective laser sintering (SLS), selective laser melting (SLM), and electron beam melting (EBM). The heat sources, either lasers or electron beams, provide sufficient energy to sinter or melt the metal powder. Although many types of metal parts have been successfully produced by SLS and SLM, powder-based fusion still faces some challenges. The challenges include large-scale production and a slow deposition rate. This chapter discusses the processes, materials, and challenges of metal powder bed fusion in detail.

10.2 Metal powder bed fusion process

Four powder bed fusion processes are mainly used for the processing of AM metal powdered parts: DMLS, SLS, SLM, and EBM.

10.2.1 Direct metal laser sintering (DMLS) and selective laser sintering (SLS)

DMLS and SLS differ in the type of materials that can be used to "print" the required parts. DMLS is designed for metallic materials, whereas SLS is an extension of DMLS and can be used for various metals and nonmetallic materials. This chapter focuses on metal AM. Hence, both are considered the same technologies that use the laser as the heat source to print a metallic material.

DLMS was developed by Electro Optical System (Germany) in 1995 [4]. In DLMS/SLS, the metal powder is fused and sintered via a highly intensive laser beam with a CAD file. During fabrication, the metal powder particles are not melted but joined together by sintering [5]. In certain cases, the metal powders are coated with certain substances for a melted bond to form between the metal particles [6]. Figure 10.1 shows the general scheme of the SLS technology. A high-power CO_2 laser is used to sinter the metallic powder material. Nitrogen or inert

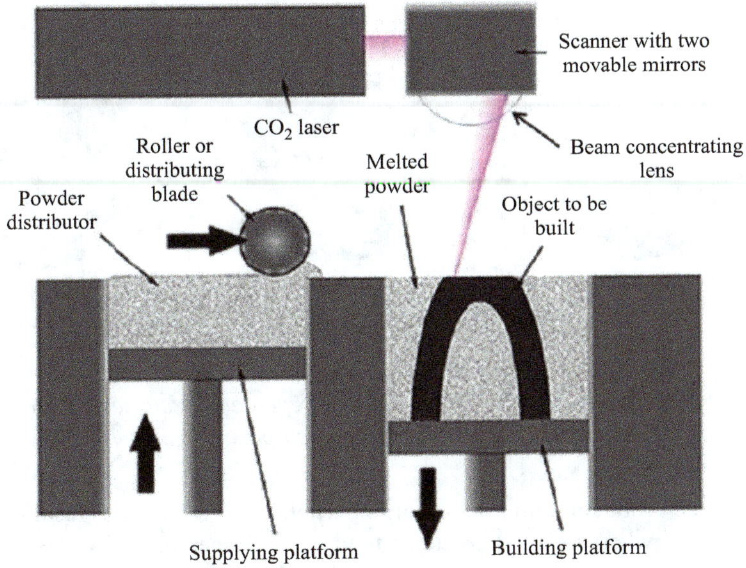

Figure 10.1 SLS process [7]

gas, usually argon, is used to control the atmosphere in the chamber. The metal powder is deposited and sintered on a layer with a typical layer thickness of 20–150 μm. The platform with models will then be lowered, and subsequent sintering will be carried out when the next powder layer is fully deposited. DLMS/SLS allows the direct printing of uniform structured metal elements from the complex design created. The relatively high power of the laser also offers high accuracy sintering.

10.2.2 Selective laser melting (SLM)

SLM is an AM technology that uses a high power-density laser to melt and fuse metallic powders. This technology is also called laser powder bed fusion or direct metal laser melting. The SLM process starts with very thin layers of metallic powders deposited at the building platform. The metallic powders are then melted by laser beams energy. By selectively melting and re-solidifying the metallic powders in each layer, a cross-section area of the designed 3D part can be built. Figure 10.2 shows the concept of the SLM building process. Once a particular layer solidifies, the building platform is lowered and a new layer of powders is deposited. The laser beam(s) can be directed to and focused on a selected layer according to the 3D design. Hence, the powder bed can be selectively melted to form the desired 3D shape.

The process parameters that require adjustment to optimize the process include laser power, scanning speed, hatch spacing, and layer thickness [9]. The metal layer thickness usually ranges from 20 μm to 100 μm, and the size of the metal powders

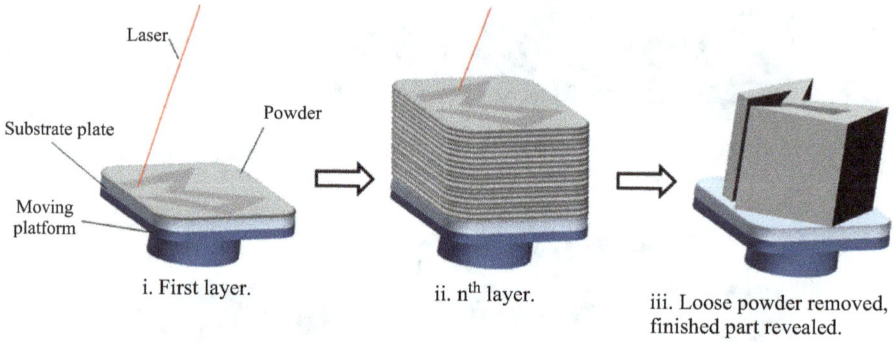

Figure 10.2 Concept of SLM process. (i) High-power laser melts selective areas of the powder bed. (ii) Process is repeated for successive layers. (iii) Loose powder removed and finished part revealed [8]

depends on the product requirement. Large particles will cause poor resolution and build tolerance, whereas small particles tend to agglomerate easily. Usually, the combination of low laser power, high scanning speed, and large layer thickness cause balling due to the lack of wetting of the molten pool with the preceding layer [9]. In contrast, high laser power and low scanning speed may cause extensive material evaporation [10]. Porosity defects occur due to poor hatch spacing.

10.2.3 Electron beam melting (EBM)

Similar to the other metal additive processes, electron beam melting prints using a 3D CAD file. EBM is also known as selective electron beam melting because of its fabrication mechanism [11] The three-dimensional parts are fabricated via focused electron beam through the layer-by-layer addition of materials. In this context, the electron beam used is selectively melted and solidifies the metal powder layer according to the intended CAD design.

Figure 10.3 presents the schematic of a standard EBM system. High-speed electrons with a kinetic energy of approximately 60 keV are emitted via the heated tungsten filament. The two magnetic coils located at the lower column are used to control the electron beams. The focus coil focuses the beam into the desired diameter, whereas the defection coil directs the focused beam to the intended location at the building platform. In the middle of the chamber, hoppers supply fine metal powder to build a thin layer through a raking mechanism. Typically, the layer thickness is in the range of 0.05–0.2 mm [12]. The electron beam with a controlled speed is directed to heat and melt the metal layers. When a metal layer solidifies, a new metal layer is laid on top, and the electron beam scanning process is repeated until all layers are completed. The EBM process takes place in a vacuum chamber. To prevent electron beam dissipation and electrical charges in the powder, low-pressure helium gas is added. Helium also assists in cooling. Once the part cools down, it is removed from the chamber. The EBM process parameters that significantly affect the quality of parts are the cathode current and the scanning speed.

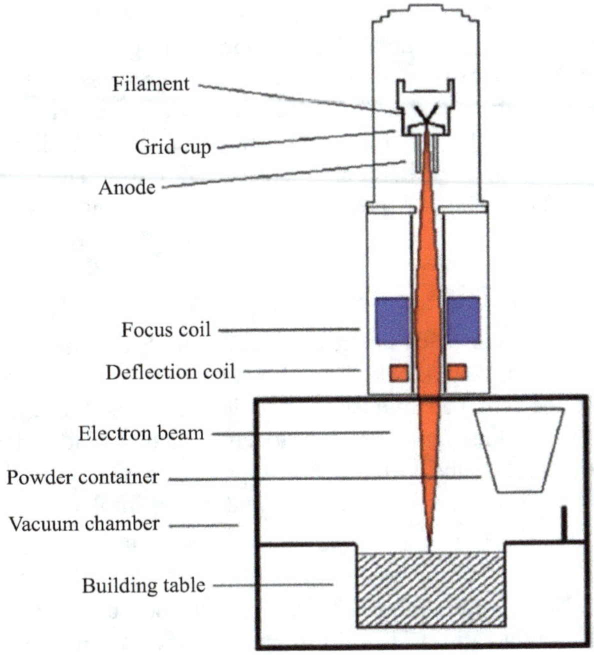

Figure 10.3 A standard EBM system [12]

A defect, such as the interstices between un-melted particles, easily forms due to the insufficient electron beam energy density [13].

10.3 Materials used in metal powder bed fusion processes

In general, metals or metal alloys that can be processed via AM are considered hard-to-machine and high-temperature creep-resistant metals. The main criteria that influence the selection of metal for AM processing are the good weldability and availability of raw materials as micron-sized spherical powders [14]. Good weldability can prevent cracking defects, while micron-sized spherical powders improve the packing density and the homogeneity of the powder deposition. The most common metallic alloys used in AM technologies are steel, titanium alloy, aluminum alloys, nickel–chromium alloys, and cobalt chrome alloys.

10.3.1 Steel

The two most popular steels used in additive technologies are martensitic steel (MS1) and stainless steel (316L). MS1 has high mechanical strength and hardness, making it suitable for fabricating building elements. The high hardness of MS1 at the 52–54 HRC level is comparable to that of many standard materials used in building injection molds [7]. The forming inserts with cooling channels, such as conformal channels, can be fabricated via DMLS. For stainless steel, 316L is widely used in producing biomedical parts due to its good corrosion resistance. Good thermal conductance also makes it a suitable process via DMLS to obtain high-dimensional accuracy parts. Different variants of steel have been tested for additive technologies, including Inox904L stainless steel, AISI Maraging 300 steel, 15-5 and 17-4 precipitation hardening steels, H20 tool steel, and tool steel X110CrMoVAl 8-2 [7]. The additive process for other variants of steel is still in the research stage.

10.3.2 Titanium alloys

Titanium alloys have many advantages, including high corrosion resistance, biocompatibility, and lightweight. Titanium alloys have been used widely in aircraft, shipbuilding, and the aerospace industry. In additive manufacturing, titanium alloys are mainly used to produce industrial components, such as blades, fasteners, rings, discs, hubs, and vessels. Due to their biocompatibility, titanium alloys have also become a good option for medical scaffolds and dental implants.

10.3.3 Aluminum alloys

Similar to titanium alloys, aluminum alloys exhibit good corrosion resistance and high strength-to-weight ratios. The additive manufacturing of aluminum alloys is particularly useful in industrial, aerospace, and automotive applications. An example of aluminum alloys is AlSi10Mg silumin that is used in the DMLS technology [7]. Many industrial components subjected to high loads are produced using AlSi10Mg silumin. A key advantage of aluminum alloys is that their excellent fusion characteristics facilitate the sintering process.

10.3.4 Nickel–chromium alloys

Nickel–chromium alloys, such as Inconel 718 and Inconel 625, are used to produce parts working in high-stress, high-temperature, and high-corrosion-resistance environments [7]. The AM parts of nickel–chromium alloys possess good pitting and cracking resistance when exposed to chlorides. Inconel 718 is the age-hardened version of 625, and it is highly resistant to corrosion. In general, AM can produce nickel–chromium alloy parts with excellent mechanical properties that could work in harsh environments. The printed parts can also be subjected to post-operation, like machining, welding, and heat treatment.

10.3.5 Cobalt–chromium alloys

Cobalt–chrome alloys, like Co28Cr6Mo, exhibit good corrosion and wear resistance and can work in high-temperature environments. In medical and dental applications, cobalt–chrome alloys serve as an alternative material when nickel-free components are required. Three-dimensional printed cobalt chrome alloy parts are comparable to hot isostatic pressing (HIP) parts. The high processing temperature enables metal powders to diffuse completely, producing fully dense metal parts.

10.4 Key challenges

Metal AM, particularly powder bed fusion, offers unique advantages, namely, it reduces materials usage and shows the capability of producing customized and complex parts. Nevertheless, the AM process shows some generic limitations, such as relatively low production and expensive operating systems. Table 10.1 outlines the key challenges of various metal powder bed fusion processes.

In general, DMLS/SLS and SLM require high-power laser beams to obtain sufficient energy to sinter or melt metal powders, thereby increasing the manufacturing cost of the parts. In the context of EBM, the high manufacturing cost is due to the expensive machine or equipment needed to produce a high-power electron beam. Table 10.1 also indicates that the limited availability of materials is one of the common pitfalls for the three processes. Many metals or metal alloys for powder bed fusion are yet to be fully explored for commercialization. The production of newly developed metal alloy powders is still in the small scale or the laboratory stage. The next challenge for these AM technologies involves the printed part quality. Some part defects, such as porosity, residual stress, and surface roughness, are usually observed. To eliminate these defects, post-processing, like polishing and heat treatment, is required. However, post-processing will increase the manufacturing lead time. Finally, the build size for these metal fusion technologies is limited compared to other AM technologies, i.e., FDM. Given that distortion occurs during the high-temperature processing of SLM, build structures are required to hold and retain the part shape.

10.5 Size of the global market and future trend

As shown in Figure 10.4, between 2018 and 2026, the market for AM or 3D printing materials is expected to grow by 12% annually to reach just under four billion US dollars by 2026. AM industry includes 3D printers, materials, software, and related

Table 10.1 Key challenges of various metal powder bed fusion processes [15–17]

DMLS and SLS	SLM	EBM
An expensive laser beam with high power and good beam quality	An expensive laser beam with high power and good beam quality	Expensive machine
Limited availability of material	Limited availability of material	Limited availability of material
Rough surface finish, parts usually post-processing	Melt pool instabilities and higher residual stresses	Rough surface finish, parts usually post-processing
Small build size compared to other AM technologies	It requires to build structures because of the metal weight and distortion at high temperature	Limited build size

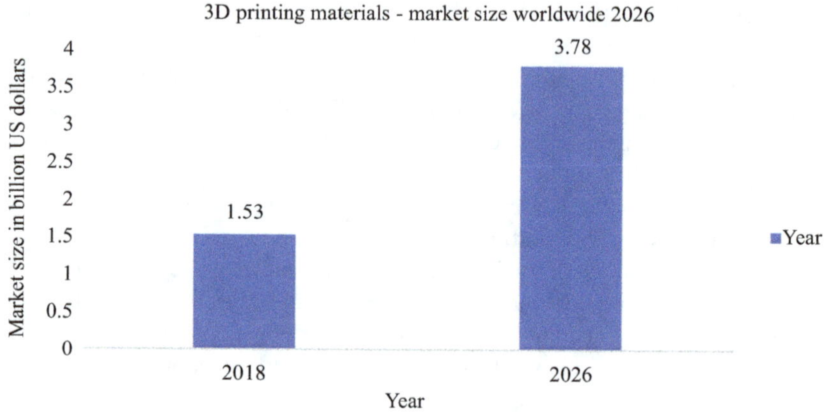

Figure 10.4 3D printing materials—market size

services. It is estimated that printers will account for about half of the market revenue, but software and related services will also likely experience significant growth.

What sounds even more impressive is that the markets for on-demand parts services and computer-aided design (CAD) software are expected to almost triple. Concurrently, discrete manufacturing is seen as the dominant industry for 3D printing, but the technology is also predicted to play a significant role in the healthcare sector. Plastics and polymers will still be among the materials that are most commonly employed in the field of AM technology. However, a wide array of other materials such as metals or even human tissue can be used to produce prototypes, finished parts, or organ transplants.

10.6 Conclusion

In line with the fourth industrial revolution and the digitalization era, AM, particularly metal powder bed fusion processes, is gaining competitive advantages against other conventional subtractive processes. Metal powder bed fusion processes are suitable for RP and customizing parts production, meeting industry, and customer demand. From the overview of the processes, material, and challenges, the following conclusions can be drawn:

(a) Both DMLS and SLS fabricate parts by sintering via high laser power. High laser power provides high accuracy sintering.
(b) Selective laser melting (SLM) uses high laser power to melt and fuse metal powder. Laser power and scanning speed are the two important processing parameters affecting the quality of parts.
(c) EBM melts and fuses the metal powder with a high density of electron beams. Cathode current and scanning speed are among the crucial factors affecting the quality of parts.

(d) Except for aluminum alloys, the metal materials for the powder bed fusion process possess a high melting point. The metal printed parts are applied in medical, automotive, aeronautical, and petrochemical industries.
(e) The main challenges of the AM powder bed fusion process are low production rate, high cost, unsuitability for mass production, rough surface finish, and limited availability of materials. Various compatible materials for powder bed fusion must be explored.
(f) The AM industry especially metal AM is expected to experience substantial growth in the coming decades.

References

[1] Sharma, V., Singh, S.: 'Rapid prototyping: process advantage'. *International Journal of Computational Intelligence Research*. 2016;12(1): 55–61.
[2] Alkahari, M. R., Furumoto, T., Ueda, T., *et al.*: 'Melt pool and single track formation in selective laser sintering/selective laser melting'. *Advanced Materials Research*. 2014;933:196–201.
[3] Habeeb, H. A., Alkahari, M. R., Ramli, F. R., *et al.*: Strength and porosity of additively manufactured PLA using a low cost 3D printing'. *Proceedings of Mechanical Engineering Research Day*. 2016:67–70.
[4] Khaing, M. W., Fuh, J. Y. H, Lu, L.: 'Direct metal laser sintering for rapid tooling: processing and characterisation of EOS parts'. *Journal of Materials Processing Technology*. 2001;113(1–3): 269–72
[5] Wong, K.V., Hernandez, A.: 'A review of additive manufacturing'. *ISRN Mechanical Engineering*. 2012(4):1–10 2012:208760-1–208760-10.
[6] Kruth J. P., Wang, X., Laoui, T., *et al.*: 'Lasers and Materials in Selective Laser Sintering'. *Emerald Group Publishing Limited, Bristol*. 2003;23(4): 357–71.
[7] Legutko, S.: 'Additive techniques of manufacturing functional products from metal materials'. *IOP Conf. Series: Materials Science and Engineering*, 2018;**393**(2018):012003.
[8] Yap, C. Y., Chua, C. K., Dong, Z. L., *et al.*: 'Review of selective laser melting: materials and applications'. *Applied Physics Reviews*. 2015;2 (2015): https://doi.org/10.1063/1.4935926.
[9] Li, R., Liu, J., Shi, Y., *et al.*: 'Balling behavior of stainless steel and nickel powder during selective laser melting process'. *The International Journal of Advanced Manufacturing Technology*. 2011;59(2012):1025–35.
[10] Krakhmalev, P., Yadroitsev, I.: 'Microstructure and properties of intermetallic composite coatings fabricated by selective laser melting of Ti–SiC powder mixtures'. *Intermetallics*. 2014;46:147–55.
[11] Nouri, A., Sola, A.: 'Electron beam melting in biomedical manufacturing, metallic biomaterials processing and medical device manufacturing', in Wen, C. (ed.), *Metallic Biomaterials Processing and Medical Device Manufacturing*, Woodhead Publishing, Sawton, 2020, pp. 271–314.

[12] Gong, X., Anderson, T., Chou, K.: 'Review on powder-based electron beam additive manufacturing technology', *Manufacturing Review*, 2014;1(2):1–12 https://doi.org/10.1051/mfreview/2014001.

[13] Gong, H., Rafi, H. K., Karthik, T., *et al.*: 'The effects of processing parameters on defect regularity in Ti-6Al-4V parts fabricated by selective laser melting and electron beam melting', *Proceedings of the Solid Freeform Fabrication Symposium*, Austin, TX, 2013, pp. 424–39.

[14] Gorsse, S., Hutchinson, C., Gouné, M., *et al.*: 'Additive manufacturing of metals: a brief review of the characteristic microstructures and properties of steels, Ti-6Al-4V and high-entropy alloys', *Science and Technology of Advanced Materials*, 2017;18(1):584–610.

[15] Rajesh, R., Sudheer, S., Mithun, V. K.: 'Selective laser sintering process—a review'. *International Journal of Current Engineering and Scientific Research (IJCESR)*. 2015;2(10):91–100.

[16] Bartolo, P., Kruth, J., Silva, J.: 'Biomedical production of implants by additive electro-chemical and physical processes'. *CIRP Annals*. 2012;61 (2):686, https://doi.org/10.3390/met10050686.

[17] Buj-Corral, I., Tejo-Otero, A., Fenollosa-Artés, F.: 'Development of AM technologies for metals in the sector of medical implants'. *Metals*. 2020; 10(5):635–55.

Chapter 11

3D processing for human–machine interaction and additive manufacturing

Mohankumar Palaniswamy[1]

The process of converting raw materials or parts into a usable or consumable products through tools, labor, machinery, chemical processing is called as Manufacturing. There are three major stages in manufacturing. They are the following: pre-production, production, and post-production. Pre-production is a preparatory stage which includes planning, scripting, designing, storing raw materials, and prototype development. Production is the actual manufacturing stage which includes process like molding, cutting, casting, machining, joining, shearing, and forming. Post-production is the stage which is in between production and dispatch. It includes assembling, inspecting, and packing. Manufacturing does indeed have a rich history, and we have transitioned from man-made to man and machine made. At times, the man-made products and goods were produced through conventional manufacturing process.

11.1 Manufacturing process

Conventional manufacturing is the traditional way of manufacturing goods. It requires a supervisor to supervise the large number of labors employed, who directly uses hard source tools over the workpiece or work material. There is always a physical contact between the tools and workpiece. Turning, broaching, grinding, slotting, and boring are to name a few. Since the seventeenth century, there have been four industrial revolutions. Steam engines, textile manufacturing, and mechanical engineering were all part of the first industrial revolution in 1780. The second industrial revolution in steel industry came in 1840. Electricity and automobile vehicles were the focus of the third industrial revolution in 1900. The fourth industrial revolution or Industry 4.0 on the other hand is focused on the information technology, and it is widely acknowledged that the fourth industrial revolution is just getting started [1].

[1]School of Allied Health Sciences, Hindustan Institute of Technology and Science, Chennai, Tamil Nadu, India

With the development of science and technology, now, the manufacturing process became more precise and accurate via unconventional manufacturing. In unconventional manufacturing process, there is no physical contact between the source tool and the workpiece or work material. Techniques like laser beam, plasma arc, abrasive jet, water jet, and chemical milling are used to get a fast output. Such technologies do not require large labors or hard tools. With a minimal supervision and less time, the output performance of the industries has increased a lot.

Internet of Things (IoT), Internet of Services (IoS), Cyber Physical Systems (CPS), interaction and exchange of data between humans (C2C), between machines (M2M), and between humans and machines (C2M) form the foundation of Industry 4.0 [2]. Since Industry 4.0 is still in its early phases, numerous research groups, universities, and corporations, such as Siemens in Amberg, Germany, are attempting to build a completely automated Internet-based industry or factory. This 108,000-square-foot plant can run and manufacture components without the need for human intervention. Cloud Computing, IoT, CPS, System Integration, Additive Manufacturing (AM), Augmented Reality (AR), Virtual Reality (VR), Data Analysis, AI Robots, and other elements contribute in the making of Industry 4.0.

11.2 Human–machine interaction

Despite the fact that the phrase "machine learning" was created in 1959, machine learning as a field of artificial intelligence did not develop until the 1980s [3]. Machine learning does not have any global definition. However, one of the widely accepted definitions proposed by Mitchell in 1997 is, "*A computer program is said to learn from experience E with respect to some task T and some performance measure P, if its performance on T, as measured by P, improves with experience E*" [4]. Machine learning is a branch of artificial intelligence that has grown out of the continuing effort to create intelligent computers that can learn. Artificial intelligence is an assortment of various fields comprising of engineering, computer science, neuroscience, psychology, mathematics, and statistics. The term machine learning was first proposed by Samuel in 1959 itself. But it emerged as an elective of artificial intelligence only in 1980. Machine learning is commonly referred to be a branch of artificial intelligence concerned with detecting patterns in data and using those patterns to generate predictions about data that have not yet been seen. In simple words, machine learning is all about constructing a computer program which can automatically enhance itself with every new experience.

Computers, learning with the help of algorithm and techniques, are termed as machine learning. It is a set of algorithms that can do multivariate, nonlinear, nonparametric regression, and classification [5]. The history of AI and ML starts from the 1950s.

- 1950: Alan Turing created the Turing Test. To pass Turing test, an intelligent computer has to persuade a human that he or she is interacting with a human and not a computer.

- 1951: The first artificial neural network, the Stochastic Neural Analog Reinforcement Computer, was created by Minsky and Dean Edmonds, which is composed of 40 neurons.
- 1952: Arthur Samuel, an American pioneer in the fields of computer gaming and artificial intelligence, developed the first computer software that could learn while running.
- 1956: Dartmouth conference organised by John McCarthy, Claude Shannon, and Nathan Rochester, where the name Artificial Intelligence was first coined.
- 1958: Frank Rosenblatt devised the first pattern and shape recognising artificial neural network called Perceptron.
- 1959: Widrow and Hoff designed a binary pattern recognizing and telephone line echo eliminating ANN models called ADALINE and MADALINE. Both the models function using Multiple ADAptive LINear Elements.
- 1962: Sebestyen came up with a Proximity Algorithm rule.
- 1965: Nilsson created Minimum Distance Classifier.
- 1967: Cover *et al.* upgraded the Minimum Distance Classifier to Nearest Neighbour rule.
- 1982: Japan declared that their fifth-generation computing uses artificial intelligence. Hopfield suggested to use bidirectional lines in machine learning instead of neurons.
- 1985: An annual meeting of Neural Networks in Computing was launched by the American Institute of Physics.
- 1987: The Institute of Electrical and Electronic Engineers launched the first International Conference on Neural Networks.
- 1997: Mitchell proposed a more specific definition for machine learning. He defined machine learning as any computer algorithm that gains experience and enhances its efficiency at a given task.
- 2012: Kevin Murphy upgraded the definition for machine learning. Kevin Murphy described machine learning as methods for detecting patterns in data and using them to predict future data or other outcomes of interest.

Machine learning can be categorized into four types. They are supervised learning, unsupervised learning, semisupervised learning, and reinforcement learning [6]. In supervised learning, a sample data is provided to the algorithm by a human which acts as a reference for the algorithm to foresee the output accurately. Here, the sample data acts as an input which is tagged with the specified set of results. When a situation or scenario is presented to the algorithm to predict its outcome, the algorithm takes reference of the sample data and determines the result based on the sample data. For example, distinguishing a spam mail from our inbox. This sample data or input acts as a supervisor and supervise the algorithm. Hence, this type of learning is called supervised learning. It is one of the most widely used machine learning algorithms. Supervised learning is further divided into classification and regression. Unsupervised learning is a process of an algorithm which analyse, categorise, and discover a pattern in the given situation. Here, the algorithm does not contain any data specified with outcomes or tagged with a predetermined result. There is no supervision in this type of learning. Unsupervised learning is further classified into clustering and dimensionality reduction. Semisupervised learning is a process where the sample data provided to the algorithm has only partial set of results tagged with it. This allows the algorithm to tag results for the remaining untagged data through its supervised learning. For example, research studies scrutinising the progression of an illness require a several years of follow up, which is a practically difficult and time consuming.

During such instance, semisupervised learning can be used. The algorithm examines the existing tagged results of the patients which were obtained during the initial years and tries to foresee the results for the untagged data. Reinforcement learning is a type of machine learning where the algorithm is set to learn from interacting with the environment. In this learning, the researcher provides a set of rewards and penalties to the algorithm when its decision towards a predefined objective is correct or incorrect. Through this way, a researcher can shape the behaviour of an algorithm. In supervised learning, the algorithm refers to the input data provided by the researcher to make a decision. Whereas, in reinforcement learning, the algorithm refers to its own decisions from the past and discovers which type of decision would acquire more rewards and less penalty on a basis of trial and error. A detailed illustration of machine learning and its sub-categories are mentioned in Figure 11.1. Based on machine learning principles, there are several machine learning algorithms available. They are presented in Table 11.1.

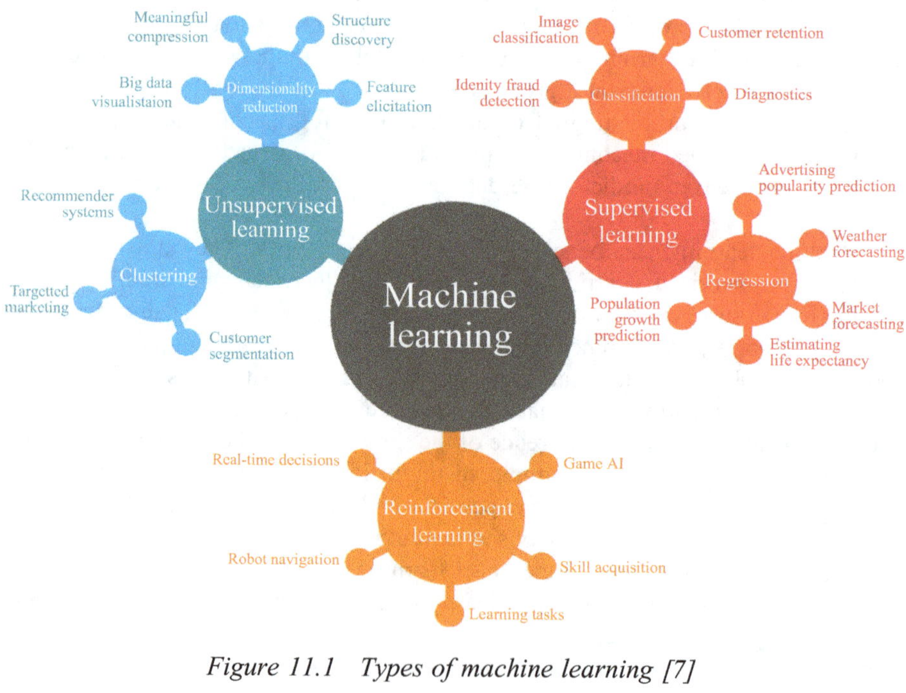

Figure 11.1 Types of machine learning [7]

Table 11.1 Types of machine learning algorithms and its functions

Machine learning algorithms	Function
Artificial Neural Network	Primitive function networks capable of receiving many weighted inputs and being assessed on their effectiveness
Bayesian Network	A supervised learning classification algorithm with a graphical model that blends graph-theoretic and probability-theoretic techniques to depict the probability connections among a set of random variables
Decision Trees	A supervised learning classification algorithm in which a hierarchical tree-shape model that sorts instances based on feature values to classify them
Deep Belief Network	An unsupervised learning classification algorithm. A neural network made up of several layers of limited Boltzmann machines, each of which is connected to the others but not to the units
Dimensionality Reduction	An unsupervised learning technique with the process of decreasing the number of features in a data
Gaussian Mixture Model	A semisupervised learning clustering algorithm. A technique for clustering data that assumes that each data point is a sample from a mix of Gaussian distributions clustering
Gaussian Process	A supervised learning regression algorithm. A group of random variables with a joint Gaussian distribution that is entirely defined by the mean and covariance functions
K-means Clustering	A semisupervised learning clustering algorithm which creates a partition in a provided dataset into a disjoint group
k-Nearest Neighbors	A supervised learning non-parametric algorithm used for classification and regression
Multilayer Perceptron	An effective algorithm for solving the problems of supervised learning
Naïve Bayes Classifier	Statistical learning algorithm
Random Forest	A supervised learning classification and regression algorithm. To achieve a more precise and consistent forecast, Random Forest algorithm creates several decision trees and combines them together
Support Vector Machine	A supervised learning classification and regression algorithm. A model that transforms the input points into a high-dimensional feature space in order to identify a separation hyperplane that increases the difference between the two classes

Human–machine interaction is the two-way communication between humans and computers, robots, and artificial intelligence [8]. Few decades earlier, users were supposed to undergo training to operate machines. For example, early computers had punch card as an interface. But with several research and invention in the recent decades, several interfaces have been invented. Not only a normal ambulatory human can use these interfaces, but even a wheelchair-bound human with multiple disabilities can use these interfaces to communicate with the machine. The following are the common human–machine interfaces used in the wheelchair propulsion [9].

- Brain–computer interface: It works based on the ECG/EEG signals. A system captures the ECG/EEG signal of the user, analyse the wave pattern, and take necessary actions like turning left or right, wheelchair speed, as per the pre-programed instructions. This type of technique is particularly used for amyotrophic lateral sclerosis (ALS). This interface has some limitations as well. The error rate in analysing the ECG/EEG signals and performing the necessary movements is between 30% and 40%. This error happens because of the noise in the signal.
- Sip and puff interface: This technology works on the basis of air inhalation and exhalation. The sensors sense the air pressure and determines the type of action to be performed. This type of technique is particularly useful for the patients who have lost control over their hands. The downside of this interface is the frequent inhalation and exhalation. This voluntary, forceful air exchange may make the patient feel tired. Patients suffering from breathing difficulties or asthma cannot use this type of interface.
- Voice-based interface: Speech recognition is a type of interface which is commonly used for the severely disabled people. It works based on voice commands. The only drawback to this interface is the surrounding noise.
- Tongue-based interface: Tongue movements are captured and processed. Based upon the direction of the tongue movement, the pre-determined actions take place. The stabilization of this interface device on the head may make the patients feel uncomfortable.
- Touchscreen-based interface: It is one of the most common interfaces used by majority of the population. The actions available are displayed in the screen and the user can select whichever is required.
- Head movement-based interface: When the patient turns their head to explore the environment, the signal processor can misjudge the information and make the wheelchair move. This could be a potential drawback of this interface.
- Gaze-based interface: This interface depends upon the visual contact with the patient. Based upon the eye movement, the pre-determined actions take place. The eye movement is captured by the infrared. As the infrared led is constantly on, it can possibly damage the eye.
- Facial movement-based interface: This interface is a combination of gaze and head movements. The sensors present in front of the face captures the eye and face movements. The eye and face movements are converted into signals and provide the basic wheelchair movements like turning left and right, and going forward and stop.
- Gesture-based interface: Using gloves like DataGlove and Z-Glove to perform gesture-based interaction [8]. These gloves contain infrared camera, accelerometers, Bluetooth, gyroscope, and magnetic sensor. Such kind of interface is widely used in gaming platforms like Nintendo, Wii, and also in military applications like bomb disposal.

The current popular technologies which use human machine interaction are Virtual Reality, Augmented Reality, Haptic interaction technology, Voice interaction technology, and Gesture recognition technology. Whereas, a Digital Twin is a new comer which enhances the human–machine interaction [10]. In human–machine interaction, sensors provide the necessary data input to the processor. Based on the placement, sensors can be classified into on-board sensors and off-board sensors [11]. On-board sensors are integrated within the board or machine and provide data. For example, gyroscope, camera, laser range finder, and ultrasonic sensor. Whereas off-board sensors are opposite of the on-board sensors. They can be placed anywhere outside the board or machine and still can pass the input data to the processor, such as movement detection security camera, lidar USB devices, and grid sensors. These automated sensors, once installed, can be accessed and regulated through IoTs which require less human interpretation. Modern industries and factories use different types of sensors, processors, and machine learning algorithms to get the job done quickly, accurately, and more importantly with minimal human intervention [12]. Some of the departments and the machine learning algorithms used by them are provided in Table 11.2.

Table 11.2 Machine learning algorithms used in industries for different purposes

Topic	Purpose	Algorithm
AI in factories	Object recognition	Neural Network
Assistance	Assembly assistance	Neural Network
Logistics	Scheduling	Neural Network Q-learning Deep Q-learning Random Forest
Maintenance	Care of produced goods	Neural Network K-nearest neighbour Principal Component Analysis Decision Tree
Optimization	Fine tune in production	Neural Network Support Vector Machine K-Nearest Neighbour
Process control	Production line	Gradient Boosted Trees Neural Network Random Forest Support Vector Machine
Production	Scheduling	Q-learning Random Forest Decision Tree
	Cost prediction	Neural Network Support Vector Regression Gradient Boosted Trees Random Forest
	Modelling	Random Forest Naïve Bayes

(Continues)

Table 11.2 (*Continued*)

Topic	Purpose	Algorithm
Quality control	Cost reduction	Neural Network Support Vector Machine
	Line quality	Naive Bayes Decision Tree
Robotics	Human–Robot collaboration	Hidden Markov Model K-Nearest Neighbour Neural Network
	Path Planning	K-Nearest Neighbour Neural Network Deep Belief Network

11.3 3D processing

Computer-aided design (CAD) systems are very essential in the development of an industrial product. Few decades earlier, before introducing 3D processing, designers and researchers use the traditional way of hand-drawn 2D orthographic projections to develop a 3D model. The orthographic projections are placed in their corresponding orthogonal planes, and the designer manually edits and redraw the lines to generate a 3D wireframe model. Researchers Markowsky and Wesley were the first to come up with a bunch of guidelines to convert a 2D image into a 3D CAD model [13]. With recent updates in the field of 3D processing, the process of retrieving a 3D model from a 2D image became automated through major software developers like Dassault Systemes, Autodesk, and Altair. There is also some open-source software in 3D modelling namely 3D Slicer, ImageJ, InVesalius, ITK Snap, Osirix Lite, Seg 3D, and few more [14]. With the help of software's, converting a 2D image or a solid model into a 3D CAD model is now a manageable task. Whereas, converting a small, irregular complex-shaped object into a 3D CAD model is still a challenging task. In order to overcome the difficulty in 3D processing of small parts with complex surfaces and shapes, Galantucci *et al.* [15] invented a powerful scanning methodology in 2016 called photogrammetric scanning system with rotary table (PSSRT). This PSSRT is capable of developing a precise accurate 3D model of objects which are even few millimetres long possessing a complicated surface. The PSSRT consists of DSLR camera with 10 megapixels, a sensor size of APS-C 22.2 × 14.8 mm, Canon EF 50 mm 1:1:8 II lens, and Kenko extension tubes. In contradiction to the traditional method of photographing a fixed object in 360°, in PSSRT, the object is placed in a rotatory table while the DSLR camera is fixed on a platform which can be moved in a single axis to tune the distance between camera and object for a better focus. This way, a precise 3D model of a very small object can be easily obtained. On the other hand, the conventional 3D processing consumes a lot of computer memory, as it has to retain several properties. To overcome this hurdle, in 2020, Qiangqiang *et al.* [16]

conducted a study on morphing-based 3D point cloud reconstruction. Maintaining topological links in the mesh for the 3D model rebuilt from the surface and volume models necessitates a lot of computer memory and puts a lot of strain on the machine. Furthermore, when reconstructing the new mesh, some tiny and flat units will invariably be created, reducing the system's stability significantly. Whereas, the point cloud model has the following properties in this context: no complicated topology, high flexibility, no mesh formation in the numerical calculation, low computing complexity, offer a large number of point clouds, and good detail reproduction capacity.

11.4 Additive manufacturing

As shown in Figure 11.2, among the formative production processes like forging, casting and subtractive production processes like turning, milling and additive manufacturing form a separate manufacturing technology. Additive manufacturing or 3D printing is also called as solid freeform fabrication [17]. It is a controlled robotic process of depositing materials layer by layer in 2D form, which eventually results in a 3D model. 3D printing can be performed using metals, polymers, and ceramics. 3D printing is standardised in the United States by ISO/ASTM 52900:2015 [18]. 3D printing is used in several sectors like

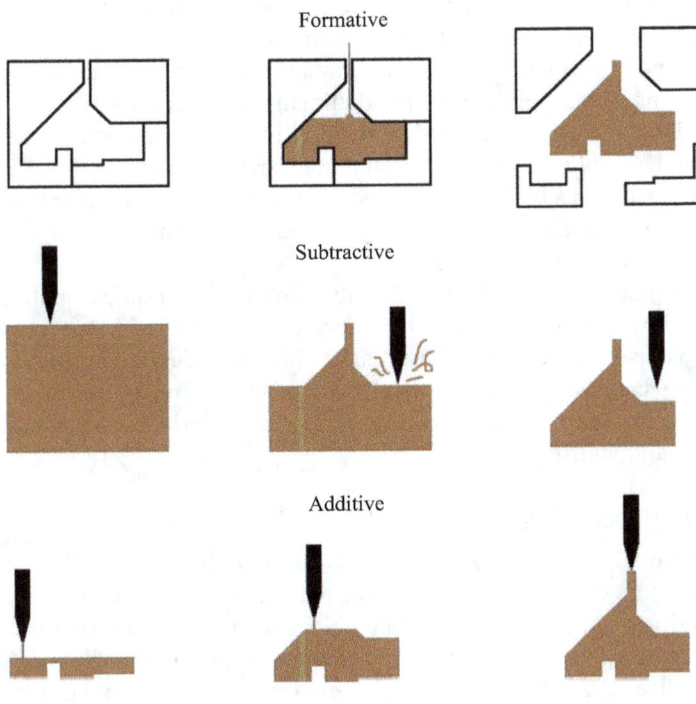

Figure 11.2 Formative, subtractive, and additive manufacturing

aerospace, automotive, architecture, consumer goods, dental, food, furniture, jewelry, medicine, toys industry and several sectors incorporate it depending upon their need [19].

Depending upon the need and requirement, a 3D printer must fulfil certain criteria [20]. They are the following:

- Resolution – The higher the level of detail corresponds to the smaller the resolution. The horizontal resolution of a Cartesian-axis system (i.e., *X*, *Y*, *Z* coordinate systems) is the shortest distance the printer can move in the 2D plane specified by the *X* and *Y* axes, whereas the vertical resolution is the smallest layer thickness the printer can generate via *Z*-axis motion.
- Size – A 3D model's size is restricted by the capacity of the printer build platform, which includes the base area *X* and *Y* dimensions as well as the height *Z* dimension.
- Material – The ability to process the material alone or in a multi-material modality, as well as the raw material forms such as solid, powder, or liquid, greatly determines the choice of a printing technique. The procedure for processing method used has an impact on the final material characteristics and functional performance. Mechanical resistance to static or dynamic stresses, biological environment endurance, and surface roughness are only a few examples.
- Finishing – Post-processing is frequently required for 3D-printed products. Support material that is required for geometries with overhangs must be removed after printing. To fully develop the desired qualities, the material may require a post cure under UV light or in an oven. Painting, surface levelling, and cleaning, for example, require additional equipment and time.
- Time – The time it takes to finish a model is highly dependent on the approach used and the resolution. A finer resolution necessitates not only more 2D layers but also greater machine precision, which may necessitate a slower layer patterning speed. Some machines can give you an estimate of how long it will take to manufacture something.
- Cost – A platform service like 3D hubs can provide a quick quote and production time for commercial machinery and materials. A specialised contractor is usually required for a more bespoke design or material. The beginning cost would be the total cost of the software package and the printing gear. More money has to be spent on materials, maintenance, trash disposal, and staffing.

11.4.1 3D processing to 3D model

A computer model is used to create each 3D-printed object. A CAD model is the name given to this digital file. An end user may create the CAD model using 3D modelling software, a scan of an existing object such as CT and MRI imaging, or a digital model acquired from a service provider. Regardless of the file's origin, users can utilise a CAD software package to make future changes during an iterative design process. Free open-source and professional versions of CAD software are available in the market. Though each CAD platform may save the geometry in a different file format, the majority of them can convert it to the common stereolithography (STL) file format, which consists of a triangular mesh representing the geometry surface in 3D Cartesian space. A support material may be necessary if the 3D geometry has layers that are greater than the layers below to prevent the layer from collapsing under gravity. The finished CAD model is turned into G-code, a numerical control programming language that includes the item surface geometry, fill pattern, and essential support material. The G-code file is transmitted to the 3D printer, which begins printing immediately. There are several 3D printing techniques available right now. They are laser stereolithography, micro stereolithography, polymer jetting, digital light processing, laser sintering, laser melting, electron beam melting, powder bed fusion, fused deposition modeling, laminated object manufacturing, selective deposition lamination, controlled metal buildup, direct metal deposition, extruding and milling, aerosol printing, and bioplotter. The most commonly followed 3D printing technologies are the following:

Fused deposition modelling (FDM): It works under the principle of melting materials. A solid material or a filament is pulled into the nozzle head, where the filament gets heated and melted by heating elements present in the head [21]. The molten material is pushed through a nozzle orifice, continued by natural cooling and solidification. The nozzle head also contains a secondary orifice to print the support material. The nozzle head moves only in the X and the Y direction, whereas the build platform moves only in the Z direction. 1.75 mm and 3.0 mm are the diameter of filaments, which is spooled into the head. If a material can be melted, then it can be used in FDM to print. The resolution ranges between 50 and 200 μm [22]. The FDM is commonly used for filament printing. Cost of FDM can range anywhere between 1,000 USD and 20,000 USD. Drawbacks are the followings: model resolution, material selection, and smoothness.

Direct ink writing (DIW): This is similar to FDM. A viscoelastic fluid is pushed through a nozzle, followed by solidification through solvent evaporation or stress reduction [21]. Drawbacks are the following: 3D models with hanging structures cannot be printed.

Selective laser sintering (SLS): A process which uses a high-power laser to irradiate a powder bed made up of either cement, ceramics, metals, or polymers [21]. During irradiation, it is made sure that the material does not reach its melting point, but hot enough to fuse into a solid mass. The laser beam is projected through a lens or mirror. This method has a very less solidification time when compared to other methods. As the model is always immersed in the powder, hanging structures can be easily printed. Commonly used for printing powders. The resolution ranges between 80 and 250 μm. Drawbacks are the following: limited material choices with high thermal conductivity, powder size, roughness in the 3D model.

Electron beam melting (EBM): This method is more similar to SLS. EBM uses an electron beam for irradiation [21]. Since the material processing is done by electron beam, a total sealing and vacuumed build chamber is necessary. Drawbacks are the following: similar to SLS.

Inkjet or multi-material jetting: A method of placing micro droplets on the build plate containing the printable material which is solidified by applying an UV light [21]. Resins are commonly used for printing 3D models. This method provides a smooth surface and hanging structures can be printed too. The resolution ranges between 5 and 200 μm Drawbacks are the following: requires an expensive high-end printer and resins with specific properties are required.

Stereolithography (SLA): A light-based 3D printing technique which shares the same methodology with other techniques such as two photon polymerization, continuous liquid interface production, digital light processing, and volumetric 3D printing [21]. Stereolithography is one of the oldest and most detailed methods of printing. A photo-curable resins that harden through a photochemical reaction are used in this light-based 3D printing technology. Galvo type scanning device is used to guide the laser beam. A typical stereolithography machine consists of build volume which is filled with build material, laser unit at the top capable of moving in *X–Y* direction, and a *Z* direction movable platform. The resolution range of Stereolithography is 10 μm.

11.4.2 Materials used

Hydrogels like microfluids – biological applications, thermoplastics like ABS, PLA, EPU in biological applications, metals like aluminum, copper, stainless steel, titanium in surgical implants, aerospace, and electronic devices, ceramics in abrasion resistive applications and optical devices, and cements in construction [23].

The types of printing process and technologies are tabulated in Table 11.3 and the depiction of each process is portrayed in Figure 11.3.

11.4.3 Post processing

Removal of the support structure and further curing procedures, coating of surfaces for functionalization and or protection, polishing and surface roughness improvements, and treatments for modifying material characteristics and structure forms are the four basic post processing approaches. Nearly all 3D printing technologies involve the removal of support structures, cleaning, and extra curing stages [20, 23]. In polymer chemistry, secondary curing procedures are used after the primary polymerization event, which is usually UV curing or heating, results in the creation of polymer chains. To attain their final shapes, these subsequent heating stages can also be employed to change the material properties or shape of the 3D printed structures. Various deposition processes, such as conformal parylene AF4 deposition, electropolymerization, and a microfluidic reactor, are used to cover 3D printed surfaces. For 3D printed metal objects, traditional polishing processes such as grinding and sanding have been used. Enclosed channels and waveguides, on the other hand, necessitate innovative polishing techniques. One example is a magnetically driven abrasive polishing system that uses electron beam melting of oxygen-free high conductivity copper to polish stainless steel WR-10 waveguides. Vibratory bowl abrasion and ultrasonic abrasion finishing are two other traditional options.

Table 11.3 Types of printing process and technologies

Process	About	Technology	Material	Pros	Cons
Powder bed fusion	Thermal energy selectively fuses regions of a powder bed. Once a layer is formed, the powder bed platform gets lowered in the Z axis	Electron beam melting (EBM) Selective laser sintering (SLS) Selective laser melting (SLM) Selective heat sintering (SHS) Direct metal laser sintering (DMLS)	Stainless steel, titanium, aluminium, cobalt chrome, copper, zirconia polyamides, polymers, alumide, silica glass nanopowder, nylon powder, polystyrene atomized metal powder, ceramic powder	Additional support structure is not required, large variety of material choices	Size limitations, high power consumption, surface smoothness depends upon the powder grain size
Directed energy deposition	Nozzle containing the material is fixed to a multi axis arm which can move in any direction. Once the material is deposited, a thermal energy is applied to fuse materials by melting upon extrusion	Laser direct metal deposition Electron beam welding (EBW) Laser engineered net shaping (LENS) Directed light fabrication (DLF) 3D laser cladding	Molten metal powders like steel, titanium alloys, cobalt chromium, ceramics, polymers	Flexibility of multi axis arm, more widely used in repair works	Limited material choices, post processing

(Continues)

Table 11.3 (Continued)

Process	About	Technology	Material	Pros	Cons
Material extrusion	The heated material is dispensed through a nozzle layer by layer.	Fused deposition modelling (FDM) Fused filament fabrication (FFF) Contour crafting (CC) Direct ink writing (DIW)	Thermoplastic polymers, plastic, nylon, nano fibres, concrete, soil, aqueous slurries and dispersions, polyamide, acrylonitrile butadiene styrene, polylactic acid Cellulose-based hydrogels	Low cost, easy maintenance, large construction	Not accurate, printing a complex structure is difficult
Binder jetting	A liquid bonding agent is selectively deposited to join powder materials and then product is baked in an oven for final curing	Powder bed Inkjet heat (3DP) Plaster based	Polymer powder ceramics, metallic oxide, bentonite, Stainless steel, bronze, plaster, resin, polymers, gypsum	Low cost, easy maintenance, quick printing	Weak mechanical properties, cannot print large models, limited material choices
Material jetting	Droplets of build material are selectively deposited through nozzle. These droplets get solidified and forms a layer. The droplets are either cured with UV light or heat or milled sometimes	Multijet modelling (poly/ink) (MJM) Material jetting (MJ) Drop on demand (DOD) Nano particle jetting (NPJ)	Thermally stable or unstable plastics, resin, Visijet crystal, ceramics, metals, polymers, photopolymer wax	Can create complex structures, accurate, quick, effective	High maintenance

Sheet lamination	Sheet of materials are cut using laser or cutter and bonded together to form an object	Laminate object manufacturing (LOM) Ultrasonic consolidation	Thermoplastic polymers, wood, metals, ceramics, polyamide, acrylonitrile butadiene styrene, paper, metallic sheet	Cheap, high speed, can work in low temperature	limited material choices, may require post processing
Vat Photopolymerization	Liquid photopolymer in a vat is selectively cured by light activated or UV polymerization or laser where the platform moves downwards after curing a layer	Stereolithography (SLA) Digital light processing	Photopolymers, ceramics, resin, epoxies, acrylates, alumina, zirconia	Highly accurate	limited material choices, expensive, need several support structures

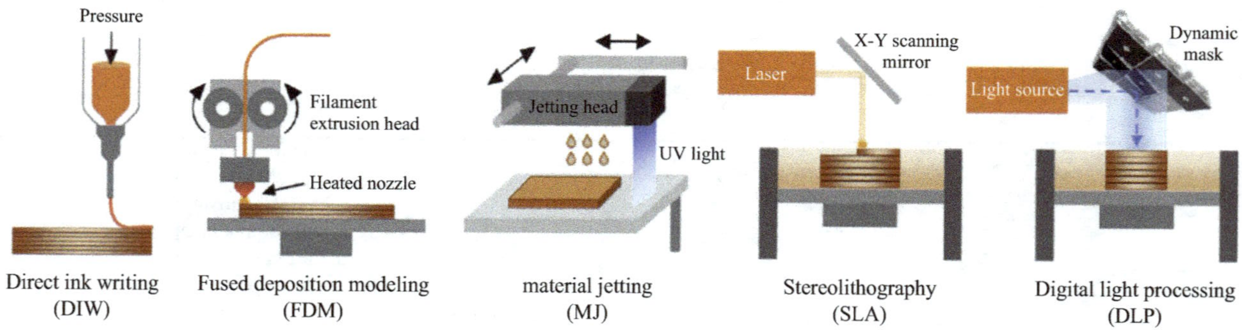

Figure 11.3 Depiction of different printing process [24]

11.5 3D printing in medical healthcare

3D printing technology is used in medicine and healthcare, particularly in drug designing and manufacturing [25], and implant manufacturing [26]. There are several drugs and medicines designed to deliver a specific outcome. Some are discussed below: a custom 3D printed tablets for personalized demand, controlled release doughnut shaped tablet for drug release at different sites in different time with different dosage, controlled release layer tablet with custom double layer drug, polypill tablet to reduce the difficulty in swallowing, tablets like organ on a ship, liver on a ship, organ on a chip, heart on a chip are used to induce a hepatotoxic reactions, microneedle used for transdermal drug injection, nanofluids for immunotherapy, and mucojet for non-invasive vaccine delivery. In regard to the implant manufacturing, a set of procedures has to be followed. They are mentioned in Figure 11.4 and 3D printed sample models are shown in Figure 11.5. Some commonly used materials, software's, and file formats in the medical industry [25, 27] are provided in Table 11.4. Using these materials, software's and technology, several companies and enterprises work in the field of medical healthcare and come up with the products approved by FDA. Some are mentioned in Table 11.5.

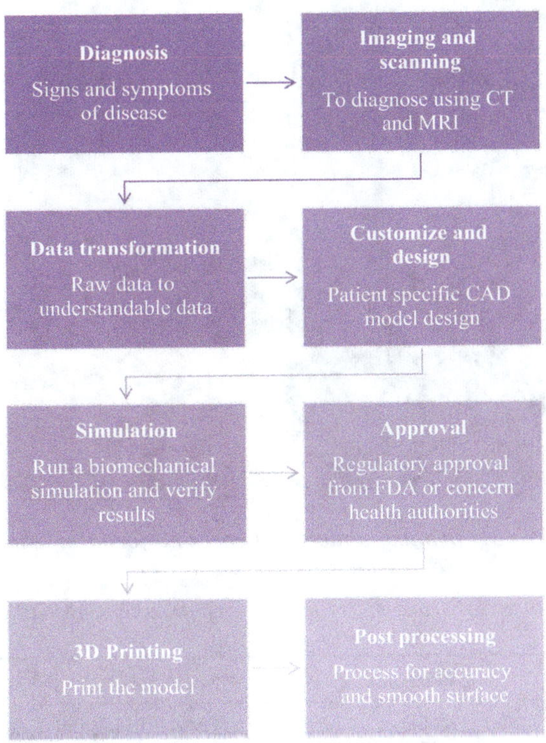

Figure 11.4 Procedures to be followed in implant 3D printing

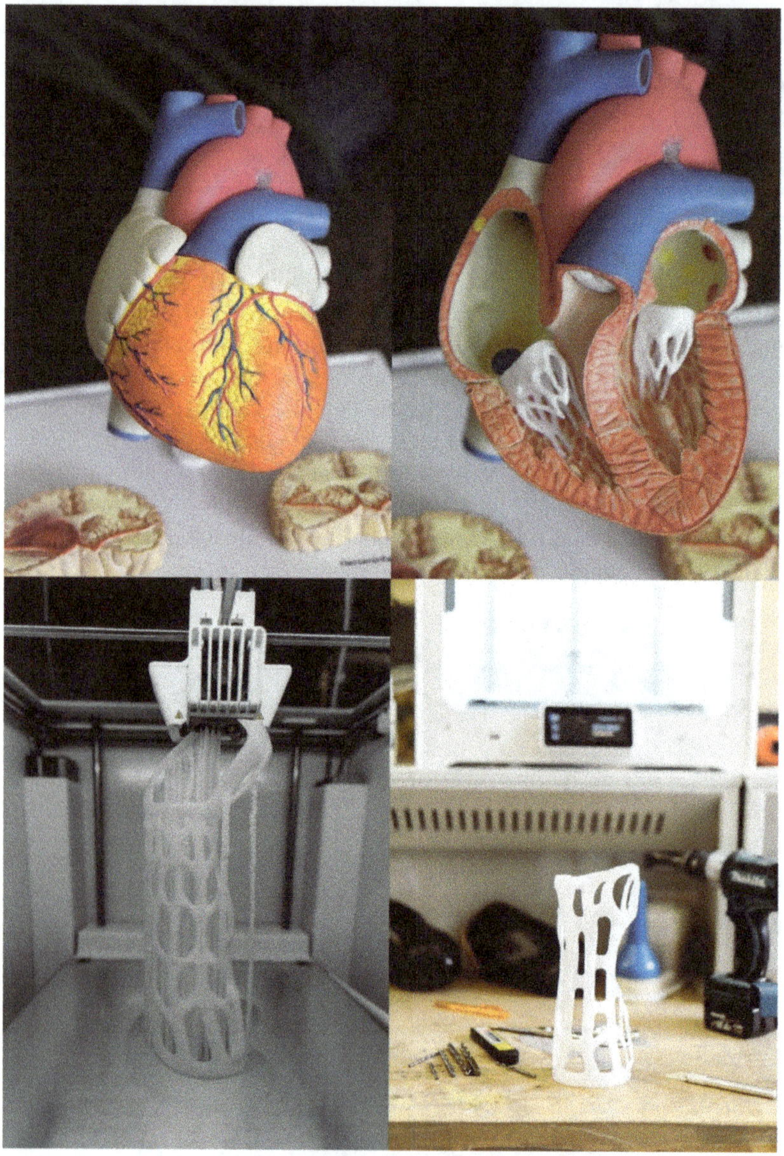

Figure 11.5 3D printed models of heart and forearm orthosis [28]

Table 11.4 Common software's, 3D model format, and material used

Software	Format	Material
InVesalius	STL	Poly Ethyl Ether Ketone
OsiriX	DICOM	Polyacidic acid
MeshLab	OBJ	Polyamide 11

Table 11.4 (Continued)

Software	Format	Material
3D Slicer		Polyhydroxyalkonates
FSL		Polyethylene
Blender		Polycarbonates
Inventor		Polyurethanes
DAZ		Silicones
Studio		Gold
LaTeX		Platinum
Meshmixer		Titanium
Materialise		Steel
		Cobalt
		Chromium
		NiTiNOL
		Calcium phosphate salts
		Glass
		Oxides of aluminum and Titanium

Table 11.5 FDA approved medical products and devices

Enterprise	Merchandise	Application
3D Systems	VSP system	Software
4WEB Medical	Lateral spine truss system	Spine
Additive Orthopaedics	3D printed bone segments, lattice planting system, osteotomy wedge system	Fractures, osteotomies, arthrodesis
Align Technology	Teeth aligners	Orthodontics
Bio Architects	Titanium cranical, craniofacial implant	Cranium
Camber Spine Technologies	SPIRA open matrix ALIF	Spine
Candid	Teeth aligners	Orthodontics
Centinel Spine	FLX devices	Spine
Choice Spine	HAWKEYE vertebral body replacement system, Tiger shark porous interbody system	Spine
Dyna Flex	Teeth aligners	Orthodontics
EIT	Cellular titanium cervical cage	Spine
Envision	E-denture, E-guard, E-shell, E-dent	Dentistry
Joimax	EndoLIF on-cage	Spine
K2M	CASCADIA lateral interbody system	Spine

(Continues)

Table 11.5 (Continued)

Enterprise	Merchandise	Application
Lantos Technologies	3D ear scanning system	Ear canal
Materialise	Mimics, 3D printing of patient specific radius and ulna guide	Software, surgical guide
Med Shape	Fast forward bone tether plate	Hallux valgus
Medicrea	IB3D TLIF banana	Spine
Mighty Oak Medical	FIREFLY pedicle screw navigation guide	Surgical guide
Onkos Surgical	ELEOS limb salvage system	Oncology
Oventus Medical	O2Vent device	Sleep apnea
Oxford Performance Materials	OsteoFab patient specific facial device	Cranium
Phonak	Virto B-titanium hearing aids	Hearing aid
Renovis Surgical Technologies	Posterior lumbar tesera porous titanium interbody fusion system	Spine
SI-BONE	iFuse 3D implant	Sacroiliac joint
Stryker	Titanium covered posterior lumbar cage	Spine
Taulman 3D	Guideline for 3D printing	Surgical tools and implants
Zimmer Biomet	Unite3D bridge fixation system	Ankle and Foot

11.6 3D printing in food science

Rapid prototyping, the method of modelling, constructing, and fabricating objects using computer-aided design, was invented in the late 1980s by Kodama of Japan [29]. Rapid prototyping technology was developed in Additive Manufacturing, a more sophisticated version that can build complicated 3D models layer-by-layer out of plastic polymer filaments, metal, and, more recently, edible materials like sugar and chocolate [30]. The materials used in the food printing can be broadly classified into three groups, namely: liquid, powder, and culture of cells. Liquid-based materials are used in extrusion and inkjet process. Powder-based materials are used in particle binder or heat source application. Whereas, the culture of cells is used in bioprinting. There are seven categories of additive manufacturing as per the American Society of Testing and Materials. They are binder jetting, directed energy deposition, material extrusion, material jetting, powder bed fusion, sheet lamination, and vat photopolymerization. Whereas there are only three main technologies used in 3D printing. They are extrusion, sintering, and liquid binding [29].

Extrusion: The extrusion process was developed by Crump in 1991 using a fused deposition modelling method [17]. This method uses a moving nozzle to expel or extrude a hot molten filament polymer as a continuous layer which fuses with its preceding layer while cooling. This method is commonly used in printing a molten material using a semi-solid viscous system or a temperature control. Based on the temperature, extrusion can be classified into room temperature extrusion (RTE), hot melt extrusion (HME), and hydrogel forming extrusion (HFE). RTE is a very basic method which is used in traditional cooking and baking for several decades. Pressing the dough, butter, cheese, and cream through a piston are some classic examples. However, with the assistance of 3D printing technology, the accuracy of the expected model is easily obtained. HME is particularly used in printing a 3D model using molten chocolate. Depending upon the material used, heat is applied to the material and kept at a temperature ranging between 30°C and 40°C using a heating block to maintain its viscosity. With the assistance of an automated syringe or piston, the molten material or a semi-solid material is pushed out of the nozzle. Once the material is extruded out of the nozzle, it has to immediately solidify and fuse to the previous layer. This concept of HME is also used in manufacturing rubber and plastic models. This method was first introduced in 2007 by a group of researchers from Cornell University. Later in 2010, Hao *et al.* proposed certain ground rules like diameter of the nozzle, height between the nozzle and bed, and extrusion axis movement to get a better 3D model of chocolate. In HFE, the hydrocolloid solution is printed into a polymer setting using syringe, jet cutter, vibrating nozzle or any other similar apparatus. In HFE, the hydrocolloid is prepared in a way that it possesses a viscoelastic character in the beginning of extrusion and latter changes into a self-supporting gel for the subsequent layers. This method is particularly used to prepare soft foods for elderly people with swallowing difficulty. There are few parameters to be considered in extrusion. They are the following: viscosity of the material to extrude through nozzle, temperature requirement, mechanical integrity of the material to withstand the layers, and stability of the printed layer in providing the accuracy [31].

Sintering: This is the second most popular 3D food printing method next to extrusion. Sintering is classified into two types. They are selective laser sintering (SLS) and hot air sintering (HAS) In SLS, a solid structure is created by sintering with the help of a 3D software which directs the infrared laser from scanner towards the printer bed containing the powdered material. Once a single layer of 3D model is sintered, the powder bed is lowered by one layer of thickness, so that the next layer can be applied on top of it. This lowering process is continued until the whole 3D model is completed. In HAS, a stream of hot air is blown towards the powder in the printer bed to form a 2D image. Once a full 2D image is obtained, the powder bed is slightly lowered so that the second layer is applied on the top. This process is repeated continuously until a 3D model is obtained. Both the SLS and the HAS are limited to a powder-based materials and faster than any other 3D printing process. 3D printed candies with different geometries are shown in Figure 11.6.

Binder jetting technique: This technology was invented by Grood *et al.*, in 2011. Inkjet printers rely on piezoelectric and thermal nozzle heads. Piezoelectric heads contain piezoelectric crystals which generate an acoustic wave that separates the droplets from the liquid at regular intervals. Whereas, in the thermal head, the droplets are forced out through the pulse of pressure which is generated by the electrically heated head. It is a repeated process of bonding a powdered material

Figure 11.6 Complex 3D food printing [32]

using a liquid binder to develop a 3D model. The powdered material is spread on the platform where the nozzle or printer head sprays the liquid agent. Here, the liquid acts as an essential bonding agent. These drops, altogether form a structure as developed in the CAD software. This technology is particularly used for powder-based materials and also has advantages as low material cost and fast fabrication. However, the 3D printed model created through this method always has a rough surface finish and has to be print on a flat surface, rather than a curved or complex surface.

To obtain an accurate and precise 3D printed model, material properties, process parameters, and post processing methods are very essential. The frequently used printable food materials are sugar, gelatine, dough, and chocolate. There are only three basic material types of food are used in 3D food printing. Basic types, method used to print, its principle, and the common materials used are provided in Table 11.6.

In recent times, several raw materials and different combination of raw materials were tested and used in the field of food 3D printing. Some of those raw materials used in the researches and their findings are provided in Table 11.7.

11.7 Future aspects

Additive manufacturing or 3D printing has more opportunities in automobile industry, aerospace, aeronautics, food art, archeology, medicine, healthcare, sports, construction industry and so on [18,49,50]. 3D printing has a big impact in the environment. There are several researches done and undergoing using 3D printing in sustainable manufacturing, air filter monitoring, air pollution monitoring, and water treatment [51]. Results have shown that replacing natural sand with 25% of recycled sand caused higher green strength and no obvious decrease in hardened mechanical properties in 3D concrete printing [52]. Adding to the clean environment and recycling of sand, 3D printing helps in preserving cultural heritage as

Table 11.6 Common methods used in food printing

Basic type	Method	Principle	Material
Liquid	Hydrogel forming extrusion	Extrusion and deposition	Gelatine, Xanthan gum
	Inkjet printing	Drop on demand deposition	Chocolate, liquid dough, jam, gel, cheese
	Melting extrusion	Extrusion and deposition	Chocolate
	Soft material extrusion	Extrusion and deposition	Processed cheese, dough
Powder	Hot air sintering and melting	Hot air	Sugar
	Liquid binding	Drop on demand deposition	Chocolate
	Selective laser sintering	Laser	Sugar
Cell	Bioprinting	Drop on demand deposition	Meat

Table 11.7 Raw materials used in food printing in recent years

Raw material	Finding	Reference
Mashed potato	In the 3D printed samples, the authors looked at the infill levels, infill patterns, and the number of shell perimeters. Firmness and Young's modulus were simply affected by infill percentage, but hardness and gumminess were highly related to infill levels and perimeters	[33]
Corn, potato, wheat starch, lemon	The researchers investigated the effects of different starch sources on the rheological properties of lemon juice gel for use in 3D food printing in this study. They used the POLYFLOW program to model the fluid properties of inks in the flow channel of positive displacement 3D printers under a variety of technological conditions, including viscosity, relaxation time, inlet volume flow rate ratio, and nozzle diameter	[34]
Cheese	Untreated and melted cheese had poorer hardness and meltability than cheese created using the 3D printing technology	[35]

(Continues)

Table 11.7 (Continued)

Raw material	Finding	Reference
Baking dough gel	Parameters like extrudability, gel strength, ductibility, and elasticity play a vital role in obtaining a better shape when printing a baking dough gel	[36]
Milk powder, oat, starch, bean protein, cellulose	According to the researchers, the semi-skimmed milk powder-based paste had the best printing precision and shape stability. The yield stress of the paste was linked to shape stability after printing	[37]
Avocado, broccoli, kiwi, carrot, pear	The antioxidant capacity of the printed samples remained stable during storage	[38]
Potato, pea	With the growth in pea protein concentration, the structural and physicochemical features changed on a regular basis	[39]
Egg, rice flour	Rice flour was used as a filler agent, and it improved the stability and strength of printed materials when combined with egg yolk and egg white	[40]
Fish, sodium chloride	Salt was used to improve slurry flow during extrusion, as well as the deposition and stability of the printed sample	[41]
Wheat flour dough, Tenebrio molitor	A more soft wheat dough was obtained when the Tenebrio molitor (insect powder) is added	[42]
Wheat starch, orange	The researcher studied the addition of wheat starch to orange concentrate for 3D printing in this study. They discovered the ideal 3D printing and ink formulation conditions for producing a printed sample with improved resolution and textural qualities	[43]
Tomato	According to the researchers, this research looked into a possible link between the printability of a food-ink formulation and its rheological properties. The pressure required to facilitate extrusion increased linearly with increasing flow stress	[44]
Potato flakes	In this study, two distinct extrusion methods for 3D printing were evaluated: the fluid parameters of the screw-based 3D food printer were complex, whereas the fluid characteristics of the syringe-based 3D food printer were simpler and easier to change	[45]

Table 11.7 (*Continued*)

Raw material	**Finding**	**Reference**
Soy protein, hydrocolloids	The researchers tested the rheology, printability, and texture of soy protein isolate-sodium alginate gelatin combinations. At 35 °C, the addition of gelatin in the mixes reduced viscosity and elastic modulus. It was also discovered that adding these chemicals to 3D printed shapes enhanced their hardness and chewability	[46]
Spinach, xanthan gum	Based on 20% spinach powder in 10% xanthan gum gel, this study investigated the effect of particle size on rheological characteristics and ink printing performance. All of the formulae could be printed	[47]
Corn starch, potato starch, rice starch	During 3D printing, each starch supply displayed a distinct rheological behavior, resulting in printed samples with varying self-supporting properties. The yield stress, flow stress, and storage modulus of the gel increased as the starch concentration increased	[48]

well [53]. Since the space 3D printing is going to be the fundamental in future space missions, NASA recently conducted an acrylonitrile butadiene styrene (ABS) 3D printing experiment using fused deposition modelling (FDM) at International Space Station to understand the effect of microgravity on ABS [54]. 3D modelling and printing shall play a major part in medical research. Instead of using real bones, implants, and arteries for blood circulation analysis, 3D digital models or printed models can be used for research purpose [55], as the availability of cadaver models with specific attributes is rare and hard to find. In the recent year literatures, several medical research analyses were also done using the 3D modelling [56] [57].

11.8 Limitations

With a decade of history and continuous improvement in scientific technology, 3D printing still has certain limitations. Those limitations can be funnelled into five categories [58]. They are the following: *Time* – the 3D printing process is still slow and takes time to print a large, complex part, excluding post processing. *Materials* – a set of materials are repeatedly used throughout the time. New materials are in need. *Material properties* – certain materials have to be stored in a moisture-free environment. If not stored properly, it may not print accurately. *Cost* – high-end printers which print accurate models are way too expensive. *Size* – large or big models cannot be printed in the printers.

References

[1] M. Palaniswamy and W. Y. Leong, "Cloud computing in Industrial *Revolution 4.0," in* The Nine Pillars of Technologies for Industry 4.0, Kuala Lumpur, IET Digital Library, 2020, pp. 245–262.

[2] H. S. Ning and H. Liu, "Cyber-physical-social-thinking space based science and technology framework for the Internet of Things," *Science China Information Sciences*, vol. 58, pp. 1–19, 2015.

[3] P. W. Langley, "The central role of cognition in learning," *Advances in Cognitive Systems*, vol. 4, pp. 3–12, 2016.

[4] T. M. Mitchell, *Machine Learning*, 1st ed., E. M. Munson, Ed., McGraw Hill, New York, NY, 1997, p. 432.

[5] K. E. Bouchefry and R. S. de Souza, "Learning in big data: introduction to machine learning," in *Knowledge Discovery in Big Data from Astronomy and Earth Observation*, 1st ed., P. Skoda and F. Adam, Eds., Elsevier, New York, NY, 2020, pp. 225–249.

[6] S. Vieira, W. H. Lopez Pinaya and A. Mechelli, "Introduction to machine learning," in *Machine Learning Methods and Applications to Brain Disorders*, 1st ed., Academic Press Inc., London, 2019, pp. 1–20.

[7] "pngfind.com," pngfind.com, [Online]. Available: https://www.pngfind.com/mpng/owmimi_machine-learning-areas-hd-png-download/ [Accessed 25 September 2021].

[8] Q. Ke, J. Liu, M. Bennamoun, S. An, F. Sohel and F. Boussaid, "Computer vision for human–machine interaction," in *Computer Vision for Assistive Healthcare*, 1st ed., L. Marco and G. M. Farinella, Eds., Academic Press, London, 2018, pp. 127–145.

[9] A. Ghorbel, N. B. Amor and M. Jallouli, "A survey on different human-machine interactions used for controlling an electric wheelchair," in *3rd International Conference on Knowledge-Based and Intelligent Information & Engineering Systems*, 2019.

[10] X. Ma, F. Tao, M. Zhang, T. Wang and Y. Zuo, "Digital twin enhanced human-machine interaction in product lifecycle," in *11th CIRP Conference on Industrial Product-Service Systems*, 2019.

[11] A. Gardecki, M. Podpora and A. Kawala-Janik, "Innovative Internet of Things-reinforced human recognition for human-machine interaction purposes," in *15th IFAC Conference on Programmable Devices and Embedded Systems*, Ostrava, 2018.

[12] S. Fahle, C. Prinz and B. Kuhlenkotter, "Systematic review on machine learning (ML) methods for manufacturing processes – Identifying artificial intelligence (AI) methods for field application," in *53rd CIRP Conference on Manufacturing Systems*, 2020.

[13] L. Governi, R. Furferi, M. Palai and Yary Volpe, "3D geometry reconstruction from orthographic views: a method based on 3D image processing and data fitting," *Computers in Industry*, vol. 64, no. 9, pp. 1290–1300, 2013.

[14] R. T, R. Ranganathan, A. Pugalendhi and S. Arumugam, "3D printed patient specific models from medical imaging – a general workflow," in *2nd International Conference on Materials Manufacturing and Modelling*, Vellore, 2020.

[15] L. M. Galantucci, M. Pesce and F. Lavecchia, "A powerful scanning methodology for 3D measurements of small parts with complex surfaces and sub millimeter-sized features, based on close range photogrammetry," *Precision Engineering*, vol. 43, pp. 211–219, 2016.

[16] Q. Cheng, P. Sun, C. Yang, Y. Yang and P. X. Liu, "A morphing-based 3D point cloud reconstruction framework for medical image processing," *Computer Methods and Programs in Biomedicine*, vol. 195, no. 105495, 2020.

[17] F. C. Godoi, S. Prakash and B. R. Bhandari, "3d printing technologies applied for food design: status and prospects," *Journal of Food Engineering*, vol. 179, pp. 44–54, 2016.

[18] S. A. Tofail, E. P. Koumoulos, A. Bandyopadhyay, S. Bose, L. O'Donoghue and C. Charitidis, "Additive manufacturing: scientific and technological challenges, market uptake and opportunities," *Materials Today*, vol. 21, no. 1, pp. 22–37, 2018.

[19] P. A. Perumal and K. P. Untol, "Additive manufacturing toward Industry 4.0," in *The Nine Pillars of Technologies for Industry 4.0*, IET Digital Library,Kuala Lumpur, 2020, pp. 233–243.

[20] R. H. Awad, S. A. Habash and C. J. Hansen, "3D printing methods," in *3D Printing Applications in Cardiovascular Medicine*, 1st ed., S. J. Al'Aref, S. Dunham, B. Mosadegh and J. K. Min, Eds., Academic Press, New York, NY, 2018, pp. 11–32.

[21] I. Karakurt and L. Lin, "3D printing technologies: techniques, materials, and post-processing," *Current Opinion in Chemical Engineering*, vol. 28, pp. 134–143, 2020.

[22] T. D. Ngo, A. Kashani, G. Imbalzano, K. T. Nguyen and D. Hui, "Additive manufacturing (3D printing): a review of materials, methods, applications and challenges," *Composites Part B: Engineering*, vol. 143, pp. 172–196, 2018.

[23] J. A. Arrieta Escobar, D. Derrien, S. Ouvrard, E. A. Yazdi, A. Hassan, V. Boly, A. J. Tinet and M. F. Dignac, "3D printing: an emerging opportunity for soil science," *Geoderma*, vol. 378, p. 114588, 2020.

[24] D. Han and H. Lee, "Recent advances in multi-material additive manufacturing: methods and applications," *Current Opinion in Chemical Engineering*, vol. 28, pp. 158–166, 2020.

[25] X. Zhu, H. Li, L. Huang, M. Zhang, W. Fan and L. Cui, "3D printing promotes the development of drugs," *Biomedicine & Pharmacotherapy*, vol. 131, p. 110644, 2020.

[26] M. Javaid and A. Haleem, "Additive manufacturing applications in medical cases: a literature based review," *Alexandria Journal of Medicine*, vol. 54, no. 4, pp. 411–422, 2018.

[27] G. Poologasundarampillai and A. Nommeots Nomm, "Materials for 3D printing in medicine: metals, polymers, ceramics, hydrogels," in *3D Printing in Medicine*, 1st ed., D. M. Kalaskar, Ed., Woodhead Publishing, Sawston, 2017, pp. 43–71.

[28] T. Claes and R. Weermeijer, "unsplash.com," unsplash.com, 26 June 2018. [Online]. Available: https://unsplash.com/photos/NIuGLCC7q54 [Accessed 26 September 2021].

[29] S. Mantihal, R. Kobun and B. B. Lee, "3D food printing of as the new way of preparing food: a review," *International Journal of Gastronomy and Food Science*, vol. 22, no. 100260, 2020.

[30] D. Sher and X. Tuto, "Review of 3D food printing," *Temes de Disseny*, vol. 31, pp. 104–117, 2015.

[31] A. Le Bail, B. C. Maniglia and P. Le Bail, "Recent advances and future perspective in additive manufacturing of foods based on 3D printing," *Current Opinion in Food Science*, vol. 35, pp. 54–64, 2020.

[32] L. Carolo, "3D PrintedFood: All You Need to Know," all3dp.com, 11 January 2021. [Online]. Available: https://all3dp.com/2/3d-printed-food-3d-printing-food/ [Accessed 25 September 2021].

[33] Z. Liu, B. Bhandari, S. Prakash and M. Zhang, "Creation of internal structure of mashed potato construct by 3D printing and its textural properties," *Food Research International*, vol. 111, pp. 534–543, 2018.

[34] F. Yang, C. Guo, M. Zhang, B. Bhandari and Y. Liu, "Improving 3D printing process of lemon juice gel based on fluid flow numerical simulation," *LWT*, vol. 102, pp. 89–99, 2019.

[35] C. Le Tohic, J. J. O'Sullivan, K. P. Drapala, et al., "Effect of 3D printing on the structure and textural properties of processed cheese," *Journal of Food Engineering*, vol. 220, pp. 56–64, 2018.

[36] F. Yang, M. Zhang, S. Prakash and Y. Liu, "Physical properties of 3D printed baking dough as affected by different compositions," *Innovative Food Science & Emerging Technologies*, vol. 49, pp. 202–210, 2018.

[37] M. Lille, A. Nurmela, E. Nordlund, S. M. Kortelainen and N. Sozer, "Applicability of protein and fiber-rich food materials in extrusion-based 3D printing," *Journal of Food Engineering*, vol. 220, pp. 20–27, 2018.

[38] C. Severini, A. Derossi, I. Ricci, R. Caporizzi and A. Fiore, "Printing a blend of fruit and vegetables. *New advances on critical variables and shelf life of 3D edible objects,"* Journal of Food Engineering, vol. 220, pp. 89–100, 2018.

[39] F. Chuanxing, W. Qi, L. Hui, Z. Quancheng and M. Wang, "Effects of pea protein on the properties of potato starch-based 3D printing materials," *International Journal of Food Engineering*, vol. 14, no. 3, p. 20170297, 2018.

[40] T. Anukiruthika, J. Moses and C. Anandharamakrishnan, "3D printing of egg yolk and white with rice flour blends," *Journal of Food Engineering*, vol. 265, p. 109691, 2020.

[41] L. Wang, M. Zhang, B. Bhandari and C. Yang, "Investigation on fish surimi gel as promising food material for 3D printing," *Journal of Food Engineering*, vol. 220, pp. 101–108, 2018.

[42] C. Severini, D. Azzollini, M. Albenzio and A. Derossi, "On printability, quality and nutritional properties of 3D printed cereal based snacks enriched with edible insects," *Food Research International*, vol. 106, pp. 666–676, 2018.

[43] R. S. Azam, M. Zhang, A. S. Mujumdar and C. Yang, "Study on 3D printing of orange concentrate and material characteristics," *Journal of Food Process Engineering*, vol. 41, no. 5, p. e12689, 2018.

[44] S. Zhu, M. A. Stieger, A. J. v. d. Goot and M. A. Schutyser, "Extrusion-based 3D printing of food pastes: correlating rheological properties with printing behaviour," *Innovative Food Science & Emerging Technologies*, vol. 58, p. 102214, 2019.

[45] C. F. Guo, M. Zhang and B. Bhesh, "A comparative study between syringe-based and screw-based 3D food printers by computational simulation," *Computers and Electronics in Agriculture*, vol. 162, pp. 397–404, 2019.

[46] J. Chen, T. Mu, D. Goffin, et al., "Application of soy protein isolate and hydrocolloids based mixtures as promising food material in 3D food printing," *Journal of Food Engineering*, vol. 261, pp. 76–86, 2019.

[47] J. H. Lee, D. J. Won, H. W. Kim and H. J. Park, "Effect of particle size on 3D printing performance of the food-ink system with cellular food materials," *Journal of Food Engineering*, vol. 256, pp. 1–8, 2019.

[48] H. Chen, F. Xie, L. Chen and B. Zheng, "Effect of rheological properties of potato, rice and corn starches on their hot-extrusion 3D printing behaviors," *Journal of Food Engineering*, vol. 244, pp. 150–158, 2019.

[49] M. Palaniswamy and W. Y. Leong, "3D modelling with CT and MRI images of a scoliotic vertebrae," *Journal of Engineering Science and Technology,* no. EURECA 2015 Special Issue, pp. 188–198, 2016.

[50] M. Palaniswamy and W. Y. Leong, "Evaluation on thoracic trunk shift and coronal balance in post-operative scoliosis patients," *ASM Science Journal*, vol. 11, no. 1, pp. 1–8, 2018.

[51] M. N. Nadagouda, M. Ginn and V. Rastogi, "A review of 3D printing techniques for environmental applications," *Current Opinion in Chemical Engineering*, vol. 28, pp. 173–178, 2020.

[52] J. Xiao, S. Zou, Y. Yu, et al., "3D recycled mortar printing: system development, process design, material properties and on-site printing," *Journal of Building Engineering*, vol. 32, p. 101779, 2020.

[53] C. Balletti, M. Ballarin and F. Guerra, "3D printing: state of the art and future perspectives," *Journal of Cultural Heritage*, vol. 26, pp. 172–182, 2017.

[54] A. Mitchell, U. Lafont, M. Holynska and C. Semprimoschnig, "Additive manufacturing—a review of 4D printing and future applications," *Additive Manufacturing*, vol. 24, pp. 606–626, 2018.

[55] M. Palaniswamy, A. S. Shuib, K. C. Ng and S. Koshy, "Verification of human lumbar vertebrae and intervertebral disc finite element models under mechanical forces," *ASM Science Journal*, vol. 15, pp. 1–10, 2021.

[56] M. Palaniswamy and W. Y. Leong, "Edge detection of the scoliotic vertebrae using X-ray images," *Journal of Engineering Science and Technology,* no. EURECA 2015 Special Issue, pp. 166–175, 2016.

[57] M. Palaniswamy, A. S. Shuib, N. Khai Ching, S. Koshy, K. Chinna and C. S. Lim, "Validation of finite element model of human lumbar vertebrae under mechanical forces," in *AIP Conference Proceedings*, Kuala Lumpur, 2019.

[58] S. M. Saptarshi and C. Zhou, "Basics of 3D printing: engineering aspects," in *3D Printing in Orthopaedic Surgery*, 1st ed., Elsevier, New York, NY, 2019, pp. 17–30.

Chapter 12

Augmented reality technology in smart manufacturing

Ranen Samer Alfakkih[1] *and Leong Wai Yie*[1]

12.1 Augmented reality technology

12.1.1 Introduction

AR technology helps a human to see a computer-generated virtual reality combined with the physical world, or even feel it. The "real world" is the place an observer perceives with his senses, that is, speak, hear, taste, smell or feel [1]. The "virtual world" is a real-world phenomenon generated by a computer, and created and stored on a storage medium or virtually experienced using computers, software, and telecommunication protocols to seamlessly allow the communication between hardwares to happen. In the AR application, a tracking program tracks the virtual reality with the physical world to combine simulated and actual information in a way that the user can understand [1].

Using any mixture of visual or aural or kinesthetic senses of the human user, AR technology basically "connects" a human user to a database or computer or partly virtual world [1]. Alternatively, as experienced by the actual person, AR technology can be seen as linking the partly simulated world with the real environment [1]. AR technology allows the human user to interpret and be directed by database knowledge accessed that is combined with the physical world to create an imaginative space where reality plays with false without any conflict. Furthermore, AR technology may allow a person to participate in an operation in a virtual world by translating the actions or behavior of the human user within given real-world coordinates (area or volume) into the desired response in the virtual world [1].

Visual AR technology comprises technologies such as "Visual See Through" AR and "Optical See Through" AR. In Visual See Using AR technologies, a camera captures real-world images and combines or incorporates simulated images electronically to create the expanded image. In Optical See Through AR technology, the visual images are displayed on a see-through monitor, allowing a human person to see through the window the reflected image of the visual entity and the images of the real-world objects [1].

By allowing individuals to access and apply visual information easily, reliably and effectively to a problem, a job or circumstance in the physical world, AR

[1]The Institution of Engineering and Technology, Malaysia

technology has major advantages in a culture that is being inundated with knowledge [1]. An AR interface can improve or strengthen the equipment's real-world experience with instructional content, sketches, and diagrams so that service technicians can operate on the equipment easily, reliably, and competently [1]. The AR program may be used to mimic real-world events, such as piloting aircraft or vehicles, and may also be used for testing, training, or rehearsals [1]. In fact, the device can be used to monitor the response of the head and eye to a stimulus by measuring and observing the reaction of either the head of the user or the pupil of the patient after a stimulus has been put to him [1]. Moreover, the wire harness manufacturing AR device implementation of the present technology has been shown to minimize cycle time by producing quality wire harnesses up to 50% faster than the current methods, and enabling aircraft, particularly wire harnesses, to be built and manufactured rapidly according to a customer's particular requirements. Therefore, a visual AR system should: (1) easily and reliably identify and trace input data such as coordinate marks or fiduciaries; (2) quickly analyze the input data to determine the relative location and orientation between the user and the target objects, and monitor the virtual world objects to the real-world objects; and (3) quickly and seamlessly incorporate the virtual world objects into the real-world objects; "Real Time" or "Near Real Time." "Near Real Time" requires an update rate of at least 60 Hz and a delay period of 16 ms or less [1].

Established trackers include optical-based magnetic, traveling infrared wave, light emitting diode (LED), ultrasonic, and mechanical-based trackers [1]. These trackers reliably calculate the position of the user with upgrade speeds of 20 Hz or more, but are complicated devices that operate within a very small volume and seriously restrict a user's mobility. Light source-based trackers use a camera to track different light source objects, such as clusters of LED, which are mounted on a target material surface [1].

12.1.2 Brief history

1960s: The first head-mounted monitor was developed in 1968 by Ivan Sutherland and Bob Sproull, and they named it The Sword of Damocles. Clearly, this was a rugged tool that had rudimentary machine graphics projected [2].

1970s: Myron Krueger established Videoplace in 1975 – a facility for artificial reality. The scientist envisaged human activities engaging with the interactive material. This idea was later used in several projectors, video cameras, and silhouettes on-screen [2].

1980s: In 1980, Steve Mann created EyeTap, the first handheld device intended to be placed in line with the front of the eye [2]. It later recorded the scene to transpose effects on it, and showed it to a user who could also play with it through head moves. In 1987, the concept of a heads-up display (HUD) was created by Douglas George and Robert Morris. It displayed real-life astronomical data [2].

1990s: The 1990s saw the coining of the term "augmented reality." It first emerged in the work of researchers Thomas Caudell and David Mizell-Boeing [2]. In 1992, the US Air Force's Louis Rosenberg developed an AR device titled "Digital Fixtures." In

1999, Frank Delgado and Mike Abernathy led a group of scientists to test new navigation tools, which produced runways and street data from a helicopter photo [2].

2000s: In 2000, Hirokazu Kato, a Japanese scientist, created and distributed ARToolKit – an open-source software development kit (SDK). It was later adapted to work for Adobe [2]. In 2004, Trimble Navigation introduced an AR system installed on an outdoor helmet. Wikitude created the AR Travel Guide for Android mobile devices in 2008 [2].

In 2013, Google beta tested the Google Glass equipped with Bluetooth connectivity. Microsoft launched two innovations in 2015: Windows Holographic and HoloLens (an AR goggles with lots of sensors for viewing HD holograms). Niantic released the mobile app game Pokémon Go in 2016 [2], and the software saw the gaming market explode, earning $2 million merely in the first week [2].

12.1.3 How does AR work

For many of us, the question what is Augmented Reality (AR) translates to asking how does AR work, that is how to explain it technically? A variety of data (images, animations, videos, 3D models) can be used in AR and users can see the effect in both natural as well as synthetic light. However, unlike in virtual reality (VR), users are conscious that computer vision is a projection of the real world only [2].

AR can be viewed on various devices: computers, glasses, handheld devices, smartphones, and monitors with head mounts [2]. This involves technologies such as S.L.A.M. (simultaneous localization and mapping), distance detection (in short, a sensor data measuring range to objects), and the following parts:

Sensors and cameras: Collecting and submitting details about the user experiences for analysis. Cameras on sensors search the surroundings and a computer locates physical items and creates 3D models with this data [2]. Just like Microsoft Hololens, this may be special-duty cameras or prevalent smartphone cameras to take photos and videos/visuals [2].

Processing: Finally, AR systems will behave like little machines, something that even current smartphones do [2]. They do need a CPU, a GPU, flash memory, RAM, Bluetooth/WiFi, a GPS, etc. [2] to determine distance, angle, direction, space navigation and so on [2].

Screening: a miniature projector on AR headsets, this takes digital content (processing result) data from sensors and projects onto a surface for display [2]. However, the use of predictions of AR has not yet been entirely developed for use in consumer goods or services [2].

Reflection: Any AR app includes mirrors that facilitate the visualization of visual representations through human eyes. Most have a "small curved mirror array" and others have a dual-sided mirror which reflects light to a camera and to the eye of a user [2]. The objective of such reflection paths is to execute a perfect fit of the image [2].

12.1.4 Application of AR technology

AR technology has been used in many areas covering tourism, archaeology, architecture, business, industrial and restoration sectors, education, emergency services, entertainment and recreation, and medical care [3].

Applying VR technologies to visitor attractions in the field of tourism will use smartphone cameras, computer apps and other technical tools to add actual scenes from history to recreate ancient sites. Additional details can be collected by observing scenes [3].

In archaeological research, AR equipment is also used to zoom in on artifacts in actual environments to make sure that archaeologists can locate the exact place where an artifact has been discovered [3].

AR technology gives users the opportunity to see every minute detail of any product without having to open the packaging [3].

By analyzing the product image, you not only can show certain design option details but also obtain other item image information [3].

AR technology can visually guide product owners during designing products and remind workers of the timing. The 3D imaging gives in-depth techniques to product designing, and makes products more attuned to user requirements and therefore more efficient [3]. Taking the automotive industry as an example, AR technology will allow car designers to enhance vehicle structure and make better choices between designing options by a visual presentation [3].

The use of AR technologies in the world of media has made it possible for people to get more points of view and perceptions of reality [3].

Indeed this fusion of reality and virtual is now a form of art [3].

AR technology is used and has played a significant part in public safety approaches [3]. The VR search and rescue program, for example, is fitted with an aerial camera that can combine the actual scene with the name and location of the forest road identified by geography, so as to locate the missing person more effectively [3].

The doctor may use the AR technology to identify the surgical site of the patient most precisely [3]. The AR technology can more effectively observe the developing fetus in real-time [3]. The AR technology may very well inform the patient to take the medicine on time by allowing the patient to wear the equipment that will substitute for the doctor [3].

12.1.5 Future of AR

Experts believe the AR/VR market will be worth more than $25 billion by 2025 and will be gradually growing [4]. But if you want to invest in something that does not go from $19,000 to $6,000 in 5 months, it could be a smart call to invest in AR/VR! [4].

AR has implementations in nearly every field; thus, there is scope for applying AR from healthcare services to property investment [4]. In medical procedures, it can be used on patients to exhibit their anatomy, or even during surgeries, doctors can use it for guides that require the muscle, nerve, and skeletal system displayed while operating using AR. In property investment, it will give you a preview into what a future house as your new home will look like, or how you would fit into an overpriced jacket without having to leave your room [4].

Let's not ignore the most critical breakthrough: the smart glasses that display answers when you are writing a report or review (patent pending) with immense accuracy [4].

While AR is in its teen years, it evolves quite quickly [4]. With even more professionals working on it, and other billion-dollar companies like Facebook investing in it, this sector will likely be a multi-billion dollar industry during our lifetime [4]. With things like smart glasses evolving, AR will most certainly be the next big thing to take the people in thrall changing our lives and our minds not to mention our future toward a better tomorrow [4].

12.1.6 Advantages

- Users can interact with the real world in real-time via AR technology [5].
- AR technology also improves human perceptivity [5].
- AR technology enables users to access required information Omni presently and in a timely manner [5].
- It saves time for users (doctors, surgeons, students, etc.) to search for particular topic information [5].
- AR technology recognizes free-hand gestures by users and saves large amounts of electricity [5].
- Information becomes more realistic, thanks to AR technology [5].

12.1.7 Disadvantages

- While it provides consumers and organizations with various benefits, certain drawbacks which should be removed have also been identified in [5].
- The introduction of VR technologies poses multiple security and privacy problems [5]. Because the AR technology software architecture is more vulnerable to cyber-attacks and risks [5], this is expected.
- Deployment of VR, and deployment of hardware and software is very complex [5].
- In the use of VR, real-world interaction becomes limited in the life of the person [5].
- The vision of humans is narrowed-down to sets of possibilities without real-world validation with the VR capabilities [5].
- AR applications can have an effect on human health and can cause poor vision with their continued use [5].
- Using VR will diminish the reality of the physical world [5].

12.1.8 Statistics of AR

AR technology brings visual content, lived and in real-time, into the physical world [6]. AR can merge real life with a superimposed image or projection using the camera on a handheld device or a special headgear by integrating graphics, sounds, haptic input and/or smell into the natural world as it happens. The AR development market is rising, with 2023 estimates valuing it at over USD 18 billion [6].

Unlike VR, which creates an artificial environment, AR uses the existing landscape essentially by overlaying new information on top of that [6]. In AR, the data about the surrounding real world is made available to the user via the use of headsets for information and/or interaction. By 2023, global AR headset shipments are expected to reach more than 30 million units, which is more than 12 times the amount expected in 2020 [6].

Globally, the number of smartphone AR apps is projected to exceed 2.4 billion by 2023, partially fueled by an increasing demand for AR technologies to improve digital advertising and entertainment experiences [6]. Rock festivals and sports activities are only two examples where AR technology is most widely used to enhance the consumer experience [6].

AR technology is used across a range of industries, including healthcare, public safety, gas and oil, tourism, marketing, etc. Everyone from tourists to soldiers to someone who is looking for the nearest subway stop can take advantage of this ability to put computer-generated graphics in their field of vision [6].

Expenditure on extended reality (XR) technology, including AR, is estimated by the manufacturing and construction industries alone to reach USD 35 billion by 2023 [6].

As technology continues to improve, there is no doubt that the use of XR technologies, including AR, will expand and develop as both businesses and consumers realize their full potential, with technology becoming as prominent as mobile devices are on the market today [6].

Here are the statistics of AR which are represented in Figures 12.1, 12.2, 12.3, 12.4, and 12.5.

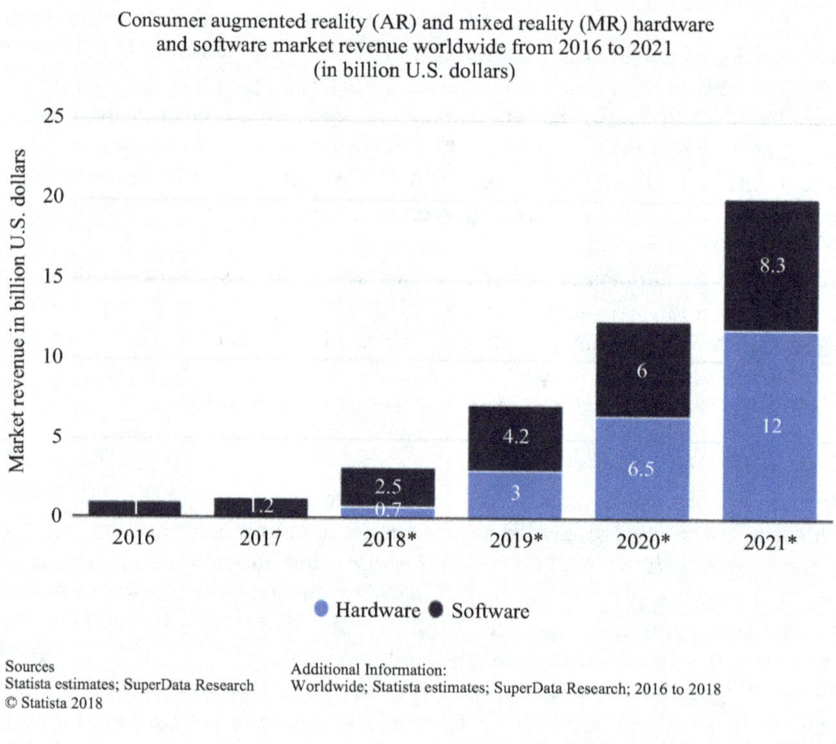

Figure 12.1 Augmented-reality-industry-statistics-by-hardware-and-software-market-size [7]

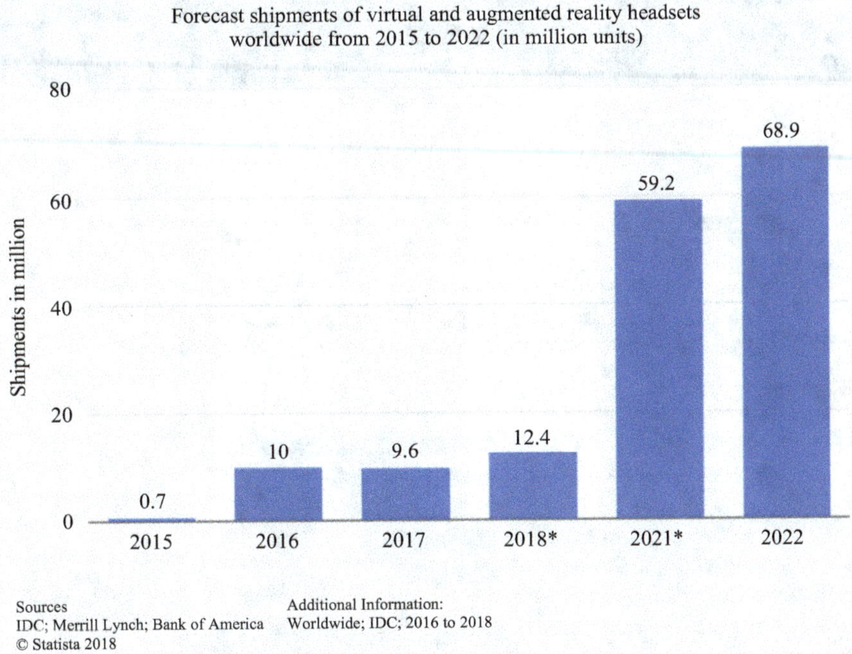

Figure 12.2 Augmented-reality-industry-statistics-by-headsets-market-size-and-forecast [7]

12.1.9 Conclusion

AR technology has gained much interest from researchers in recent years. AR technology, powered by computer vision and artificial intelligence technologies, has been showing a strong traction in progress [3]. Both the accuracies of the tracking record show equipment significantly increased efficiency and the complexity of human–computer interaction [3]. However, it can be seen that the VR System also has many challenges to address [3]. As far as tracking registration technology is concerned, the existing method of monitoring registration will use just a limited amount of information in the situation, such as feature point information, which leads to poor comprehension of the device and its context [3]. So far as computer technology is concerned, the size and price of VR glasses that can offer consumers a strong sense of submergence cannot satisfy the demand of the market. The more realistic and multi-user AR interface technology is yet to be analyzed in terms of the interaction style [3]. The use of AR technologies will make their appearance in large numbers over the next few years, especially in the implementation of mobile intelligent terminals [3]. While mobile devices are less submersible than helmet-mounted displays, they are highly common. Around the same time, the arrival of technology platforms ARKit and ARCore is theoretically

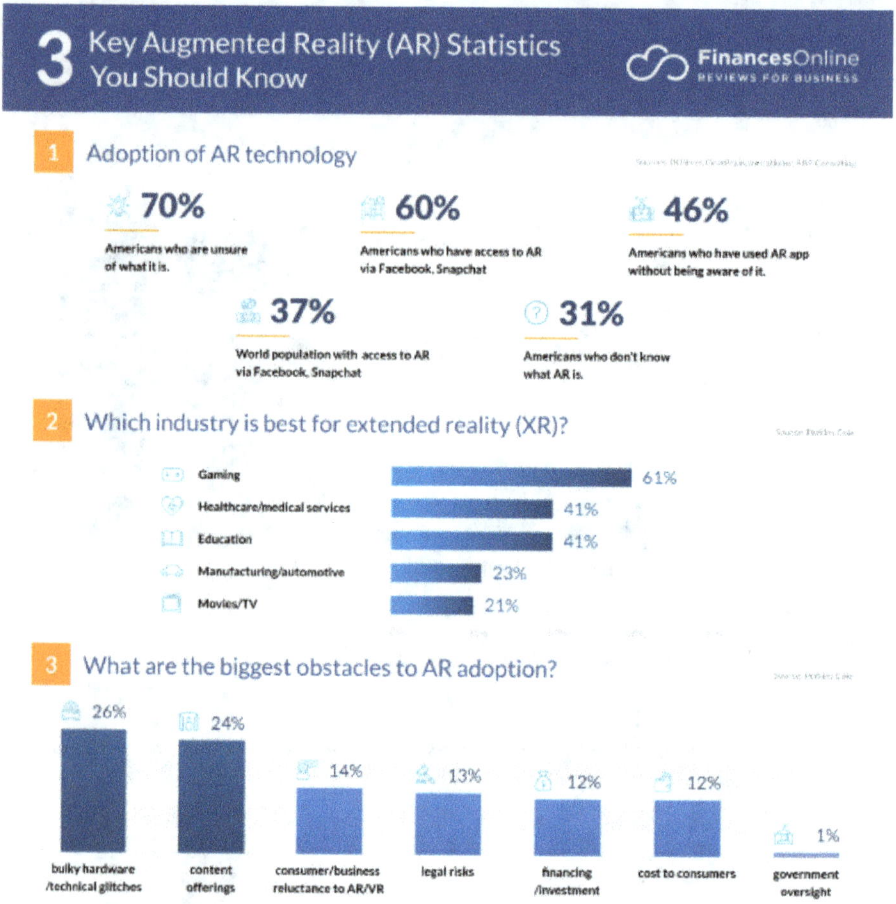

Figure 12.3 Augmented-reality-statistics [8]

contemplating the convergence between VR and smart mobile apps. Smart apps, which will offer maximum play to the advantages of AR technologies, can create a more practical convergence environment for humans in future [3]. People can communicate more easily with the machine in a human-computer context [3].

12.2 Literature review

12.2.1 Introduction

AR is an immersive simulation of a real-life environment where objects that exist in the physical world are augmented by computer-generated cognitive content, often through several sensory appliance types, including visual, auditory, haptic, somatosensory and olfactory. AR can be described as a device that fulfills three basic

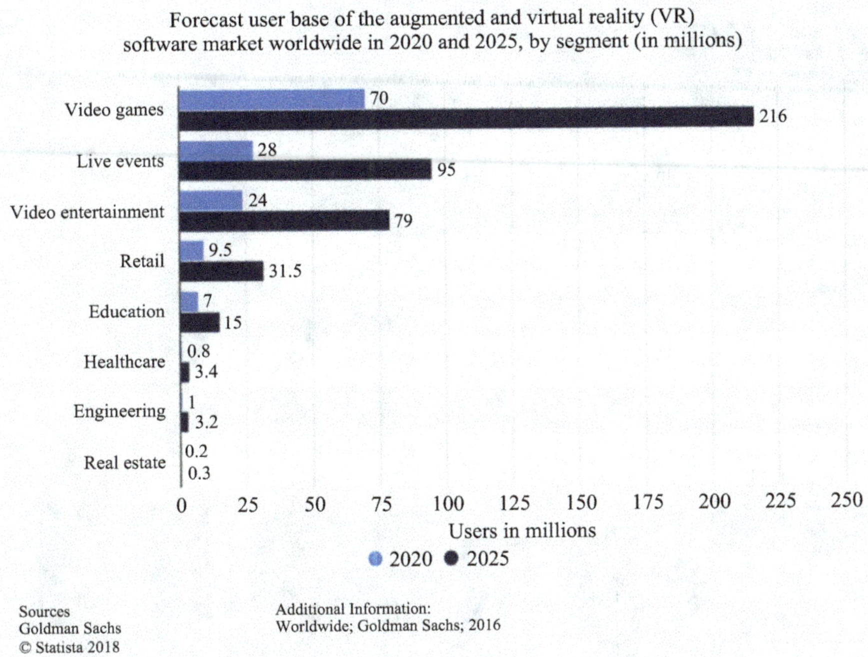

Figure 12.4 Worldwide-forecast-augmented-and-mixed-reality-software-users-by-segment [9]

features: a mixture of real and virtual environments, real-time interaction, and precise 3D visualization of virtual and physical objects. Overlaid sensory perceptions may be constructive (i.e. additive to the natural world) or destructive (i.e. masking the natural world). This perception is effortlessly interwoven with both the physical world in such a way that it is viewed as an immersive part of natural life. In this way, AR changes one's current experience of a real-world environment, while VR totally substitutes the user's real-world environment with a digital one. Increased reality is linked to two largely synonymous terms: mixed reality and computer-mediated reality [11].

The primary benefit of AR is the manner in which the elements of the physical world merge into a person's understanding of the real world, not as a static representation of details, but by the incorporation of sensory stimuli that are experienced as natural aspects of the environment. The first practical AR systems to provide consumers with immersive mixed-reality environments were developed in the early 1990s, beginning with the US Interactive Fixtures program: the Armstrong Workshop of the Air Force in 1992. Commercial AR experience was first used in television and gaming companies. Consequently, virtual and AR technologies spanned consumer markets such as technology, media, medicine, and entertainment. In school, content can be viewed by downloading or displaying a picture using a mobile device or by using marker-less AR methods. [11].

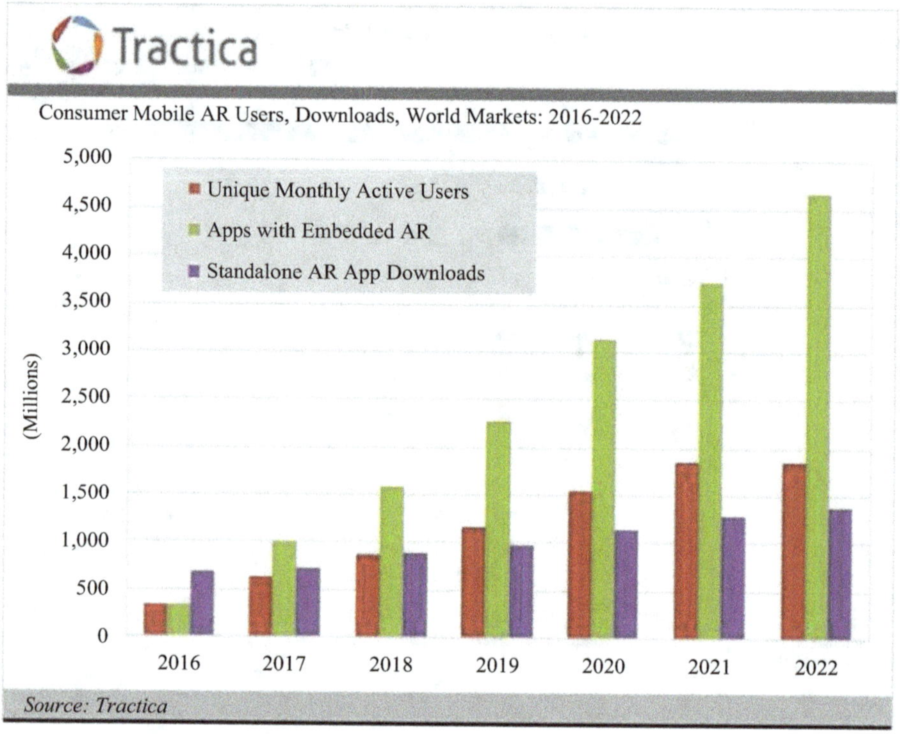

Figure 12.5 Consumer mobile AR users, downloads, world market [10]

12.2.2 AR in education

The importance of AR in the sphere of education is closely related to how AR is structured, incorporated and integrated into academic and nonacademic environments. An important consideration is how AR technologies support and provide meaningful learning. Taking into consideration AR as a term rather than just a particular form of technology would be beneficial for educators [2]. The participation of educators is critical to promote the creation of favorable AR technologies for teaching, which increases the opportunity for AR to be integrated into education. Implementations for AR have been established in many areas of education [12].

A few of these application areas for AR were used in previous research. Gopalan *et al.* studied the effect of improved AR science textbooks on lower secondary school students in Malaysia. Chiang *et al.* examined the use of the AR-based virtual learning program for natural science testing practices for fourth-grade students in Taiwan. The system guides students to the specific areas of ecology and displays the correlating learning process or related study guides. Akçayır *et al.* tested the need for an enhanced AR laboratory manual in science laboratories for first-year students in Turkey [2]. The research investigated the effect of the Anatomy 4D smartphone application, mostly, on learning satisfaction of undergraduate college students in medical sciences at the University of Cape Town (UCT) [2]. Figure 12.6 depicts AR in education.

12.2.3 Advantages of using AR in education

AR opens up unique opportunities to communicate with the physical world and can construct interactions that would not bc available with either a truly physical or a virtual environment. Given AR has a unique ability to build immersive hybrid learning environments which combine real and virtual objects [2], thus it follows

(a)

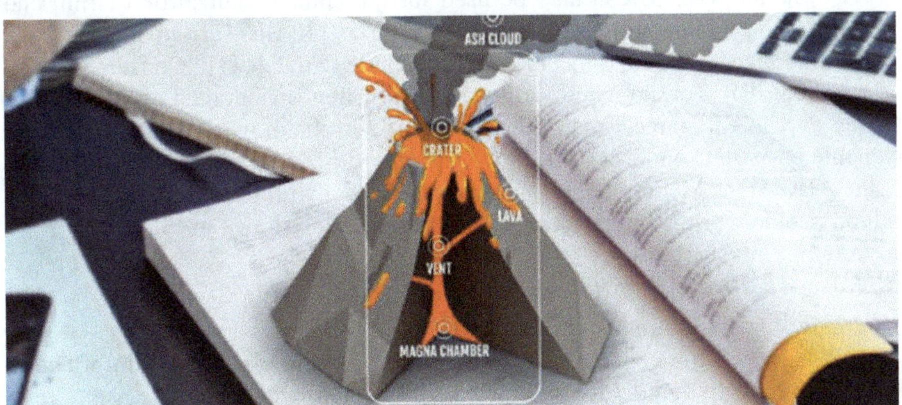

(b)

Figure 12.6 The implementation of AR in teaching and learning (a) and (b) [13,14]

AR technology enables participants to perceive scientific occurrences that are not realistically possible, with the exception of certain chemical reactions, making unreachable subjects available to students. The simulation of abstract structures and the detection of events that are impossible to detect in the physical world can be enabled by AR [2]. This method of learning environment will promote cognitive skills and improve the intellectual awareness of anomalies that are either intangible or impossible to detect, as well as clarify misconceptions [2]. AR tackles cognitive problems often faced by visualizing unknowable phenomena [2].

The expertise, knowledge, and experience that students build through technology-enhanced learning experiences can be more effectively built by AR technology [2]. The cognitive responsibility can be whittled down by combining different sources of data. The integration and communication features offered by AR could well encourage students to participate in active learning and might even strengthen the motivation and interest to learn [2]. AR offers highly immersive experiences that can produce genuine learning progress, interactivity and a high degree of realism. Communication with the world is significant to this study and, other than reality, AR is also one of the best ways to facilitate all these interactions [2].

12.2.4 AR in video game

The gaming industrial sector has already implemented AR technologies. A variety of games have been created for engineered confined areas, such as AR air hockey, Titans of Space, multiplayer battles against simulated opponents, and AR-enhanced pool table games [11].

AR has encouraged video gamers to witness digital gaming experiences in a real-world environment. Niantic has launched the AR video game Pokémon Go [1]. Disney has collaborated with Lenovo to develop the AR game Star Wars: Jedi Challenges, which functions with the Lenovo Mirage AR headset, the monitoring system and the Light saber joystick, and was slated to be released in December 2017 [11].

AR Gaming (ARG) can also be used for the commercialization of film and television amusement properties. On 16 March 2011, Bit Torrent introduced a freely authorized edition of Zenith in the United States [1]. Consumers who accessed the Bit Torrent client software were often encouraged to access and convey Part 1 of its three parts of the film. Part 2 of the movie was made publicly available on VODO on 4 May 2011 [1]. The episodic video release, complemented by the ARG Trans Media Publicity Strategy, caused a viral impact and more than a million people worldwide downloaded the film [11], as shown in Figure 12.7.

12.2.5 AR in healthcare

AR and its application in healthcare [16]

AR is indeed the concept of superimposing computer-generated information across a live world view. AR combines digital content with the participant's real-time experience which is becoming more available and inexpensive for medical training and photography, dental care and nurse training. AccuVein, VIPAR, ARnatomy, and VA-ST will be among the AR possibilities for healthcare [16].

Figure 12.7 AR in video games [15]

Dentistry – AR software installed into smart glasses – replicates real-time data immediately from the dental scanner, allowing a dentist to construct accurate crowns or caps [16].

Training nurses – AR-enabled, device-based simulation models of different patient circumstances – allow nurses to effectively communicate with patients (e.g. SimMan) and manage daily job situations that require a mixture of social, practical and team skills [16].

Medical imaging – CT or MRI data – which AR improves the visualization of by overlaying stereoscopic representations during surgery [6]. This information is crucial in surgical procedures requiring precise accessibility of the particular organ. For example, AR can also be used for pre-operative preparation to improve the precise location of tumors and surrounding structures in processes such as minimally invasive partial nephrectomy or radical prostatectomy in which challenging the anatomy of the vascular or nervous system may exacerbate the disposal of the tumor [16].

Medical education – to teach medical students about fundamental anatomy and notions – led professors to use AR; doctors use these to learn about experimental treatments and medications. For example, ARnatomy uses optical characters detection software to view classroom photographs and add visual data on individual body components to enable students know the names of bones and muscles. The aim is to replace cumbersome textbooks and charts with more live feed [6].

Pediatric MRI Evaluation – that tests children 's capacity to lay down for a period of time before an MRI test – has put Current Studios to work and who have created an AR tablet game. Physicians use it to decide if an infant may need an anesthetic for an MRI treatment [16].

Helping the visually impaired – to improve the optical perception of ordinary objects by individuals – VA-ST SmartSpecs has utilized 3D detection tools to do the requisite. It assists legitimately people who are blind or those with severe visual disabilities to recognize familiar faces, find lost objects, and traverse their environment easily [6].

Peripheral Vasculature Visualization – enabling clinicians to check for venous disease – has led to AccuVein AV400 electronically displaying a real-time vasculature map on the exterior of the skin so that patients benefit from less pain throughout venipuncture operations.

Remote Surgical Expertise – VIPAR (Virtual Interactive Presence in AR) – is a video assistance approach that exceeds telemedicine. During the operation, a surgeon directly advises a peer by putting his hands onto an AR monitor [16].

12.2.6 Healthcare-focused AR apps

EyeDecides: Use a digital camera to show the effect of different conditions on a person's sight to inform patients experiencing cataracts and age-related macular degeneration [6].

DoctorMole: Helps physicians to assess suspected moles throughout the body and provide immediate feedback that is useful in the identification of malignant lesions [6].

Healthcare App by pixelbug: Helps the doctor to immediately understand the operating or mechanisms involved of a modern medical device [6].

Anatomy 4D: Provides an immersive, 4D interpretation of human anatomy to medical practitioners, students, and teachers with respect to the spatial connections of internal organs, skeletons, and muscles [6].

MEVIS Surgery App: Created by Fraunhofer MEVIS, it provides virtual 3D rebuilding of real organs. Surgeons use this to execute liver enucleation using iPad overlay images that assist to locate blood vessels or tumors [6].

MedicAR: Google Glass software for surgeons to better locate incision points with reduced damage to the patient. It is intended for use in complex surgeries and MRI studies. [16] [12].

12.2.7 AR technology is in nascent stages of market penetration

The AR market is at an early phase of advancement but is expanding quickly with marked progress significantly in the last two years. Numerous AR materials from different fields and, thus, more accessible user solutions are predicted within the next five years. AR Enterprise applications and memberships will increase the revenue of start-ups [2].

Several more industries are bound to take to AR only as a major source of multisensory training and regular functioning guidelines for the protection of the labor force. Medical education and training of personnel are, by the needs of the medical fraternity, to remain the primary use of AR in healthcare [12].

Behavioral and psychological medical complications are tackled by AR innovations that create exciting and empowering environments for medical intervention purposes; apart from interactive innovations also promoting the understanding between patients and their clinicians, thereby helping to identify and manage diseases [7]. For example, the USC Center for Body Computing has developed a Virtual Care Clinic method that integrates AR, analytics, and artificial intelligence technology with mobile apps, wearable sensors and virtual human health providers to give patients complete and accurate access to medical treatment and information from anywhere in the world [12]. The program features an interface for linking patients to their primary physician's persona, which directs patients through the various courses of their medical treatment [17], as shown in Figure 12.8.

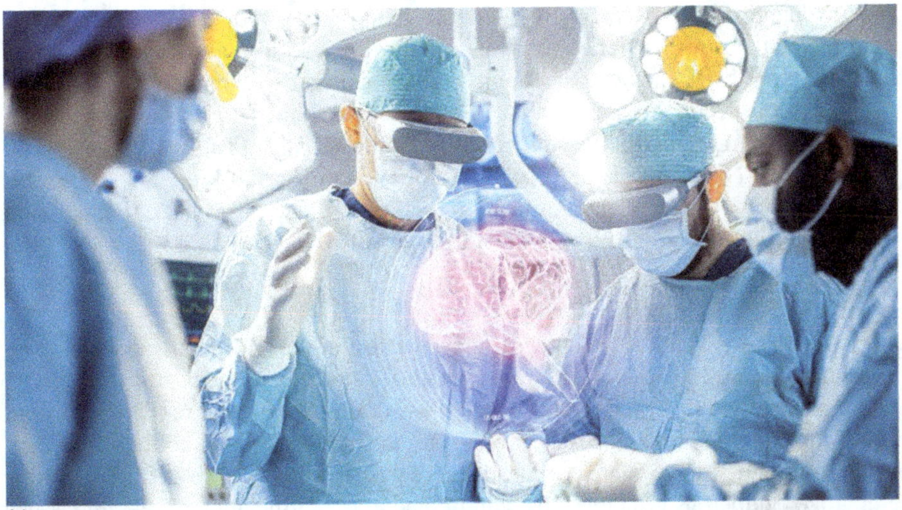

(a)

(b)

Figure 12.8 (i) AR in healthcare [18]. (ii) AR in healthcare [19]

12.2.8 AR in glasses

AR is mainly experienced through wearable glass devices, head-mounted devices, or smartphone applications. AR superimposes digital information above the real world. Thus, AR enhances the user's perspective in the real world instead of just replacing it [6].

Not to be misunderstood with VR or fused reality, although for reasons they both come within the extended spectrum of reality, AR is considered to also have the largest potential for widespread use [20]. Here's our selection of some of the best AR glasses available in the market, as shown in Figure 12.9.

12.2.9 AR in business

The introduction of AR to industries is likely to shift the importance from how much production to how many workers are being employed to achieve that produce. Specialists in this field should be able to provide live assistance to remote personnel who will show signs, point out problems, superimpose templates on objects such as vehicle engines and the like, and more [7]. AR erases noise and supposition from manual labor and simplifies processes [7] (Table 12.1).

Example: NASA used Microsoft's HoloLens to help build its new spacecraft, using mixed reality implementations to erase any need for user manual and sophisticated instructions [7], as shown in Figure 12.10.

12.2.10 Conclusion

This research aims to understand the significant effect of the AR mobile application on the academic performances of UCT undergraduate students in medical sciences. The literature suggested that there is still a lack of work on the effect of mobile AR use in education, and there is scope to investigate the ability of AR to boost student engagement and lead to increased academic achievement [2]. The literature review summarized numerous principles that contributed to the formulation of research issues with a need to know the focus, importance, optimism and fulfillment (ARCS) motivational architecture model. AR was described as integrating physical and virtual realities, complementing the real world with computer-generated virtual images in real-time, and how AR was clarified in the sense of education. Mobile AR has been addressed because AR can be conveniently used across mobile devices. The concept required the use of the Anatomy 4D smartphone device as an interactive AR aid [12].

The literature review looks into the use of AR in education supplemented by a summary of several of the prior research using AR applications [11]. Various benefits and drawbacks for the use of AR in education have also been discussed. Encouragement in the learning context and the underlying motivational theories used to understand the encouragement of learning were addressed. This spurred on ahead the main research question (RQ) (RQ 1): What are the points of distinctiveness in ambition for learning students before and after using the AR mobile application? The ARCS model has been used to react to the RQ 1. The ARCS model and each ARCS feature were used to fully comprehend the effect of AR technology on student encouragement for learning. The instructional materials motivation survey (IMMS) was used to create tools in this research [12].

(a)

(b)

Figure 12.9 (i) AR in glasses [21]. (ii) AR in glasses [22]

The methods and design of the research were mentioned, and also the methodology of data collection and analysis were delineated. Statistical analysis was used to describe the research results to address the research questionnaire. The findings of this study have shown that the use of a smartphone AR software has improved the learning fulfillment of undergraduate medical sciences students at the UCT [2].

Table 12.1 Gap analysis of different Augmented Reality transformer

References	Author	Data transformer	Gap analysis
[23]	Martín-Gutiérrez *et al.*	The study looked at the emergence of technology developments in education from 2004 to 2014 via a systematic review of the Horizon AR Reviews and several other technology-boosting learning topics	Positive effect findings include: improved understanding of the concepts, learning spatial systems, language organizations, long-term memory retention, enhanced collaboration and encouragement. The detrimental effect findings are: publicity parallelization, problems with accessibility, unproductive classroom convergence, disparities between learners
[24]	Saggio and Ferrari	Normal visualization and communication. Spatial resolution and photo-reality	More pragmatic, need to merge scenario with real world
[25]	Garrett *et al.*	Study of pilots, mixed techniques.-Exploration of emerging smartphone AR tools has the ability to improve clinical skills learning in the laboratory	Medical test equipment definitions by the use of bar codes / QR codes.AR tools have been built into a clinical modeling situation
[26]	Frank L. Baum's	The essential concept of this article is that smart glasses can become a means of creating corporate value for businesses and also for social system as a whole	There is a need to clearly differentiate between public and private value creation variables. External value formation variables are linked to aspects
[27]	Wang *et al.*	Evolving the beauty market and providing interactive support to consumers who do not want to leave their own homes to check out the best in fashion and makeup	This precision of AR makes it possible to research it more methodically

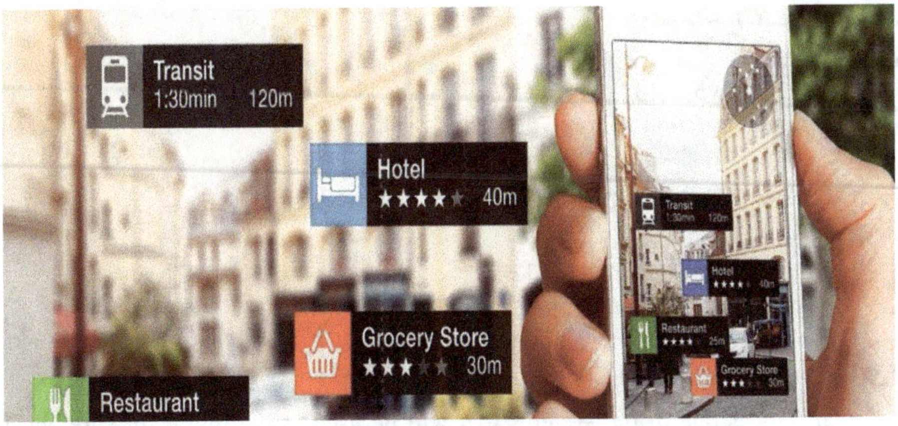

Figure 12.10 AR in business [28]

The findings are associated with prior research performed by Di Serio *et al.*, Chiang *et al.*, Gopalan *et al.*, and Solak and Cakir. The overall mean value acquired for the recent method of teaching was 3.05 and the overall mean value for the use of Anatomy 4D mobile app was 3.49. Increased attention, confidence and sense of fulfillment factors were identified following the use of the Anatomy 4D mobile app, whereas the pertinence factor decreased [11,12].

12.3 Industry Revolution 5.0

12.3.1 Introduction

The fifth industrial revolution will be centered around the co-activity between man and machine, as human knowledge works in agreement with cognitive figuring [29], by returning people to modern creation with collaboration robots [29]. Workers will be upskilled to provide esteem-included assignments in production, prompting mass customization and personalization for clients [29].

With the ongoing headway in smartphone and wearable AR headset showcase, a splendid future for AR is guaranteed in the Tourism Industry [29]. Likewise, AR has just begun to throw the Sports Industry into a tizzy as new experiences arrive fast and strong [29]. As per latest news about AR innovation, a market insight organization, named ABI research, says that by 2020, AR specialty showcase is evaluated to arrive at a value of US$ 100 billion [29].

So that means this technology may enhance more in the future and be one of the technologies widely used in 2030 [29].

12.3.2 AR technology characteristics

- Overlay of the genuine world with the computerized world [30].
- Ongoing association [30].
- Enlistment and Alignment in 3D [30].

12.3.3 AR features

- Enhance the real world [31].
- Realize the world [31].
- Solution suggestion to some problems [31].

12.3.4 Future system standards include

- Functional requirement [32].
- Performance characteristics [32].
- Test evaluation criteria [32].

12.3.5 5G smart city

5G smart city is one of the systems that will be used in the future for sure and I think they will start working on it from now [33]. Because smart city uses different technologies like IoT (Internet of Things), AR and AI [33], as shown in Figure 12.11.

12.3.6 Future of commercial transportation – Toyota e-Palette

Toyota was available at the CES 2018 Expo with another idea called Toyota e-Palette Concept [34]. As per the vehicle creator, this electric vehicle and programmed guiding framework is a dream of things to come from independent portability administrations [34].

The new idea of the Toyota e-Palette Concept has been created from a joint effort between specialists from a few organizations, for example, Amazon, DiDi,

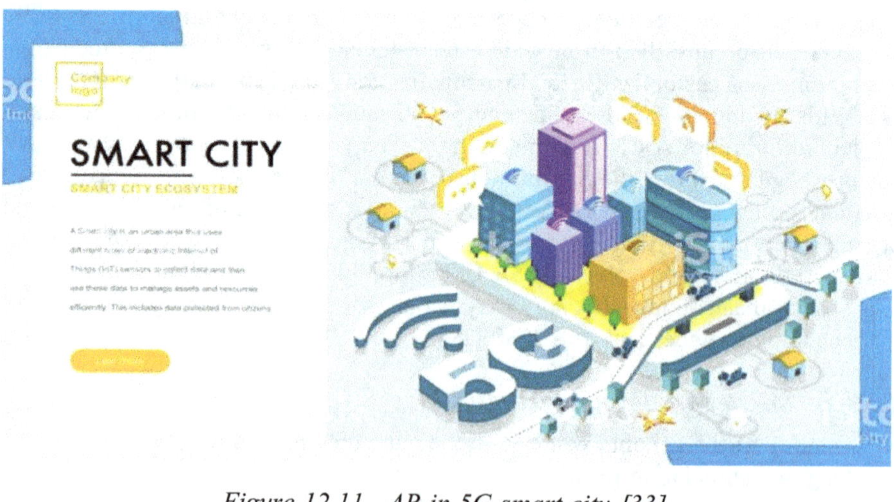

Figure 12.11 AR in 5G smart city [33]

Figure 12.12 AR in commercial transportation [34]

Mazda, Pizza Hut and Uber. The model brought to CES has the state of a rectangular "box" on wheels [34], as shown in Figure 12.12.

12.3.7 Applications

12.3.7.1 Pokémon Go

One of the famous applications that captured people's attention and I was one of them [35], this game gives you the reason to go out and walk looking for the Pokémon and catch them [35]. It uses GPS to know your location to move your avatar game in the right place while the smartphone camera is on to show the Pokémon in the real world [35]. So, it gives you the feeling that you are really playing in the real world in your home or town and that's what is special about using AR. You feel the experience [35], as shown in Figure 12.13.

12.3.7.2 Amikasa

One of the apps that save your money and time by offering you the best layout for your bedroom, kitchen or any other place in your house [37], this app is using your phone Cam to give you a clear vision of what you want including the color of the room and its 3D model [37], as shown in Figure 12.14.

12.3.7.3 YouCam Makeup

Who doesn't love makeup?! No one. This application allows you to try makeup on from any brand to see if it fits you, or not, and then to go and buy it if it satisfies your requirements. Especially now, with COVID-19, when we cannot go out and try any makeup product due to the safety requirement, this app reads your skin type using AR technology and gives you the best lip color or concealer and so on [37], as shown in Figure 12.15.

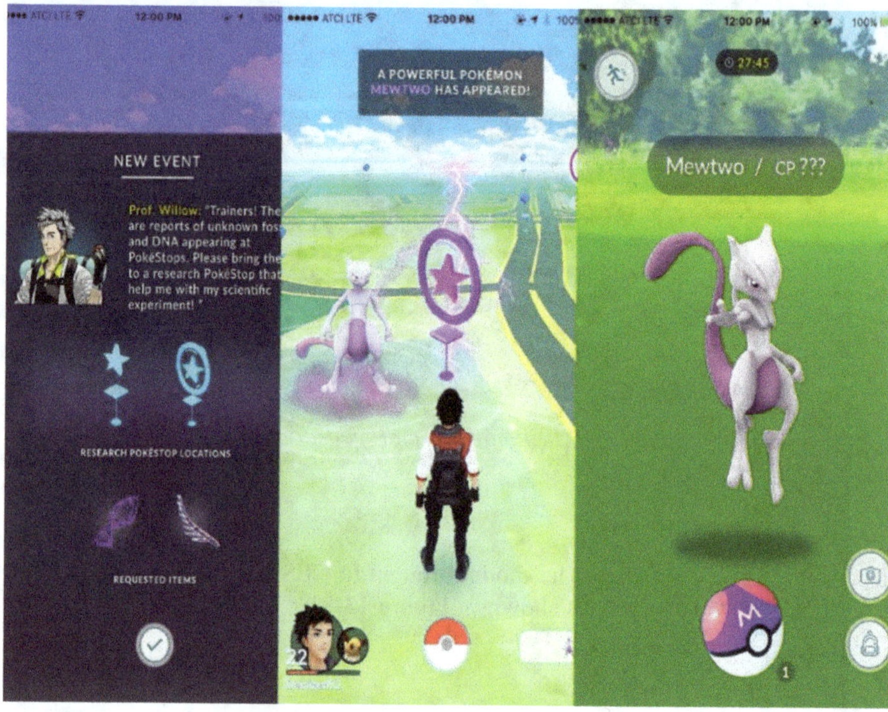

Figure 12.13 AR in Pokémon Go [36]

Figure 12.14 AR in Amikasa [37]

12.3.7.4 Houzz

Again another great app for home layout and furniture sellers, this app is one of the top ones if you want to plan interior layout design [37]. The main thing in this app is it gives you the option to see what your home may look like and to buy what you

Figure 12.15 AR in YouCam makeup [37]

Figure 12.16 AR in Houzz [37]

have designed by allowing the user to browse and buy the products [37], as shown in Figure 12.16.

12.3.8 Complete anatomy

Visually, the human body is complex detailed and to learn more about every part of your body [38], This app will be very helpful to the physicians and also the patients. It will help them to get more information about the recovery of the disease-afflicted human body [38], as shown in Figure 12.17.

12.3.9 Conclusion

AR is regarded by many technology experts as the next big thing on the block because it is seen to be achieving popularity in nearly every area it is used [32].

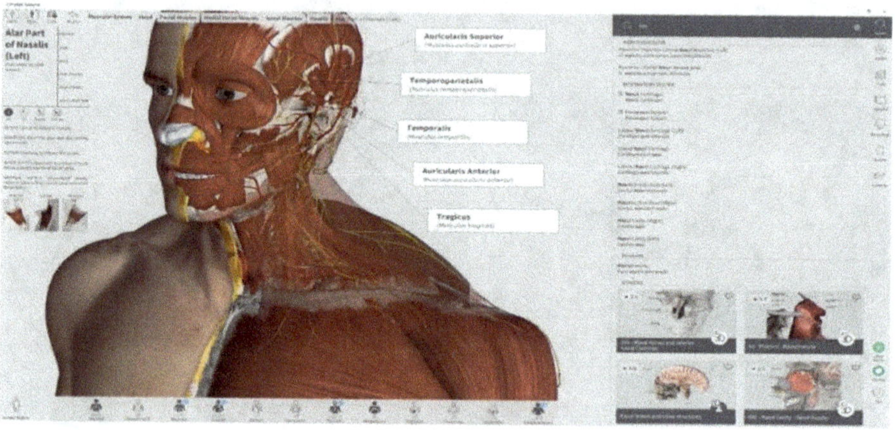

Figure 12.17 AR in anatomy [38]

With the aid of AR solutions, almost any problem can be solved readily [32]. When it comes to application development using AR, it is clear that it provides such great opportunities to delight the user experiences [32]. By incorporating the use of AR, we can bring more innovation in multiple fields, particularly given the possibility of enhanced production of applications [32].

The UX designers have already begun working to improve their designs using developments in AR technology [32]. Without a doubt, this AR development can give UX experts a lot of advantages. UX professionals are trying to improve the devices' ability to make use of their advances in the IT world [32].

So, this year is full of surprises waiting to show up with an increase in any object's overall performance, reliability and performance using AR technology [32]. Brace yourself to see AR technology's development on its own or in combination with AR and artificial intelligence [32].

References

[1] W.P.D.R.R. Jeffrey Meisner, Augmented Reality Technology, 6 April 2019. [Online]. Available at: https://patents.google.com/patent/US6625299B1/en [Accessed 13 May 2020].

[2] N. Kovach, What is Augmented Reality (AR) and How Does It Work, 24 January 2020. [Online]. Available at: https://thinkmobiles.com/blog/what-is-augmented-reality/#:~:text=Augmented%20reality%20is%20the%20technology,real%20with%20a%20virtual%20one [Accessed 13 May 2020].

[3] Y. Chen, Q. Wang, H. Chen, *et al.*, An overview of augmented reality technology, *Journal of Physics: Conference Series*, 2019;1237(2): 6,.

[4] A. Islam, The Future of Augmented Reality, 11 December 2018. [Online]. Available at: https://medium.com/predict/the-future-of-augmented-reality-90143b98f7a3 [Accessed 14 May 2020].

[5] J. Charles Wilson, What are the Advantages and Disadvantages of Augmented Reality?, 5 February 2019. [Online]. Available at: https://assignmenthelp4me.com/discussions/what-are-the-advantages-and-disadvantages-of-augmented-reality.html [Accessed 14 May 2020].

[6] Statista Research Department, Augmented Reality (AR) - Statistics & Facts, 9 March 2020. [Online]. Available at: https://www.statista.com/topics/3286/augmented-reality-ar/ [Accessed 14 May 2020].

[7] B. Gaille, Augmented Reality Industry Statistics, Trends & Analysis, 19 July 2018. [Online]. Available at: https://brandongaille.com/13-augmented-reality-industry-statistics-trends-analysis/ [Accessed 14 May 2020].

[8] N. Gilbert, Vital Augmented Reality Statistics: 2020 Market Share & Data Analysis, 12 November 2019. [Online]. Available at: https://financesonline.com/augmented-reality-statistics/ [Accessed 13 May 2020].

[9] S. Greene, 7 Magento 2 E-Commerce Trends That Add Values To Sites In 2020, 11 November 2019. [Online]. Available: https://bsscommerce.com/blog/ecommerce-trends-2020-with-magento-2/ [Accessed 15 May 2020].

[10] TRACTICA, Mobile Augmented Reality Market to Reach 1.9 Billion Unique Monthly Active Users by 2022, 17 April 2017. [Online]. Available at: https://tractica.omdia.com/newsroom/press-releases/mobile-augmented-reality-market-to-reach-1-9-billion-unique-monthly-active-users-by-2022-according-to-tractica/ [Accessed 14 May 2020].

[11] D. Weibel and B. Wissmat, Immersion in computer games: the role of spatial presence and flow, *International Journal of Computer Games Technology* 2012;2011(05):14.

[12] T. Khan, K. Johnston, J. Ophoff, The impact of an augmented reality application on learning motivation of students, *Advances in Human-Computer Interaction* 2019;2019(03):14.

[13] Alamy, Industrial 4.0, Augmented Reality Concept. Hand Holding Tablet with AR Service, Thermal Monitoring Motor Application for Check Destroy Part of Smar, 25 November 2016. [Online]. Available at: https://www.alamy.com/stock-photo-industrial-40-augmented-reality-concept-hand-holding-tablet-with-ar-141527334.html [Accessed 26 May 2020].

[14] blippar, AR for Education, 29 December 2018. [Online]. Available at: https://www.blippar.com/build-ar/augmented-reality-education [Accessed 26 May 2020].

[15] D. Herrera, A 3D Game In Augmented Reality, October 2018. [Online]. Available: https://www.udemy.com/course/3d-game-development-learn-to-build-duck-hunt-ar/ [Accessed 25 May 2020].

[16] U.-V. Albrecht, U. von Jan, J.Kuebler, *et al*. Google glass for documentation of medical findings: evaluation in forensic medicine, *Journal of Medical Internet Research*, 2014;2014:16.

[17] T. Rai, Augmented Reality Examples: 10 Industries Using AR to Reshape Business, 3 September 2019. [Online]. Available at: https://www.clickz.com/augmented-reality-examples-10-industries-using-ar-to-reshape-business/214953/ [Accessed 28 may 2020].

[18] P. Dialani, Future Prospects of Augmented Reality Inhealthcare, 30 November 2019. [Online]. Available at: https://www.analyticsinsight.net/future-prospects-of-augmented-reality-in-healthcare/ [Accessed 27 May 2020].

[19] rnr, Qualitative Report on Augmented Reality in Healthcare Market Growth, Trends, Demand, Share, Analysis and Market Survey 2026 by Google, Microsoft, Daqri, Psious, Mindmaze, Firsthand Technology, 19 February 2020. [Online]. Available: https://www.bccourier.com/qualitative-report-on-augmented-reality-in-healthcare-market-growth-trends-demand-share-analysis-and-market-survey-2026-by-google-microsoft-daqri-psious-mindmaze-first-hand-technology/ [Accessed 27 May 2020].

[20] Interaction Design Foundation, Augmented Reality – The Past, The Present and The Future, 25 september 2019. [Online]. Available at: https://www.interaction-design.org/literature/article/augmented-reality-the-past-the-present-and-the-future [Accessed 26 may 2020].

[21] Brochesia, Augmented Reality, 20 September 2019. [Online]. Available at: https://www.brochesia.com/what-is-augmented-reality/ [Accessed 20 May 2020].

[22] J. Horwitz, Apple's AR Glasses Deserve an Apple Watch-Style 'early access' launch, 31 January 2020. [Online]. Available at: https://venturebeat.com/2020/01/31/apples-ar-glasses-deserve-an-apple-watch-style-early-access-launch/ [Accessed 26 may 2020].

[23] J. Martín-Gutiérrez, C. E. Mora, B. Añorbe-Díaz, and A. González-Marrero. Virtual technologies trends in education. *Eurasia Journal of Mathematics, Science and Technology Education*, 2017;13(2): 469–486.

[24] G. Saggio and M. Ferrari. New trends in virtual reality visualization of 3D scenarios, in Virtual Reality – Human Computer Interaction. London, United Kingdom: IntechOpen, 2012 [Online]. Available at: https://www.intechopen.com/chapters/38742 doi: 10.5772/46407.

[25] B. M. Garrett, J. Anthony, and C. Jackson. (2018) Using mobile augmented reality to enhance health professional practice education, *Current Issues in Emerging eLearning*, 2018;4(1) , Article 10. Available at: https://scholarworks.umb.edu/ciee/vol4/iss1/10.

[26] L. Frank Baum. The Wonderful Wizard of Oz, CreateSpace Independent Publishing Platform; Anniversary edition (February 3, 2014), ISBN-10: 1495421864.

[27] Y. Wang, E. Ko, and H. Wang. Augmented reality (AR) app use in the beauty product industry and consumer purchase intention. *Asia Pacific Journal of Marketing and Logistics*, 2022;34(1), 110–131.

[28] IQUII, Augmented Reality and Brand Experience: Examples and Business Opportunities for Enterprises, 22 August 2018. [Online]. Available at: https://medium.com/iquii/augmented-reality-and-brand-experience-examples-and-business-opportunities-for-companies-6dc6e2ad7f38 [Accessed 26 may 2020].

[29] Manufacturing Business Technology, 5 Reasons Why AR Is the Future of Smart Factories, 13 March 2019. [Online]. Available at: https://www.mbtmag.com/home/article/13250129/5-reasons-why-ar-is-the-future-of-smart-factories [Accessed 11 June 2020].

[30] S. Kumar, Amazing Facts About Augmented Reality: Your TechWorld, 12 March 2016. [Online]. Available at: https://www.augrealitypedia.com/7-amazing-facts-augmented-reality-tech/ [Accessed 12 June 2020].

[31] D. Moon, 'AR as a feature' is the Best Use of AR Tech Right Now, and Amazon Gets It, 9 January 2018. [Online]. Available at: https://venturebeat.com/2018/01/09/ar-as-a-feature-tech-amazon-view/ [Accessed 12 June 2020].

[32] S. Ratnottar, Augmented Reality (AR) Trends: The Past, Present & Future Predictions For 2019, 19 Febuary 2019. [Online]. Available at: https://towardsdatascience.com/augmented-reality-ar-trends-the-past-present-future-predictions-for-2019-8e1148345304 [Accessed 12 June 2020].

[33] ZooNo3,5G Smart City Future, 15 October 2019. [Online]. Available at: https://www.shutterstock.com/image-vector/5g-smart-city-future-abstract-metropolis-1544137442 [Accessed 10 June 2020].

[34] A.J. Hawkins, Toyota's 'e-Palette' is a Weird, Self-Driving Modular Store on Wheels, 18 January 2018. [Online]. Available at: https://www.theverge.com/2018/1/8/16863092/toyota-e-palette-self-driving-car-ev-ces-2018 [Accessed 11 June 2020].

[35] M. Jansen, The Best Augmented Reality Apps for Android and iOS, 9 June 2019. [Online]. Available at: https://www.digitaltrends.com/mobile/best-augmented-reality-apps/ [Accessed 12 June 2020].

[36] A. Choi, Pokémon GO!—Mewtwo Event Concept, 17 January 2017. [Online]. Available at: https://medium.com/@AlbertChoi92/pok%C3%A9mon-go-mewtwo-event-concept-6dfa72789cb2 [Accessed 13 June 2020].

[37] A. Sheehan, Best Augmented Reality Apps in 2020, 24 October 2019. [Online]. Available at: https://www.oberlo.com/blog/augmented-reality-apps [Accessed 15 June 2020].

[38] Pradeep, Popular iOS App Complete Anatomy now Available for Download from Windows Store, 15 March 2017. [Online]. Available at: https://mspoweruser.com/popular-ios-app-complete-anatomy-now-available-download-windows-store/ [Accessed 13 June 2020].

Chapter 13
Extended reality on smart manufacturing

Dhasaani Raj and Leong Wai Yie

Extended reality (XR) is expected to form brand new world with a completely different business experience [1]. The main reason is that most people carry their smartphones with them [1].

"Extended Reality" (XR) depicts a comprehensive range of enhanced digital experiences and physical experience: covering augmented reality (AR), virtual reality (VR), and mixed reality (MR) [1]. AR, in particular, has been obtained from a simple "take a photo, get information" application (such as Wine Searcher), developed into a sports game digital messages and markers super imposed on the screen during the event relay [1].

By 2020, 3.4 billion mobile phones will provide more advanced AR functions. This means that billions of shoppers can use digital integrated content to understand physical properties products, make smarter shopping decisions [1]. Currently, nearly 1/3 (32%) of consumers use AR applications, mainly concentrate on games and social media [1]. However, according to analysis agencies such as IDC forecast, consumer AR/VR spending is expected to grow steadily in the next 5 years. The compound annual growth rate will exceed 52% [1].

For example, the leading online platform for interior design Houzz recently launched an AR feature that enables shoppers to use virtual technology at a real scale lay the ground with tiles to preview the effect of the decoration [4]. Houzz CEO Adi Tatarko pointed out that when customers use AR-enabled applications, significant increase up to 2.7 times in the past [4]. More importantly, customers which used with AR functionality are more likely to shop than customers who do not use AR 11 times [4]. Houzz's example fully demonstrates the broad development prospects of AR [4].

In fact, we think AR is a huge shock in the retail industry nothing worse than Web or mobile technology [8]. The business and brand influence brought by AR and related XR technologies directly impacts companies' core of the industry's digital reshaping strategy [8]. The success of XR technology depends not only on the company industry 's infrastructure, data and cloud capabilities, a clear vision is needed to design a better customer and employee experience [8].

It is not just retailers who need to be aware of this change. The wide application of AR is also available in companies that can influence the definition and expression of brands in all walks of life, as well as company find ways to interact with users of products and services [8]. In fact, it is expected that in 2023, the financial services industry is expected to become the strongest AR/VR consumer growth field (a compound annual growth rate of 133.9%) [8]. However, retailers as the first adopters, by implementing retail XR, can help consumers build expectations for future experiences in other industries.

13.1 Difference between VR/AR/MR/XR

As the old drivers of the technology security circle, we must have a history of VR or AR exposure. Even many enthusiasts have purchased VR/AR equipment for a large price, but what are the new varieties of MR and XR? Do not worry, let us talk about the difference between VR/AR/MR/XR [3].

At a glance, XR is a collection of VR/AR/MR, so some people have defined XR as “all real and virtual environments generated by computer graphics and wearable devices” according to the definition of XR (Figure 13.1) [3]. The “X” is understood as a variable. In other words, there are AR/VR/MR now, and there may be Baffle Reality (BR), Cinematic Reality (CR), Deceive Reality (DR), etc. in the future [3]

To be honest, the XR is a general term that includes three technologies, and VR/AR/MR is different:

First of all, VR is also called virtual reality, to put it bluntly is false reality [3]. Specifically, it refers to simulating a three-dimensional space with a computer, simulating a sense of immersion and presence through the senses, just like bringing people to a virtual simulation world [3]. However, if it is pure video content, it is no different from watching ordinary 3D movies; so the difference between VR content and traditional 3D video is the interaction between people and video content [3].

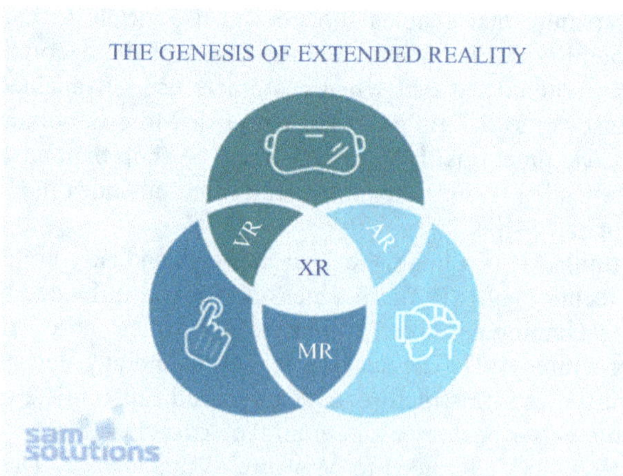

Figure 13.1 Genesis of XR [3]

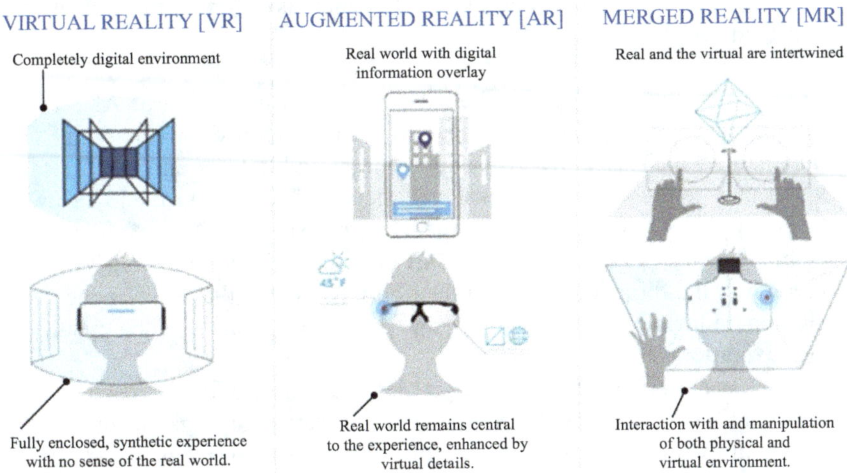

Figure 13.2 Types of XR [3]

As for AR, it is the Pro version of the real world. To put it simply, it is to implant false content in the real space and map the virtual information (objects, pictures, videos, sounds, etc.) in the real environment [3]. Obviously, VR takes you through the spirit, and AR is the reality of tall buildings, reproducing the mirage [3].

Finally, mediated reality (MR) can be directly regarded as a combination of VR/AR, because it refers to the new visualization environment created by merging reality and virtual world [3]. As a simple example, when you go home from work, you wear a MR device to watch a violent American drama, Arnold Schwarzenegger will jump out of the traditional screen, blowing a beard in front of your eyes and staring, no matter where you move, he will only play against you (Figure 13.2) [3].

13.2 What "R" technology can do now?

Among the several technologies of XR, it can be said that the AR people have the largest foundation, but what can they do? The light AR application can open your eyes [3].

1. AR games

   The integration of virtual and real entertainment to death has spawned the ultimate experience of AR games, which has also contributed to the popularity of AR games worldwide [3]. Technically speaking, AR games born out of AR technology, with the addition of multi-dynamic Marker, HDR reflection, occlusion collision and GPS technology, finally broke the virtual and real boundaries [3].

   In 2016, "Pokémon Go" jointly developed by Nintendo and Niantic became popular in various countries, further accelerating the development of

AR in the field of games, and classic IPs such as Harry Potter began to AR gamification [3].

2. AR camera

Outside of the game, under the lens, there is also a world of AR. In 2019, ByteDance's subordinate laboratory AI Lab developed a technology called Landmark, which uses 3D vision technology and detects and calculates outdoor scenes to present such AR special effects in the fastest time [3]. Now, this technology has been applied in the Douyin App [3]. When shooting city landmarks, you can experience this AR effect. In addition to some AR camera research in the direction of the selfie, you have to say that byte bounce brings AR camera technology entered the new generation of short video creation [3].

3. AR navigation

Traditional navigation is no longer new, AR navigation is premeditated to replace it [3]. In 2019, Google launched GoogleMaps AR navigation, the AR scene directly in front of the field of view brought a new experience of "head-up" navigation and opened a new trend of reversing the "low head." In addition to Google Maps AR, "AR navigation" such as AR City and Moovit broke out and even cars began to develop AR navigation applications [3]. Bus application developer Moovit has also introduced AR into its mobile application to help users find stations visually so that users can more easily find stations and stations visually [3].

4. AR military

In addition to entertainment and leisure civilian development such as games and cameras, AR technology has gradually penetrated into the military field [3]. As early as 2018, the US Army once purchased 100,000 HoloLensAR glasses for combat missions and soldier training [3].

Since then, AR technology has gradually developed a military-oriented market, and six AR military markets including North America, Europe, Asia-Pacific, Middle East, Africa and South America have rapidly emerged [3]. Among them, the North American AR military market alone has gathered a large number of companies such as Google LLC and Applied Research Association Inc. [3].

From games to tools, from civilian to military, "R" technology can do much more than it sees [3]. Careful research will also reveal that the application of technology is inseparable from the revival of the sensory experience brought by the XR [3].

13.3 What sensory experiences can XR simulate in the future?

When it comes to sensory physical examination brought by technology, the three technologies of XR and VR/AR/MR, the most prominent is vision, followed by hearing, touch, and smell [3].

1. Visual experience

The eyes are the most intuitive and the most difficult to deceive, and the visual experience brought by the XR is also different according to the difference of VR/AR/MR [3]. First of all, the images that VR sees are completely virtual and cannot see the real environment, so there are two perspectives of personal experience and bystanders (Figures 13.3 and 13.4) [3].

2. Auditory experience

   Music enthusiasts know that Dolby sound, lossless sound quality, and the hearing experience under XR technology bring a 3D version of immersive audio experience [3]. To put it simply, the sound comes from all directions, standing at the center of the stage to listen to music in minutes [3].

3. Tactile experience

   The so-called tactile sensation refers to the reaction caused by contact [3]. The tactile experience of XR technology is to simulate the feeling of human contact. For example, if you put on equipment, you have nothing in your hands, but you feel like you are tearing down the table [3].

4. Olfactory experience

   In the field of XR technology, although the simulated smell is relatively new, there are also a few R&D institutions [3]. The principle is not complicated to

Figure 13.3 Personal experience [3]

Figure 13.4 Bystanders [3]

explain, that is, by changing the smell of equipment, according to what you see, you can feedback the closest taste. It sounds like the olfactory experience is nothing, but it is different if it is placed on the actual experience [3]. You "saw" someone handing over a bunch of flowers, reaching for them and actually touching the weight of the flower, and then you smell the fragrance of the flower [3].

13.4 Current statistics of XR simulate

According to research and consulting company Tractica, this growth will be justified mainly by the increasingly advanced and affordable software development kits (SDKs) allowing more developers to include AR in their apps [6,7]. About 75% of the forecast AR revenues will originate from equipment sales, advertising, and ecommerce (Figures 13.5 and 13.6) [6].

This revenue model suggests that AR mobile apps will focus on an ad-based monetization model, rather than premium subscriptions [6]. As the technology becomes more widespread, sales of headsets and smart glasses will increase as well [6].

PwC's 2017 Global Digital IQ Survey has found 10 industries where AR-related investments are expected to pick up in the next 3 years (Figures 13.7 and 13.8) [2].

XR is changing the landscape in a number of industries. It is expected to grow eightfold, reaching an estimated market size of more than $209 billion by 2022 (Figure 13.9) [5].

A glance at current use cases shows the potential for XR across industries [3] :

- Entertainment

  XR brings immersive experiences to the entertainment world and offers consumers an opportunity to virtually experience live music and sporting events from the comfort of their VR headset [3]. While a majority of market share leans heavily toward entertainment, it is not the only one gearing up for a virtual expansion [3].

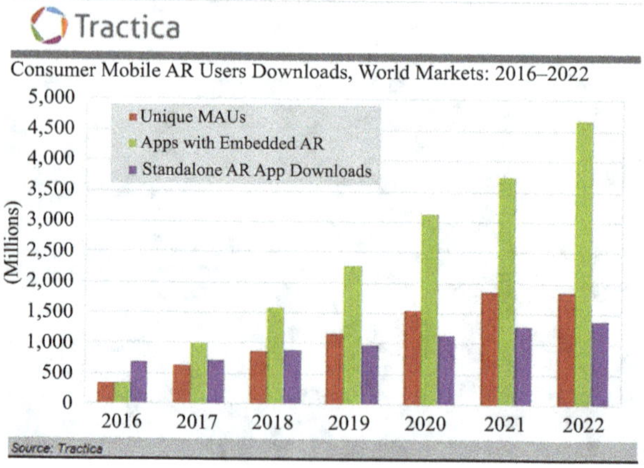

Figure 13.5 Consumer mobile AR users from 2016 to 2022 [6]

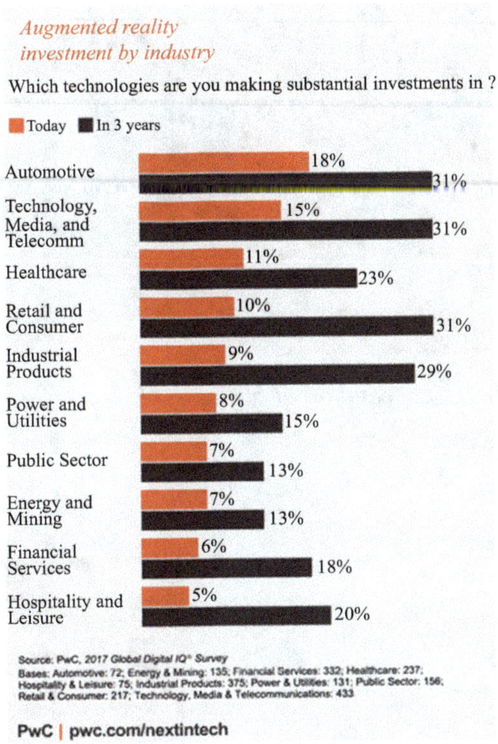

Figure 13.6 AR investment by industry [2]

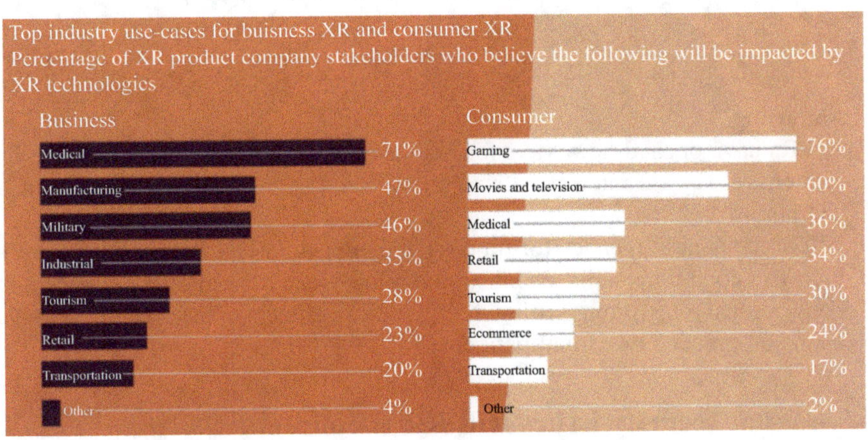

Figure 13.7 Data of top industry use cases for business XR and consumer XR [5]

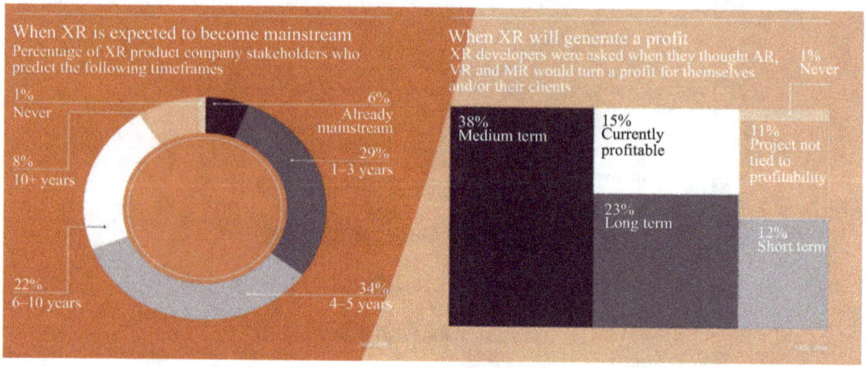

Figure 13.8 Expectation of business XR in a few years [5]

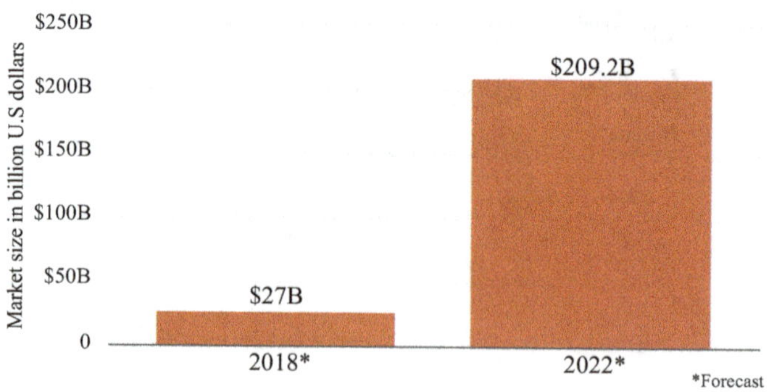

Figure 13.9 Market size of business XR in US [3]

- Marketing
  VRs have opened new ways for brands to engage with consumers, offering immersive ways to interact with new products [3].
- Training
  XR opens new avenues for training and education [3]. People who work in high-risk conditions—like chemists and pilots—can train in safety from a more conventional classroom setting [3]. Medical students, meanwhile, can get hands-on practice on virtual patients [3].

13.5 What security issues will XR encounter?

In fact, everyone also understands the truth of the advantages and disadvantages of technology. Since the core of XR is to simulate human senses, its security risks, or threats to the ultimate victim are the human itself [5]. Next, from the perspective of security, talk about the threat of XR to people in two aspects of data security and vulnerability attacks:

1. Data threat

   Compared to conventional electronic devices and applications that have plunged us into the vortex of information leakage, XR devices not only collect regular data but also collect a large amount of detailed privacy data such as user's voice, habits, interaction, appearance, and attention [5]. Once these more personal data are stolen by intentional people, a little analysis can generate images of the characters and strike them accurately [5].
2. Hidden risks

   Vulnerabilities are everywhere, not alarmist. This old problem also exists in XR technology [5]. As XR technology is widely used in professional fields such as games, shopping, architecture, medical care, and education, hackers come through the "hole," the first destruction is the visual screen, which may be displaying wrong data, or it may be tampering that the threat level is probably the same as current cyber attacks [5].

13.6 Concept of XR

The concepts of augmented reality (AR for short) and virtual reality (VR for short) have been around for decades. However, it is the latest thing that VR/AR has appeared in the technology media to attract the attention of all parties. In this section, we briefly introduce these two concepts and their history, and, at the same time, clarify the difference between them.

First of all, let us think about: What is reality? Very philosophical and profound question, the ancient Greek thinker Plato said, oh, far away, back to the computer world on which everyone depends on to support their families: "seeing is believing." More specifically, the reality we care about is the perception of the real world presented to the human visual system in the form of images. Simply put, it is a real-world image.

The above definition clearly gives ambitious people a back door: if the image is generated in a certain way, as long as it is realistic enough, can it fool human eyes or even the entire brain? This VR, without suspense, is virtual Definition of reality (VR). Of course, the current technical capabilities are still far from the fascinating immersion in the Matrix. In fact, from the earliest VR prototype, the VR 3D personal theater Sensorama (Figure 2.1) invented by Morton Helig in 1962, to the recent Oculus [9].

The immersion brought by VR technology is always in a state that requires users to actively believe. Of course, this state is not a problem for many applications, so there are now all kinds of VR products and hot-selling VR predators and rookies (Figure 13.10) [9].

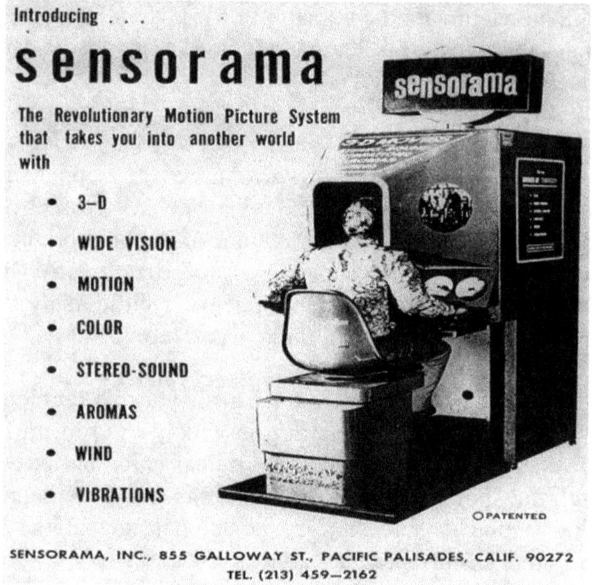

Figure 13.10 Sensorama [9]

13.7 AR in education

The educational value of AR is closely linked to the way in which it is designed, implemented, and integrated into formal and informal learning environments [10]. An important consideration is how AR technologies support and afford meaningful learning [10]. Considering AR as a concept rather than a certain type of technology would be productive for educators [10]. The involvement of educators is important to facilitate the development of favorable AR applications for teaching, which increases the potential for AR to be incorporated in education [11]. AR applications have been developed for many areas of education [12].

Some of these AR applications have been used in previous studies [13]. Gopalan *et al.* [13] tested the impact of AR enhanced science textbooks on lower secondary school students in Malaysia. The system guided students toward target ecology areas and displayed the corresponding learning tasks or related learning materials [13].

13.8 Advantages of AR in education

AR provides new ways of interacting with the real world and can create experiences that would not be possible in either a completely real or virtual world [14].

1. Reduce costs

   With AR technology, the physical materials needed for many courses are not needed. For example, when going to human anatomy, a 3D virtual human body can be presented through AR. The user can control and watch the 3D stereoscopic image, and the teacher does not need to bring a human body model. In the biology class, the AR card teaching method is adopted, and the biology structure can be more intuitively and vividly understood in the biology class, which can greatly reduce the procurement cost of biological specimens.[14]

2. High interest

   Because the content presented by AR is all 3D, it is very intuitive and vivid, which is helpful for students to understand and remember. With the help of AR technology, students' classroom experience jumped from 2D to 3D. It is no longer the flat content presented in books or blackboards, but the vivid 3D content. For animals, plants, daily necessities, etc., which are originally visible 3D objects, students do not need to fill the 3D image from the planar 2D image; for the abstract or invisible content such as radio waves, magnetic fields, atoms, geometry, etc. [14]

   AR can be visually displayed, and the teaching of images replaces boring theories, which is very attractive to students. In addition, AR's visualization and interactivity can naturally design very attractive gamified teaching content, which is entertaining and entertaining, so as to greatly increase students' willingness to learn, stimulate learning interest, and improve learning effects [14].

3. Teamwork

   When students use AR technology to learn, they no longer learn by rote, but to experience the learning content, and can participate in teaching in a teamwork manner. AR can give learners a special space to make them feel co-located with other people or things. This sense of presence can strengthen students' awareness of the learning community. Collaborative AR has already had an impact on the business operations of enterprises. [14]

   It also has important uses in the education industry. A huge benefit of AR is that it allows groups with different levels of perception and experience to be at the same starting point. By fusing virtual digital images and superimposing them into the physical world, AR education makes it easy for students scattered in different locations to collaborate as if they were studying in the same classroom [14].

   In a paper in the academic journal "Computer Collaboration", educational researchers studied collaborative learning based on AR. According to the paper, AR has been used to allow scientists from different countries and regions to work together in a building excavation address, allowing educators to create a virtual lecture hall with a highly immersive 3D environment [14].

4. Save time

   In the past teaching, teachers often used virtual presentations such as PPT and video, and then took out the real objects for observation. With the integration of virtual and real AR technology, many digital teaching contents can be directly integrated into physical teaching items, which brings great convenience to teachers. There is no need to repeatedly switch between virtual and reality, thereby saving a lot of time and improving teaching efficiency. Taking the AR globe of Phantom Technology as an example, students can see the encyclopedic knowledge of animals, buildings, country introductions, meteorology, etc. on the maps of countries around the world by scanning the physical globe [14].

In addition, in traditional teaching, the learning level of different students varies greatly, and teachers need to spend a lot of time to achieve individualized education based on their aptitude. AR technology can easily help teachers achieve personalized teaching [14].

Taking the AR cube as an example, the physical cube is placed in front of a mobile phone or tablet, and the current state of the cube can be scanned and identified immediately, and then analyzed to automatically solve the best algorithm, and then show the step-by-step operation process. Even if you do not understand the formula of the Rubik's Cube at all, you can restore the Rubik's Cube quickly by following the operation steps. In this way, students with strong learning ability can do self-study, students with average learning ability can make a little allocation, and then make time to focus on explaining to students with poor learning, so as to improve the teaching effect [14].

5. Remote support

   AR allows teachers and students from different regions to gather in a virtual classroom and achieve real-time and real-time interactions. Therefore, many high-quality educational resources from around the world can be tilted to the third and fourth lines, rural areas and other underdeveloped areas at very low cost, so that remote mountainous schools can also enjoy the personal guidance of famous teachers [14].

13.9 Disadvantages of AR in education

Users of AR technology may experience usability issues and technical problems, and some students may find this technology complicated.

1. Usability

   There is no evidence to suggest that usability issues are directly related to AR technology and may instead stem from inadequate technology experience, interface design errors, technical problems, or negative attitudes [15].

2. Hard to differentiate reality or fantasy

   The combination of real and virtual objects may cause confusion as students may face difficulty navigating between fantasy and reality [15]. The use of AR technology within a learning environment requires multitasking, as students need to engage with large amounts of information and multiple technological devices to accomplish complex tasks [14]. This may result in a cognitive overload and a feeling of being overwhelmed or confused [15]. The confusion indicates the authenticity of an AR system; however, this may be unproductive in a learning environment as students may lose track of the real environment [15].

3. Stability

   The stability of mobile AR technology is not guaranteed, and difficulties may be encountered if the technology lacks well-designed interfaces and guidance as this may result in the technology being too complicated [15]. Users may also need time to get familiar and comfortable with AR technology [15].

13.10 Previous technologies on XR

1. Microsoft HoloLens

   HoloLens is Microsoft's MR headset display that came out in January 2015. Nadella has repeatedly mentioned the importance of HoloLens in public: MR, artificial intelligence, and quantum computing will be the three major technologies in the future [16].

   Because Microsoft's hybrid display business has always focused on the B-end market and vertical industries, many consumers have not yet had the opportunity to contact HoloLens, which is called "black technology" products. But in fact, the technology behind HoloLens has been touched by ordinary consumers, and the carrier is the prestigious Kinect in the consumer electronics world [16].

   Kipman, the father of Kinect, once said that the development team of HoloLens and Kinect are the original team. Identifying and tracking human movements in HoloLens actually extends the experience of the game console XBox to another platform [16].

   In 2010, Kipman began to develop another project, Project Baraboo, a collection of holographic glasses: projecting virtual images in the real world, users interact with gestures and voice. In January 2015, the results were officially published, Microsoft named it HoloLens, and the developer version cost is $3,000 [16].

   This is an all-in-one device, with its own computing processing unit, equipped with the Cherry Trail CPU + GPU of Intel 14nm process, and the self-developed Holographic Processing Unit (HPU). HPU adopts TSMC's 28-nanometer process and integrates 24 cores. Approximately 1 trillion calculations can be performed per second, and the data transferred to the CPU requires little additional processing [16].

   HoloLens has four environment perception lenses, one depth lens, and one video lens, which are in the same line as Kinect's depth perception (Figure 13.11) [16].
2. LyteShot

   In order to bring video games into the real world, the startup LyteShot just launched a crowdfunding campaign on the crowdfunding platform Kickstarter. They hope to attract more people's attention and bring their mobile gaming platform into real life [17].

   The four members of this entrepreneurial team have extensive experience in the 3D and entertainment industries. They not only create a very interesting platform for gamers but also leave room for makers and developers in their hardware and software design. So that they can further develop on this system and add their own creation: "You can print out swords/sticks/guns and any other equipment you want with a 3D printer, and then combine our Lyter and Puck to achieve plug-and-play games Experience." [17].

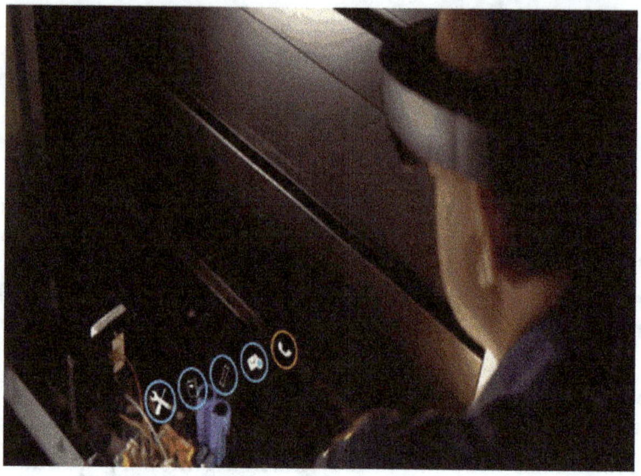

Figure 13.11 Hololens [16]

In addition to an open and customized game platform, LyteShot has also built an online market where users can share and even sell their various equipment designed for games [17].

LyteShot's peripherals are designed to be compatible with various weapons, objects and tools created with 3D printers. In addition, the Arduino-based hardware platform allows makers to interact with 3D printed designs in the gaming environment.[17]

Similar to simulation games that use laser marking, LyteShot's game mode is very simple: users can carry a cartoon gun with a "Puck" shooting sensor (you can also design and 3D print it yourself) and find other "laser marking." The player will pull the trigger, and the "hit" user will be judged to have lost blood or died like the traditional video game. Using Bluetooth technology, all players can stay in sync via the smartphone app and use infrared to shoot [17].

Players can track the in-game logistics, leaderboards, geographic tracking, complete tasks, and communicate in the game in real time, and can enhance these functions through their own game design, or play an existing game provided by LyteShot [17]. Perhaps LyteShot's more impressive feature is its ability to digitally track game statistics, game maps, guide storylines, and rely on developers to choose to modify their designs. This game can provide gamers with a fully immersive open world gaming experience through a central controller, or no controller at all [17].

The first game level announced by LyteShot includes assassination, siege, and invasion. These three scenes have different story development main lines, ranging from political thrillers to zombie themes. Players can play alone or in teams, and there are various levels of social interaction in the game environment (Figure 13.12) [17].

3. Pokémon Go

What kind of game is Pokémon Go? In addition to the strong IP that lasted 20 years ago, it incorporates AR elements and expands based on geographic

Figure 13.12 LyteShot [17]

location [18]. AR stands for AR technology, which allows virtual Pokémon to appear more realistically in real life, successfully allowing a group of otakus and otakus to walk out of their homes and seek "treasures" everywhere [10]. It also has a strong social attribute: it allows players to exchange pets, and also allows elves to play against each other; its promotional film has scenes where handsome men and beauties accidentally gather in the same place to fight, and then look at each other with a smile, which is very imaginative [18].

This game is also Nintendo's "big move" after the death of former president Satoshi Iwata and the change of company structure. A technology commentator said that AR technology has been deeply cultivated for a long time, and this game shows its huge profit potential [10]. Since 2012, the amount of investment in related fields has been increasing. Just over half of 2016, the global AR/VR investment has reached 1.3 billion US dollars, an increase of 85% over the total amount last year [18].

AR games, referred to as AR games, also known as AR interactive games, refer to games that use AR (expanded reality, expanded reality, AR) technology [18].

Since such games use AR technology to superimpose the real environment in reality and the virtual things in the game to the same scene in real time, they can achieve strong interaction. The game needs to have camera support. And with the support of the satellite positioning system, you can also interact with things in the game in a very large real world.[18]

Since mobile terminals are more convenient to move in real time and realize the interaction between games and reality, most existing AR games are games on mobile terminal platforms (mobile phone games).[18]

In addition, AR games is not equal to VR games. Some people confuse AR games with VR games, which are different types of games. The VR game is a completely virtual game environment, and does not interact with the real-world environment, but uses an immersive display device (such as a game helmet) to bring the vision into the virtual environment (Figure 13.13) [18].

Figure 13.13 View of Pokémon Go game [18]

4. Dream Glass

   DreamGlass is an immersive AR headset focused on 3D content. DreamWorld has created a Unity-based software development kit (SDK). Unity is generally used in all ways of video game development, and there are many tools for creating AR and MR content, and supports Windows and Android [19].

   DreamGlass AR can be connected to a smartphone and can use a USB Type-C connector to complete network sharing [19]. This connector allows users to use a smartphone as a controller to interact with AR content displayed on the headset. Similarly, the device can also be connected to a PC, which is useful for the development and prototyping of AR and MR content [19].

   DreamWorld has now launched DreamGlass Air, an AR headset focused on portability that provides a 90° field of view with an 18:9 aspect ratio, providing impressive visual effects [19]. DreamGlass Air is connected to an independent Android-powered computing unit, which provides approximately 5 hours of continuous runtime with the headset [19].

   It can copy files (videos, photos, documents, etc.) to the computing unit via USB to view the content on the wearable device. Audio is provided via a 3.5-mm jack or Bluetooth headset connection, and the device supports Wi-Fi. DreamGlass Air can also be used to mirror tablets, PlayStation 4 game consoles and even drones (Figure 13.14) [19].

5. North Focals

   Canadian company North launched the first AR smart glasses-North Focals, priced at 999 US dollars. It is reported that this glasses not only follow the fashion appearance of spectacles but also combine some functions of Google Glass [20].

   North said that this is the first everyday "smart glasses," using retinal projection technology, can browse mobile phone information (including SMS,

Figure 13.14 DreamGlass [19]

Figure 13.15 North Focals [20]

calendar, weather forecast, etc.), and supports taxi, navigation and calling Amazon Alexa and other functions [20].

Focals consists of two parts: smart glasses and a loop controller similar to the "ring" style [20]. Loop is a plastic ring with a button like a joystick. In terms of battery life, North said that the battery life of each Focals glasses can reach 18 hours, but can only be charged through the matching glasses case. The Loop is also charged through the glasses case [20].

In terms of operation, the wearer can browse or select the displayed information by toggling the joystick to the left or right and pressing a button. You can also use it to activate Alexa, the Amazon smart assistant built into the glasses. In simple terms, the role of this ring is similar to the VR/AR hardware controller [20].

Focals has built-in microphone and speaker, so it supports voice control function. Users can issue instructions to Alexa and hear the system's response when necessary, such as texting, calling Uber, map navigation, checking the date and weather, and other functions (Figure 13.15) [20].

Table 13.1 Gap analysis and benchmarking

No.	Study	Research objective and context	Research methodology and design	Gap analysis
[25]	Herz and Rauschnabel (2018)	To develop a comprehensive framework to study consumer reactions to wearable glasses	• $N = 661$ • Germany • Survey • Structural equation modeling	It is a set of potential benefits and risks associated with VR glasses are identified
[26]	Shin (2018)	To test a VR experience model that integrates presence, flow, empathy, and embodiment	• $N = 661$ • South Korea • Experiment and survey • Structural Equation Modeling	High immersive devices affect the users' perceived presence and make them feel flow
[27]	Yim and Park (2018)	To examine the role of consumers' perceived body image in their reactions to AR-based vs. web-based product presentations	• $N = 406$ • USA • Survey • Analysis of variance	Consumers who perceive their body image as unfavorable evaluate AR more favorably than traditional web-based product presentations
[28]	Huang and Liu (2014)	To examine the extent to which presence, media richness, and narrative experiences yield the highest experiential value	• $N = 344$ • Survey • Structural Equation Modeling	Compared to features such as presence and media richness, using the narrative perspective to design ARIT creates experiential value for online consumers
[29]	Nunes and Filho (2018)	To analyze consumer behavior in relation to Google Glass	• $N = 8$ • Netnography	Three categories (socially satisfied, socially constrained, and early adopters) and two categories (enthusiasts and visionaries) of consumers are found

13.11 What is the concept of XR in 2030?

In 1999, no one would think that 20 years later, an electronic device the size of a cigarette box, the computing power display effect, will be close to a serious computer [21].

Twenty years ago, the Internet had just appeared, and there were only 9 million people online. Twenty years ago, we simply could not imagine that 20 years later, without the Internet, we could not do anything [21].

Some people ask, does it make sense to discuss things 20 years later?

Let me remind you: 20 years ago, Alibaba and Tencent were just established.

After 20 years, VR will reach its ultimate state:

- Retina resolution
- Field of view above 180°
- Somatosensory input and output
- Wearing comfort similar to ordinary glasses

By then, VR will not only make people's lives more convenient, it will create a new parallel virtual world [21].

Everyone has their own new identity in the virtual world and leads a new life. People will spend a lot of activities in VR, visiting attractions, dating, studying, and working [21].

Because the economic system in the virtual world can communicate with the real world, new economies will appear in the virtual world, and you can get a job that can really feed your family in the virtual world. For example, an architect in a virtual world scene, or an interactive actor in a virtual world [21].

Of course, new jobs will also be created. How new? Just like the workers during the Industrial Revolution, they could not imagine that there would be a new job like a programmer in the future [21].

Twenty years later, because a large number of people work in virtual offices at home, people no longer need to squeeze into big cities, housing prices will not be hot, and the travel industry will also be affected [21].

Twenty years later, VR, which has reached its ultimate form, has an impact on human society, similar to the impact of the Internet on human life, even deeper than that. After the material needs are greatly satisfied, the pursuit of spiritual needs will be realized in all dimensions in the virtual world [21].

Bill Gates said that people will always overestimate the changes after 2 years, but also underestimate the changes after 5 years. In the 20-year time period, there will be a lot of incredible new technology revolutions and new trends of thought [21].

Standing at the starting point of 20 years, the general direction is clear, but on the way to the goal, timing and nodes are very important. We can only grasp a few signs of the future and depict as far as possible the image of the future of the virtual world [21].

13.12 The future of VR, livable VR technology

At the 2018 MWC conference, the concept of "Vive Reality" also believed that 5G and AI will give VR more vitality. It is not difficult to imagine that in the future world, VR will become a way of life, playing an increasingly important role in all aspects of work and life. So what are the characteristics of our ideal VR system in the future? [22]

When discussing the future of VR, we will admit a hypothesis: "Users will spend more and more time in the VR system, and the frequency of use will become higher and higher until they become an important part of daily work and life." This assumption means that in the future, part of people's lives will be completed in virtual spaces, so we have borrowed the view of "livable cities" to create the concept of "Livable Virtual Reality." This concept can be simply understood as a VR technology that can be used continuously for a long time, can be used at a high frequency for a long time, and supports the basic survival maintenance system [22].

Livable VR is divided into three main functions:

1. Can be used continuously for a long time.
2. Can be used at a high frequency for a long time.
3. Support the basic survival and maintenance system.

13.12.1 Can be used continuously for a long time

The term "continuous use for a long time" here means that the user continuously uses the VR system for work or entertainment without leaving the immersive environment. The basis is the "comfort" and "health" of the head-mounted display device itself [22].

"Continue to use for a long time without leaving the immersion state" is actually a necessary and extremely demanding requirement. Generally, even if a display screen with an eye protection function is used, the user needs to remove his eyes frequently to relax eye muscles and relieve fatigue. However, in the VR system, as long as the head-mounted display is not removed, all the content that is visible is presented by the display, so the task of relaxing the eye muscles must be achieved by the display itself without leaving the immersion state. This is a feature that has not yet been perfectly implemented in all displays. But from another perspective, this demand is very necessary. Compared with the existing human-computer interaction system, the VR system provides an unprecedented sense of separation from reality [22]. When the user is immersed in the virtual environment, their feelings have completely left the real world, so whenever they leave or enter the virtual environment, they will bring a huge gap to their overall feelings. According to previous research, such switching can easily cause user fatigue and loss, which should be avoided as much as possible in the process of constructing a highly immersive virtual environment [22].

13.12.2 Can be used at high frequency for a long time

"Long-term high-frequency use" means that the VR system can be used daily. This requirement not only puts forward requirements on the comfort and health of the system hardware itself but also puts forward strict requirements on the quality of the software system that builds the virtual environment and the basic guidelines for building the virtual world. This is because long-term high-frequency use may affect people's "common sense" perception, and poor design may cause users to form a wrong perception, leading to perception conflicts [22]. Imagine that users are immersed in a VR game that simulates basketball every day. In order to reduce the difficulty, the developer intentionally shortened the feeling of space in the virtual environment, so that the user can more easily put the ball into the basket. Over time, when one day this user wants to play a real basketball game, He will be sad to find that because of the indistinct sense of distance, he cannot score at all [22].

When these perception confusion affects the user's judgment related to life safety (such as the judgment of the distance), it will put the user in a very dangerous situation. And this confusion will become more and more serious and stubborn as the time and frequency of using VR increase [22]. Of course, the erroneous design that caused this confusion is not necessarily intentional by developers. In most cases, they are caused by developers inadvertently. After all, it is really a VR environment to create a variety of common sense rules that are exactly the same as the real world. It's too difficult. Therefore, in order to achieve the goal of "long-term high-frequency use," the VR system must undergo rigorous inspection to ensure that its "common sense" perception of users will not cause fatal harm. In the future, it is extremely necessary to establish mandatory standards to ensure this aims [22].

13.12.3 Support basic survival maintenance system needs

The third requirement of livability is: "Support basic survival and maintenance system needs" [22]. To put it simply, users are required to eat, excrete, clean the body, sleep, etc. without leaving the immersive environment. This seems to be a critical request. But in many practical applications, these requirements are rigid. For example, in VR cinema applications, researchers hope to replace the large screen with high-performance head-mounted displays and become the main playback device in cinemas [22]. Due to the high immersion, high display performance and other characteristics of VR, this idea itself has great appeal. But the researchers have to face a grim fact: a considerable portion of the revenue in the cinema industry comes from the cost of eating popcorn and drinks while watching movies, and VR cinemas that are unable to eat will find it difficult to reach a break-even. The same example is when using the VR office system, it is difficult to imagine the user to work for four hours without drinking a cup of coffee. These seemingly excessive demands actually determine whether the VR system can really be used in this field [22].

Based on the above discussion, we have listed the minimum demand system that constitutes livable VR. There are three main points in terms of hardware:

1. The livable VR system must be able to support the integration of virtual and real, so that users can interact with the virtual and real environment at the same time.
2. Livable VR requires the use of highly human-headed reality devices and must ensure the health of users' continuous use of head-mounted displays.
3. In response to different business needs, Livable VR must redesign efficient input devices and methods to match the completion of relevant tasks under the premise of wearing a real-world device.

In terms of software, Livable VR can at least meet the following functions:

1. *Survival needs*: The software system can realize basic functions such as drinking and eating.
2. *The need for security*: The system should have the ability to sense the environment and map the real environment to the virtual environment to ensure that the user will not accidentally touch the real environment objects during the use of the system.
3. *Emotional needs*: The livable VR system should provide corresponding functions to construct a pleasing virtual space and should be designed to use functions to reduce the loneliness caused by long-term exposure to the virtual space.
4. *Existence requirement*: The system is used to ensure that users can observe their limbs in a virtual environment, to avoid separation of the body. The system should allow users to feel the real time lapse in the virtual environment and ensure the consistency of the virtual and real-world time.

Livable VR will be an important part of people's daily life and work in a few years [22]. This concept can help the industry accelerate the realization of idealized VR equipment and spawn related long-term immersive content, thus making VR truly necessary for people's daily life device. Correctly understanding its technological development trends and possible problems will help us better meet the arrival of the real VR era [22].

13.13 Current application of XR technology

1. VR smart medical

   The application of VR in medicine has very important practical significance. In a virtual environment, a virtual human model can be established. With the help of a trackball, HMD, and sensory gloves, in medical teaching, students can easily understand the internal organ structure of the human body, which is more effective than the current way of using textbooks [23].

   At present, VR medical applications are mainly divided into several aspects: first, VR technology can be used for various mental illnesses or rehabilitation training. VR technology can be used to virtualize various scenarios that may not exist in reality, which can help patients undergo psychological treatment and let students complete various health projects [23]. Second, the visual immersion and interactivity brought by VR technology can allow doctors to "enter" human observation and learning, simulate human anatomy and other activities [23].

   With the help of VR technology, doctors can learn in virtual scenes. In addition, VR technology has very important significance in remote surgery, planning of complex operations, information guidance of the surgical process, prediction of surgical consequences and improvement of the living conditions of the disabled, and even the development of new drugs (Figure 13.16) [23].

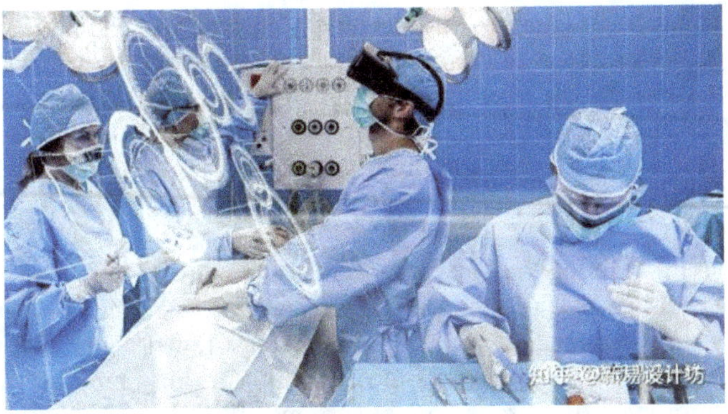

Figure 13.16 Medical equipment with VR technology [23]

Figure 13.17 Education with smart glasses [23]

2. VR cultural education

   Promote the application of VR technology in higher education, vocational education and other fields and experimental and demonstrative courses in physics, chemistry, biology, geography, etc., construct virtual classrooms, virtual laboratories and other educational teaching environments, and develop virtual lesson preparation and virtual teaching, virtual exams and other new methods of education and teaching, promote learner-centered personalized learning, and promote the transformation of teaching and learning models [23]. Really displays the overall campus environment and infrastructure through VR panorama, providing students with a more intuitive and comprehensive choice when choosing a school (Figure 13.17) [23].

3. VR entertainment

   The rich sense ability and 3D display environment make VR an ideal video game tool. For the development of games, the current VR technology is more suitable for the development: role-playing games, action games, adventure puzzle games, racing games, and its advanced graphics engine is no less than the image performance of the current mainstream game engines [23].

And the integration of supporting dynamics and AI systems has also provided convenience for game development. Since the reality requirement for VR in entertainment is not too high, in recent years, VR has developed most rapidly in this area (Figure 13.18) [23].

4. VR military

Simulation training has always been an important issue in the military, allowing soldiers to personally experience the battlefield. Arne Odland, BEMR's chief technical expert, said: "In the past few decades, VR technology has been used as a simulation training, mainly for those projects that are very expensive or even dangerous on real systems" [23].

Now BEMR is exploring more cases to improve the visualization of the battlefield and strive to provide better information for each soldier on the battlefield. The system can connect more than 200 simulators. In addition, the use of VR technology can simulate a zero-gravity environment to replace the current non-standard method of training astronauts underwater (Figure 13.19) [23].

5. VR design

Taking interior design as an example, using VR technology can not only perfectly represent the indoor environment but also walk freely in the 3D indoor space [23]. At present, VR technology is commonly used in the industry for indoor 360° panoramic display and indoor roaming, which is unanimously praised, and not only can roam indoors but also can use VR technology as a pre-decoration system, which can realize the dynamic change or paste of the wall color in real-time wallpapers of different materials can also change the color of the ground or paste different wooden floors, ceramic tiles, etc., which can move the placement of furniture and replace different

Figure 13.18 Gaming by using VR [23]

Figure 13.19 AR technology used in planes [23]

Figure 13.20 House being designed by using VR technology [23]

decorative objects. All this is perfectly represented under VR technology (Figure 13.20) [23].

6. VR traffic

   No matter in the transportation planning simulation of air, land or ocean rivers, VR technology has its unique advantages. It can not only use 3D GIS technology to show various traffic routes in place but also dynamically simulate various natural disasters happening (Figure 13.21) [23].

13.14 AR glasses in the future

Waking up early in the morning, we can use AR glasses to get a variety of the latest personalized notifications. The AR glasses cooking app provides us with customized dietary recommendations and informs the health monitoring data [24].

Figure 13.21 Traffic being shown on map by using VR [23]

On the way to the airport, in self-driving cars, we can use AR glasses to interact with friends in the virtual world, create various artworks, and even turn the carriage into a part of the virtual world [25–28].

After arriving at the airport, the AR glasses will switch to the navigation mode and guide us to the boarding gate by showing arrows [24].

After arriving at the hotel, we can activate the travel mode of AR glasses, understand the information and related data of nearby historical buildings and cultural relics, and obtain information on nearby restaurants, flavors, reviews, promotions, etc.[24]

When resting at night, we can browse photos and videos taken during the day using AR glasses and share them with family members through the eye-tracking control interface.

This sci-fi scene is an enhanced world in 2030. Although many enterprises are currently expanding the application scenarios and application scope of AR devices, a wide range of consumer application scenarios will be an important field of AR device applications in the future. With the rapid development of hardware and data connections, product prices are more affordable, AR glasses will become a part of daily life.[24]

Cloud computing and 5G coverage will allow AR products to be infinitely scalable, ultra-fast, and mobile. Ultimately, brain implant technology will allow AR to be directly embedded into our nervous system through a brain–computer interface.

The future of AR is full of infinite possibilities.

13.15 Conclusion

Due to the characteristics of VR technology itself, it is very convenient, fast, and true to engage in the simulation of the above-mentioned related work. VR technology has been recognized by more and more people. Users can experience the most real feelings in the VR world. The authenticity of its simulated environment and the real world are difficult to distinguish between true and false, making people feel immersive.

At the same time, VR has all the perceptive functions possessed by human beings, such as hearing, vision, touch, taste, smell, and other perception systems. Finally, it has a super simulation system that truly realizes human–computer interaction and makes people operate in the process. In particular, you can operate the VR at any places and get the most realistic feedback from the environment. Its existence, multi-perception, interactivity, and other characteristics of VR technology that make it popular with more and more people.

References

[1] Martechseries (2018). 32 Percent of Consumers Use Augmented Reality According to a Study by Artillry Intelligence and Thrive Analytics. https://martechseries.com/predictive-ai/augmented-reality/32-percent-of-consumers-use-augmented-reality-according-to-a-study-by-artillry-intelligence-and-thrive-analytis (Accessed 10th NOV 2021).

[2] Dhaval Sarvaiya (January 2019). Augmented Reality: Market Potential and Practical Applications. https://hackernoon.com/augmented-reality-market-potential-and-practical-applications-71f0d8faef56 (Accessed 10th NOV 2021).

[3] Jenny Scribani (January 2019). What is Extended Reality (XR)? https://www.visualcapitalist.com/extended-reality-xr/ (Accessed 10th NOV 2021).

[4] Houzz Company News. (April 2019). Virtually Cover Your Floor with Tile Using the Houzz App's Updated View in My Room 3D Tool. https://www.houzz.com/press/657/Virtually-Cover-Your-Floor-with-Tile-Using-the-Houzz-Apps-Updated-View-in-My-Room-3D-Tool (Accessed 10th NOV 2021).

[5] Pandolph, Stephanie (2019) AR is now a Must-Have in Retail. Business Insider. https://www.businessinsider.com/ar-is-now-a-must-have-in-retail-2017-10 (Accessed 10th NOV 2021).

[6] Statista Research Department (March 2020). Extended Reality (XR) – Statistics & Facts https://www.statista.com/topics/6072/extended-reality-xr/ (Accessed 10th NOV 2021).

[7] AR Post (November 2017) Key Augmented Reality Statistics You Need to Know. https://arpost.co/2017/11/15/key-augmented-reality-statistics/ (Accessed 10th NOV 2021).

[8] Armando Ortiz and Mark McGiffin (January 2019) How Extended Reality will Reshape Commerce: Reinventing Brand and Customer Experience. IBM. https://www.ibm.com/downloads/cas/GAZR2Z69/ (Accessed 10th NOV 2021).

[9] Chris Payatagool (September 2008) Theory and Research in HCI: Morton Heilig, Pioneer in Virtual Reality Research. http://www.telepresenceoptions.com/2008/09/theory_and_research_in_hcimor/ (Accessed 10th NOV 2021).
[10] H. Wu, S. W. Lee, H. Chang, and J. Liang (2013) "Current status, opportunities and challenges of AR in education," *Computers and Education*, vol. 62, pp. 41–49.
[11] X. Wei, D. Weng, Y. Liu, and Y. Wang (2015) "Teaching based on AR for a technical creative design course," *Computers and Education*, vol. 81, pp. 221–234.
[12] J. Martín-Gutiérrez, P. Fabiani, W. Benesova, M. D. Meneses, and C. E. Mora (2015) "AR to promote collaborative and autonomous learning in higher education," *Computers in Human Behavior*, vol. 51, pp. 752–761. https://www.sciencedirect.com/science/article/pii/S0747563214007110?via%3Dihub (Accessed 10th NOV 2021).
[13] V. Gopalan, A. N. Zulkifli, and J. A. A. Abubakar (2016) "A study of students' motivation using the AR science textbook," *AIP Conference Proceedings*, vol. 1761, no. 1, pp. 27–35.
[14] T. H. C. Chiang, S. J. H. Yang, and G. J. Hwang (2014) "An AR-based mobile learning system to improve students' learning achievements and motivations in natural science inquiry activities," *Journal of Educational Technology and Society*, vol. 17, no. 4, pp. 352–365.
[15] M. Akçayır and G. Akçayır (2017) "Advantages and challenges associated with AR for education: a systematic review of the literature," *Educational Research Review*, vol. 20, pp. 1–11. https://www.sciencedirect.com/science/article/pii/S1747938X16300616?via%3Dihub (Accessed 10th NOV 2021).
[16] Maria Aleksandrova (May 2018) Augmented Reality in Education. https://dzone.com/articles/augmented-reality-in-education (Accessed 10th NOV 2021).
[17] A. Jana, M. Sharma, and M. Rao (July 2017) Digital Reality - Under the Hood : Introduction to Microsoft Hololens. https://hub.packtpub.com/introduction-hololens/ (Accessed 10th NOV 2021).
[18] S. Hendershot (May 2015) Lyteshot : Aims to turn Augmented Reality into reality. https://www.chicagobusiness.com/article/20150305/ISSUE01/150309934/lyteshot-s-mobile-game-tech-dazzles-augmented-reality-headset-makers (Accessed 10th NOV 2021).
[19] G. Skarre (June 2018) Dream Glass is an Interesting Augmented Reality Glass. https://skarredghost.com/2018/06/14/dreamglass-is-an-interesting-augmented-reality-glass/ (Accessed 10th NOV 2021).
[20] J. Crook (June 2018) Focals by North Review : The Future is Here. https://techcrunch.com/2019/06/18/focals-by-north-review-the-future-is-almost-here/ (Accessed 10th NOV 2021).
[21] M. C. Aslan (Jan 2019) Virtual World 2030: The New World with Virtual Reality. https://arvrjourney.com/outlook-2030-the-virtual-world-69f3772c56bc (Accessed 10th NOV 2021).
[22] D. Ruggeri, C. Harvey, and P. Bosselmann (Nov 2017) Perceiving the Livable City : Livable Virtual Reality. https://www.tandfonline.com/doi/abs/10.1080/01944363.2018.1524717 (Accessed 10th NOV 2021).

[23] Tech Industry (April 2020) Virtual Reality : Uses and Application for Virtual Reality in Different Fields. https://www.fdmgroup.com/5-exciting-uses-for-virtual-reality/ (Accessed 10th NOV 2021).

[24] J. Benton (Nov 2019) AR Glasses: Future is a Few More Years than Expected. https://www.niemanlab.org/2019/11/our-ar-glasses-future-is-a-few-more-years-away-than-some-thought/ (Accessed 10th NOV 2021).

[25] Leong, Wai Yie (Nov 2005) Implementing Blind Source Separation in Signal Processing and Telecommunications. Diss. University of Queensland.

[26] U. Mehrotra and W. Y. Leong (2009) NSEEAR: A energy adaptive routing protocol for heterogeneous wireless sensor networks, 2009 35th Annual Conference of IEEE Industrial Electronics, pp. 2647–2652, doi: 10.1109/IECON.2009.5415255.

[27] W. Liu, E. Lupito, Y. L. Sum, B. Tay and W. Y. Leong, (2009) Ultra wideband antenna for real time location system application, 2009 35th Annual Conference of IEEE Industrial Electronics, pp. 2738-2742, doi: 10.1109/IECON.2009.5415424.

[28] Han Fei Min and Leong Wai Yie (2012) Investigating target detection and localization in wireless sensor network, Procedia Engineering, Volume 41, pp. 75–81, ISSN 1877-7058, https://doi.org/10.1016/j.proeng.2012.07.145. (https://www.sciencedirect.com/science/article/pii/S1877705812025313).

Chapter 14
Intelligent transportation systems

Mohsen Sahal Mohsen Fadhl[1] *and Wai Yie Leong*[1]

In the few last decades, intelligent transportation systems (ITS) have been built and implemented to enhance transport protection and accessibility, rising transport network environmental effects of road accidents on road safety, encourage sustainable transport growth and increase productivity. ITS blends high-tech and upgraded computer networks, communications, sensors, controls and sophisticated mathematical methods with the modern transport network environment. As an interdisciplinary study area, it is challenging to provide a good view of the whole program for an inexperienced researcher.

14.1 Introduction

Connected vehicles models the smart transport system on the pivot of wireless communications with expectations to improve the transportation system efficiency and traffic safety. The communication taking place without wires between the connected vehicles and the transport infrastructure, that among vehicles, or that between the vehicles and the wireless devices is also known as vehicle to infrastructure, vehicle to vehicle, or custom vehicle networks, and car to device communication, respectively. During the past decade, the connected vehicle communication connections attracted a great deal of interest from a wide range of academics, industries (telecommunications, vehicle manufacturers, etc.), governments (transport authorities), and standards organizations (ASTM, ISO, ETSI, IEEE, SAE, etc.) [1]. The synergy among these actors has caused numerous concurrent research and development projects worldwide, such as Intellidrive, CVIS, AHS, and several consortia (VIIC, C2C_CC, etc.). All this action is expected to expand, enhance and demonstrate the viability of the Connected Vehicle Communication systems. The principal inspiration driving the development of the Connected Vehicle Communication system is to improve road safety of the cars fitted with applications. This system provides the intelligence and support the driver's needs to prevent accidents on the road and have a smooth ride home [1]. Taking into consideration that public safety must have a scalable, robust, and

[1]The Institution of Engineering and Technology, Malaysia

cost-effective core technology program backboning the connected vehicle communication applications in cars required to perform at the Quality of Service, provided the latency of the wireless signals connecting the cars reliably provide the quality required, not to mention coverage, bandwidth, and also availability. Although many other-than-safety applications may be enabled, the service requirements in terms of latency and reliability of the safety applications must preside above the quality requirements of these other technologies [1].

14.2 Intelligent transportation system

Intelligent transport is the use of advanced means like information controllers to manage traffic efficiently, use of advanced and integrated applications of information transfer from sensors and software, and how the transport management add information to the system [2].

14.3 Types of ITS

Intelligent transport systems are classified into five classes that are common to the concerned parties as follows:

1. Advanced traffic management systems.
2. Advanced mobile information systems.
3. Commercial vehicle operations systems.
4. Advanced public transportation systems.
5. Advanced vehicle control and safety systems [2].

14.4 The necessities of this system

1. Information gathering: This includes, firstly, the placing of sensors in particular places on the roads in order to collect data and gain information on traffic to facilitate traffic management and to help reduce traffic congestion. The sensors are linked to smart traffic signs which, added to the information collected from sensors, help in evaluating the system in the short term and strategy-wise to roll out plans for the development of mass transportation. The system is considered costly and an important part of such studies. Secondly, on the traffic side, the roads have cameras fitted at strategic locations, and the cameras are used in evaluating road characteristics for driving cars, the security of roads, traffic violations and its monitoring, etc. [2].
2. Data processing techniques: the software and devices to convert data into information for the driver to use, such as do occurrences of traffic jams make it necessary to reduce speed limit or how many changes in the track of vehicles due to the traffic jam [2].
3. Control techniques: Traffic signs, indicative and warning signs, coordination with public transportation using agencies based on various transportation activities, establishing control rooms and traffic and transportation information banks [2].

14.5 Statistics of the car connections

Annual production count of motor vehicles is expected to reach 800,000 units worldwide by 2030 as shown in Figure 14.1. Motor vehicles are light vehicles equipped with advanced auxiliary systems for level 4 or level 5 drivers (very little human intervention or no human intervention is required). This high degree of automation can be used for the purpose of providing transportation to passengers without the need for a human driver as shown in Figure 14.1 [3].

While we head towards what many expect the future of driverless transport to be, in this technological competition to build a completely autonomous vehicle, many companies are currently involved in it as shown in Figure 14.2. In recent years, driverless cars are no longer science fiction because artificial intelligence has developed so rapidly. Indeed, in the United States, the autonomous vehicles have traveled millions of miles on public roads, and, in recent years, with a markedly small number of tragic incidents recorded [4].

This graph compares the statistical data from the National Safety Council from two polls. One poll happened in January and the other happened during the end of March/early April 2018. 36% of respondents reporting in January said that they found self-driving cars to be less safe than a human-controlled vehicle. But, in the beginning of April, this percentage had increased by 14% by the same anxiety index. This increase in concern is the result of closing in of the fantasy to reality's

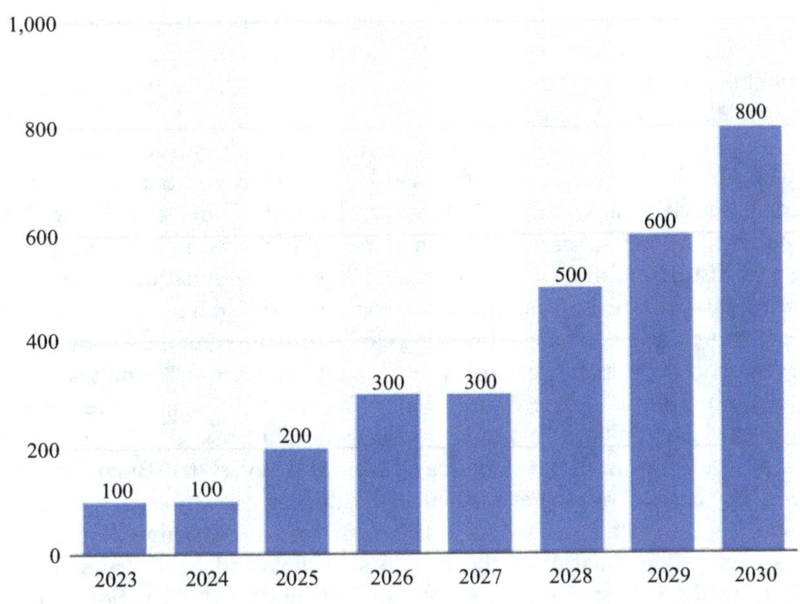

Figure 14.1 The expected production level of motor vehicles between 2023 and 2030 [3]

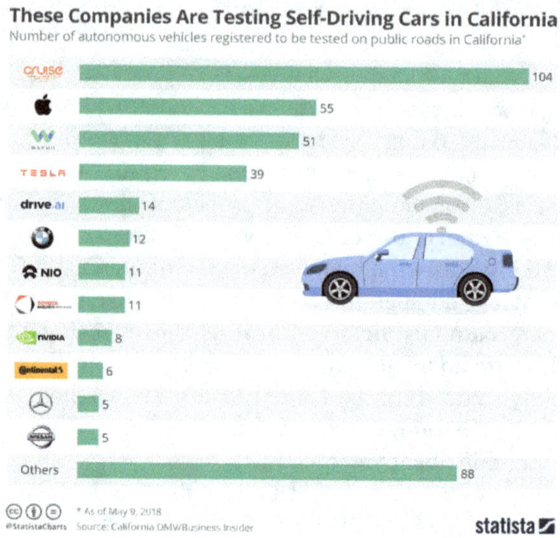

Figure 14.2 The number of autonomous vehicles registered to be tested on public roads in California [4]

door as some companies, such as Google spinout Waymo, plan to offer self-driving cars to the public as shown in Figure 14.3 [5].

The chart below may indicate to some that the United States is not yet ready for self-driving cars. However, everyone should want one because they will be safer than the alternative. After all, there are over 30,000 deaths from car accidents in the U.S., alone, each year as shown in Figure 14.4 [6].

Self-driving cars are in the focus after two fatal tech crashes in the U.S. on 16-4-2018. The first included a pedestrian killed by an Uber autonomous car; while the second was a Tesla Model X autopilot that crashed into a highway divider before it caught on fire. Ipsos conducted a major study on people's ideas about self-driving cars just before these couple of incidents and found that the preference for self-driving cars was the highest in the developing world. Indian participants in the survey are particularly keen on self-driving cars with 49% of those surveyed saying that they support the idea and cannot wait to use it. Another 46% said that they are unsure but found the idea interesting, while 5% said they would never utilize it. China experienced similar levels of optimism, while in Russia, 33% were supportive and 59% unsure. In South America, three out of every ten Brazilian respondents said they cannot wait to get a ride in a self-driving car, while six out of ten are still unsure but interested. In the U.S., skepticism was already high before the two incidents occurred. A quarter of the people said they will never try self-driving technology while 54% said they are not sure but interested. Twenty-two percent said they support the idea and cannot wait for the driving test to start. India showed an enthusiasm much higher than anywhere else as shown in Figure 14.5 [7].

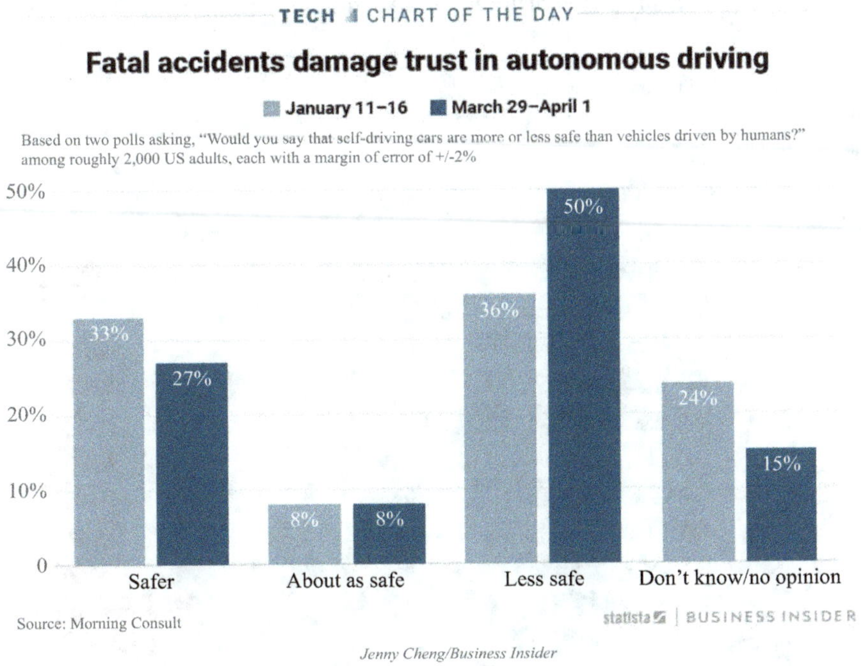

Figure 14.3 Fatal accidents damage trust in autonomous driving [5]

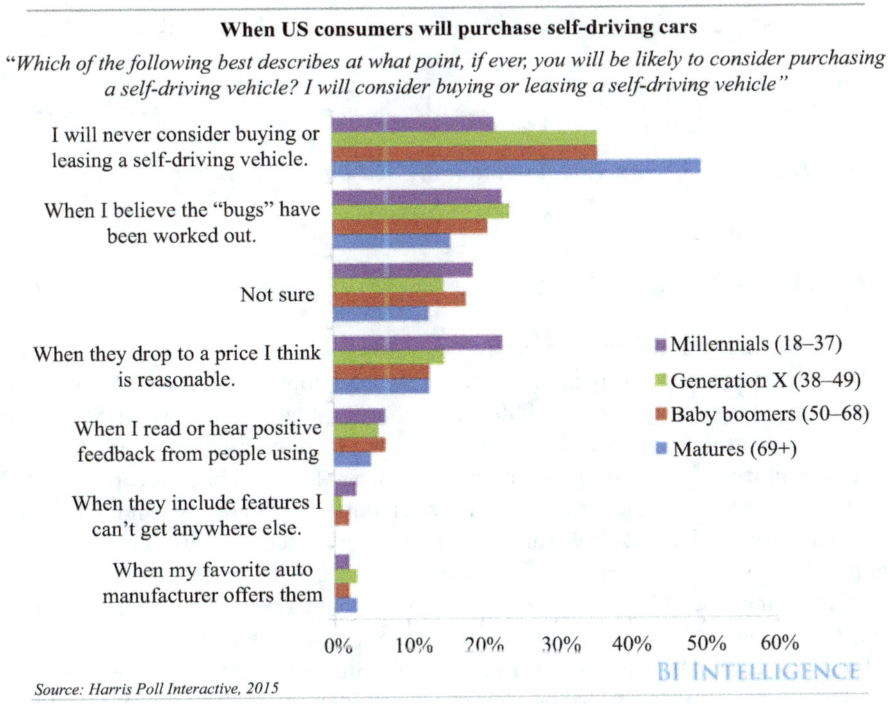

Figure 14.4 When US consumers will purchase self-driving cars [6]

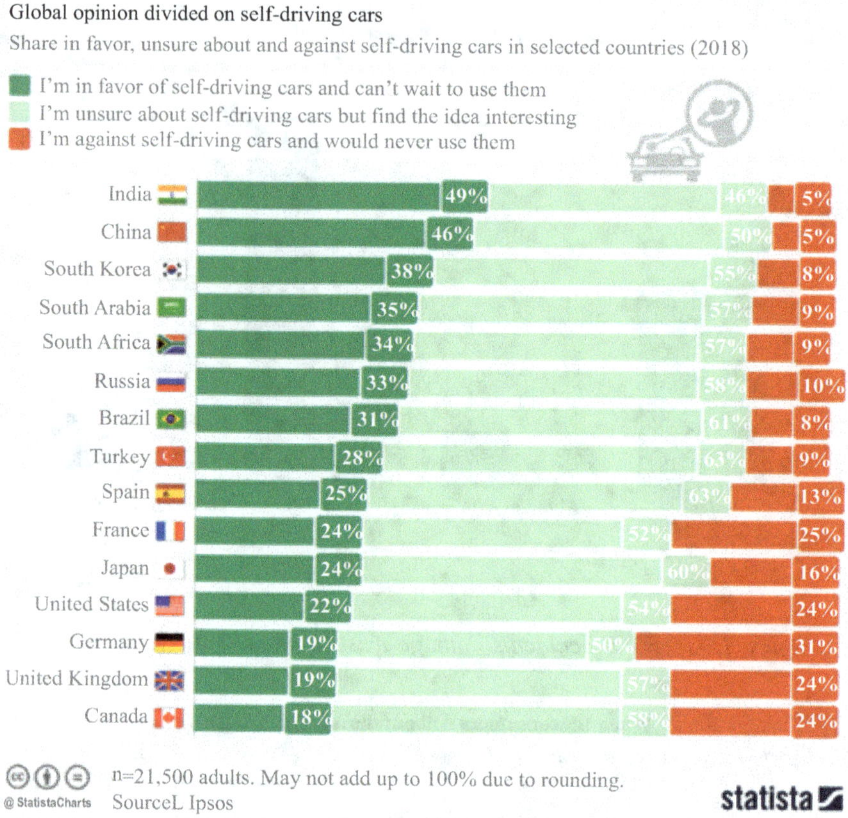

Figure 14.5 The global opinion divided on self-driving cars [7]

14.6 History

Cars began to communicate primarily with other sources for safety in the late 1990s. For emergency notifications, the cars were initially equipped with cellular connections. Beginning in year 2000, the technology has been developed for car security and comfort, and included remote sensing technology associated with cell phone monitoring [8]. During a similar period, G + D Mobile Security started making associated vehicle protection arrangements. Beginning around 2010, the expanded network had taken chance with cutting-edge security driving highlights, stopping/review, self-driving, and remote-control capacities. In the early years, specifically the late 1990s, in the early long periods of telematics-related car benefits, the highlights were basically crash warnings to crisis responders, vehicle finding and emergency aides. In 1996, the OnStar simple cell framework was declared. As the airbags spread, their frameworks interfaced with an OnStar advisor who passed on data to crisis responders. In 1997, Mercedes-Benz presented the principal remote key called "Key – less Go" for the W220 S 1998 classification. BMW propelled data administrations with BMW help. In 1998, the GM improved OnStar's administration by evolving plants and introduced models that permit sans hands calls and voice acknowledgment [8]. In 1999, the Mercedes-Benz dispatched TeleAid remote help with emergency aides and found acceptance with vehicles. In the 2000s, the services propelled in by this agreement incorporated propelled diagnostics, route, web associations, taken vehicle log jam, and cell phone applications with constrained remote highlights. TeleAid highlights a Mercedes S-Benz model that opens the entryway remotely. In 2001, the OnStar started giving ongoing traffic data and opened the entryway remotely with the fourth era with about 8 million connections. Vehicle remote demonstrative gadgets propelled by Continental arrived. In 2003, the OnStar GM Good wrench Remote Diagnostic Service became accessible. Mainland offered vehicle well-being and versatility gadgets bit by bit through a computerized organized get to gadget. In 2005, the Navigation was turned on by running each progression independently. In 2007, the slow vehicle administrations were taken from OnStar. The first iPhone was introduced. BMW Assist administration propelled in the United States with 4 years of free help remembered for vehicle purchase. Mainland had invented advanced information technology, otherwise called "machine-to-machine data advancements," which are presently being adopted widely around the world [8]. In 2008, the dispatch of the main Android cell phone was completed. Avoid Chrysler and Pocket Uconnect Web, fueled via Autonet Mobile, furnished uninterrupted web association with convenient 3G Wi-Fi hotspots. In 2009, the Mercedes-Benz dispatched mbrace administration that permits proprietors to interface by means of iPhone and BlackBerry by remotely bolting an entryway or opening and finding the vehicle. In 2010, and after more applications, traps and hacks, in the multi-decade, there were more vehicle associations through various gadgets. The advancement of electric vehicles just invigorates the requirement for remote applications. A few vehicles have programming refreshes over the air. New administrations require more associations and more information. The main vehicles show up in V2I and V2X innovation. The expansion of remote applications and administrations in new structures, like the Nissan LEAF was propelled with the CarWings iPhone application that gave access to charging, driving reach and atmosphere control. The OnStar Mobile application was unveiled in 2010 CES, which permitted Chevy Volt proprietors to set charging time. In 2011, the OnStar tested FamilyLink, which finds relatives and sends instant messages and email alarms. Propelling Nissan LEAF applications came BlackBerry and Android gadgets. In 2012, the Mercedes-Benz embraced Facebook, Yelp, Internet browser, geo approach, alarms, and added capacity to accommodate companions. The New York Auto Show displayed Audi 7 debut with 3G web association with Wi-Fi hotspots, Google Earth route. The primary Tesla Model S was presented, including 3G availability and the capacity to refresh programming over the air. In 2013, the Audi A3 presented the principal vehicles with a 4G LTE information association in Europe. In 2014, the Wi-Fi 4G

LTE hotspots became accessible for most GM 2015 vehicles in the U.S. Tesla S models were outfitted with gadgets that empowered autopilot included after a future programming update. In 2015, the weak spots implied that ADAC (Allgemeiner Deutscher Automobil – Club) analysts can copy BMW servers and remotely open vehicles. It is called e-Call. Mercedes-Benz Remote Parking Pilot application permits the driver to control the leaving outside of the vehicle. Charlie Miller and Chris Falasic hack a Cherokee jeep remotely, from a PC 16 km away. It controlled brakes, radio, windshield and quickening agent wipers. Sami Kamkar has made a gadget that can catch the BMW/Mercedes-Benz mbrace, Uconnect and Viper remote control framework associated with vehicles remotely using iPhone iOS applications. Mercedes-Benz uncovers the new E-Class with the first creation model with Car-to-X correspondence. Tesla Motors discharges over-the-air 7.2 programming update with self-driving innovation for Autopilot. In 2016, the Hacks uncovered various vulnerabilities for different models empowering programmers to assume control over auto capacities, even drive them away. Tesla discharged 7.1 with the Summon highlight to control leaving outside of the vehicle. Audi America reported V2I innovation (Traffic light data) for select 2017 Audi models. BMW Connected Version 3.0 for iOS includes Alexa, first mile route, vehicle administration alarms and an Apple Watch application. BMW Connected 1.0 for Android is accessible. In 2017, the G + D Mobile Security turned into a trusted e-SIM the board accomplice for BMW and their Connected Drive Services [8].

14.7 How connected vehicles work

The associated vehicle framework of connected vehicles system is still being worked on, and much research remains to be finished. The security-related frameworks of associated vehicle innovation are probably going to be founded on Dedicated Short-Range Communication (DSRC), an innovation like Wi-Fi. DSRC enables vehicles to communicate with each other and other road users directly, without involving cellular or other infrastructure. Unreliable applications may rely upon various kinds of remote innovations [9]. Private vehicles, trucks, public transports, and different vehicles will have the option to "talk" to one another inside the vehicle or via post-retail gadgets that continually share significant security and transportation data with one another. Associated vehicles can likewise utilize the "talk" remote association with traffic lights, work zones, expense stalls, school territories, and different sorts of frameworks [9]. The vehicle data being sent is encrypted and, so, vehicles cannot be followed and the framework is safely giving associated vehicles' drivers 360-degree consciousness of likewise connected vehicles within a scope of approximately 300 m. This safe framework keeps individual data hidden and does not follow your vehicle drivers who will be admonished if they stray out of the restricted precincts of driving lane that advise them regarding potential risks through a visual presentation, seat vibration or tone. These are just alerts taking care that the driver stays in charge of the vehicle consistently.

Anyway, the innovation can likewise bolster constrained robotized works in which the driver just gets halfway control of the vehicle and alerts can assist drivers with reacting rapidly to dodge potential crashes; for instance, the convergence development help give application cautions to drivers when it is hazardous to enter a crossing point, where the do not pass application cautions drivers. When it is not safe to pass a more slow moving vehicle, the crisis electronic brake light application tells the driver when a far out vehicle a few vehicles ahead is slowing down and it envisions by estimation this car as vulnerable. The admonition application lets this business driver essentially observe what's going on in their vulnerable side associated vehicles, which can speak with sensor-fitted side of the road. At rail level intersections, to put drivers on alarm when a train is approaching regardless of whether the driver can see or hear the moving train, the associated vehicles application can help. With climate-related traffic, and considering wellbeing issues too given this is particularly evident with hostile natural conditions: for example, dark ice is a condition where the climate may not give an impression of being perilous, yet, the streets are apparently safe, data may be gathered from numerous associated vehicles, which can help decide when a potential danger exists. For example, frosty streets can be cautioned to drivers before they experience them. Road climate information from the vehicles can be sent to traffic board habitats or TMCs giving the minutest details of what is ongoing to help screen and oversee transportation risks. The framework executes and takes care of activities like altering traffic signals and changing speed limits telling safety upkeep groups, dispatching support vehicles to the streets, and broadcasting admonitions for drivers about safety who can get continuous road climate data [10].

14.8 Literature review (introduction)

This research presents a review of the smart car as the main technology of the smart transportation system, and presents the work of one of the many high-tech integrated transport organizations. Connected vehicles is a general term referring to comprehensive vehicle technology to complete one or more driving tasks in whole or in part. An important field for robotics application, it has emerged in the last 20–25 years where it revolves around cars and smart vehicles [11]. The car was among the twentieth century's most important products. It generated a huge industry and gave free movement to individuals which completely changed our lifestyles. Yes, with the way our modern economies are organized, the automobile was a huge influence and a significant transition from old to modern lifestyle. More than 800 million automobiles are present on the planet today, and this figure is predicted to increase in the next decade. This major challenge really has led to the development of an active research field that aims to automate the typical tasks that people perform while driving. The smart car is defined as a vehicle optimized for perception and reasoning, and the operating devices that make it possible to automate driving tasks such as following the safe lane, obstacle avoidance, to negotiate heavy traffic, to pursue the car in front of you, to evaluate and avoid hazardous conditions, and to decide the lane to drive in. The overall motivation for building smart vehicles has been to improve driving safety, comfort, and efficiency [11].

14.9 Fundamental autonomous vehicle technology

Smart vehicle technology is a concept usually associated with developing autonomous vehicle functionality. Smart vehicles have the main features of:

- The capacity to sense the vehicle's state and its environment.
- Environmental communications abilities.
- The capacity to plan and execute the most suitable exercises.

Smart vehicle technologies is a fast-growing field that the automotive industry, academia, and government agencies are seeking actively. Numerous unmanned ground vehicle (UGV) competitions around the globe are also fueling the general curiosity in smart car technology [12].

Although the self-driving car is not currently a target for car companies, components of this technology are rapidly making their way into passenger vehicles to assist the driver at critical moments. In the automobile industry, the implementations of connected car technology are also used as the next wave of vehicle protection systems. Specifically for Automobile uses, Richard Bishop describes "smart vehicle systems" as devices detecting the driving situation and supplying knowledge or vehicle power to support the driver in the optimal vehicle activity [12].

14.10 Intelligent driver model (IDM)

We address the IDM in this section [13]. The IDM is the most frequently utilized model for smart vehicle simulations because it is one of the easiest and most accident-free models that produce a realistic acceleration profile in a single lane situation. The IDM is similar to the Adaptive cruise control (ACC) vehicles than to the characteristics of the human-driven vehicle since it has no specific reaction time and is given in a continuously differentiated acceleration function [28]. By changing certain parameters, we can use the Internet Download Manager (IDM) itself as an ACC or a model of a human-driven vehicle. In addition, the amended IDM is used to simulate connected autonomous vehicles (CAVs) [13,29]. The basic function of IDM is as follows:

$$a_{IDM}(s, v, \Delta v) = \frac{dv}{dt} = a\left[1 - \left(\frac{v}{v_0}\right)^{\delta} - \left(\frac{s^*(v, \Delta v)}{s}\right)^2\right] \tag{14.1}$$

$$s^*(v, \Delta v) = s_0 + vT + \frac{v\Delta v}{2\sqrt{ab}} \tag{14.2}$$

In most situations, the IDM results in a plausible rate of acceleration and deceleration. But when the current gap in vehicles is significantly lower than the desired gap, the rate of deceleration becomes unrealistically high. In fact, when it comes to human-driven vehicles, drivers assume that with the hardest deceleration, the preceding vehicle will not suddenly stop without cause. Therefore, the present difference greater than the ideal difference distance is called a comparatively mild-critical situation [30]. To fix this problem, Arne Kesting and Martin Treiber merged the IDM and the Constant Acceleration Heuristics (CAH) to restrict excessive deceleration speeds. The basic expectation of the CAH model is that the previous vehicle would not unexpectedly adjust its acceleration within seconds [13].

The CAH has three underlying conditions: (i) the acceleration of the vehicle under consideration and the preceding vehicle will not change in the applicable future (generally, a few seconds); (ii) no safe time forward or minimum distance is needed at any time; and (iii) the car owners will respond without delay (zero response time) [13,31].

For actual gap, current speed, preceding speed of the vehicle, and acceleration values given, the maximum acceleration which helps to prevent crashes is provided by:

$$a_{CAH}(s, v, v_1, a_1) = \begin{cases} \dfrac{v2\widetilde{a}_l}{v_1^2 - 2s\widetilde{a}_l} & \text{if } v_1(v - v_1) \ \leq -2s\widetilde{a}_l \\ \widetilde{a}_l \dfrac{(v - v_1)^2 \theta(v - v_1)}{2s} & \text{otherwise,} \end{cases} \tag{14.3}$$

Where effective acceleration = min is being used to prevent artifacts that may be affected by preceding automobiles with higher acceleration abilities. The statement is satisfied if the automobiles stopped at the time the minimum gap $s = 0$ was reached. Otherwise, negative approach rates cannot make any sense to the CAH and therefore are removed by the Heaviside step function Q. By merging IDM and CAH acceleration, Kesting and Treiber [31] have proposed the ACC model as laid down in (14.4). Depending on the following factors, the ACC model generates various acceleration speeds depending on the IDM or the CAH. The ACC model produces the same acceleration if both the IDM and the CAH achieve the same speed outcome. If the IDM causes the unrealistically strong deceleration, whereas the deceleration of the CAH is in a reasonable deceleration range, the condition becomes slightly dangerous, and the acceleration of the ACC stays above the acceleration of the CAH minus the deceleration. If either the IDM or the CAH results in acceleration well below $-b$, the problem is seriously critical, and the acceleration of the ACC must not be higher than the IDM and CAH acceleration maximum. The acceleration of the ACC should be a continual and differentiable function of the acceleration of the IDM and CAH [13]:

$$a_{ACC} = \begin{cases} a_{IDM} & a_{IDM} \geq \ a_{CAH} \\ (1 - c)\ a_{IDM} + c\left[a_{CAH} + b \tan h\left(\dfrac{a_{IDM} - \ a_{CAH}}{b}\right)\right] & \text{otherwise,} \end{cases} \tag{14.4}$$

Similar to IDM, the ACC model includes one extra parameter. It is called a consideration for coolness. Then, the ACC model returns to the IDM even when, under tiny distances, the intensity of the distance varies, and no difference in velocity occurs. Kesting and Treiber [23] assumed that c is = 0.99.

Zhou and Qu [29] established the cooperative intelligent demand model (CIDM) utilizing IDM as the benchmark model and analyzed the performance of CAVs in the system. Communication of the CAV is implemented in the human driver model (HDM) utilizing the concept of spatial anticipation.

The HDM anticipation is applied to the CIDM dividing the IDM into (14.5) based upon (14.1):

$$a_n(\Delta x, v_n, \Delta v) = a_n^{free} + \sum_{m}^{n-1} a_{nm}^{int}(\Delta x_{nm}, v_n, \Delta v_{nm}) \tag{14.5}$$

The foundation IDM (14.1) consists of two main parts: one is the term for acceleration comparing the current velocity to the desired velocity, while the other is the term for breakage comparing the current distance to the expected. In (14.5), with acknowledgment of V2V activity, it is the same description of, which is the same as in (14.1) [13].

14.11 Vehicle networking

Networks may function in more than one context to prevent accidents between individual vehicles. If vehicles can communicate with each other, they can serve as a way to decrease personal automobile accidents or numerous car crashes [14]. Accident avoidance may also involve independent autonomous systems and integrated systems, to produce the best performance. The researchers have begun to combine a self-control framework with machine learning algorithms to improve the performance of a possible dual system that incorporates autonomous and combined systems [32]. Desjardins and Chaptra Draa are turning their attention to adaptive cruise control (CACC). They achieved success in simulation development, which proved the importance of utilizing autonomous and integrated CACC systems. Figure 14.6 shows the conceptual degradation of the CACC definition for readers. The knowledge obtained from the atmosphere is transmitted to the GPS, which is filtered through it so that the device can make the changes needed. With certain changes to the infrastructure and technologies, this can be extended to a practical application [14].

14.12 Enabling technologies

1. Environment state:

   Sensing the world surrounding the automobile is an essential feature in smart car systems [11]. Knowing the traffic scenario is the most challenging task in smart cars. It involves defining the principal characteristics: lane, other cars, people, traffic lights, road signs, and other unorganized obstacles. The toughest problem is pace regulation when an incident is detected in a road scene. Popular sensors include ultraviolet, ultrasound, radar, laser rangefinders, and

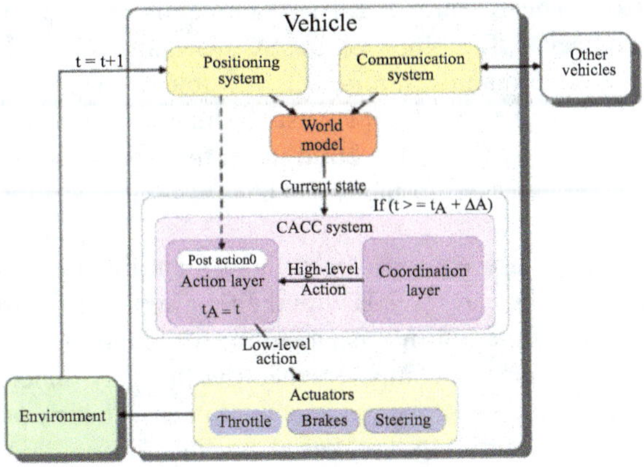

Figure 14.6 CACC layout [14,31]

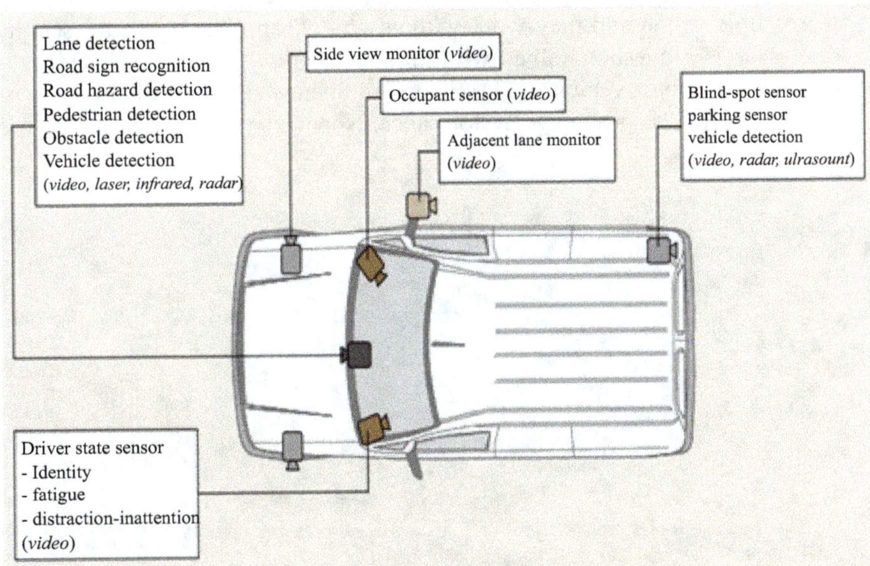

Figure 14.7 Environment state sensing [11,32]

computer vision, which scans the world as seen in Figure 14.7. Radar is typically used to remotely locate obstructions, while infrared and ultrasound are utilized to identify obstructions in the area. Under varying environmental situations, laser spectrum and image analysis are used to provide more energy over the road scene. Any aspects of the road scene such as road signals and traffic indicators can only be noticed by using a vision sensor. In smart vehicle applications, sensor fusion is widely used, particularly in both monovision and radar/laser sensors. This research is carried out mainly by level one automotive suppliers (Bosch, Denso, Del phi, Visteon, Siemens, and NEC) in collaboration with major automotive manufacturers. It is possible to map the terrain around the vehicle using these sensors, then use techniques such as simultaneous localization and mapping [33], then build the complicated maneuvers required to park the vehicle or escape obstacles [11].

2. Tracking path/lane:

   Several car designs are equipped with lane identification and monitoring devices, beginning in the early 1980s with the first implementations. Computer vision plays a crucial function in this situation. Although the path can also be scanned with laser scanners, computer vision is the only general technology capable of detecting lane structure and lane markings with high precision. Most approaches to lane monitoring are based on identifying lane signals and leveraging the layout in the area, such as comparisons between left and right lane markers, set road width, or expectation of a commonly traveled flat path. These assumptions were used mainly to fix the issue of providing a single camera (a cost-effective option). For identifying track lines, certain devices use stereoscopic vision and may work without any constraints. In road cases, the track issue is a question solved with the introduction of industrial devices in travelers and motor vehicles [11]. Figure 14.8 shows an indication of a commercial lane tracker's average performance. Nonetheless, these systems cannot

Figure 14.8 Lane detection [11,33]

promise 100% reliability for lane recognition systems; such systems typically run with 95–99% reliability. Thus, lane detection devices are used mainly in-lane deviation alert systems where no malfunctions can be allowed for autonomous driving. Efforts are being made to build algorithms that accommodate a range of driving environments and drive efficiency levels of 100% [11,34].

3. Detecting traffic signals:

   Another application of machine vision, specifically, is identifying and recognizing the road signs. Road signals are specifically built to support individual users. Road signs utilize a well-defined mix of forms, colors, and designs. Banners are placed against the road at appropriate heights and locations. The interpretation of road signs is also a workable job for machine vision. Recognition is achieved utilizing a mixture of type detection systems and/or light detection. The identification is rendered during the recognition and localization point. This function is generally accomplished by pattern matching strategies such as cross-linking between images and neural networks or vector supporting machines due to the small and well-specified possible range of road signs. Figure 14.9 demonstrates the idea of a speed warning device based on the velocity signal detection. The obstacle for research work in this field resides in the robustness of mark recognition identification and reliability. Many automotive makers are in the midst of designing technologies [15,35].

Figure 14.9 Street sign identification for transmission of level warning[11,35]

4. Detection of cars:

   Detection of cars has been approached utilizing a broad variety of sensor methods, varying from vision to lidars, from radars to sonar. Notwithstanding the different shapes and colors, the substances possess the same properties and are distinguished by a broad scale and a transparent surface. The location of the vehicle may be predicted until there is an estimated indicator of the direction of the path/street. Indeed, several separate sensors can autonomously identify vehicles successfully (Figure 14.10). A vision-based device exists for car identification. While the answer to this question appears straightforward, every sensor has its own area of operation and its own challenges. Vision is usually high but can struggle as vehicles clutch each other in low visibility and bad lighting situations (night or tunnels) or in intense traffic conditions. Infrared vision (thermal imaging) can identify automobiles with great confidence because automobile tires and mufflers typically display elevated temperatures and can thus be readily identified in the image. However, stationary cars, trailers, and even cars only beginning to travel are cooler than moving vehicles and hence less visible. Lidars are generally solid but have lowered sensitivity under adverse weather conditions. While radars are inexpensive, due to the existence of certain surrounding reflecting objects, they can exhibit bias inside measurements. The sonar can only be used for very short periods, at least. The work goal is to introduce a successful combination of multi-sensors. A growing solution is to incorporate radar visibility [11,36].

Figure 14.10 Vehicle identification-location of the lane and the length [11,36]

Figure 14.11 Advanced cruise control (ACC) [11,37]

5. Adaptive cruise control:

   Adaptive cruise control (ACC) is the functional expansion of the traditional cruise control to provide a reasonable distance from the previous car. If there are no cars in front of the smart car, it maintains a fixed pace, as per automatic cruise control. When a slower-moving vehicle is in front, the vehicle can be detected by an ACC-equipped automobile using radar or lidar, and will fail to retain a reasonable distance (typically set to a range of 1.5–2 s). Figure 14.11 shows an example of the ACC principle. The sensing task for ACC is somewhat simpler than the rear-end impact countermeasure problem, as ACC devices are built primarily for other driving cars. The greatest sensing complexity for rear-end collision countermeasures is the separation of stopped cars on the road from off-road objects; for ACC, this complexity is overlooked by neglecting all stopped items. Moving items are marked as in-lane or out-of-lane depending on a variety of heuristics [16]. Also, the smart vehicle's own steering range is used as a street bend gauge to assess if the cars ahead are on a specific road. Because the devices are specifically marketed as a comfort rather than protection system, they can only handle normal situations with typically low-speed contrasts and leave the more difficult circumstances to the human driver. The human being is always aware, monitoring the course, and observing the traffic. Both big automotive manufactures are introducing such schemes [16,37].

14.13 Modulation system in intelligent vehicles

Smart vehicles used the V2V and V2I communication system which used the spatial modulation (SM). SM is a hybrid Multiple-Input Multiple-Output (MIMO) and modulation technology initially developed for RF communication that increases spectral network performance and decreases device variability at the same time.

14.14 Advantages and disadvantages of SM

1. Benefits:
   - SM absolutely evades both Inter-. Channel interference (ICI) and Inter Antenna synchronization (IAS) [17].
   - It needs just one RF chain on the transmitter. This is attributed to the SM system of motion. A Single transmitting antenna for data processing transmission while all other antennas stay silent [17].
   - 3D telescope diagram in SM offers double spatial gain by growing the logarithmic field with the number of the transmission antenna [18].
   - SM offers a highly effective, spectrum-based symbol with an equal symbol rate greater than one [18].
   - Receiver architecture is easy since interference de-mining algorithms are not needed to manage ICI as compared to traditional spatial multiplex methods for MIMO systems [17].
   - SM will access ML decode with a single current receiver [17].
   - SM will function effectively if the $Nr < Nt$ because the reception of the antenna used for diversity only receives.
   - SM is capable of operating in multiple control situations because of the separate pairs of transmitters and receivers. It typically occupies various spatial positions. When all receivers use the channel pulse array, responses of all transmitters is to detect data (e.g., several users are detected); other users can use the same wireless services for telecommunications [18].
   - SM has higher power than traditional low-complexity encryption methods for MIMO systems [17].
2. Disadvantages:
   - At least two transmission antennas are needed to allow the use of the SM principle [18].
   - If the radio transmits and receives where wireless links are not sufficiently distinct, the SM model may not be utilized or may not contribute to satisfactory results, and this constraint is very similar to conventional spatial multiplexing strategies, which allow a rich dispersion setting to achieve a substantial improvement in the rate realizable data [18].
   - The receiver needs complete channel awareness to identify data, which can place specific restrictions on the channel estimator [17].
   - SM only produces a logarithmic (rather than linear) rise in the data rate with the transmitting antenna level. It can restrict SM to obtain very large spectral efficiencies for the realistic amount of transmitter antennas [17].

14.15 Gap analysis and benchmarking

Most people usually do not consider the car sitting in their driveway to be on the leading edge of new technology. However, for most people, the personal automobile has now become their initial exposure to new intelligent computational technologies such as fuzzy logic, neural networks, adaptive computing, voice recognition and others. In Table 14.1 we analysed the gap and benchmarked with various intelligent vehicle systems that are now being deployed into motor vehicles. These intelligent system applications impact every facet of the driver experience and improve both vehicle safety and performance. We also described recent developments in autonomous vehicle design and demonstrate that this type of technology is not that far away from deployment.

14.16 Industry Revolution 5.0

There is a strong need for smart mobility schemes because of the high expense of urban traffic jams, but the introduction of self-driving vehicles would make it much more necessary in the future. Intelligent mobility services would still need to build and incorporate technology such as cloud infrastructure, 5G, the Internet of Things, sensors and artificial intelligence in order to enable and operate the fleets of automated vehicles at the city level, leading to the complete reliance of self-driving cars on IT networks. Therefore, within 10 years, smart transit networks will effectively act like a core nervous system for sustainable cities, and cases like Lisbon and Toronto give us an idea of what such networks feel like. Smart transit networks would have a huge effect on emerging towns, but they would still have a significant part to play in transforming the culture. As long as smart transport systems are common, many sectors will also be changing. To start with, most types of insurance will not be needed by a scientist with entirely autonomous vehicles. Real-estate prices will now be drastically altered because infrastructure cost needs to be within control, and police and other public protection departments will be forced to spend their resources on non-car related problems (because with the connected vehicles system in place, other forms of income, such as traffic fines, may disappear). In order to achieve this, though, you would require a robust information and communication technology system where innovations such as 5G networks, sensor networks, artificial intelligence and cloud computing would be required to capture, process, preserve, and interpret these data [19].

14.17 Important technologies in intelligent transportation system

1. Advanced tracking system: Today, the main part of vehicles is fitted with GPS in-vehicles. The GPS framework gives two-path correspondence to help traffic experts find vehicles, screen vehicle speed and allot to crisis administrations. Cell phones, versatile applications, and Google maps have become helpful instruments for following errant drivers, for finding out about street quality, traffic thickness and for finding various courses and areas [20].
2. Advanced sensing technologies: These give intelligent sensors to two or more vehicles and street systems. Radio frequency identification (RFI) and keen signal sensor frameworks ensure the wellbeing of walkers in networks the

Table 14.1. Gap analysis and benchmarking

References	Authors	Intelligent vehicle technology	Gap analysis
[28]	Treiber and Kesting	Traffic flow dynamics, data, models and simulation	It has no clear reaction time and is given in a continuously differentiable acceleration function
[29]	Zhou, Qu, and Jin	The impact of cooperative autonomous vehicles in improving freeway merging	Unrealistically high deceleration rate
[30]	Treiber, Kesting, and Helbing	Understanding widely scattered traffic flows	Ability decreases, and platoons as results of variance-driven time gaps
[31]	Kesting, Treiber, and Helbing	Enhanced IDM to access the impact of driving strategies on traffic capacity	Royal Society's Philosophical Transactions A: Mathematical, Physical & Technical Sciences
[32]	Desjardins and Chaib-Draa	Cooperative adaptive cruise control: a reinforcement learning approach	Due to possible network errors and/or delays, collision prevention systems do not rely solely on networking to operate
[33]	Sebastian Thrun and John J. Leonard	Simultaneous localization and mapping	The SLAM question is the issue of measuring the world and at the same time deciding the location of the robot relative to the map
[34]	Apostoloff and Zelinsky	Robust vision-based lane tracking using multiple cues and particle filtering	The systems cannot promise that the lane identification systems would operate with 100% accuracy
[35]	Piccioli, De Micheli, Parodi, and Campani	Robust method for road sign detection and recognition, Image Vis	It has a problem associated with the difficult conditions of image acquisition, and therefore the problems of noise, gradient
[36]	Kato, Ninomiya, and Masak	An obstacle detection method by fusion of radar and motion stereo, IEEE Trans	Radars, although inexpensive, can suffer from bias in lateral measurements due to the effect of other nearby reflective artifacts;eventually, sonars are not valid for very small distances
[37]	Kesting, Treiber, Schönhof, and Helbing	ACC design for active congestion avoidance	Some sensors may not work well (or do not work at all) in bad weather like rain or fog

world over. Street reflectors, inductive circles are based out and about, assisting with controlling traffic and to drive securely, particularly during the night. They will likewise keep the traffic network updated about the traffic number in a given time length and can recognize both drowsy and rapid vehicles [20].

3. Advanced video vehicle detection: Outside cameras or CCTV checking will address various issues for traffic chiefs. Video film of key areas and prime intersections may assist administrators with observing traffic streams, recognizing any crisis circumstances and street blockage. In worked in car controls, robotized number plate acknowledgment permits to follow vehicles for wellbeing purposes [20].
4. Advanced traffic light system: RFID recognition is normally utilized in the rush hour gridlock light system. Innovation offers the correct calculation and database likewise when reaching out to a few streets, street intersections and cars. Such lights might be changed during fundamental and pinnacle hour traffic conditions with no manual interventions [20].
5. Emergency E-call vehicle service: In case of a crisis, for example, a mishap or any disaster, the in-vehicle sensors can reach the close-by crisis jog. The e-call can permit the driver to speak with the certified administrator and will likewise transfer essential subtleties legitimately to the jog, for example, date, place, position of the vehicle and portrayal of the vehicle. The e-call administration has been made compulsory across Europe for all new endorsed vehicles [20].
6. Route information: In the event that drivers have early information about the course that is most appropriate to their excursion. Their excursion will be straightforward and unwinding, especially for inter districts. Constant data on traffic conditions, travel frameworks, sharp turns, stop signs, street conditions, moves and different guidelines on the courses can make travel simpler. The driver may get this information from their PC, versatile or media communications organized beforehand to proceeding with the outing. They may likewise screen drive time, traffic stream level, elective streets, street building tasks, transport ways, cost rates, stopping offices by means of variable message signs (VMS), correspondence and remote contraptions [20].
7. Safety and vehicle control: The administration offers wellbeing help to vehicle administrators through alert and vehicle control subtleties. Drivers will survey their driving aptitudes, street conditions and vehicle effectiveness. They will be alert of any front or backside mishaps as they move headings or switch at crossing points by observing the way of different vehicles. Propelled cameras in vehicles may frequently help drivers during low natural condition perceivability situations because of hostile climate or poor night vision by taking photos of the characteristic zone. The objective is to take cognition of the accident or episode condition by sending notice to drivers and rescue vehicle laborers of looming mishaps [20].

14.18 Smart transportation architectures

The core principle of smart transport is to have the right details in the right location and on the right computer to make the transport-related choices confidently and to make it easy for people to use quicker and better ways at road safety. We need software architecture to achieve this goal. Such systems tackle different roles such as parking, flow, flow data exchange, public cars, routing, energy management, etc. [21].

1. Graph-oriented transportation utilizing IoT generated Big Data: Street sensors are mounted at every street corners to quantify the traffic subtleties for all vehicles on the course, including the amount of vehicles, normal vehicle speed, traffic volume, size, and so on. The vehicle arrangement is utilized to acquire data about the individual vehicle, for example, area and speed (Figure 14.12). The associated two frameworks (Road Sensors Network and Vehicular Network) give information to the web through hand-off hubs, organizers, entryways, total and classifiers, and so on [21].

   Navigation is the core feature discussed in this architecture. This gadget might be used to choose the most safe course from source to goal based on existing traffic conditions, or it might be utilized to alarm the specialists viewing traffic, for example, street blockage, higher traffic volume, car accidents, etc. [21].
2. Transportation system architecture for a multi-station shared vehicle system (Figure 14.13): In fact, a community car network usually consists of a number of cars utilized many times a day by multiple people [21].

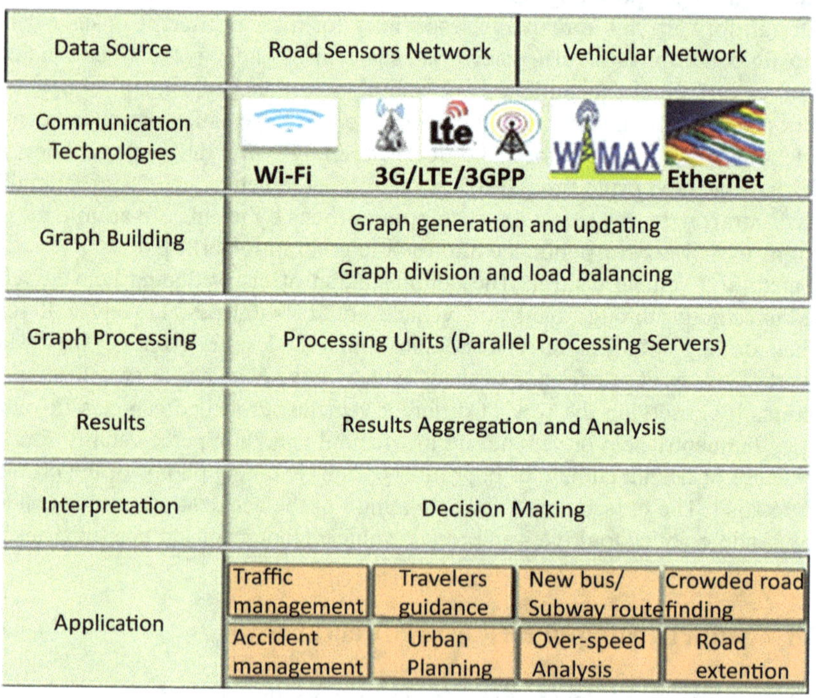

Figure 14.12 Graph-oriented smart transportation [21]

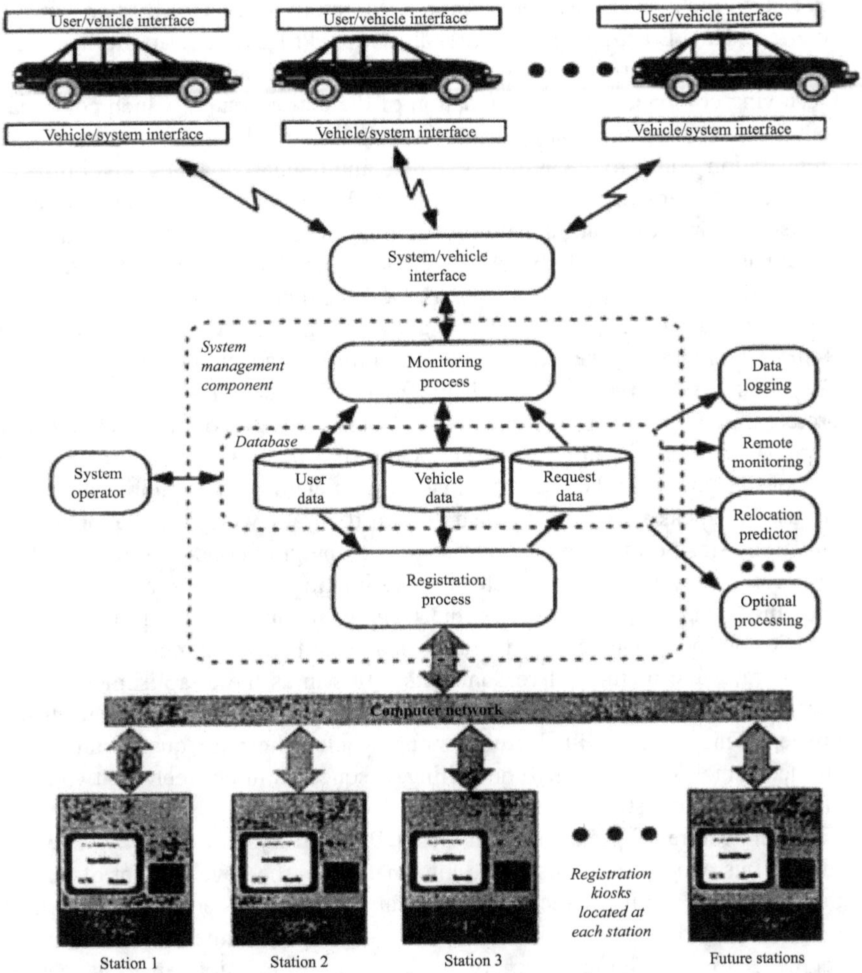

Figure 14.13 Architecture for a multi-station shared vehicle system [21]

There are three general components of this architecture:

1. The utilizer trip-registration component
2. The system management component
3. The vehicle component [21]

The primary goal of the User Trip-Registration Component (lower some portion of the figure) is to enroll a vehicle application that requires recording. To demand a vehicle, the client should arrive at the enrollment work area situated at each station with a contactless card per-user and a touchscreen. The subsequent segment is the system management component (the centerpiece of the figure) which is based on a constant database containing data about clients, vehicles and solicitations. Shrewd Agents are ceaselessly refreshing this database. These keen specialists include: Registration Process, Monitoring Process, System Operator Interface, Data Logging, and Additional Remote Processing. Data from access to the enrollment corners is overseen by an electronic enlistment process. Observing component is the foundation of the entire system, which continually tracks information originating from the vehicle armada and furthermore sends "registration" signs to the assigned vehicle and regularly gauges travel time and separation toward the finish of the outing. The gadget administrator interface causes the framework administrator to follow the situation of every vehicle in a guide-based condition. Both vehicle subtleties and enlistment booths might be concentrated so as to get full appraisal of customer driving conduct, vehicle action and the dependability of the gadget from the executives' procedures. Remote techniques give programmed control of gadget movement, charging clients as they arrive at a pre-characterized timespan, following vehicles to protect if they do not outperform pre-set cutoff points, recording vehicle trouble warnings, vehicle time of appearance and traffic request expectation. The third part of this gadget is the UCR Intellishare Vehicle Module. Radio transponders are used to speak with the administration of the framework. It collaborates with numerous signs on the vehicle to survey the charging condition of the battery, assistant battery voltage, odometer beat estimations, entryway open/shut signs, charging sign, and sign scope of the rigging; it interfaces with a few of the vehicle control signals (e.g. entryway lock and start enact/murder hardware); and incorporates the method of reasoning that fills in as the establishment of convention. The central component of this framework is common armada upkeep. This system will decrease stopping clog, help air quality and render surface vehicles increasingly compelling. It squeezes in between the two clients and gadget administrators [21].

3. The architecture of Transportation-Cyber Physical System (T-CPS): The main idea of this concept is to distribute information on the development of physical things, and the status of the intelligent transportation system to the digital framework (Figure 14.14), and afterwards to join digital frameworks and physical segments to one single assembly made up of calculation, interchanges and control advances so as to accomplish data trade, gadget synchronization and effective vehicle dynamics. Its model executes a collective programming building philosophy concentrated on Service Oriented Architecture (SOA) [21].

   The first layer in this system architecture is the Perception layer. This layer utilizes hubs of sensor and actuators assortment. The key capability is to sense the world's physical resources, including vehicles and utilities. Next level is the Communication layer that incorporates every one of the contact stations and system hubs. The key capability is to transfer the first information deciphered by the layer below to the data center by taking for granted the power of the transmission lines, networking productivity of the contact (gateways, base stations and network nodes). The third segment is the Computational layer. The key assignment of this layer is to continue ongoing information processing and classification with storage examination and re-enactment proficiency. The fourth layer of this structure is the Control layer with its impact on final output for data

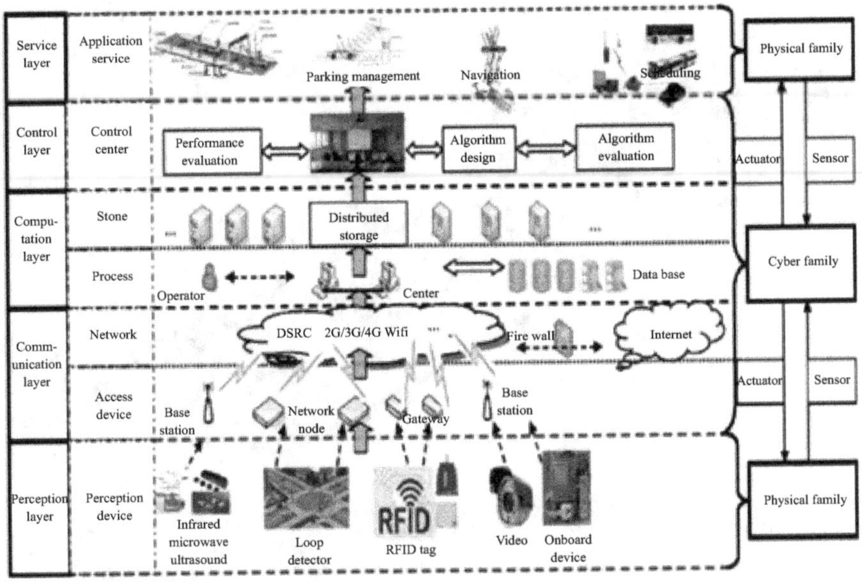

Figure 14.14 Architecture of T-CPS [21]

integrity and verifiability. Based on traffic mass information, this machine can give scientific control calculation and send directions to the control hubs or transmit a crisis ready sign through the actuators. Next up is the Service layer which produces the final output on screen or paper for visualization. The application program (parking management system, navigation system, and scheduling system), would provide information to the shopping terminals and traffic data office. This system interfaces genuine physical components to the web/digital web, yet there are a few issues with this gadget, for example, keeping up very proficient availability, guaranteeing the dependability and consistency of this framework to accomplish sufficient control effectiveness, adapting adequately to irregularities and conveyance, and improving the accuracy and practicality of the conveyance of traffic data [21].

14.19 Intelligent transportation applications

1. Electronic toll collection (ETC): ETC allows vehicles to drive at traffic speed through toll gates, reducing toll traffic congestion and automation of toll collection (Figure 14.15). Originally, ETC systems were utilized to automate the collection of tolls, but more recent innovations have been utilized by ETC to enforce congestion pricing through cordon zones in city centers and ETC lanes. Until recently, most of the ETC systems were focused around the usage of radio equipment in automobiles that would use patented protocols to classify the vehicle when it went through the gantry on the roadway [22].

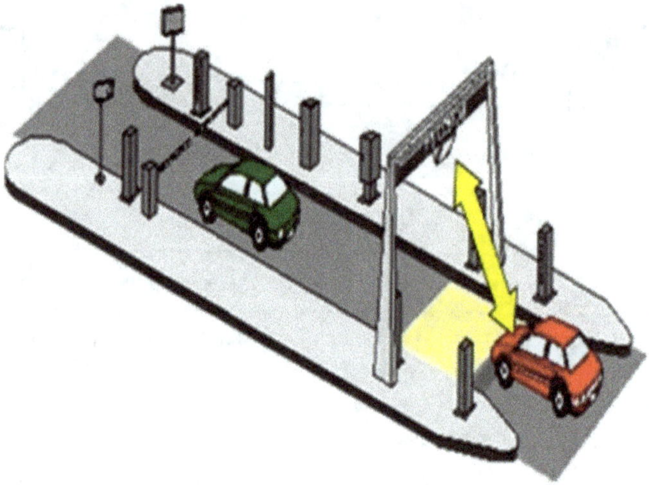

Figure 14.15 ETC system [22]

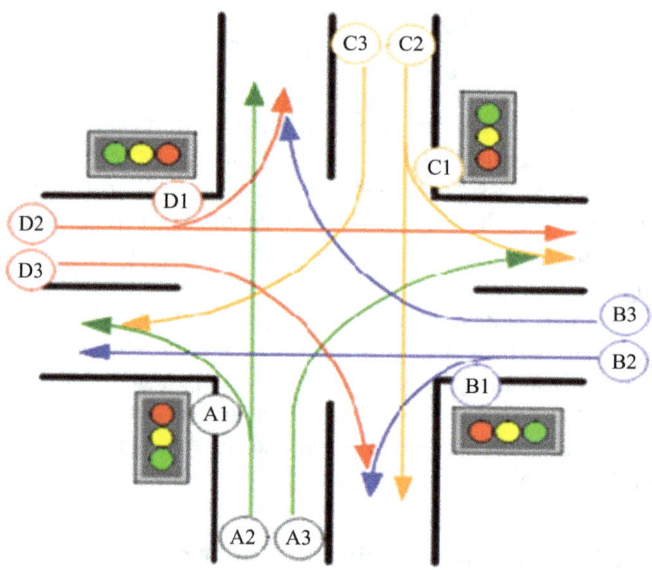

Figure 14.16 Multilane traffic sequence flow [23]

2. Dynamic traffic light sequencing (Figure 14.16): Utilizing RFID for dynamic traffic light sequence bypasses or to avoid problems that usually arise with systems that utilize image processing and beam interruption techniques, the RFID system with an effective algorithm and database is introduced to a multi-vehicle, multilane and multi-road intersection area to provide an optimal time-management scheme. For the passing of each board, a complex time

schedule is worked out. The dynamic sequence algorithm adjusts itself with the presence of some extreme cases. The program may imitate the decision of a traffic police officer on-duty by considering the number of vehicles in each column and the routing characteristics [23].

3. Emergency vehicle notification system (Figure 14.17): An in-vehicle e-call is generated manually by the vehicle occupant(s), or automatically via an in-vehicle e-call generated manually by the occupant(s) of the car or by activation of the in-vehicle sensors after an accident. The e-call carries both voice and data directly to the nearest emergency point (usually the public safety response point). The voice contact allows the occupant(s) of the vehicle(s) to connect with the qualified e-call operator and, at the same time, a minimal collection of data including vehicle identity, accident detail, date of event, and then the exact position and path of travel are transmitted to the e-call operator to take the required action. Mostly found in developed countries, e-call is a common choice for all licensed types of vehicles [24].
4. Automatic road enforcement: Automatic road enforcement system detects and identifies vehicles that disobey a speed limit or some other legal requirement. This then charges the defendant on the basis of the number tag [25]. Traffic passes are then submitted to the owners via e-mail. The program shall include:

   - Speed cameras to identify over-speeding vehicles (Figure 14.18). Many systems also use radar or electromagnetic loops buried in each lane of the road to detect vehicular speed.
   - Traffic light cameras to detect vehicles that cross a stop line while a red traffic light is showing.
   - Bus lane cameras to identify vehicles traveling in lanes reserved for buses.
   - Double white line cameras that identify vehicles crossing these lines.
   - High-occupancy vehicle (HOV) lane cameras to identify vehicles violating HOV requirements [25].

5. Collision-avoidance systems (Figure 14.19): Any of the vehicles already utilize proven technology to prevent accidents. Since car makers are beginning to add

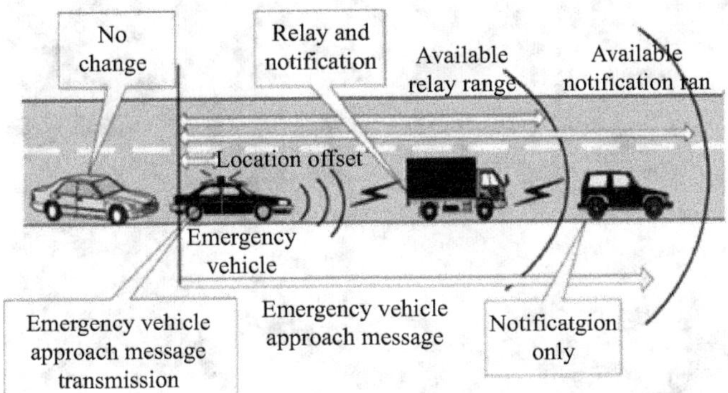

Figure 14.17 Broadcast area for emergency-vehicle-approach information [24]

Figure 14.18 The ASE camera at Putrajaya [25].

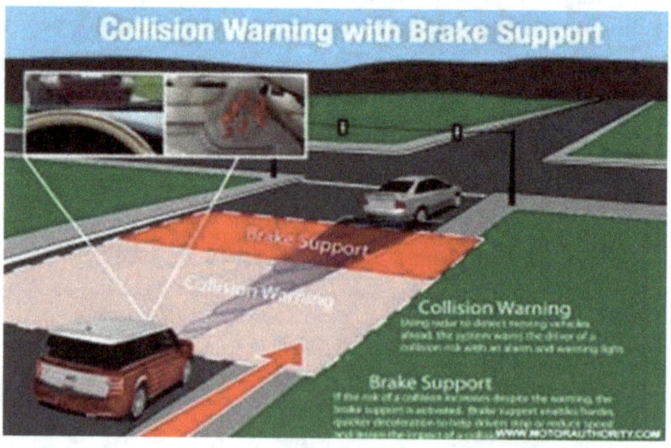

Figure 14.19 Collision avoidance areas [26]

front-facing sensors to locate a vehicle for us, thus when collecting data from front-facing sensors, the collision-avoidance device determines if there is any possible obstacle present. If the speed difference between the vehicle and any other object is too large, the system may be capable of performing tasks that result in a safe driving behavior. The best collision-avoidance mechanism would simply warn the driver, and ideally the driver can touch the brakes and slow down or step away from the barrier [26, 38–41].

14.20 Conclusions

The total population will progressively get familiar with smart systems, like sensors and vehicle communications. The only proof of extensive usage of robotics technology in the automotive sector is the industry 's willingness to produce innovations on total protection, concentrating emphasis on and reliance on the problems of secure systems. The rail and avionics enterprises have tackled these issues. In these ventures, the security cost per vehicle can be a lot higher than a vehicle or transport as experts work and look after request. Producers will gradually create ability in unwavering quality and cost building, and governments will continuously work to determine obligation issues. Therefore, frameworks that put the driver in a pattern of new robotization advances will be the essential focal point of improvement [27].

References

[1] K. Jadaan, S. Zeater, and Y. Abukhalil, "Connected vehicles: an innovative transport technology," *Proc. Eng.*, vol. 187, pp. 641–648, 2017, doi: 10.1016/j.proeng.2017.04.425.

[2] H.A. Idan, "Towards the use of smart transportation system in Iraq," Faculty of Engineering, University of Kufa, 2018. http://eng.uokufa.edu.iq/archives/5061 (Accessed May 03, 2020).

[3] I. Wagner, "Projected production of robo-cars worldwide 2030 | Statista," *Statista*, 2020, https://www.statista.com/statistics/1068697/projected-robo-car-production/ (Accessed May 03, 2020).

[4] F. Richter, "• Chart: These Companies Are Testing Self-Driving Cars in California | Statista," Statista, 2018, https://www.statista.com/chart/13868/registered-autonomous-vehicles-to-be-tested-in-california/ (Accessed May 03, 2020).

[5] "Public worries about self-driving cars are spiking after a pedestrian was killed by a robo-car | Business Insider," Business Insider Australia, 2018, https://www.businessinsider.com.au/public-concern-increase-safety-self-driving-vehicles-charts-2018-4 (Accessed May 03, 2020).

[6] Tehrani.com – Comm & Tech Blog, "Yes, Consumers Will Buy Self-Driving Cars – Tehrani.com – Comm & Tech Blog," 2015, https://blog.tmcnet.com/blog/rich-tehrani/technology/yes-consumers-will-buy-self-driving-cars.html (Accessed May 03, 2020).

[7] N. McCarthy, "• Chart: Global Opinion Divided On Self-Driving Cars | Statista," 2018, https://www.statista.com/chart/13531/global-opinion-divided-on-self-driving-cars/?nr_email_referer=1&utm_source=Sailthru&utm_medium=email&utm_content=COTD&utm_campaign=Post Blast %28sai%29: Roughly half of all Americans expect self-drivi (Accessed May 03, 2020).

[8] "A Brief History of Car Connections | G+D Spotlight," G+D Mobile Security Smart , 2017, https://www.gi-de.com/en/spotlight (Accessed May 03, 2020).

[9] United States Department of Transportation, "Intelligent Transportation Systems – How Connected Vehicles Work," https://www.its.dot.gov/factsheets/connected_vehicles_work.htm (Accessed May 03, 2020).

[10] United States Department of Transportation, "Intelligent Transportation Systems – Connected Vehicle Basics," https://www.its.dot.gov/cv_basics/cv_basics_benefits.htm (Accessed May 03, 2020).

[11] A. Broggi, A. Zelinsky, M. Parent, and C.E. Thorpe, "Intelligent Ve 51," *Springer Handbaook of Robotics*, pp. 1175–1198, 2008.

[12] O. Gusikhin, D. Filev, and N. Rychtyckyj, "Intelligent vehicle systems: applications and new trends," *Lect. Notes Electr. Eng.*, vol. 15, pp. 3–14, 2008, doi: 10.1007/978-3-540-79142-3_1.

[13] W. Do, O. M. Rouhani, and L. Miranda-Moreno, "Simulation-based connected and automated vehicle models on highway sections: a literature review," *J. Adv. Transp.*, vol. 2019, 2019, doi: 10.1155/2019/9343705.

[14] D. Elliott, W. Keen, and L. Miao, "Recent advances in connected and automated vehicles," *J. Traffic Transp. Eng. (English Ed.)*, vol. 6, no. 2, pp. 109–131, 2019, doi: 10.1016/j.jtte.2018.09.005.

[15] M. Kuga, "Simultaneous," *Sen'i Gakkaishi*, vol. 40, no. 4–5, pp. P393–P395, 1984, doi: 10.2115/fiber.40.4-5_p393.

[16] A. Kesting, M. Treiber, M. Schönhof, and D. Helbing, "Adaptive cruise control design for active congestion avoidance," *Transp. Res. Part C: Emerg. Technol.*, vol. 16, no. 6, pp. 668–683, 2008, doi: 10.1016/j.trc.2007.12.004.

[17] K.B. Ekta balotra1, *Spatial modulation,* vol. 2, no. 3, pp. 1–10, 2013.

[18] M. Di Renzo, H. Haas, A. Ghrayeb, S. Sugiura, and L. Hanzo, "Spatial modulation for generalized MIMO: challenges, opportunities, and implementation," *Proc. IEEE*, vol. 102, no. 1, pp. 56–103, 2014, doi: 10.1109/JPROC.2013.2287851.

[19] https://www.nec.com/, "How Will Smart Transportation Systems Work in Ten Years?," *NEC/ orchestrating a brighter world,* 2020; https://www.nec.com/en/global/insights/article/2020022504/index.html (Accessed Jun. 10, 2020).

[20] Smartcity, "How Smart Cities can Benefit from Intelligent Transportation Network," 2017, https://www.smartcity.press/intelligent-transportation-system-for-smart-cities/ (Accessed Jun. 09, 2020).

[21] P. Parmar and T. Champaneria, "Study and comparison of transportation system architectures for smart city," *Proc. Int. Conf. IoT Soc. Mobile, Anal. Cloud, I-SMAC 2017*, pp. 675–680, 2017, doi: 10.1109/I-SMAC.2017.8058264.

[22] K. Sampoornam, "Electronic Toll Collection and Gate Automation," January, 2019.

[23] K. Al-Khateeb and J.A.Y. Johari, "Intelligent dynamic traffic light sequence using RFID," *Proc. Int. Conf. Comput. Commun. Eng. 2008, ICCCE08 Glob. Links Hum. Dev.*, December, pp. 1367–1372, 2008, doi: 10.1109/ICCCE.2008.4580829.

[24] S. Drakatos, N. Pissinou, K. Makki, and C. Douligeris, "A future location-prediction replacement strategy for mobile computing environments," *IEEE Wirel. Commun. Netw. Conf. WCNC*, vol. 4, no. April, pp. 2252–2260, 2006, doi: 10.1109/WCNC.2005.1424866.

[25] F. Khairuddin, M. Hamzah, M. Yusof, and C. Ng, "The automated speed enforcement system—a case study in Putrajaya," *J. East. Asia Soc. Transp. Stud.*, vol. 10, February 2019, pp. 2133–2146, 2013.

[26] Roboauto, "How it works? Collision avoidance system – Roboauto – blog," 2017, https://blog.roboauto.cz/how-it-works-collision-avoidance-system-d05bd9807f1a (Accessed Jun. 10, 2020).

[27] A. Eskandarian, "Handbook of intelligent vehicles," *Handb. Intell. Veh.*, vol. 1–2, pp. 1–1599, 2012, doi: 10.1007/978-0-85729-085-4.

[28] M. Treiber. *Traffic Flow Dynamics*. Springer, 2013. 10.1007/978-3-642-32460-4.

[29] M. Zhou, X. Qu, and S. Jin. "On the impact of cooperative autonomous vehicles in improving freeway merging: a modified intelligent driver model-based approach," *IEEE Transactions on Intelligent Transportation Systems*, vol. 18, pp. 1422–1428, 2017.

[30] M. Treiber, A. Kesting, and D. Helbing. "Understanding widely scattered traffic flows, the capacity drop, and platoons as effects of variance-driven time gaps." *Physical Review E, Statistical, Nonlinear, and Soft Matter Physics*, 2006. 74. 016123. 10.1103/PhysRevE.74.016123.

[31] A. Kesting, M. Treiber, and D. Helbing. (2010). "Enhanced intelligent driver model to access the impact of driving strategies on traffic capacity." *Philosophical Transactions. Series A, Mathematical, Physical, and Engineering Sciences*, vol. 368, pp. 4585–605, 2010. 10.1098/rsta.2010.0084.

[32] C. Desjardins, and B. Chaib-draa. "Cooperative adaptive cruise control: a reinforcement learning approach." *Intelligent Transportation Systems, IEEE Transactions on*, vol. 12, pp. 1248– 1260, 2012. 10.1109/TITS.2011.2157145.

[33] C. Stachniss, J. Leonard, and S. Thrun. *Simultaneous Localization and Mapping*. Pp. 1153-1176, 2006. 10.1007/978-3-319-32552-1_46.

[34] N. E. Apostoloff, and A. Zelinsky. "Robust vision based lane tracking using multiple cues and particle filtering." In *Proceedings IEEE Intelligent Vehicles Symposium*, Columbus, OH, USA, 2003.

[35] G. Piccioli, E. De Micheli, P. Parodi, and M. Campani. "Robust road sign detection and recognition from image sequences," In *Proceedings of the Intelligent Vehicles' 94 Symposium*, pp. 278–283, 1994. IEEE.

[36] T. Kato, Y. Ninomiya, and I. Masaki. "An obstacle detection method by fusion of radar and motion stereo," SICE 2003 Annual Conference (IEEE Cat. No.03TH8734), 1, 689–694, 2003.

[37] A. Kesting, M. Treiber, M. Schönhof, and D. Helbing, "Adaptive cruise control design for active congestion avoidance," *Transportation Research Part C: Emerging Technologies*, vol. 16, pp. 668–683, 2008. 10.1016/j.trc.2007.12.004.

[38] W. Y. Leong. *Implementing Blind Source Separation in Signal Processing and Telecommunications*. Diss. University of Queensland, 2005.

[39] U. Mehrotra and W. Y. Leong, "NSEEAR: A energy adaptive routing protocol for heterogeneous wireless sensor networks," 2009 35th Annual Conference of IEEE Industrial Electronics, pp. 2647–2652, 2009. doi: 10.1109/IECON.2009.5415255.

[40] W. Liu, E. Lupito, Y. L. Sum, B. Tay and W. Y. Leong, "Ultra wideband antenna for real time location system application," 2009 35th Annual Conference of IEEE Industrial Electronics, 2009, pp. 2738-2742, doi: 10.1109/IECON.2009.5415424.

[41] M. Han Fei and W. Y. Leong, "Investigating target detection and localization in wireless sensor network," *Procedia Engineering*, vol. 41, pp. 75–81, 2012, ISSN 1877-7058, https://doi.org/10.1016/j.proeng.2012.07.145. (https://www.sciencedirect.com/science/article/pii/S1877705812025313).

Chapter 15
Optical fibres for data interoperability and real-time production tracking in medical manufacturing

Bandar Khalid Al-Kudaini[1] *and Wai Yie Leong*[1]

15.1 Current technology

15.1.1 Optical fibre in modern technology

Optical fibre is an elastic clear fibre made by drawing glass (silica) or plastic to a breadth marginally thicker than a human hair. Optical fibre normally contains a centre encompassed by a straightforward cladding substance. It works on the principle of total internal reflection. When light ray strikes at the internal surface of optical fibre cable called such the incidence angle is greater than the critical angle, then incident light ray reflects in the same medium, and this phenomenon repeats. The illumination is preserved in the centre by using the phenomenon of all-out inside reversal [1]. The innovation of fibre optic in the present day is transmission media, such as twisted-pair copper and wireless technology. An optical transmitter changes an electrical sign into an optical sign and sends it through the optical fibre line, which is a cable packed with various optical filaments and then steered through underground canals and structures, and used to send information transmitted regularly as advanced content created by PCs, phone industries and satellite TV corporations [1]. In 1990, Bell Labs transmitted a 2.5 Gb per second signal over 7,500 km with no regeneration, where the system utilised a laser and an erbium fibre speaker, known as the Erbium Doped Fibre Amplifier (EDFA), that permitted the illumination wave to keep up its formation and thickness. Today Dense Wavelength Division Multiplexing (DWDM) (thick frequency division doubling) innovation has paved the way for information transmission capacity increments [1]. Narinder Singh Kapany who is an Indian physicist, is broadly recognized as the father of fibre optics [1].

15.1.1.1 Applications

Applications of optical fibre in communication and medical field:

[1]The Institution of Engineering and Technology, Malaysia

Optical fibre is widely utilised in web links since it transmits a lot of information at rapid speed. It is utilised in telephones and it is broadly utilised in medical procedure and dentistry [1].

Applications of optical fibre in PCs, TV and decoration:

It is utilised in PC-to-PC communication where the PCs are connected with optical fibre links for high speed information transmission requirements. Broadly, it's utilised in TV's link, and, additionally, utilised in lighting and design [1].

Applications of optical fibre in space:

NASA's ongoing projects is using optical fibre and it shows that the U.S. space organization is utilising cutting-edge products for progressive space exploration. NASA is, as of now, using Made in Space fibre optics, a product of the Made in Space company proficient in developing optical fibre for space journeys, which is made of ZBLAN optical fibre, more evolved than conventional silica based optical filaments [1].

15.1.1.2 Advantages of optical fibres

Unlimited transfer speed [1].

- Long transmission distance [1].
- Low expense [1].
- Small size and lightweight [1].
- Easy establishment [1].
- Secure transmission [1].
- Not alluring to misrepresentation [1].

15.1.1.3 Disadvantages of optical fibres

- Installation costs are high [1].
- Susceptibility to physical harm [1].
- Fragility [1].
- Need unique abilities [1].
- Require more care and unique gear [1].

15.1.2 Optical fibre communication in the twenty-first century

There is no doubt that telecommunication has assumed a critical role in the shaping of the modern world. Without the existence of the telecommunication system and the electronic foundations behind it, advanced life would be incomprehensible, even it is difficult to imagine how life would be for modern man [2]. After all, it is the telecommunication that shapes us as individuals and makes the world comprehensible to us, our day-to-day life is highly interlaced with the telecommunication and its signals. To put it simply, under the impact of telecommunications, the world is contracting from segregated terrains isolated by vast oceans to an interconnected town [2]. Communication in the twenty-first century is showing the way to a friendly integration of disparate nations different by language, culture, or history. Communication empowers individuals to share thoughts, express their sentiments and add to conversations and discussions generally, and language is a huge hindrance to this expression of common interests.

Communication between different communities and different cultures [2], for example, is being simplified by the arrival of Internet technology using cable lines to connect computers and phone lines to access the Internet. Therefore, there is a sharp increase in the demand for fast broadband 'connectivity' for community and the economy to expand, with a credible, affordable, and scalable state-of-the-art telecommunication infrastructure to back it. During this time of quick innovative and technological change, the fibre optic innovation has developed in the past two decennia and is predicted to be the backbone for universal communication for most of this century, if not beyond [2]. The interest for optical fibre systems is expanding quickly, customers are requesting for further developments and enhanced administrations are taking over along these lines. The fibre to the home and the enterprise has the potential to change our communication prowess and communication environment [2]. The innovations related to the enterprise are very helpful in the remote control and monitoring systems [2].

15.1.3 Statistics and graphs

As seen in Figure 15.1, the upload speeds for different types of connection media for businesses has a slowly increasing graph, and it shows that the optical fibre media has the highest connection speed with a high deference value among the various comparable media types. The download speeds for a majority, other than fibre, is usually higher than the upload speed [3].

From the graph in Figure 15.2, it shows the number of subscribers by technology in the period spanning 15 years from 2000 until 2015, and we can say that for 14 years, the copper had the highest number of subscribers until in 2015 the fibre became the one with the highest number of subscribers. Cable connection was more than the fibre for 12 years and in 2013 it became equal with the fibre connection after that it became the lowest [4].

As seen in the graph in Figure 15.3, which shows that the optical signal to noise ratio (OSNR), or fidelity of connections, for different capacity connections

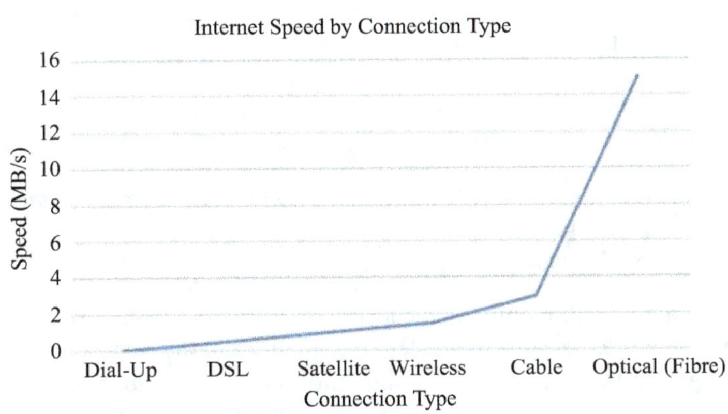

Figure 15.1 The data transfer speed of the Internet using optical fibre is the highest with 15 MB\S [3]

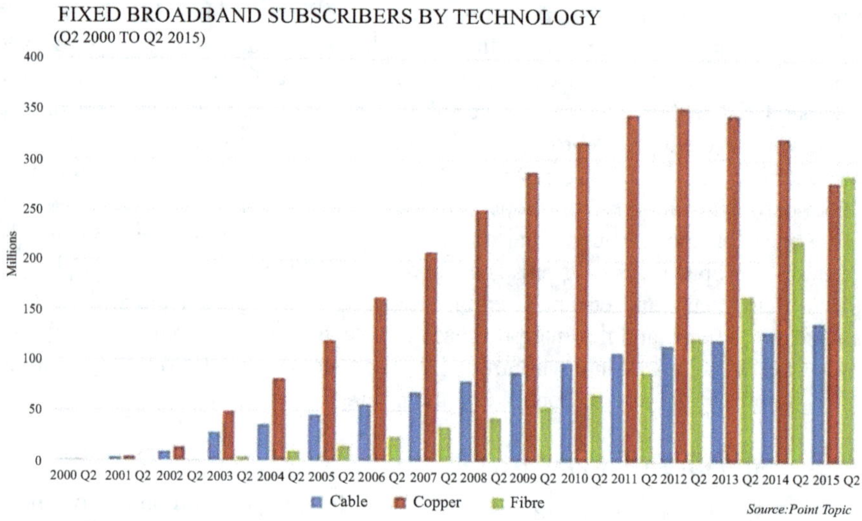

Figure 15.2 The number of subscribers by technology of the period of 15 years from 2000 until 2015 [4]

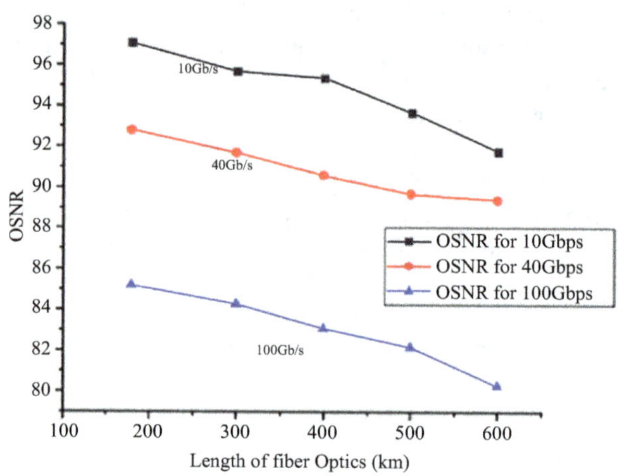

Figure 15.3 The optical signal to noise ratio (OSNR) against different length of line in DWDM [5]

vis-à-vis the length of fibre optic line; capacity of different length goes into DWDM, and it seems the more the capacity the less the fidelity and the less the distance pulled of the lines. A significant distance and high information rate transmitted system with power per channel equal to 7 dBm and $0.36E^{-18}$ nonlinear effective area [5].

As seen in Figure 15.4, the revenue of the optical fibre industry is increasing significantly from 2017 and is expected to increase in the future, with the

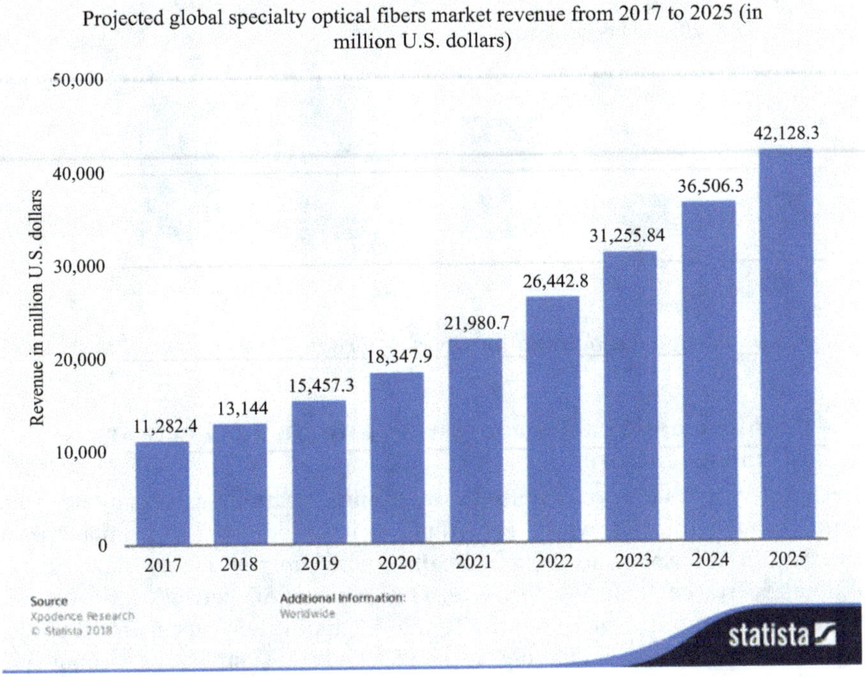

Figure 15.4 The projected revenue of optical fibre industry between 2017 and 2025 [6]

expectation that in 2025, it will be over 40 million U.S. dollars, especially with the increases in demand worldwide [6].

As seen in Figure 15.5, the global demand for optical fibre in 2019 compared to 2018, in quarterly buckets, that prove to us that the demand of fibre optic is growing and will keep growing in the future [7].

15.1.4 Trends of optical fibre

The future of the fibre optic cable seems filled with promise, with commerce foreseeing a 12% complex yearly development rate in the year 2024. In this paper, some of the future trends for fibre optic cables is depicted [8].

15.1.4.1 Global demand for greater Internet bandwidth

As the world wide web (WWW) keeps on developing at a fast rate, so does the interest for expanded bandwidth as well. A tremendous quantity of data is passed over the web daily, with data structures, for example, multimedia, sound and video on the rise. This requires a telecommunication infrastructure that can deal with the tremendous quantities of information with quality and convey this information effectively across the world [8].

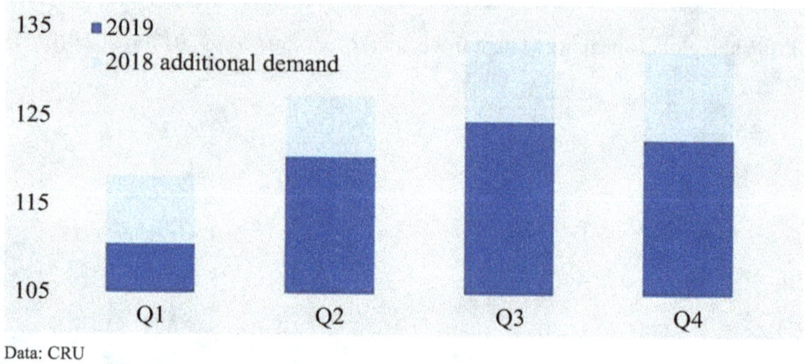

Figure 15.5 The global demand for optical fibre in 2019 compared to 2018 [7]

15.1.4.2 Fibre optic cables are essential within a 5G network infrastructure

5G technology vows to revolutionise the telecommunication business by providing superior bandwidth rates where everything will be digitized – from homes to businesses, from hospitals to shopping malls – with superior quality of service and real-time experience of closing distances. The launch of 5G network substructure is set to be essential in the progress of VR and AR innovation, just as for driverless vehicles and distributed computing. Moreover, 5G transmitters are a supplement for the Internet of Things (IoT) hubs which are at present working alongside Wi-Fi routers [8].

15.1.4.3 Increasing number of applications within medical industries

As fibre optic innovation has progressed, various use cases have been distinguished over a wide variety of businesses. The medical field has generally used fibre optics for detecting and sending pictures of internal organs at very fast rate during surgeries. Looking to the future, an expanding number of applications will possibly become available – for instance, applications facilitating the curing of wounds and tumour [8].

15.1.5 History search and who developed it

As for the invention of optical fibres, it went through several stages starting in 1840, when Daniel Colladon, the Swiss physicist, discovered that light could be used to glow along the inside of a water pipe, where water carrying light jets out while trapping light within the pipe using total internal reflection (TIR). In 1870, John Tyndall, an Irish physicist, had demonstrated to the Royal Society in London that light used internal reflection along a specific path . When he poured some water from the jug the sunlight that shone in a water jug followed the curved path of the water trail, then he observed and followed the trail of the water [9].

The idea of light curvature is exactly the same as what is happening in optical fibres. In 1930, German students Heinrich Lamm and Walter Gerlach tried to utilise

Figure 15.6 Charles Kao at work in his laboratory at Harlow, England in 1966 [10]

illumination tubes to make a gastroscope, a tool to look into a person's stomach [9]. In 1950, in England, London Narinder Kapany, an Indian physicist, with Harold Hopkins, a British physicist, managed to transmit a simple photo under the light tube made of thousands of glass fibres. Kapany gained fame with his publications of articles on fibre optics and was nicknamed the father of optical fibres [9]. Then, in 1957, three Americans scientists at the University of Michigan, Lawrence Curtis, Wilbur Peters and Basil Hirschowitz succeeded in using fibre optics to make the first telescope ever; and, in 1960, Charles Kao (Figure 15.6) and his colleague George Hookham, who are American physicists, realised that admixed glass cannot be used to a far-reaching distance, as Kao proposed that the fibre optic cable is made of finely unmixed glass and can be used to convey phone signals over far-reaching distances. For this unprecedented discovery, Kao was awarded the Nobel Prize in Physics in 2009 [9].

15.1.6 Types of optical fibres:

The two main types are:

1. Single-mode fibre which allows just a single illumination signal to be transferred in each fibre of the fibre bundle. This kind of fibre is distinguished by the smaller radius of the core's vitreous where it reaches about 9 μm and passes through the infrared laser wavelength of 1.3–1.55 nm [11].
2. Multi-mode fibre which allows a lot of light signals to be transferred through a single optical fibre, which makes it the best fibre for use in PC networks. This

kind of fibre has a larger radius, reaching 62.5 μm and infrared can be transferred through it [11].

15.1.7 How optical fibres work and how it conducts light

It is well known that when light travels in straight lines, it will reach the other side with no problems. But what if there is a curvature of the track? This problem can easily be solved by placing a mirror at the curvature to reflect the light back into the path again. In the same way, the problem is solved if the way that has a lot of curvatures is installed with mirrors along the way to continuously reflect light from side to side to the remainder of its way [11]. This is exactly the idea of fibre optic functioning, where the light is transferred by continuous reflection from the wall next to the glass heart (cladding), a complete internal reflection because this wall does not absorb any of the light falling on it. The light signal can travel far distances but sometimes it might happen that part of the light is lost as it is absorbed by the impurities present in the glass heart [11].

The optical fibre system is formed of three main parts which are:

1. Transmitter is the part that creates and encodes the illumination signal while the heart is the illumination source, which may be a laser or an LED. Modulation of the optical signal may be altered by density up and down analogue modification or lit and turned off in a series which is what is called digital modification [11].
2. Fibre optic is the part that delivers the illumination signal over range [11].
3. Receiver is the part that receives the optical signal and decodes it to change it into an electrical signal transmitted toward the user [11].

15.2 Literature review

15.2.1 Introduction

The highly important need of communication between individuals since the antiquated times has consistently grown. As the days passed, numerous types of communication apparatus have been created. The inspiration behind each new type of communication was essentially either to make it feasible for additional information or data to be sent by expanding the information transfer rate, to upgrade the transfer fidelity or signal exactness, or to expand the transfer separation between various hand-off stations. Until the nineteenth century, all communication frameworks worked at an uniform rate and they included just acoustical or optical methods, typically drums, signal lights, horns and fire. Individuals were utilising particularly fire signals for making declarations, to ask for help or sending an alert. Presently, there are better approaches for speaking with individuals far and close, and one of them is the utilisation of the WWW, despite the fact that the essential applications and rules that make the web conceivable had existed for just about two decades, yet it is only in the 1990s that the system gained public attention. On 6 August 1991, CERN, a European association for nuclear research inquired about how to connect to the new Internet invention. There are various methods for accessing the web, and some of them are wireless, dial-up, Digital Subscriber Line (DSL), cellular broadband and satellite Internet. Every type of technology has its own advantages and disadvantages [12].

15.2.1.1 Wireless

Wireless communication is the interchange of information over a distance without the use of electrical channels or 'wires', where the distance could be either short (a couple of metres such as TV remote control) or long (radio communications distance could be thousands of kilometres). When the Wifi setting is clear at that gateway point, the term is usually shortened to 'wireless' [12]. Here in Figure 15.7, it depicts how the Wifi works.

15.2.1.2 Dial-up

Dial-up Internet is a type of Internet connection to the utilisation of phone lines. User's computer or switch utilises an appended modem associated with a phone line to dial into an Internet Service Provider organization's (ISP) hub to set up a modem-to-modem interface, then uses it to route Internet Protocol packets between the user's equipment and hosts on the Internet. Dial-up associations have dissimilar velocity. For instance, Modem 110 baud has a velocity of 0.1 Kilobit per second, a Modem (103 Bell or 21 V), with data transmission speed of (three hundred baud), has a velocity of 0.3 kilobit per second, and 9600 Modem (two thousand and four

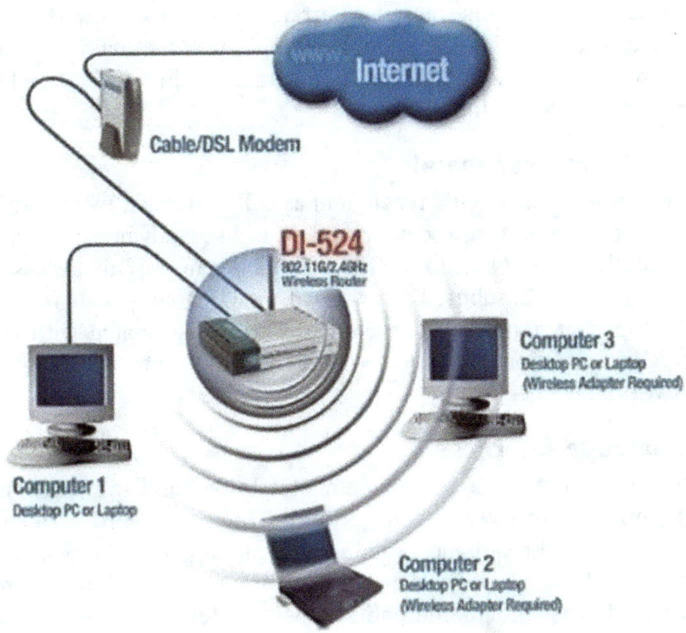

Figure 15.7 Simple block diagram of how wifi works [13]

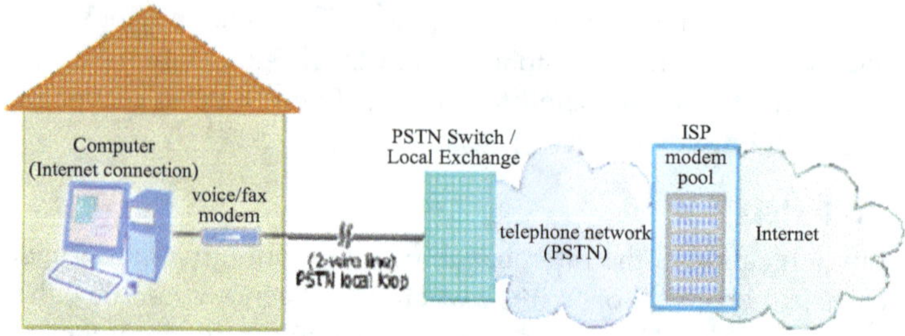

Figure 15.8 Dial-up Internet access through the telephone network [14]

hundred baud) (32 V) has a velocity of 9.6 kbit/s (ITU, 2008). Dial-up is generally the main decision-making criterion given inaccessibility of countrysides or remote regions where broadband organizations are rare owing to small populations and requests for connections. Dial-up time to build a usable phone association (a few seconds, depending on the area) is expected, and performing handshaking using protocols sets up synchronized data moves to occur [12,13]. As seen in Figure 15.8, the Internet access happens as depicted.

15.2.1.3 DSL

The DSL is wired to the Internet at all times, meaning you don't have to dial-up to connect. It does not utilise a phone line and it does not need to connect to your phone. It does anyway utilise a phone arrangement. It's viewed as broadband Internet, which means it uses a fast connection. It was the main genuine advance towards improving the information and voice transmission apparatus. DSL speeds range from 128 kbps to 8 Mbps [14,15].

15.2.1.4 Cellular broadband

Cellular phone towers have wide reach, and as cell systems move to fifth-age (5G) systems they can support fast data rates; it utilises advanced technology, for example, Evolution-Data Optimized (EVDO). This can provide access to the web using broadband, with Cardbus, ExpressCard, or USB cell modems, or with cell routers, which permit more than one computer to be associated with the web utilising one cell connection [12,15]. As an example, Figure 15.9 shows how the cellular network is connected.

15.2.1.5 Satellite Internet

Satellites in geostationary circles can transfer broadband information from the satellite organization to every client. Satellite Internet is ordinarily one of the costliest methods of broadband web access; in any case, the cost has been decreasing in the last few years that has encouraged the technology to be more competitive with other broadband choices. Most satellite Internet suppliers likewise have a Fair Access Policy (FAP). Maybe the biggest disadvantages of satellite web, these FAPs usually throttle a user's throughput to dial-up

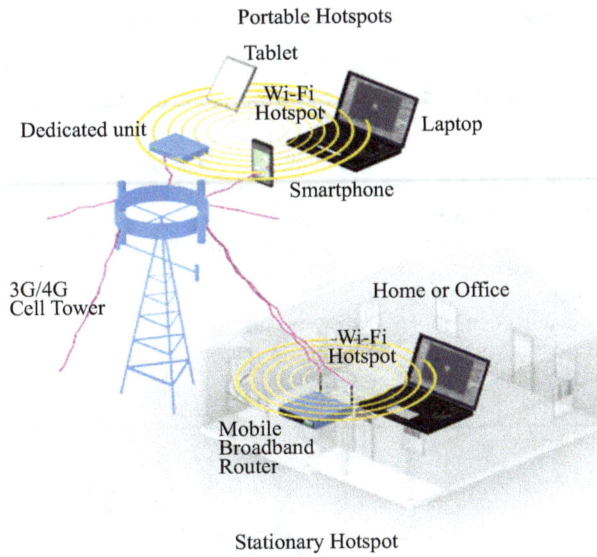

Figure 15.9 How we get the connection from the cell towers [16]

information rates after a specific 'invisible wall' is hit (for the most part around 200 MB daily). This FAP generally goes on for 24 h after the wall is hit, and then a client's throughput is re-established to whatever level they paid for. This makes data transfer capacity concentrated exercises almost difficult to finish in a sensible measure of time (examples include P2P and newsgroup binary downloading) [15,16].

15.2.2 Comparison between DSL, cable Internet lines and fibre optics

The presence of the Internet has become something very important in our lives as any home or workplace cannot do without it. Here in Table 15.1, we will make a comparison of DSL, cable and fibre optics with the specifications that they all possess and the differences between them.

15.2.3 Gap analysis and benchmarking

15.2.4 Mathematical model of the transmission of light in optical fibre

The inhomogeneous three-dimensional wave equations of the general form as [18]:

$$\nabla^2 f_e + \frac{\omega^2}{c^2} f_e = -4\pi \nabla \varepsilon .\, d\mathbf{r}, \tag{15.1}$$

Table 15.1 Some comparisons between DSL, cable and fibre optics technologies [17]

	DSL	Cable	Fibre optics
Speed overview	Slowest type and usually have no more than 35 Mbps	Fast and can be a rival to fibre's download speeds	fastest available
Uses	It uses copper phone lines to deliver Internet	It uses a coaxial cable to deliver Internet	It uses fibre optic cables to deliver Internet
Stability	Very stable connection	Very unstable	Very stable
Availability	DSL Internet is easily accessible for everyone	Cable is widely available	Available in big cities only

Table 15.2 Gap analysis and benchmarking of optical fibre

Reference	Authors	Optical fibre	Gap analysis
[12]	Khayat and Mike	Wireless local area network	Security is an issue. Wireless reception varies from area to area
[13]	MacLean	Internet connections	Believed to be the slowest but the cheapest type. The rates of connection for dial-up modems head for falling within 24–56 kbps
[14]	Cheryl Hodgkinson, Alfredo Terzoli and Ingrid Brandt	Models of Internet Connectivity in previouslyDisadvantaged Secondary Schools in the EasternCape	More expensive and it has a maximum distance of approximately 5 km, which limits it to use in urban areas or areas very close to an exchange
[15]	Erik Dahlman, Stefan Parkvall and Johan Sköld	4G LTE/LTE-Advanced for Mobile Broadband	Being able to access the web anytime and anywhere means that you may not know when to stop working, which will put the worker under a lot of pressure and cause stress
[16]	Alessio Botta and Antonio Pescap	New generation satellite broadband Internet services: should ADSL and 3G worry?	Including high response time, exigency to employ middleboxes and accelerators, effect of the weather conditions

where c is the speed of light, w is the angular frequency, dr is the differential of the radius vector, $\nabla\varepsilon$ is the gradient of the spatial distribution of the absolute permittivity and $\nabla\varepsilon.dr$ is the permittivity difference function [18]:

$$\left.\begin{aligned} x_0 &= nx, \\ y_0 &= ny, \end{aligned}\right\} \tag{15.2}$$

where n is the refractive index of the given medium in the case of an insulator and x_0, y_0 are the 'air' values of the linear coordinates [18]:

$$n = \sqrt{\frac{1}{2}\left(\sqrt{\mu^2\varepsilon_r^2 + \mu^2\varepsilon_i^2} + \mu\varepsilon_r\right)}, \tag{15.3}$$

$$nk = \sqrt{\frac{1}{2}\left(\sqrt{\mu^2\varepsilon_r^2 + \mu^2\varepsilon_i^2} - \mu\varepsilon_r\right)}, \tag{15.4}$$

where ϵ_r and ϵ_i are the real and imaginary parts, $\epsilon\hat{}= \epsilon_r + i\,\epsilon_i$. The relative permeability μ for optical frequencies is virtually always unity [18]:

$$a(\Delta z) = \exp\left(-\frac{2\pi i}{\lambda_0} nC_z\Delta z\right) \tag{15.5}$$

for a dielectric medium, and

$$a(\Delta z) = \exp\left(-\frac{2\pi i}{\lambda_0} nC_z\Delta z\right)\exp\left(-\frac{2\pi}{\lambda_0} nkC_z\Delta z\right) \tag{15.6}$$

For a conductive medium, where the frequency λ_0 is in vacuum, while the bearing of the cosine C_0z is determined as an element of the comparing locative frequencies and the frequency λ_0 in this medium is $c_0z = \sqrt{1 - \lambda^2 v_x^2 - \lambda^2 v_y^2}$ if the discriminant is negative, which matches to undamped waves, or $c_0z = i\sqrt{1 - \lambda^2 v_x^2 - \lambda^2 v_y^2 - 1}$ if the discriminant is less than zero, which matches to damped evanescent waves [18]. The expression for the scalar potential has the form of:

$$f_e(\mathbf{r}) = \sum_i\sum_j \tilde{f}_{eij}u_{ij} = \sum_i\sum_j \tilde{f}_{eij} \times \exp[2\pi\imath(v_i x + v_i y)]\tilde{f}_{tij} \tag{15.7}$$

where $\widetilde{f}_{tij}$is the transmission factor that explains the propagation of plane waves in space [18]:

$$f_e(x, y, z = 0) = \sum_i \sum_j \widetilde{f}_{eij} u_{ij}(x, y) = \sum_i \sum_j \widetilde{f}_{eij} \times \exp[2\pi i(v_i x + v_i y)], \tag{15.8}$$

whence it follows that

$$\widetilde{f}_{eij} = \Delta x \Delta y \sum_x \sum_y f_e(x, y) \exp[-2\pi i(v_i x + v_j y)]. \tag{15.9}$$

As easily can be observed, the discrete inverse Fourier transformation of$\widetilde{f}_{eij}$ is Eq. (15.7), while the discrete forward Fourier transformation of the scalar potential $f_e$taken in the $z = 0$ plane is Eq. (15.9) [18].

The following diagram in Figure 15.10 shows the explanation of optical fibre when it is used to transfer an illuminate signal (Sun, flames, incandescent filament lamps, fluorescent tubes and other discharge lamps) over a far distance. Optical fibre consists of materials with various refractive indicators. The centre has a refractive index higher than the cladding. At the point when illumination falls on the centre, it is refracted at the limits of the essential cladding. In this manner, contingent upon the edge of the dense medium used (such as plastic or glass), some portion of the light is refracted outside the medium and the fundamental part is totally reflected inside the medium. Therefore, the aggregate impact everywhere (between media, air or water) is the loss of light vitality because of refraction through cladding [19].

15.2.5 Snell's law

Snell's law was named after Willerbrord Snellius, the Dutch mathematician, and the law states that the proportion of the sines of the angles of incidence θ_1 and

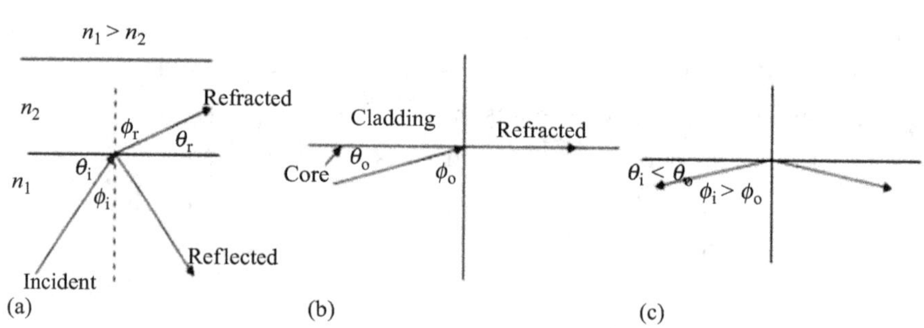

Figure 15.10 Ray diagram of fibre optics that is used to transfer an illumination signal over a long range [19]

refraction θ_2 is equivalent to the proportion of light velocity in the two materials, or equivalent to the opposite proportion of the refractive index of refraction [20]:

$$\frac{\sin\ \theta 1}{\sin\ \theta 2} = \frac{v_1}{v_2} = \frac{n_2}{n_1}$$

Snell's law is also written as: $n_1 \sin\theta_1 = n_2 \sin\theta_2$

where v is the velocity, n is the refractive index, θ_1 is the angle of incidence and θ_2 is the angle of refraction [20].

15.2.6 Modulation system in optical fibre

Fibre optic communication system transmission uses pulse code modulation which is a type of digital modulation. So, data is transmitted as a progression of pulses. The computerized pulse code-modulated signal is coupled into a fibre. The fibre end is situated by a connector to maximize the input power [21].

15.2.6.1 Pulse code modulation advantages

- High efficiency for significant distance communication [22].
- Its transmitter effectiveness is high [22].
- Its noise immunity is high [22].

15.2.6.2 Pulse code modulation disadvantages

- Requires huge transmission capacity when contrasted to an analogue system [22].
- Has a very complex circuit for encoding, decoding and quantizing [22].

15.3 Beyond 2030

15.3.1 Industrial Revolution 5.0

The analysis of Industry 5.0 indicates there are many doubts about what it will bring and how it will disturb business in detail, yet it will remove separations acting as obstructions between this present reality and the virtual one. The Universal Robots Chief Technology Officer calls attention to the fact that the subsequent stage of the Industrial Revolution will be necessary as indicated by the purchasers' high-demand of individualization in the items they will purchase. In this manner, Industry 5.0 will expand joint effort among humans and smart systems like robots, particularly in the manufacturing sectors. With this progression, machines assume control over all monotonous, repetitive errands while humans take an inventive side to assume greater liability and expanded management of systems to hoist the quality of production across the board. However, optical fibre is an essential element of the industrial revolution 5.0 because it is developing to keep up with the new technologies, especially, the virtual world which the industry 5.0 is based on. Optical fibre is called the future of the communication system and, in 2030, it will be the best communication system since it is increasingly competent to pass on information considering the fact that optical strands are of higher bandwidth than standard wires, which can put a large number of data inside one bundle, which increases the telephone wire capacity or the quantity of TV communication stations in a single rope. It's enough to realise that the optical fibre transfer speed arrives at 50 THz (Terahertz) while the biggest data transmission needs to communicate TV content doesn't surpass 6 MHz, which is comparatively little as the cable radius is lower than the radius of conventional copper wire, copper wire is lighter at 94.5 kg and can be supplanted by optical fibre weighing only 3.6 kg, with less signal loss. The signals are transmitted through adjacent fibres whether for phone discussion, transmission or television broadcast [23].

15.3.2 Additional technology, feature and characteristics beyond 2030

Fibre optics communication is the inevitable fate of information technology. The progression of fibre optic communication had been impelled by the development and extensive eagerness for fibre optic communication. It is required to go on into the future, with the improvement of new and progression. Coming up next are a segment of the visualize future samples in fibre optic communication [24].

15.3.2.1 All optical communication networks

An all fibre optic communication network is visualized which will be absolutely located in the optical space. In such networks, all signals will be set up in optical media, which has no electrical control setup. Before the transacting with and trading of signals in the territory of electrical, optical signals shall initially be converted over to electrical banners before they can be coordinated. After getting ready and controlling, then the signals are re-changed over to optical signals, which are transferred over long detachments to their goals. This optical to electrical change, and the reverse, results in included inertness of the system and subsequently is a constraint to fulfilling high information rates [24].

15.3.2.2 Multi-terabit optical networks

DWDM prepares for multi-terabit transition. The precondition for expanded transition limit availability has led to creating multi-terabit optical systems. Long before there were four terabit systems utilising 40 Gb per second information rate, there were 100 DWDM existence channels. Developers are looking at achieving fundamentally higher information transitions with 100 Gb per second. With the persistent diminishing of fibre optic segments cost, the accessibility of substantially much noteworthy transmission capacity later on is conceivable [24].

15.3.2.3 Intelligent optical transmission network

Regular optical systems are not ready to adapt to the rapid adaptation to changes of online data advantages because of the nontraditional assignment of element of information transfer. Standard optical networks rely for the most part upon the manual arrangement of system network, which is time consuming, and not ready to totally change in accordance with the requirements of the consumer. Intelligent optical network is an example of future optical network development and will have the accompanying applications: development structuring, dynamic resource course allocation, phenomenal control shows for framework organization, versatile hauling capacities, information transmission on demand, wavelength rental, wavelength rebate, isolated organizations for a grouping of Quality of Service levels, etc. It will require some investment before the intelligent optical system can be associated with all degrees of the system; it will initially be associated with entire arrangement frameworks, and consistently be associated with the system edge [24].

15.3.2.4 Ultra-long haul optical transmission

In the area of ultra-optical transition, the hindrances due to transition medium are still under research. Cancellation of scatter sway has affected analysts to think about the likely focal points of soliton spread. More understanding of the relationship within the electromagnetic illumination wave and the transition medium is critical to proceed to a structure with the greatest conditions for an illumination pulse to increase [24].

15.3.2.5 Laser neural network nodes

The laser neural network is a novel development for the of optical system. A committed gear course of action working in the optical space and the utilisation of ultra-quick photonic portions is required to furthermore develop the breaking point and velocity of the media transition system. As optical networks change to be more perplexing later on, the use of optical laser neural centres can be a suitable course of action [24].

15.3.2.6 Polymer optic fibre

Polymer optical fibre offers numerous favourable situations while being different from other data communication media. For example, copper fastens, distant control communication structures, and optical glass fibre. In assessing optical glass fibre, polymer optical strands give a basic and lower cost for planning of optical flags, and are increasingly multilateral for fitting interconnections. The usage of polymer optical fibre as the transition media for aircraft is under examination by different R and D labs as a result of its preferences. The Aerospace Center in Germany has assumed that the use of Polymer Optical fibre blended-media strands looks to be serviceable for future applications of flying machines [24].

15.3.3 *Types of future systems*

There is a huge number of future technologies, this paper has identified seven broad application areas of developing and future technologies that could get operational in the close range (10 years), medium-range (10–20 years) and long term (20 years) future. This brought future innovative applications whose cost-efficient models were developing now; however, these would not be plausible until the 2030 time spans or beyond [25]. By identifying these early innovations as empowering communications, we can increasingly develop ideas created from such advancements:

15.3.3.1 Biomimetic robotics

There are as yet numerous parts of the environment that we are as yet attempting to gain from to develop our innovations and robots. One model is simply the creation of gathered robots. This stems off multitude practices and would permit a robot to alter itself to make the perfect auxiliary course of action. This additionally fetches us into organic biomimicry where analysts are attempting to utilise hereditary calculations and neural systems to make the possibility of savvy mechanical technology. Just as this there are as yet numerous materials in nature that individuals are still researching. It is obvious to see biomimetic that apply autonomy have numerous applications. It permits people to expand on what the environment has taken a long time to motivate for innovations of our own [26]. Example of an anthropomorphic robotic hand technology is shown in Figure 15.11.

15.3.3.2 Computational drug design and testing

The steps of medication disclosure are exceptionally confusing and require an interdisciplinary effort to plan powerful and economically practical medications. The goal of drug design is to create a compound of chemicals which can fit to a specific cavity on a protein target both chemically and geometrically. After passing the animal exams and clinical trials for humans, this compound becomes a drug available to patients. The traditional medication structure strategies include irregular screening of synthetic chemicals found in nature or synthesized in research centres. The matters with these techniques are long structure cycles and significant cost. Present day approach including structure-based medication plan with the help of informatics innovations and computational strategies has speeded up the medication revelation process in an effective way. Exceptional advancement has been made during the

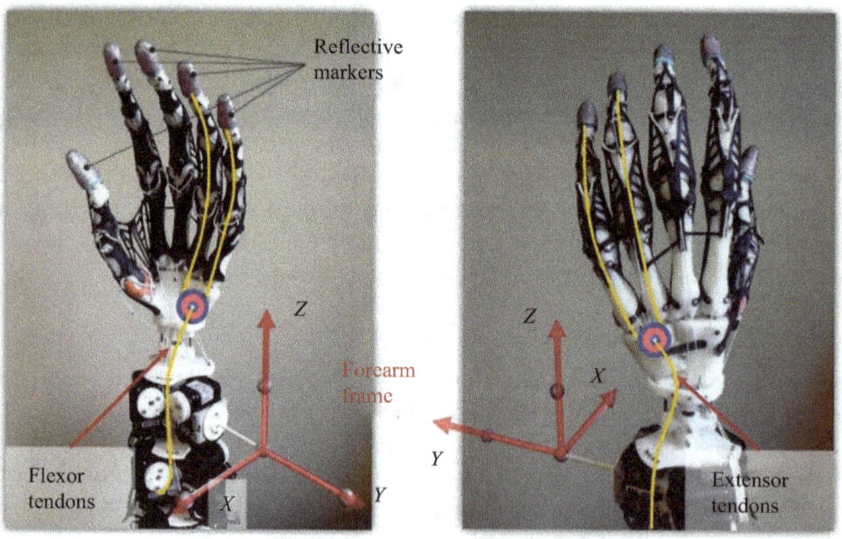

Figure 15.11 Anthropomorphic robotic hand [27]

previous 5 years in practically all the regions worried about drug design and discovery. An improved age of programming projects with simple activity and better computational devices than producing chemically steady and commendable mixes with refinement ability has been created. These apparatuses can take advantage of the information to abbreviate the pattern of medication disclosure, and subsequently make drug discovery more cost-effective. A total review of the medication disclosure process with correlation of ordinary methodologies of medication revelation is talked about here. The computational methodologies and programming projects are utilised extensively in medical planning [28]. As seen in Figure 15.12, we can see the method of Computational Drug Design.

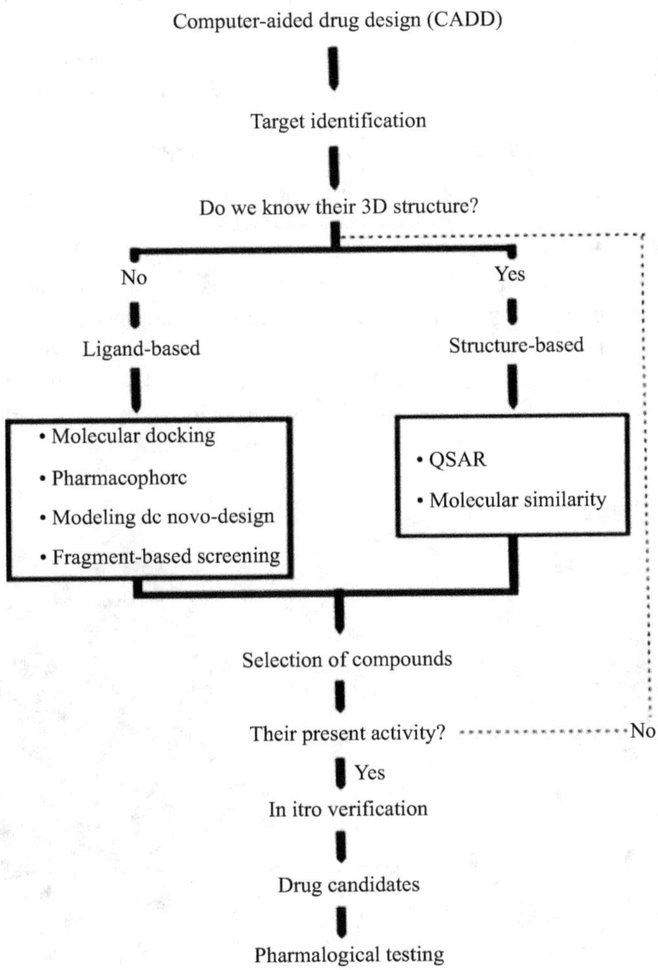

Figure 15.12 Computational drug design methods – current and future perspectives [29]

15.3.3.3 Artificial implants for improvements or recovery of biological functions

Controlling/mimicking high-order biological functions through synthetic means, artificial extensions of human capabilities, including brain repair (2020–2030), are in the offing. Long-lasting, biocompatible cochlear, optoelectronic implants for better sensing performance. Brain machine interfaces is another. Artificial implants that enable additional sensory in individuals of restricted hearing or sight are relied upon to decrease, be totally biocompatible and won't require supplanting. Such close coordination of innovation with the human body is relied upon to proceed onward from furnishing extra tangible mindfulness in individuals with incapacities to developing into brain (cerebrum) machine interfaces (BMI) by 2030. BMI could be utilised related to propels in wearable PCs and info gadgets from 2020, voice and motion acknowledgement and optical detecting are relied upon to turn into a standard method of imparting plans to registering and independent systems [25]. In Figure 15.13, mind machine interfaces in neurorehabilitation of stroke is depicted.

15.3.4 FIVE real applications of optical fibre

The application and uses of optical fibre can be seen in the following.

15.3.4.1 Medical industry

On account of the incredibly dainty and adaptable nature, it is utilised in different instruments to see interior body parts by embeddings into empty spaces in the body. It is utilised as lasers during medical procedures, endoscopy, microscopy and biomedical exploration. An example is shown in Figure 15.14 which is the endoscope and it has two fibres inside it: one is to illuminate inside the patient and the second one is for the camera to send the photo back to the doctor [31].

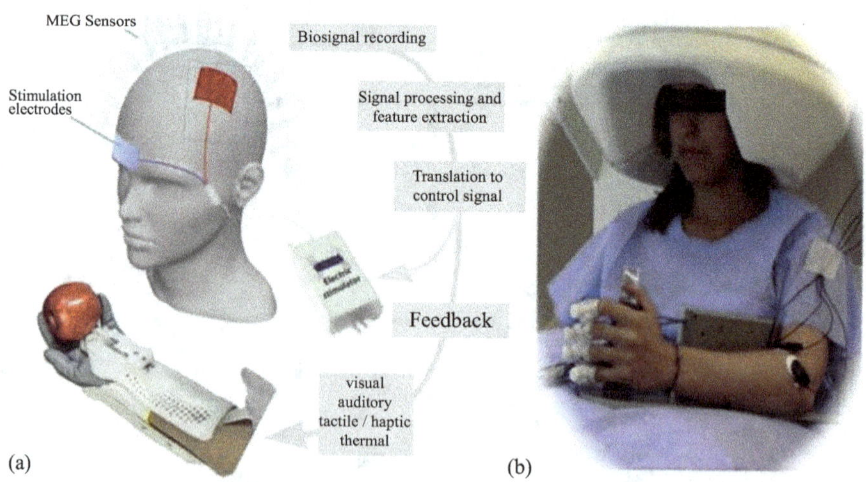

Figure 15.13 Representation of a brain machine interface (BMI) framework for stroke neurorehabilitation preparing [30]

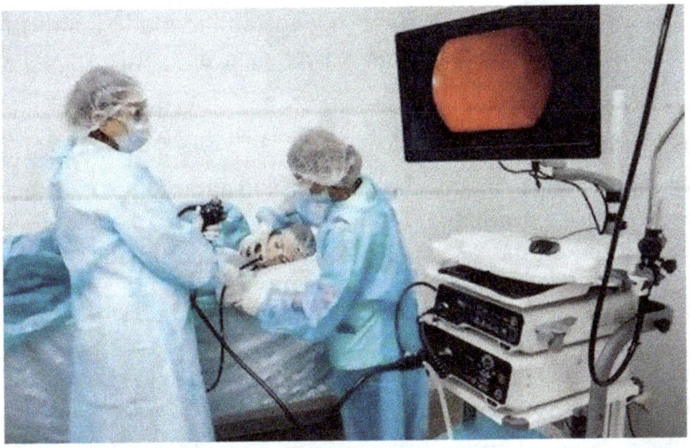

Figure 15.14 Doctor holding endoscope before gastroscopy [32]

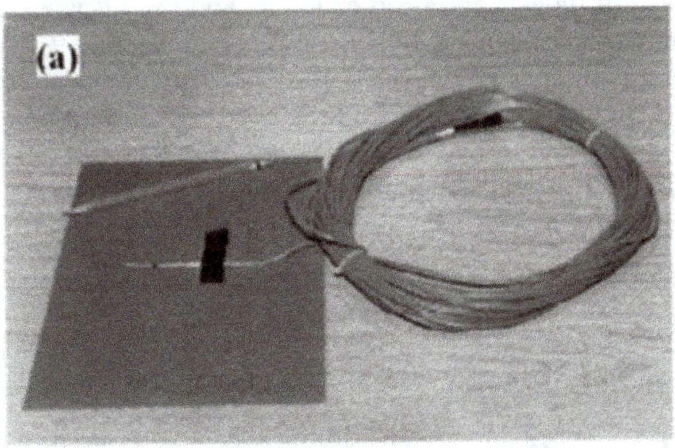

Figure 15.15 one of the first types of optical fibre [33]

15.3.4.2 Fibre optic sensors

Optical strands can be sensors used to quantify race, heat, compression and different boundaries. The small size and the way that electrical force is not desired at the distant area gives the fibre optic sensor points of interest to regular electrical sensors in specific applications. Optical fibre sensors for temperature and weight have been produced for down-gap estimation in oil wells. The fibre optic sensor is appropriate for this condition as it is working at temperatures unreasonably high for semiconductor sensors [33]. As seen in Figure 15.15, this is a sample of the sensor.

15.3.4.3 Manufacturer/commercial

Used for imaging in rugged regions, they are used as wiring where Electromagnetic interference (EMI) is a problem, as sensory devices to make temperature, pressure and different assessments, and as wiring in autos and in modern industrial settings. These include smart processing plants, where data and orders must be given or gotten continuously in real-time.

15.3.4.4 Broadcast/CATV

The wide prevalence of accessing broadband and premium video benefits along with the presentation of top-notch TV is forcing the system to redesign and progressively remold the digital TV scene. As a result, the broadcast/cable organizations presently are utilising optic fibre connectors to wire the HDTV, CATV, web, video-on-request and many other applications [34]. An example of CATV architecture is seen in Figure 15.16.

15.3.4.5 Defence/government

Defence contracts have additionally been known to utilise fibre optics. There are numerous conceivable military applications, however, some probably would not be uncovered to people in general. Among the ones that are referred to include utilising a hydrophone for seismic and SONAR, wiring in aircraft and submarines, to improve the reaction time between order inputs and the response of the vehicle. Field networking is likewise a likely zone to use optical fibre. Government capacities stretch out past this, obviously. Divisions that screen seismic movement depend on hydrophones and SONAR, which both utilise fibre optic transmission lines to send the information real-time. This involves quick analysis of the information in case of an earthquake, provides emergency services and immediate actions to the incidents [34]. In Figure 15.17, a defence radar using fibre optics is shown.

15.3.5 Conclusion

In conclusion, the fibre optic is the best option in future fast communication networks since its coverage reaches a longer distance and furthermore has a higher speed and has almost no electromagnetic obstruction, despite the fact that it has its own impediments vis-à-vis the UTP links, coaxial links and wireless. The velocities

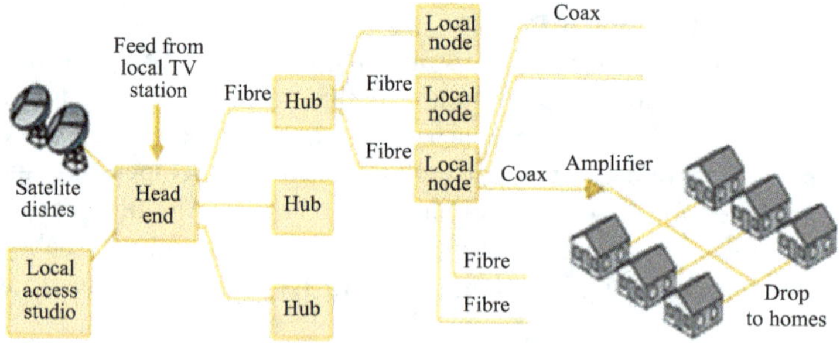

Figure 15.16 The architecture of CATV wiring from the programming centre to the consumer [35]

Figure 15.17 Defence radar technology using optical fibre [36]

of fibre optic and copper links are both constrained by length; however, copper is considerably more forcefully restricted in this regard. Fibre optic is frequently supposed to be 'future proof' on the grounds that the speed of the broadband network is typically restricted by the terminal gear as opposed to the fibre optic itself, allowing probably some speed enhancements by hardware updates before the fibre optic itself must be upgraded [15].

References

[1] Shubham, "Optical Fiber In Modern Technology By Shubham." [Online]. Available: https://www.slideshare.net/ShubhamSrivastava405/optical-fiber-in-modern-technology-by-shubham (Accessed May 12, 2020).

[2] M. Azadeh, *"Fiber Optic Communications: A Review,"* Springer, Boston, MA, 2009, pp. 1–27.

[3] "The Benefits of Fiber Optic vs DSL vs Cable for Business – Packetworks." [Online]. Available: http://www.packetworks.net/blog/the-benefits-of-fiber-optic-vs-dsl-vs-cable-for-business.htm (Accessed May 12, 2020).

[4] "The Newsletter of The Fiber Optic Association." [Online]. Available: https://www.thefoa.org/foanl-10-15.html (Accessed May 12, 2020).

[5] D.Y. Khan, "Optical Signal to Noise Ratio Against Function of Various Length Ranges... | Download Scientific Diagram." [Online]. Available: https://www.researchgate.net/figure/Optical-signal-to-noise-ratio-against-function-of-various-length-ranges-in-DWDM-long_fig329653581 (Accessed May 13, 2020).

[6] B. Gaille, "15 Optical Fiber Industry Statistics and Trends – BrandonGaille.com." [Online]. Available: https://brandongaille.com/15-optical-fiber-industry-statistics-and-trends/ (Accessed May 12, 2020).

[7] M. Finch, "Optical Fibre and Cable Industry Review | CRU." [Online]. Available: https://www.crugroup.com/knowledge-and-insights/spotlights/2020/optical-fibre-and-cable-industry-review/ (Accessed May 12, 2020).

[8] "Future Trends for the Fibre Optic Cable Market – Rebound Electronics." [Online]. Available: https://www.reboundeu.com/blog/fibre-optics-future-demand/ (Accessed May 12, 2020).

[9] "History of Fiber Optics – Timbercon." [Online]. Available: https://www.timbercon.com/resources/blog/history-of-fiber-optics/ (accessed May 12, 2020).

[10] H. Johnston, "Optical telecoms pioneer and Nobel laureate Charles Kao dies at 84 – Physics World." [Online]. Available: https://physicsworld.com/a/optical-telecoms-pioneer-and-nobel-laureate-charles-kao-dies-at-84/ (Accessed May 13, 2020).

[11] "What Is Fiber Optic Internet And How Does It Work? [GetInternet]." [Online]. Available: https://getinternet.com/fiber-optic/ (Accessed May 12, 2020).

[12] W. Is and I. Connections, "Types of Internet Connections – Webopedia.com," 2006, [Online]. Available: http://www.webopedia.com/quick_ref/internet_connection_types.asp.

[13] Ddivi, n.d, "How WiFi Works – Lab Management." [Online]. Available: https://engineering.utm.my/computing/lab/?p=801 (Accessed May 27, 2020).

[14] Conniq, n.d, "Internet Access Guide: Dial-Up (page 1 of 2) – How Dial-Up Internet works?" [Online]. Available: http://www.conniq.com/InternetAccess_dialup.htm (Accessed May 27, 2020).

[15] B. Internet, O. Credit, C. Required, and C. B. Services, "Types of Internet Connections – Webopedia.com," pp. 1–4, 2006, [Online]. Available: http://www.webopedia.com/quick_ref/internet_connection_types.asp.

[16] Yourdictionary, n.d, "Cellular Hotspot Dictionary Definition | Cellular Hotspot Defined." [Online]. Available: https://www.yourdictionary.com/cellular-hotspot (Accessed May 27, 2020).

[17] Spectrum Aothurized Retiler, n.d, "DSL vs Cable vs Fiber Internet: Which One Is Right for You?" [Online]. Available: https://www.buyinternetcable.com/blog/dsl-vs-cable-vs-fiber-internet (Accessed May 29, 2020).

[18] N.B. Voznesenskii, V.P. Veiko, N.N. Voznesenskaya, and Y.M. Voronin, "Mathematical model of the transmission of light in optical-fiber laser microinstruments," *J. Opt. Technol.*, vol. 70, no. 1, p. 35, 2003, doi: 10.1364/jot.70.000035.

[19] R. Bhansali, 20th Dec, 2017 "Please explain with the help of a ray diagram An optical fibre is used to transmit a light signal over a long distance without the loss of energy Howe – Physics – TopperLearning.com | tazt8022." [Online]. Available: https://www.topperlearning.com/answer/Please-explain-with-the-help-of-a-ray-diagram-An-optical-fibre-is-used-to-transmit-a-light-signal-over-a-long-distance-without-the-loss-of-energy-Howe/tazt8022 (Accessed May 27, 2020).

[20] Fosco, n.d, "Basic Optics for Optical Fiber – Fosco Connect." [Online]. Available: https://www.fiberoptics4sale.com/blogs/archive-posts/95048070-basic-optics-for-optical-fiber (Accessed May 27, 2020).

[21] "Fiber Communication." [Online]. Available: http://electron6.phys.utk.edu/optics421/modules/m8/fibercommunication.htm (Accessed May 27, 2020).
[22] Technical Editor, 19 May, 17, "Advantages, Disadvantages and Application of PCM (Pulse Code Modulation) – Polytechnic Hub." [Online]. Available: https://www.polytechnichub.com/advantages-disadvantages-application-pcm-pulse-code-modulation/ (Accessed May 27, 2020).
[23] D. Paschek, A. Mocan, and A. Draghici, "Industry 5.0 – the expected impact of next industrial revolution," *Management, Knowledge and Learning International Conference*, pp. 125–132, 2019.
[24] F. Idachaba, D. U. Ike, and O. Hope, "Future trends in fiber optics communication," *Lect. Notes Eng. Comput. Sci.*, vol. 1, no. 1, pp. 438–442, 2014, doi: 10.5121/ijci.2017.6203.
[25] N. Luketic, "Future Technology Themes: 2030 to 2060," 2013, [Online]. Available: https://apps.dtic.mil/dtic/tr/fulltext/u2/a591405.pdf.
[26] R. Mitchell, "Overview of Biomimetic Robotics Robotics Mechanical Systems Advanced Research Report," vol. 0.
[27] "Anthropomorphic Robotic Hand." [Online]. Available: https://www.wevolver.com/wevolver.staff/anthropomorphic.robotic.hand./ (Accessed Jun. 13, 2020).
[28] A. Baldi, "Computational approaches for drug design and discovery: an overview," *Syst. Rev. Pharmacy*, vol. 1, no. 1. pp. 99–105, 2010, doi: 10.4103/0975-8453.59519.
[29] F.D. Prieto-Martínez, E. López-López, K. Eurídice Juárez-Mercado, and J. L. Medina-Franco, "Computational drug design methods—current and future perspectives," In: *In Silico Drug Design*, New York, NY: Elsevier, 2019, pp. 19–44.
[30] S.R. Soekadar, N. Birbaumer, M.W. Slutzky, and L.G. Cohen, "Brain-machine interfaces in neurorehabilitation of stroke," *Neurobiol. Dis.*, vol. 83, pp. 172–179, 2015, doi: 10.1016/j.nbd.2014.11.025.
[31] C. Ex, S. Csio, and C. Chhabra, "Optical Fibres Beyond Telecommunications."
[32] "Endoscopy At The Hospital. Doctor Holding Endoscope Before Gastroscopy... Stock Photo, Picture And Royalty Free Image. Image 144079145." [Online]. Available: https://www.123rf.com/photo_144079145_endoscopy-at-the-hospital-doctor-holding-endoscope-before-gastroscopy-medical-examination.html (Accessed Jun. 13, 2020).
[33] M. Mahdikhani and Z. Bayati, "Application and development of fiber optic sensors in civil engineering," *14th World Conference on Earthquake Engineering*, 2008, [Online]. Available: www.iitk.ac.in/nicee/wcee/article/14_11-0118.PDF.
[34] "Applications of Optical Fiber – Communication Systems & Applications (Optical Fibers)." [Online]. Available: https://sites.google.com/site/csapgroupc/home/applications-of-optical-fibre (Accessed Jun. 14, 2020).
[35] "Fiber Plays an Increasing Role in CATV | Laser Focus World." [Online]. Available: https://www.laserfocusworld.com/fiber-optics/article/16554637/fiber-plays-an-increasing-role-in-catv (Accessed Jun. 14, 2020).
[36] "Defense Applications | RF over Fiber and Fiber Optic Delay Lines." [Online]. Available: https://rfoptic.com/defense/ (Accessed Jun. 14, 2020).

Chapter 16

Human–Machine Interface for Healthcare Technology Manufacturing

N A Rahman[1] and V Rajaratnam[2]

Healthcare is the organised provision of services to manage the diseases or health impairments and the psychosocial, economic, and spiritual concerns and needs of the individuals and their communities. It endeavours to maintain or improve their health outcomes via therapeutic, remedial, and preventative measures [1]. The rising healthcare costs, particularly for the ageing population with age-related conditions such as diabetes, have resulted in more significant spending to promote wellness and healthy lifestyles, disease prevention and early diagnosis [2]. The shift in focus to realise this benefit towards a knowledge-driven patient-centric system, the individual and the agents of the healthcare industry and all other community stakeholders must collaborate to design and implement solutions to promote healthy lives [3]. There has been a trend towards ensuring that the healthcare industry has greater efficiency, accountability and produce value for money services [4].

Traditionally, the healthcare system can be divided into five levels (Figure 16.1). The patients, their families, and communities have been prioritised as the centre of the system. In a patient-centred healthcare system, health professionals should carefully consider the needs and preferences of the patients. They should be involved, and to a certain extent, their families, in all the decisions related to their medical treatment. Next is the primary care, made up of the healthcare professionals and their network that deliver care in the frontline. Secondary care follows the institutions that provide the infrastructure, resources and processes for providing care. The overall system of financiers, institutions and regulatory bodies enables these connections and procedures among the three levels to facilitate access, monitor, evaluate and ensure the efficient and effective utilisation of these resources. The fourth level is tertiary care, which offers specialised treatment for complex conditions aided by high technology instruments, followed by public health to look after the population [5].

[1]CRE, Perdana University, Kuala Lumpur, Malaysia
[2]KTPH, Yishun, Singapore

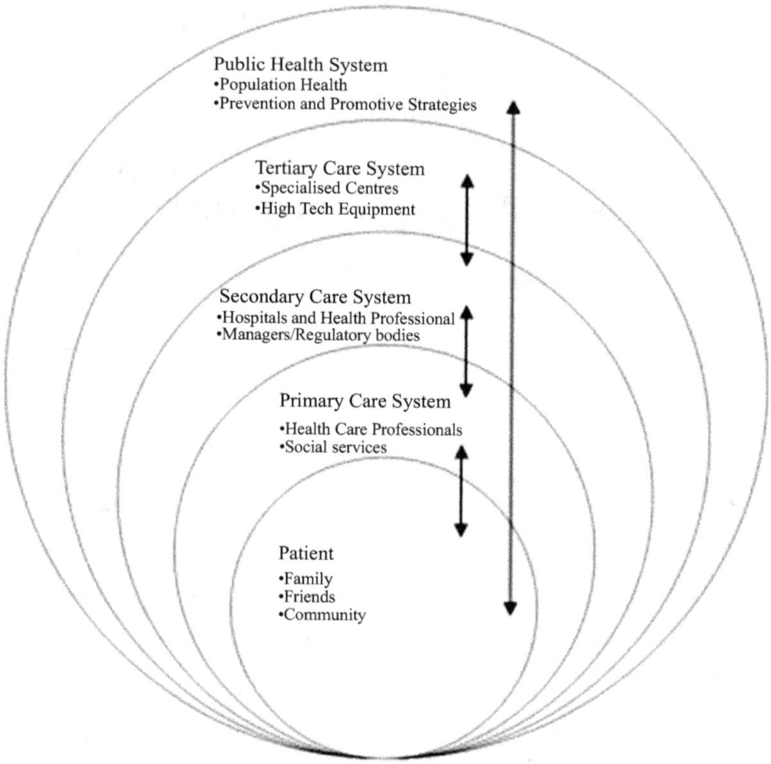

Figure 16.1 Healthcare systems (adapted from [5])

16.1 The subsystems within the healthcare system

16.1.1 Primary care system

Primary healthcare is the first point of contact when patients access the healthcare system. The aim is to provide an easily accessible route for whatever problems the patients have. It includes general practice, community, pharmacy, dental and optometry services. These healthcare professionals – the doctors, nurses and allied health professionals (e.g., pharmacist, psychologist, occupational and physiotherapist, etc.), deliver the care to patients as frontline staff. They also include social services as they provide ancillary and support services for the individual patients and their families in the community, especially for their psychosocial and economic needs.

Primary healthcare provides treatment for common illnesses, manages chronic diseases like diabetes, and prevents conditions through screening programmes, immunisations, and health advice. These professionals working in general practice are generalists dealing and caring for people rather than specific diseases. They usually have an extended long-term relationship with their patients; they act as the patient's advocate and coordinate the care of those with chronic, multiple health problems [6].

16.1.2 Secondary care system

One of the exit points from the primary care system is for the patients to be referred for more specialised care in the secondary care system. The patient may be cared for by a specialist professional with more specific expertise in the health issue, usually at a hospital with extensive resources. The specialists focus on a particular disease or body system, including cardiologists, endocrinologists and oncologists. The cost of care in the secondary care system is high. Thus, structures and procedures are in place for the judicious use of these resources through the gatekeeper function of the primary care or other bodies like financiers (insurers/health maintenance organisations (HMOs), etc.) [6].

16.1.3 Tertiary care system

These are highly specialised units that focus on complex medical conditions requiring highly skilled professionals and equipment. This level includes coronary artery bypass surgery, dialysis, complex or experimental treatments, or procedures. Thus, they are so specialised that the local hospitals are unable to offer these services. There are generally very few institutions in the country that can provide the level of service. They are not directly accessible by the patients through the public systems. For cost-effectiveness, admissions to these units have strict referral criteria and are monitored closely by regulatory bodies based on their outcomes [6].

16.1.4 Public health system

Public health is involved in population health rather than an individual, and the approach is more holistic. It is based on preventing diseases by reducing the causes and threats of ill health, improving health through health promotions, and ensuring our health services are practical and equally accessible. Their main activities are involved in monitoring and evaluating various public health policies of the population's health. It has been defined as a multidisciplinary practice engaging with a larger population and is generally the government's responsibility. However, public health may sometimes be a partnership with private, educational and research organisations [7].

16.2 The patient journey

When a patient requires the services of the healthcare systems, they will typically access the services via the primary care system (general practitioners (GP), etc.) and occasionally via secondary care (the Emergency Department in hospital). Therefore, we will need to study the patient's journey to understand how technology can enable and enhance the healthcare systems and thereby the quality and effectiveness of healthcare. We can generally summarise the contact points in a healthcare journey to the four groups, i.e., patients, doctors, nurses or allied health professionals. The interventions may be any of the following: a drug/a device/or a process (Figure 16.2).

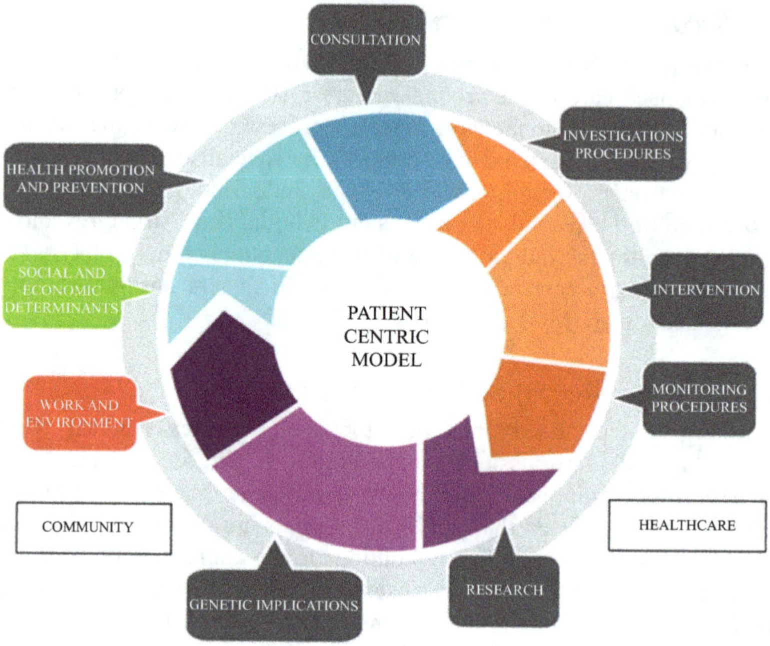

Figure 16.2 The patient journey

The patient journey navigates the sequence of events that a patient experiences within a given healthcare system, from scheduling an appointment for a consultation to receiving treatment for an illness or injury. It is dynamic and ongoing and incorporates all parts of the healthcare ecosystem, from G.P. clinics to hospitals to community centres and care and social services [8]. The introduction of technology and digital health has transformed the patient journey. These disruptive technologies have changed what used to be a "doctor knows all" model of healthcare relationships by empowering patients with access to information and expertise [9].

The patient journey is an important place to understand and learn from the patient's experiences to increase productivity, quality, and efficiency in healthcare. This allows the health stakeholders to formulate strategic goals, systems and processes to empower the healthcare professionals and patients to provide seamless sustainable health service. Understanding these touchpoints will allow digital transformation to enhance and deliver efficient, personalised care [10]. It can improve decision-making, transform care delivery and improve the relationship between health professionals and patients. In addition, it allows providers to have a more holistic understanding of patients and their concerns and expectations, thus achieving better outcomes at lower costs [11].

16.3 Healthcare stakeholders

The principal stakeholders are the patients and their families, and the secondary stakeholders are the healthcare systems and the government (Figure 16.3). The use of a transdisciplinary research process will help maximise the effectiveness and

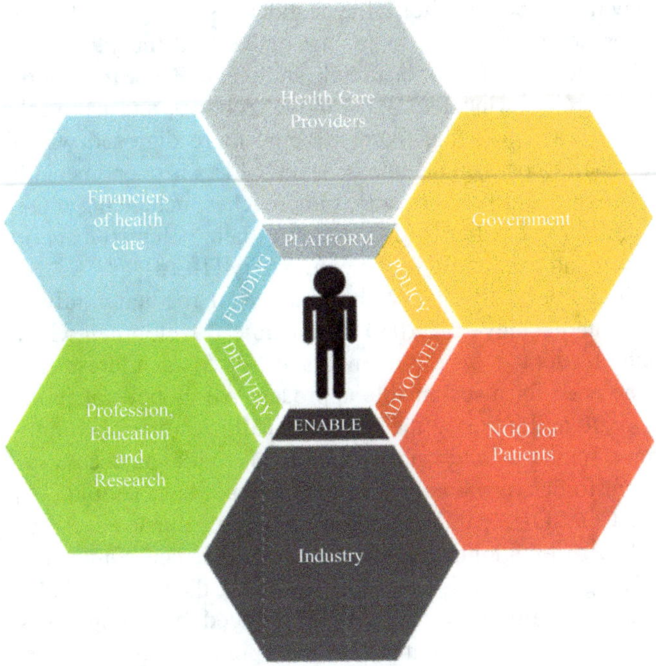

Figure 16.3 Stakeholders in healthcare [12]

impact of technology in healthcare by involving all primary and secondary stakeholders in healthcare. This approach will allow the healthcare providers and other secondary stakeholders to analyse the whole patient experience within the complex healthcare system and its processes. By mapping and analysing these processes, the secondary technology actors in healthcare can co-design and develop solutions together with the healthcare stakeholders while ensuring the patient remains the principal actor. They can create value for the patients, other stakeholders, and the community to maintain a just and equitable healthcare system [12]. This process, therefore, provides the alignment of their purpose as an enterprise.

16.4 Impact of technology on healthcare

The trending body of evidence of the positive impact of technology in healthcare is overwhelming, especially in these categories:

1. integrated management of information technology in health,
2. medical images,
3. electronic medical records,
4. development of portable, mobile devices in health,
5. access to e-health,
6. electronic telemedicine, and
7. privacy of medical data [13].

Technology has helped transform the healthcare sector through the digitalisation of the various processes and systems in care delivery. It includes digitalisation and big data, machine learning (ML), artificial intelligence (AI), the Internet of Things (IoT), additive manufacturing like 3D printing, robotics and nanotechnology. The healthcare workflows will in the future become more digitised with the focus on precision care. Their impact on healthcare stakeholders is phenomenal by improving operational efficiency and patient-centred approaches, enhancing the experience of health professionals and patients [14]. Despite their potential benefits, they pose significant threats and challenges to the community [10,15].

These technologies will change the practice of medicine, advance universal health coverage, and achieve the 2030 Sustainable Development Goals – a United Nations initiative adopted by all members in 2015, to address the inequalities globally and the need for a sustainable approach. All global stakeholders in health need to collaborate and increase their digital health readiness, comply with health data governance standards and bridge the digital divide, as access to digital technology is critical to delivering equitable health benefits to people [16].

The global COVID-19 pandemic has been the impetus in accelerating efforts to improve health care by adopting technologies to enhance connectivity and make changes to existing health services. Before the crisis, the need to deploy technology was recognised as health services were challenged by rising costs, population growth, and aging patients with chronic or multiple diseases [17]. The traction was slow as patients would typically consult their GPs if they were ill or attend the emergency department as the first-line contact. They found it difficult to break their attachments with their GPs and the health facilities and had concerns about regulations and privacy using online services. The pandemic challenged these traditional contact points due to overwhelming fears about infection. GPs offer virtual care instead of face-to-face consultations, and patients, even those with serious health problems, have stayed away to protect themselves. This led to virtual care, an online healthcare model that allows patients to consult with their GPs regardless of their location, lessening the risk of infection and empowering them to monitor themselves and manage self-care [18].

Digital health is 'the field of knowledge and practice associated with the development and use of digital technologies to improve health' [19]. The US Food and Drug Administration expands that,

> The broad scope of digital health includes categories such as mobile health (mHealth), health information technology, wearable devices, telehealth and telemedicine, and personalised medicine. … Digital health technologies use computing platforms, connectivity, software, and sensors for health care and related uses. These technologies span a wide range of uses, from applications in general wellness to applications as a medical device. They include technologies intended for use as a medical product, companion diagnostics, or adjunct to other medical products (devices, drugs, and biologics). They may also be used to develop or study medical products [20].

As healthcare demand grows, the adoption of digital technologies that enable cost-effective change is lagging compared to other industries. These innovations are helping to shift our focus from acute intervention to early-stage prevention to reduce frontline health care [10]. They are essential in preventing debilitating and life-threatening diseases. Technology alone will not make real change, but the government, relevant organisations, health professionals, and other stakeholders need to understand patient needs and expectations and have a thorough understanding of the capabilities of technology that would be beneficial to their organisations [14]. The main components of this in healthcare include AI, robotics, the IoT, autonomous vehicles, 3D printing, nanotechnology, additive manufacturing, neurotechnologies, biotechnology, materials science, energy storage and quantum computing. It has resulted in the rapid integration of these technologies with physical, digital and biological fields [21].

The complexity of the healthcare systems affords well to the integration of technology. The process by which this technology will enable healthcare delivery is in the following areas:

1. information platforms,
2. data collection technology,
3. market intermediaries,
4. services for remote and on-demand healthcare,
5. augmented and virtual reality provider,
6. blockchain-based electronic health record (EHR),
7. cloud service provider, and
8. intelligent data analysis for healthcare providers [10].

16.5 Types of technology impacting healthcare

Digital health, defined as tools and services that use technology (information and communication, software, sensors, and wearable devices) and integrate AI, big data and proteomics applications to promote and deliver the health of the population

The data acquisition from these devices and ML algorithms helps analyse the data to make predictions. This will impact the efficiency, economics and value add of care delivery [13]. These technologies can be incorporated into the care pathway as a medical product, adjunct to diagnostics, or other devices, drugs, and biologics applications [22].

One application in practice is prescription digital therapeutics (PDTs) – improving disease management by using a technology enable software program prescribed by doctors, captures patients' information from symptoms and progress, which are shared directly or remotely with providers full support to enhance compliance. They are designed to work independently, combined with a therapeutic approach or connected to a medical device. The PDTs must be proven to be safe and formally approved by the FDA [23]. Pear Therapeutics collaborated with Novartis to develop a PDT for substance use disorders [24] and Akili Interactive has developed the first prescription treatment delivered through gaming for attention deficit hyperactive disorder and an autism spectrum disorder. It will allow doctors to tailor patients' treatment according to their needs [25].

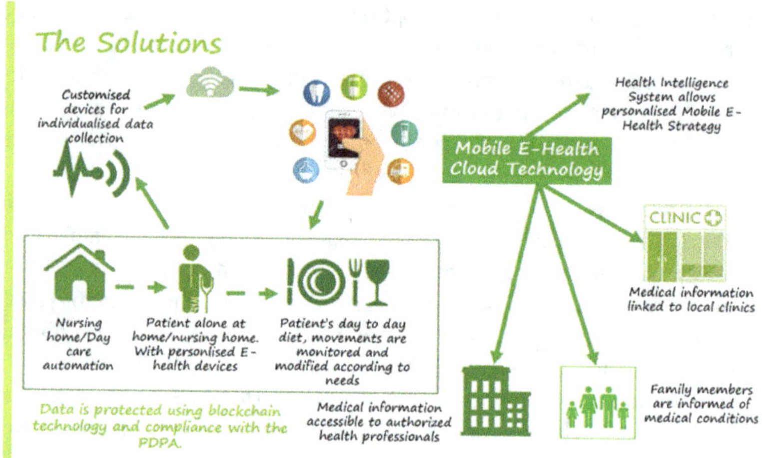

Figure 16.4 Mobile health in the community

Mobile health uses mobile devices like smartphones, tablets and wearable devices for healthcare monitoring and communication. Wearable devices are portable medical or electronic health devices that are intelligent and interactive. They are small, easy to wear, sustainable and durable so that sensors can perceive, record, analyse, regulate and intervene to maintain health and treat illnesses if necessary. However, they require wireless connectivity and biometric and cloud technologies to implement [26] (Figure 16.4). However, they need to be perceived as valuable, irreplaceable, convenient and credible for their wide adoption. Allowing users to check their health status at their fingertips and other unique functions to improve and maintain health will make them more attractive [27].

Triboelectric technology is the production of electricity due to displacement or movement between two different materials – it stands out as a promising versatile technology due to its flexibility, self-sustainability, broad material availability, low cost and good scalability. The interface includes interactive gloves, blink/movement-triggered interfaces, voice/breath monitors and self-induced wireless interfaces [28].

Virtual healthcare (aka telemedicine, mobile apps and AI-supported patient education systems) are platforms for enhanced access to healthcare professionals and services. Virtual health technologies will help bring healthcare closer to the patients through integrated and accessible care solutions. They are a practical way of providing patients with the convenience of accessing healthcare, including specialist health services from home, without commuting or taking time off work or away from their families and enjoying shorter wait times. These applications are utilised as the following to enhance healthcare systems:

- general health and fitness-based applications,
- information on medicine-based applications, and
- applications for managing the healthcare process [13].

AI is the process of developing an algorithm using the computer system to attempt to reproduce the task that would have otherwise required human intelligence. It is believed to support, facilitate and enhance human work and accelerate opportunities to realise healthcare benefits. *ML* is an application of AI that uses technology that can learn automatically without a specific programme written by humans to execute it. It identifies patterns and predicts future behaviours based on prior behaviour. As data volume has increased rapidly, ML can promptly perform these tasks by analysing large datasets due to their processing power resulting in better responses and decision-making. AI has been extensively used in healthcare in various areas of diagnosis, monitoring, and managing treatment and research [29]. Here again, for the AI system to work, it needs HMI to acquire the data from the source and display the computational results in a conducive format for consumption by the users (Figure 16.5).

Big data in healthcare is described as large and complex data sets that can come from diverse sources (e.g., EHRs, laboratories, government sources, etc.) stored on computer systems that support a wide range of medical and healthcare functions. Big data analytical software is required to harness data insights to identify patterns, trends and associations to guide decision-making, disease surveillance and planning. Data science will enable us to understand the disease more comprehensively by integrating our biophysical and clinical knowledge with the socioeconomic impact of disease and the health systems. This is a fundamental component for AI and ML to add value to the healthcare system [29].

AI/ML have impacted all areas of healthcare, from diagnosis and treatment to research and drug development and general administrative applications [30]. AI can be used in clinical decision support systems, population health and disease management (Figure 16.6).

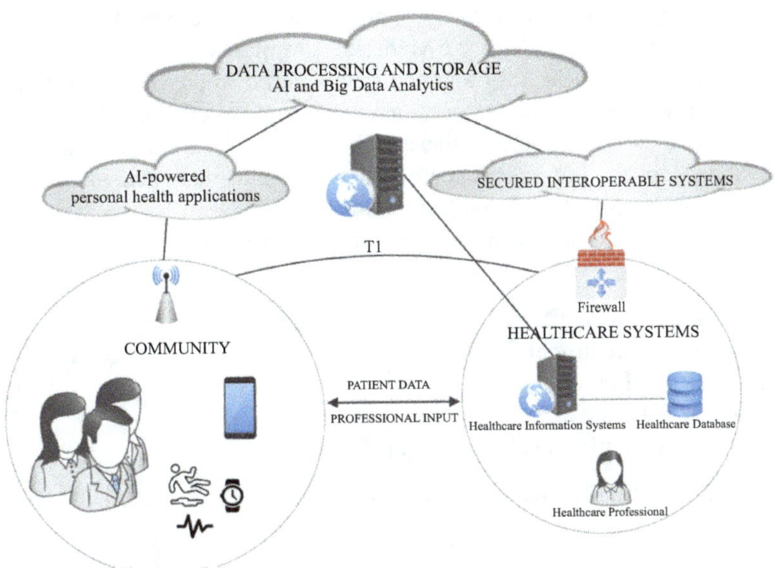

Figure 16.5 Mobile health and AI conceptual guide (adapted from [29])

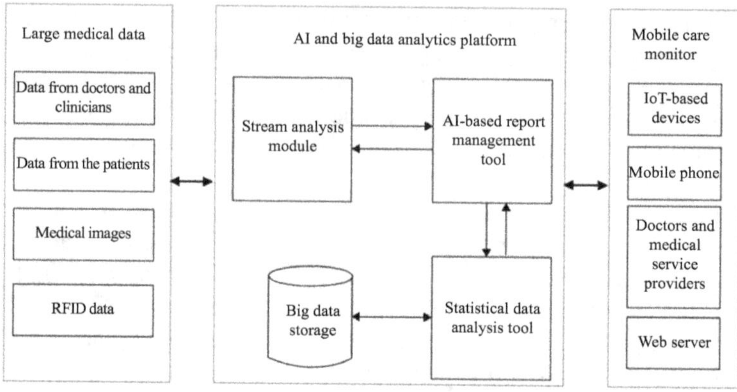

Figure 16.6 Proposed model for AI and big data for mobile health systems [29]

The popularity and trends of the impacts of technology in healthcare have not been without challenges. Adoption has not been smooth due to the resistance from the primary stakeholders of healthcare – patients and professionals. For this technology to work, it requires authentic behavioural characteristics of its users who in healthcare have significant biases and aversion for machine-driven decision-making, preferring the established models of the professional–patient interaction. This lack of trust for the machines deciding clinically significant issues has resulted in the regulators in healthcare mandating the documentation for the evidentiary and transparent decision-making process with the AI systems. Some have argued that pushing for consensus among primary stakeholders in healthcare to adopt these technologies may not be beneficial for healthcare in general [31].

In their comprehensive report, the United States Government Accountability Office and the National Academy of Medicine summarised that the use of AI has shown promise in healthcare. AI showed that it could improve patient treatment, reduce the burden on providers and increase resource efficiency at healthcare facilities. They also cautioned that these technologies '... are at varying stages of maturity and adoption, but except for population health management tools, many we described have not achieved widespread use...' [32].

They identified the following clinical areas where AI tools are applicable in healthcare:

- predicting health trajectories,
- recommending treatments,
- guiding surgical care,
- monitoring patients, and
- supporting population health management.

In the administrative areas in health, the committee identified the following areas:

- recording digital clinical notes,
- optimising operational processes, and
- automating laborious tasks.

They concluded a lack of high-quality, unbiased data to back the widespread evidence-based use of AI in healthcare. This was further confounded by the lack of transparency of the AI engine for scrutiny to ensure its safety and effectiveness. They, however, did see evidence for its consideration outside the healthcare system (hospital and clinics), in the community – like personal use AI-powered health applications. The following challenges need to be addressed to be successful: data interoperability, standardisation, privacy, ameliorating systemic biases in algorithms, reimbursement of AI-assisted services, quality improvement and integration of AI tools into provider workflows. Others have also highlighted these concerns, including the ethics of its use and the operationalisation of its use in the community. A governance model has been proposed to address these issues [33].

Virtual reality and augmented reality (VR and AR) is defined as the multisensory, immersive experiences from technology that brings together a virtual world and the real one with enhanced, 3D visuals. These are computer-generated simulations of an alternate world. AR layers the ordinary physical world into a vibrant visual stimulus, enhancing the original image with additional images using the device's camera. Augmented reality is the process of adding the VR to the user's real-world experience. It requires the use of special tools such as computers, sensors, headsets and gloves [34,35] (Figure 16.7).

The distinctions between VR and AR come down to the devices they require and the experience itself:

- AR uses a real-world setting while VR is completely virtual.
- AR users can control their presence in the real world; the system controls VR users.
- VR requires a headset device, but AR can be accessed with a smartphone.
- AR enhances both the virtual and real-world while VR only enhances a fictional reality.

Healthcare worker training

- VR and AR can be used to train surgeons in a realistic and low-risk simulated environment, avoiding trying something on a patient for the first time.
- Improving physical and mental health.
- VR and AR have therapeutic and rehabilitative potential for acute pain and anxiety disorders [10].

Figure 16.7 The continuum of the real and virtual world

All these domains relate to each other. Even though they can increase efficiency and productivity within the systems [14], it must be reinforced that the doctor–patient contact is the relational value of the medical consultation, which is rarely measured in the value-driven paradigm [36].

The Internet of medical things derived its meaning from the original *IoT,* which describes physical objects embedded with sensors, processing ability, software and other technologies that connect and exchange data with other devices and systems over the Internet or other communications networks. In healthcare, which produces large amounts of data with smart medical sensors that are dynamic and real-time based, there is a need for Cloud and IoT architectures to enable the meaningful application of these data to deliver appropriate and timely care. The HMI challenges here are acquiring the data from the patients and healthcare workers with minimal interference in care delivery. These can be categorised into wearable things (smartwatches, glasses and sensor patches), which incorporate fabric and flexible sensors, and ambient things (such as motion, door, pressure, video, object and sound sensors) [37].

The IoT empowers users with a personalised form of care through self-management, a home telemedicine system and monitoring. It allows providers to enhance the patient experience in their healthcare journey. Networked wearable devices and apps play an integral role, serving as the foundation to the ever-evolving practice of the IoT in healthcare [38]. Some of the areas of its application in healthcare are summarised in Figure 16.8.

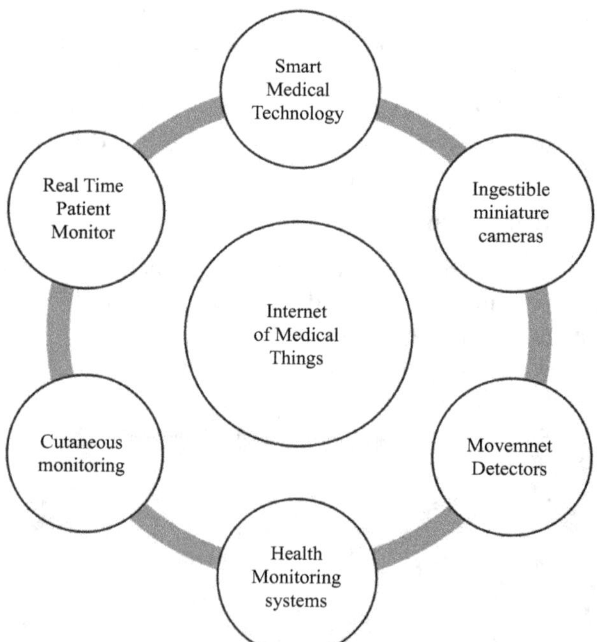

Figure 16.8 Application of Internet of Medical Things in health care (adapted from [40]

However, the current systems pose challenges to response time and data security, authentication and authorisation of application use and their breaches, apart from the general protection and privacy to facilitate the universal adoption of the Internet of medical things. As a result, there has been a recent trend in *edge and fog computing* to overcome some of these challenges.

Edge computing is the process of collecting data. Its computation occurs in the network on the organisation (local area network) and is physically close to the device, collecting the information and creating the data. *Fog computing* connects the data from the edge after filtering and manipulating it to the cloud for interoperable use by other networks. Edge computing is not dependent on the fog, but fog computing depends on the edge for its primary data [39].

16.6 Human–machine interface (HMI) in healthcare

These technological impacts on healthcare depend on effective HMI as healthcare predominantly depends on humans using various devices. Therefore, their interaction with these devices will engage the technological systems to enable optimal care delivery [41].

This HMI in healthcare is crucial for its adoption and success by its leading actors (the patient and healthcare professionals). The interface must be acceptable in the environment of healthcare without interfering with the delivery of care. They need to see the perceived ease of use of the technology based on the human interactions of the machines and their programs in the clinical environment [42]. It comprises hardware and software solutions to exchange information and communication between systems/machines and healthcare workers. This facilitates the management and/or visualisation of device processes from simple inputs on a touch display to control panels for highly complex automation systems not confined to healthcare but including manufacturing. These changes have been one of the critical processes of the Fourth Industrial Revolution I4.0 [37].

For the end user's interaction in digital health, the main technology components are the IoTs, 3D printing, advanced HMIs, AI and big data, and AR and VR (Figure 16.9). For this interaction to occur, there is a need for the interface between

Figure 16.9 Main components of digital health

the real world and the cyber/virtual world through the human-machine interface. This system is called the cyber-physical system (CPS).

16.6.1 CPS

This CPS architecture is needed for the interaction of the physical world actors and cyber world things like Smart modules (IoT sensors, IoT actuators, embedded chips, etc.), smart devices, Edge devices – (Local area communication, storage, networking system), Edge servers (local and mobile PC and servers, cloudlet servers and Cloud servers (with infrastructure as a Service (IaaS), Platform as a Service (PaaS) and Software as a Service (SaaS)).

The CPS allows for the interaction of the various hardware and software with networking systems (Figure 16.10).

They encompass the physical world of patients, medical hardware, and equipment and devices, which are bridged to the cyber world through fast, stable and secure connections to transmit health and clinical data. This allows for real-time and asynchronous analysis for optimal and efficient delivery of care and planning and prediction. In addition, the CPS provides for the integration of multiple technologies for the optimal utilisation of interconnected devices for the delivery of healthcare [43]. These healthcare technologies include IoT, Big Data Analytics, Cloud Computing, Blockchain, AI/ML, Robotics, Digital Twin, 3D printing and Nano & Biotechnology.

For the CPS to be successful, the following conditions have been identified:

Interconnection: the ability of all the available components (hardware and software) to communicate and connect among themselves and the people operating.

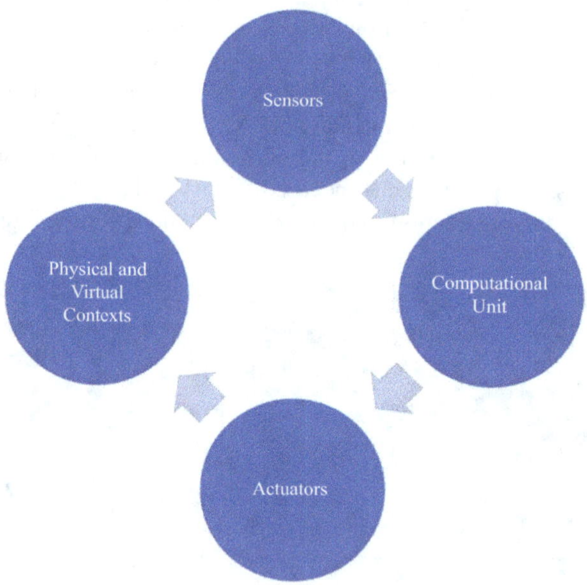

Figure 16.10 The interactions of the CPS

Information transparency: allowing all stakeholders to trust the data to help the processes in delivering and innovating services, manufacturing products efficiently and for continuous improvement program

Technical assistance: aiding all actors in the healthcare system for decision making, performing unsafe tasks, and problem-solving

Decentralised decisions: the system must be autonomous to make decisions and perform tasks themselves without requiring any assistance from an operator during routine operations [37].

One of the commonly used modalities for CPS is the smart glove. This form of HMI is equipped with triboelectric sensors, fibreoptic cables and piezoelectric haptic mechanical stimulator (providing vibratory feedback to the user using piezoelectricity) is an effective and electrically efficient solution for interaction in VR/AR [44,45].

For personalised care delivery, the HMI is important to integrate all the technological advances with minimal barriers. It facilitates the ease of use by the main actors and will ensure the successful adoption of these technologies. These HMI devices need to be non-obtrusive, allowing for ease of connectivity with sufficient security. It needs to allow for just-in-time assistance and allow automated measurement of the transaction, interaction, and outcomes allowing for unbiased and independent decision making.

One of the requirements of the HMI is the use of a multi-modal interface for the interaction of the immediate persons in the clinical environment with the cyber system. The interaction of humans with their environment depends on the sensory-motor systems, dependent on the five senses. The multi-modal system would thus use these senses to acquire data from these persons and their environment and present its computation and outcomes to be received by the same five senses of the persons. This multi-modal system currently used in healthcare include the following:

- *Eye-based control* – in these modes, the eye-tracking device uses near-infrared light to track eye movements and gaze points on a screen, whether physical or virtual. Execution of function is triggered by selection of a button mediated by prolonged gaze at the object and execution by gaze time or winking
- *Gesture-based control* – these can use signs, a position of a body part, maintained for a given time. Signs can be made from any part of the body visible to a sensor camera for computation.
- *Voice-based control* – this mode works using voice commands which are pretty standard with most operative systems currently.
- *AR interface* – this mode uses applications running on devices or augmented reality and uses positioning systems with either the hand and fingers and gestures to interact with the CPS.

The use of smart spaces within the healthcare environment will aid the integration of these technologies. The space enables the integration of data between robots, sensors, actuators and interfaces of different technologies to produce and deliver care. In addition, it would allow patients with various physical disabilities to use their residual abilities to interact with the multi-modal system.

It can offer the possibility of controlling physical objects in the environment like lighting and controlling various assistive devices for their disabilities [46]. What is needed in healthcare, therefore, is a total solution. It is orientated for healthcare that provides functionality that includes a multi-modal sensor, data analysis and management. It will also require a conversational virtual assistant with a natural user interaction that incorporates the medical history-taking model. This type of dialogue management system should accept a variety of input from patients and healthcare professionals. Touch, gestures and voice interaction will be the modes needed for this.

The HMI for the platform would need to use multi-modal verbal (dialogue) and non-verbal (sensor/camera) information management processes and include patient medical history data presented in a user-friendly experience, using computer vision displays, big data analysis, natural language processing and semantics. It should also:

- be user-agnostic, meeting the needs of authorised clinical staff and patients;
- allow for natural user interaction time-independent using conversational interaction with dialogue management, including enhanced question–answer tasks with social interactions;
- affords comprehensive patient monitoring, from multi-modal sensor data (from cameras, sleep sensors, blood sugar/pressure levels and wearable readings);
- provide a computation of aggregation, analysis and synthesis of the data collected;
- respond to complex queries by applying AI in a natural conversational way [47].

Various research projects are working on these platforms, including REA (rea-project. gr) and Suki – AI Assistant | Medical Dictation Software which has commercialised the virtual medical assistant in the US.

Another area in healthcare is the human activity recognition (HAR) interface. It is a system that uses sensor cameras in video surveillance, which can be augmented with wearable smart devices to monitor patient activity. It can recognise activities from data collected based on the actions of patients and the environmental conditions and is a form of human–computer interaction [48]. It has been used to identify early mobility activity for intensive care patients and gait characteristics in certain muscular diseased patients and can aid in predicting and preventing specific health outcomes [49].

16.6.2 Nanomedicine and genomics

Nanomedicine is technology applied to target and deliver medications/devices for diagnoses, prevention and treatment of various diseases. It is in contrast to the traditional methods of treating the whole body and getting the unwanted side effects on normal tissues and other complications of the drug therapy. Nanomedicine can deliver higher concentrations of drugs to produce targeted and better results with minimal impact on normal tissue, thus minimising the toxicity and side effects, such as cancer treatment [50,51]. Healthcare professionals can incorporate its use in devices to treat complex diseases like cancer. It has been successful in various clinical areas, including medical devices to nanopharmaceuticals. However, challenges remain in standardising the processes, evaluation and regulatory framework in multiple countries and regions. The future of this technology is promising in healthcare based on the current research [52].

An extension of this technology is Genomics, a branch of molecular biology about the gene structure and its functions, evolution and mapping for disease detection and management. Genomic medicine is impacting four main fields of medicine: oncology, pharmacology, rare and undiagnosed diseases, and infectious disease [53]. Proteomics is the study of the proteome – investigating how different proteins interact with each other and the roles they play within the organism. Many factors modulate them in addition to the expression level of the relevant gene [54].

Their contribution in medicine has been in specific areas like the following:

- *Precision medicine* – understanding the individual patient and his genetic makeup and risk for disease and drug interactions and design customised prevention programmes specific to the patient,
- Clustered Regularly Interspaced Short Palindromic Repeats (CRISPR) program to 're-edit' genetic codes and thereby rewrite the genetic code to help cure cancers and some genetic diseases and also drug designs; however, the ethical challenges include the trend for the creation of 'designer babies'.
- Targeted alteration of proteins in the cells through proteomics to manipulate the reaction of drugs, the synthesis of certain types of proteins and the alteration of metabolic pathways in the cells.
- Genetic testing in predicting and thereby preventing disease and adverse events to drugs has become standard and even available to consumers with ethical challenges. There are questions on their validity, and their results require specific knowledge to interpret and thus the need for genetic counsellors. The privacy and security of these personal data held by third-party organisations is another challenge.

Gene therapy is one of the functional areas of use of this technology. By rewriting the genetic code on the cells, it enables these cells to heal themselves. It has been used successfully in immune therapy by getting the patient's cells to find and destroy the cancer cells. Currently, gene therapy is expensive, but with advances in technology and scaling of production, it will become more affordable and accessible in time [55].

16.6.3 Robotic medicine

Robots are generally devices that assist movement by converting energy (usually electrical energy) into mechanical force. Electrical motors, sometimes assisted and augmented with hydraulics, are used to provide movements. In terms of medical use, robotics can also get feedback from the sensors within the mechanisms and record their performances.

Robots can be classified as follows:

- Active robots are those that autonomously execute a preplanned motion.
- Passive robots are unactuated and manually driven.
- Semi-active (e.g., cooperated) robots employ constrained motion.
- Teleoperated robots involve a tool held by the robot and remotely controlled by the human [56].

Robotics is not new in medicine. In 1910, Theodor Büdingen patented a 'movement cure apparatus'. It used an electric motor to guide and support the stepping movements of patients with cardiac disease. The 'Meridian', a cable-driven apparatus, a human-powered mechanotherapy machine, was used by Richard Scherb as a device for assistive activities to move joints [57].

Their evolution of robotics in medicine has been documented by Troccaz *et al.* in Figure 16.11.

Robotic surgery, or robot-assisted surgery, uses automated mechanical devices with motors and sensors to enable doctors to perform a highly complex motor task for intervention in patients with high precision, flexibility and control, not possible with conventional techniques. It is a form of minimally invasive surgery through tiny incisions. In the standard setup, the surgeons control the device with a visualising camera system with high definition and may include magnification and mechanical arms to perform a motor task. The surgeon uses this telemanipulation system to control the surgical arms. It allows for the performance of delicate and complex procedures that would have been challenging and unsafe with the conventional manual technique [58].

The evolution of surgery has been through the phases of freehand surgery, to minimally invasive surgery and the future with fully automated surgery with robots. In the process, surgery moved from open procedures with wide incisions made to enter the body to provide an intervention to minimally invasive precision techniques with small incisions, thus reducing unnecessary tissue damage and side effects and complications. However, this brings with it the constraints of tight and narrow working spaces to work. Therefore, it requires specialised instruments and devices to work with while at the same time providing sufficient haptic feedback to the surgeon to perform his tasks safely and efficiently [58].

The requirements for any robotics surgical systems would include as in Figure 16.12.

The HMI is crucial in robotics to facilitate safe and authentic clinical interactions via the cyber world. Visual imaging will include diagnostic modalities, such as medical ultrasound, computer tomography, magnetic resonance imaging and endoscopic images. High-definition camera systems further augment these with magnification to provide high-quality images of the anatomy of the structures.

Force sensing provides the haptic feedback needed for the kinaesthetic cues while performing the motor tasks by the operating surgeon. The use of strain gauges, capacity sensors in piezoelectric sensors with optical sensors helps to overcome the lack of 'feeling' of the task being performed by the surgeon. The use of optical sensing allows for force measurement of up to six degrees of freedom of the surgeon. This uses the principles of changes of intensity or phase of light passing through a flexible tube to a compliant structure [58].

Real-time position sensing is crucial for the accurate manipulation of instruments in the human body. Advanced positioning systems have been used in modern surgical robots. The use of optical and electromagnetic tracking systems allows for this without using the human touch. The ability to drive surgical instruments within the human body would require actuators that may be remotely installed. They need to be small with higher flexibility as surgical tools. They can be cable-driven or

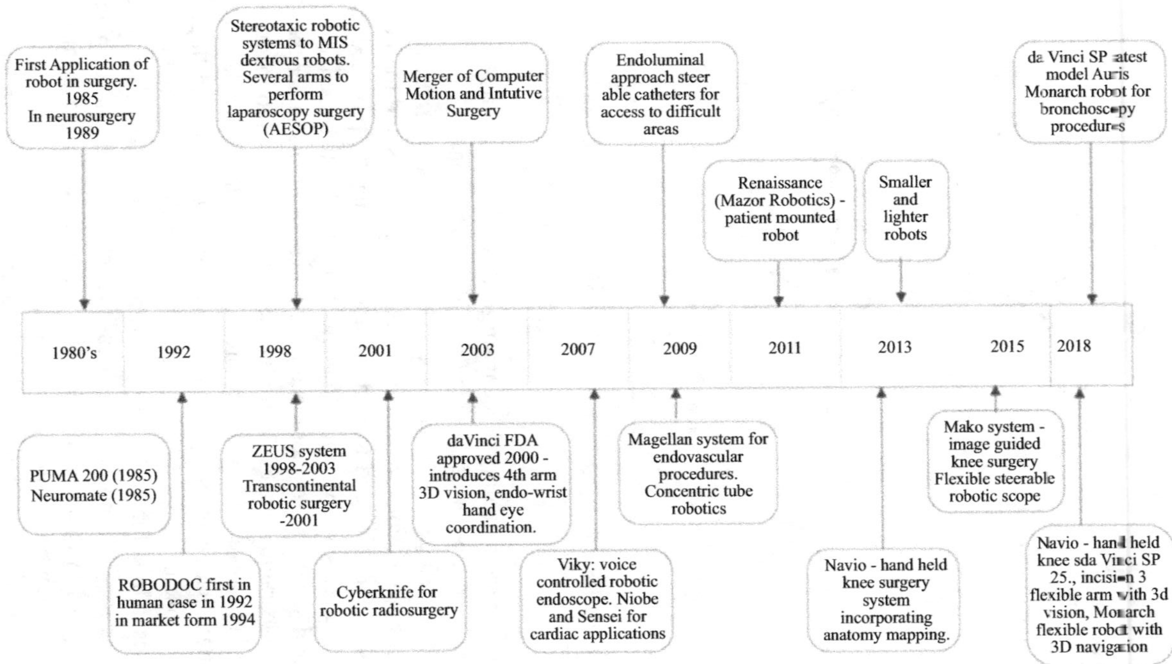

Figure 16.11 Evolution of robotics in medicine (adapted from Troccaz et al. [56])

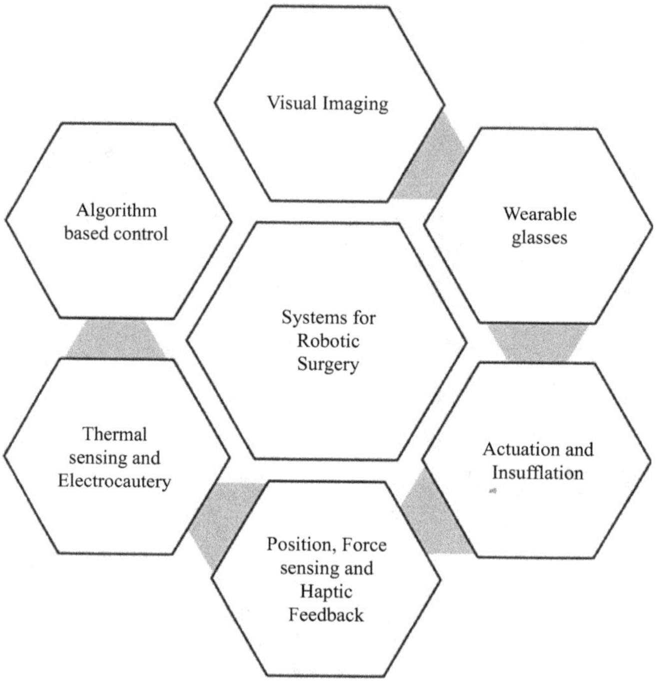

Figure 16.12 System requirements for robotic surgery [58]

flexible fluidic, or magnetic actuators. Recent surgical robotics systems have used smart materials like shape memory alloys and piezoelectric actuators [59].

For safe and effective surgery, the sense of touch is crucial (haptic). It consists of kinaesthetic and tactile perceptions. Kinaesthetic pertains to forces and torques that are sensed in the joints, muscles and tendons. These are generally provided in the systems using graspable telesurgery handheld devices. Tactile haptics such as touch, vibration, pressure and shear forces are sensed by mechanoreceptors under the skin, usually provided via wearable devices. There has been a recent trend to incorporate vibrotactile actuators into a wearable glove to produce cutaneous feedback. Another challenge is contactless haptic feedback or mid-air haptic feedback, and there has been progress with the use of pressurised air, air vortices and magnetic fields [58].

Wearable glasses allow for real-time vision and remote visual access and training. By incorporating AR and VR, it is possible to create enhanced imaging using diagnostic imaging overlaid on the real-time anatomical image during surgery, ensuring accurate identification and manipulation of tissue.

Thermal sensors allow for devices like thermal ablation catheters to locate and ablate tissue accurately. Pressure systems allow for the insufflation of cavities to allow for better visualisation through narrow portals access to structures. Electrosurgical technology allows for electrocautery (based on high frequency and alternative electric current at various voltages) in surgery. It is used for cutting, coagulating, dissecting, fulgurating, ablating and shrinking tissues through heat generation. Electrosurgical devices are used in many surgical specialities, traditionally in open operations. Still, they have been increasingly used in endoscopic procedures as they are exact and can reduce damage to the surrounding tissues with limited blood loss. There are some adverse effects associated with these devices like burns, electric shock, nerve damage and scarring [58].

A bi-directional communication exists between a master console (commands) and the slave end-effector (surgical tools) in telesurgery. For these to perform optimally, there is a need for algorithms for various movement strategies to conduct the surgery. These include optimal trajectory planning, input shaping technique and predictive control. In non-model-based controllers, input signals are regulated by real-time feedback from sensors to minimise the vibration and system delay. They include position feedback control, linear velocity feedback control, fuzzy logic control, adaptive control, and neural network (NN)-based control, which requires sufficient training data sets [60].

In the computer-controlled robotic system, the surgeon uses a computer programme to control the mechanical robotic arms. This process allows for precision control and manipulation of the instruments using ML, VR, haptics and artificial intelligence. In addition, the systems allow for the negation of physiologic tremor that is inevitable with manual surgical procedures, thus improving surgical accuracy and reducing unnecessary collateral damage to adjacent healthy tissues. It also provides high-definition visual feedback to surgeons. This is an example of an advanced form of HMI.

Another area is the use of collaborative robots in surgery – called 'cobots'. It is a collaboration between the robots and the doctors. It is recognised as co-existence, synchronised collaboration, cooperation and responsive collaboration. The robot and healthcare professional work alongside each other with or without a shared workspace synchronously or asynchronously [61].

This has paved the way for telesurgery – where a surgical procedure is performed where the surgeon is not at the immediate site as the patient. Telesurgery allows surgeons to remotely control the system from a distance and thus perform remote surgery from anywhere. The prerequisites are wireless networking and robotic technology to connect the surgeons and patients. This can address the challenges of global surgery due to the shortage of surgeons and access to these services by patients needing the benefits due to geographical, transport and safety constraints. It facilitates surgical collaboration amongst surgeons. However, latency time is a significant problem due to network issues which can cause surgical inaccuracies and risk the safety and recovery of patients [60].

Furthermore, for automated robotic surgery systems, where there is no or partial human guidance, the requirements are object recognition, navigation, human-robot interaction, manipulation, and decision-making, all summarised in Figure 16.12. It has been shown that robotic surgical systems with high-level automation may offer higher accuracy and speed for intricate procedures that mandate delicate manipulation of vital structures like nerves and vessels during surgical procedures. However, fully automated robotic surgery would not be universally accepted in practice on the grounds of ethical practice. The challenges include potential failure in the systems from mechanical, electrical and software and the potential for breach in the security and safety of the system. Supervised autonomy would be the alternative strategy for automated robotic surgery soon [58]. The current challenges for the widespread rollout of such technology systems in surgery are the availability and cost of these systems, the lack of training facilities, the need for a stable and secure global network, medicolegal, ethics and funding issues [60].

16.6.4 Rehabilitation and robots

Therapy robots are reprogrammable devices with multiple functions designed to perform different motor tasks like grasping or walking. Their use has been for sensorimotor deficits after damage to the central nervous system. It can be distally based – like assisting the hand function or proximally based – like helping the shoulder and arm position the hand. These devices can be passive, where the motors will move the limb for passive mobilisation of the joints, which would otherwise have to be done by a therapist. It can also be an active assistive exercise where the device can detect movement to augment the task based on the patient's response through a sensor within the device. The data captured allows for feedback and a personalised therapy regimen to enable the patients to build up their muscle power. It can also aid the therapist in automating the monitoring and accurate recording of progress and performance [57]. The role of an efficient and user-friendly HMI is necessary for adopting these technologies by the professionals and patients.

Robotic exoskeletons are wearable devices used to augment the user's physical performance with physical disabilities like weakness of the muscles, as an orthotic device for walking and assistance for daily activities like climbing steps and daily activities. The Food and Drug Administration (FDA) defines these devices as 'a prescription device composed of an external, powered, motorised orthosis used for medical purposes that are placed over a person's paralysed or weakened limbs of providing ambulation' [62].

Their potential in medical rehabilitation include the following:

- to increase user independence;
- secondary benefits such as improved bowel/bladder function;
- to decrease chronic pain;
- to reduce spasticity;
- to increase bone marrow density;
- the reduction of energy consumption to move joints as the robot offloads them;
- providing automated repetitive, long, and intense physiotherapy sessions freeing resources, and;
- they provide objective measurements of several kinematic and dynamic parameters of patient limb movement and performance-related indicators (e.g., range of motion, velocity, smoothness) to quantify patient progress objectively.

The robotic rehabilitation systems for the lower limbs can be classified into the following: fixed site/stationary and mobile/overground walking systems. Recent advances have made these exoskeletons available commercially, though costly.

16.6.4.1 Fixed/stationary systems

These systems have a fixed structure and moving platforms allowing for automated physical therapy freeing the therapist for other activities. The exoskeleton worn by the user over the lower limbs provides for gait training, while the fixed system suspending the patient will offload their weight to facilitate the lower limb movements for the gait training.

16.6.4.2 Mobile/overground system

These exoskeletons are mainly for enabling lower limb locomotion in individuals with spinal cord injury or paralysis of the lower limb. It provides motor power to the lower limb joints to facilitate standing, walking, climbing steps and sitting down for independent activity. These devices, however, require the ability to balance and preserve good upper limb motor function. Therefore, initial training has required the therapist to assist so that the wearable devices worn by the user can augment the control of the movements.

A systematic review of robotic neuromuscular rehabilitation showed that it has a role to play and can be integrated into clinical practice. However, it requires training and priming of the patients for effective use from training to real-world performance transition. The challenges include access to these devices, especially in low- and middle-income countries.

The robotic neuromuscular rehabilitation was used mainly for stroke, and they also found that the effect depended on the type of assistance it provided. It also has an essential function as an objective evaluation tool affording insights into the motor recovery process. Humans interact with their environment with their limbs and with the sensory feedback generated during the interactions. These were impaired during injury and disease of the neuromuscular systems. So, neuro-rehabilitation devices, especially in the upper extremity, should allow the hand and finger placement for optimal grasping and releasing function augmented with visual and haptic feedback.

To maintain a high level of motivation, the training regimens to improve the function should be functionally relevant tasks for daily living in the real world. The application of VR games is potentially helpful in improving arm function after a stroke [57,63]. The role of robotic rehabilitation can be both in assessment and decision support systems to provide rehabilitation strategies customised to the patient's needs.

Furthermore, this automated evaluation of motor function, range of movement and task performance can be used to measure and monitor the rehabilitation journey, especially in work-based assessment and aid intervention planning. They can also provide important biofeedback to enhance the rehabilitation process for the patient. These robots can position the limb optimally and eliminate gravity if needed to give automated resistance to motor strengthening for muscular rehabilitation. Repetitive manoeuvers can be programmed, and outcomes monitored and recoded for asynchronous assessment. The robotic technology will free the therapist from other activities with other patients.

Generally, robots in rehabilitation are an enabler, but it does not negate the therapist's role [64]. The future of robotic rehabilitation will require rigorous collaboration between transdisciplinary subject matter experts in medicine, rehabilitation, engineers and computer scientist.

16.6.5 3D printing

3D printing or additive manufacturing is building a 3D product using a design of a model of the product. It uses various types of materials being applied layer by layer to produce the final product. It allows for rapid prototypes, like bespoke prostheses to 3D lattices of tissue for reconstruction in healthcare. In deformity correction, this technology can build various aids and templates to assist surgeons in correcting and reconstructing the deformity precisely. The use of biological materials within this technology allows for custom-made products to be printed with ease, at less cost for the benefit of the patients.

Broadly, its use in healthcare includes customised implants, and prosthetics, medical models and medical devices. These can be used for patients as an intervention, for doctors to plan their treatment, and aids in health professional education and research [65].

This technology has been used in precision medicine and bespoke drug delivery with tailored dosages, geometries and drug release, depending on the needs of the patients. It also has significant uses in dentistry. The 4D printing method integrates the fourth dimension, that is, time. Here, printed products can change their configuration, that is, change in shape, property or functionality over time in response to external stimuli like heat and pH. This process is possible with the use of smart materials using smart design. The result will allow for targeted site delivery of medication at a specified time when the external stimulus is triggered [66].

Innovative science companies need to bridge the digital and physical environments for more personalised and higher-quality treatments using 3D printing.

16.6.5.1 3D printing delivers tremendous value to patients

Human organs are more complex than aircraft parts. They are also unique to each person. 3D printing enables the design of solutions bespoke to one's anatomy, leading to shorter recovery times due to the customisation and accuracy of treatment. This process is especially useful in medical orthosis and prosthesis, allowing for customisable devices to be produced based on the individual patient's anatomy (Figure 16.13).

These innovations have led to the open-source Do It Yourself 'DIY' production of these devices, reducing the time and cost of the production [67]. Here, the HMI is crucial to ensure a hassle-free process of data acquisition from the patient's problem via the healthcare professional to the remote fabrication of the solution. It helps care providers create better solutions.

The ability to capture the nuances of a patient's anatomy enables physicians and surgeons to conceptualise, design, and practice complex procedures before an operation. This improves the patient's experience and recovery time and allows physicians to deliver care with confidence.

16.6.5.2 Prototyping gets things done right and fast

Medical tools and solutions can be prototyped earlier due to the digital environment's ability to accommodate feedback from doctors and patients and ease manufacturing of small series. In this environment, innovative sciences companies can safely conceptualise, simulate and optimise their suite of solutions with digital twins.

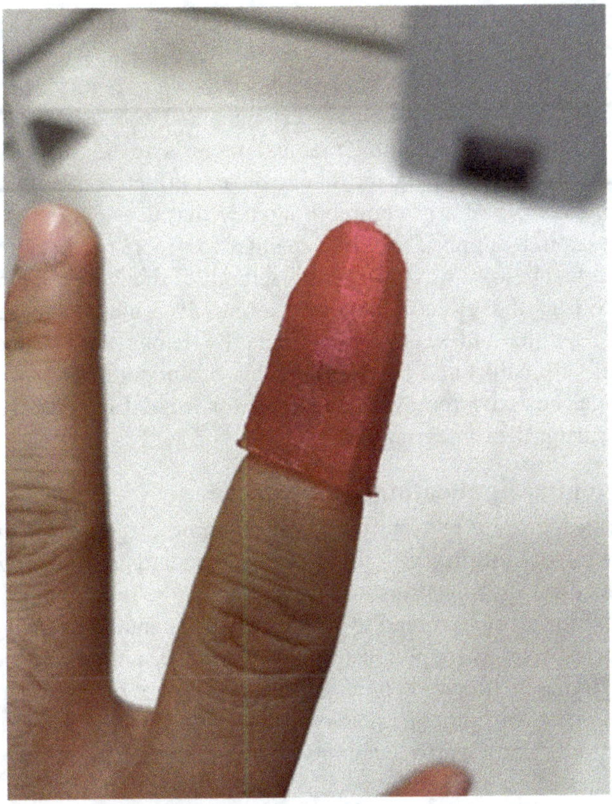

Figure 16.13 *Photo of a 3D printed orthosis for treating a common fingertip injury (from VR's practice)*

16.6.5.3 The rapid deployment of solutions

Once regulators approve digital models, 3D printing brings digital possibilities into the real world quickly and cost-effectively. When care is delivered faster, there is more time for the treatment of patients.

16.6.5.4 Blockchain technology

Blockchain is a distributed database technology that has a wide range of applications and uses in healthcare. It is a ledger technology that requires no central authority. It is mainly used in healthcare to keep patients' medical records safe and secure. The distributed ledger of time-stamped 'blocks' of data in a chronological sequence cannot be corrupted retroactively. Each new input notifies the nodes that anonymously comprise the distributed, peer-to-peer network and their replicated copies of the ledger are updated accordingly. It records and stores an audit trail of all transactions. The decentralised nature of the technology makes the ledger highly resistant to fraud and compromise [68].

The blockchain immutability and decentralisation make the distributed networks fault–tolerant and will remain connected even if some nodes fail or are compromised; the ledger remains unchanged and accessible. Blockchain's anonymous network of verifying nodes removes the need for any central authority or trusted party. It protects sensitive data by concealing the identity of the individuals with secure codes, thus making it more efficient and robust than other database technologies. It allows the patients' medical records to be shared securely and quickly, with their consent, for faster diagnosis and effective treatment plans [69]. The CDC is an example of a government agency that uses Blockchain to monitor and track diseases in a supply chain-like manner to report outbreaks in real-time.

Blockchain technology was conceptualised out of the 2008 financial crisis as a viable alternative to the global financial system. It came to the fore as a core component of Bitcoin and was first adopted by banks and financial services. Blockchain can help banks and financial service providers improve efficiency and lower operational costs by removing the need for trusted intermediaries in transactions and ensuring data integrity [70,71].

16.6.5.5 Potential applications in healthcare

Blockchain applications are relevant to almost anything that involves digital transactions and are being used to improve the prevalent standards of handling, sharing and processing of medical data and personal medical records. In addition, industries and technologies ripe for disruption tend to rely heavily on centralised networks or information repositories, such as supply chain management, insurance and cloud storage.

Many are trying to improve processes in healthcare by improving efficiency, controlling access, advancing technology, protecting privacy and managing data security. As a result, blockchain's suitability to transform healthcare has been increasingly discussed in the last few years, specifically for its ability to protect and enable the exchange of electronic healthcare information across fragmented healthcare systems.

Blockchain applications can help address concerns related to medical diagnostics and improve patient care [72].

Some of the potential uses of blockchain in healthcare are the following:

1. Supply chain management in healthcare systems.
2. Remote monitoring of patients.
3. Monitoring, processing and auditing of the healthcare 'transaction'.
4. Data analysis and research.
5. EHR [73,74].

In the Internet of medical things domain, the mobile and wearable gadgets that surround every individual enable the wireless systems to feel, activate, interact and communicate with other identical devices. They could be a person, place, device, context of time, any network, and business or service. In healthcare, the patient is the source of all data for the Internet of medical things. They have devices attached closely or remotely monitoring their health and activities and generating a large volume of data. The data are stored on a secure network. With the help of AI and blockchain, the healthcare providers can track, monitor and authenticate the healthcare interactions and deliver safe, effective and efficient care promptly with consent from the owners of the data (Figure 16.14) [75,76].

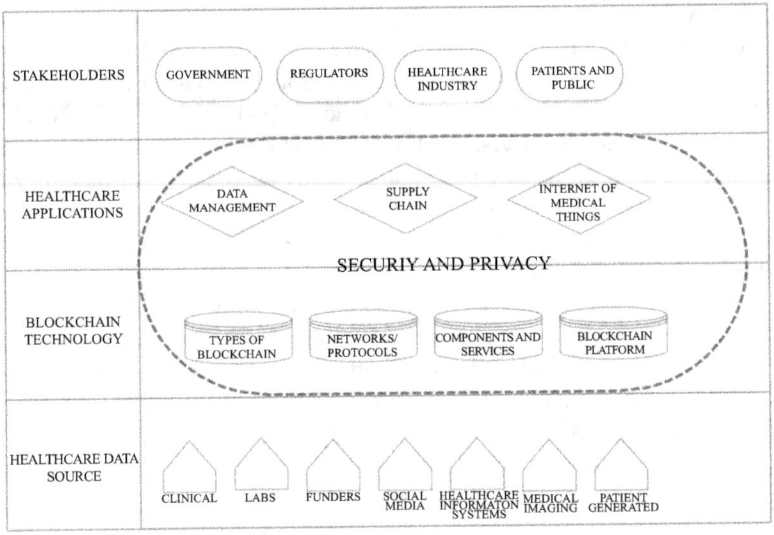

Figure 16.14 A workflow of blockchain-based healthcare applications. The workflow comprises four main layers: raw healthcare data, blockchain technology, healthcare applications and stakeholders. The blockchain as a decentralised technology enables multiple stakeholders to benefit from healthcare applications [79]

Blockchain technology in the pharmaceutical industry can manage pharmaceutical supply chain issues. It can also prevent the delivery of counterfeit or inadequate medications, which can have critical consequences for the patients. It can also benefit health insurance claims as it can provide encrypted, transparent and auditable data [77].

Many healthcare services around the world face issues with their EHRs. A particularly pertinent issue is the lack of interoperability and data exchange. The different healthcare entities (hospitals, caregivers, insurers, pharmacies) rely on various 'sources of truth' for patients' health records, leading to outdated, patchy or incorrect data. These siloed pockets of centralised information have low fault tolerance and are vulnerable to security breaches and ransomware attacks. Furthermore, different EHR providers use different workflows to store this data, making manual efforts to combine the data to get a comprehensive view of a patient's medical history extremely difficult.

If blockchain technology was integrated with EHRs, all parties could use the same shared ledger as a source of truth, improving interoperability. Given the integrity and immutability of data in a blockchain – as the active nodes must verify and confirm every input before it becomes part of the blockchain – data exchange between healthcare entities can be done securely and confidently without the need of a trusted intermediary. Integrating blockchain into EHRs could reduce administrative overheads and improve efficiency for healthcare entities [78].

No data would have to be sent or requested, in contrast to the classic 'push' or 'pull' models of medical data exchange, which do not guarantee data integrity, lack a standardised auditing trail and require ad hoc security permissions. With blockchain – integrated EHRs, the shared ledger could contain a patient's entire medical history, including prescriptions, doctor visits, ailments – stored in a comprehensive historical timeline with the most up-to-date information easily accessible. All this information would be encrypted, ensuring privacy. Being a distributed system, blockchain – integrated EHRs would also be more fault–tolerant. If nodes in the network are compromised or crashed, the data will remain unchanged and fully accessible. It would also enable standardised auditing across diverse healthcare entities, which could, for example, lead to more efficient claims processes for the insurance industry.

Furthermore, patients currently have minimal access to their medical records and information. Blockchain could empower patients by granting them access to their medical records, enabling them to see every time their records are updated and sharing their data with researchers or different healthcare entities. It could also help patients generate their data through IoT and wearables, which could then be stored securely in their records in real-time. The benefits outlined above could also be applied to data from research and clinical trials, improving the integrity of the data and enabling efficient auditing. Comprehensive digitised medical records that are shareable create a better experience for patients and healthcare providers [80].

16.6.6 Case studies

Blockchain technology in the pharmaceutical industry is capable of managing pharmaceutical supply chain issues. It can also prevent the delivery of counterfeit or inadequate medications, which can have critical consequences for the patients. It can also benefit health insurance claims as it can provide encrypted, transparent, and auditable data [81]. It manages authentication, confidentiality, and data-sharing and provides the patients' access to their medical records across providers [82].

Another example is MeDShare, which provides data provenance, auditing and control between big data entities to share medical data in the cloud. It allows for secure and reliable collaboration in monitoring and control of healthcare transactions [83].

In remote sensors collecting biomedical data, HyperLedger Fabric uses blockchain to transmit, store and share data securely [84]. IoBHealth takes this further with the use of the technology on the Internet of medical things [85]. MIStore is a medical insurance system that uses blockchain technology to ensure all healthcare transactions and healthcare insurance data for reimbursements are immutably stored and retrieved for the medical insurance industry [77].

Tokenisation is a process that allows digitalisation of healthcare data and allows for specific user rights and access to and retrieval of their healthcare data without decrypting or re-encrypting. This empowers patients to own, store and share their medical record information with anyone, thereby reducing the barriers of healthcare access and treatment and reducing the powers of the central intermediaries of health data (e.g., insurance companies, hospitals, pharmaceuticals, etc.) [76].

16.6.7 Challenges of blockchain

Blockchain is surrounded by hype, and the current outlook could be overly optimistic. Therefore, it will be essential to proceed cautiously. Regarding EHRs, outside of the issues outlined above, there are other underlying issues that blockchain might not be able to solve, such as implementation, operational and maintenance costs, standardisation, and incomplete or delayed documentation.

Despite the promise of blockchain technology in the health sector, much of the existing evidence surrounds the theoretical frameworks, architectures or models with few technical details. There have been few successful implementations and scale-up national versions. It has been recommended to be used to best benefit the healthcare systems in collaboration with other technologies with appropriate auditing measures. It is important to note that healthcare data can be high-volume and rapidly changing. It will pose significant challenges to the current existing blockchain and be inefficient and costly. Online analysis and planning with blockchain stored is challenging and thus, limiting its clinical, statistical and research use.

One significant challenge to the current blockchain technology is its lack of total conformity to the international data protection regulation, particularly the right to erasure. There has been a failure to engage stakeholders in blockchain development and incentivise the end-users of the technology [73].

A cautious approach to technology in healthcare is needed. As personalised health and wearable devices become popular and adopted by the general public, more research is needed to explore the usefulness and value of these applications in healthcare [26]. Further challenges will be posed by sharing sensitive personal health data among healthcare stakeholders and compliance with personal data protection laws [86].

The Health Insurance Portability and Accountability Act (HIPPA) in the US and General Data Protection Regulation (GDPR) in Europe established the rules, expectations and framework that determined the appropriate legal storage and dissemination of private, personal health information [86].

Despite the potential of technology use in healthcare, the difficulty in large-scale rollout and sustainability of these projects and lack of evidence of the outcomes in health improvement remains a persistent challenge. In addition to all these is the issues of data security, governance and protection. There have also been unrealistic expectations of delivering results from the integration of technology into the healthcare systems. Other risks include noncompliance with data protection policies, inadequate data governance and overestimated expectations resulting from the rapid introduction of new technologies. As with all things new, the healthcare sector needs to carefully assess the risk and benefits and ensure that technology is evidence-based [4].

16.7 Conclusion

The impact of technology on healthcare has been pervasive and relentless. Despite the tremendous benefits to outcomes, experiences and efficiency, it has also produced sufficient challenges and concerns for society. The complex nature of the healthcare system and processes requires a transdisciplinary approach to generate innovation and solutions to enhance further the impact of healthcare on the individual and the community. The current status of healthcare provides an excellent opportunity for the engineering, manufacturing, computing and information technology industry to work with the main stakeholders in healthcare to learn from each other and explore options for transformative ways to add value to society.

References

[1] Kruk, M.E., Gage, A.D., Arsenault, C., *et al.*, 2018. High-Quality Health Systems in the Sustainable Development Goals Era: Time for a Revolution. *Lancet Glob. Health* 6, e1196–e1252. https://doi.org/10.1016/S2214-109X(18)30386-3.

[2] Ferguson, B., Belloni, A., 2019. Ageing and Health Expenditure Categories: Health Profile for England, Public Health Data. Ageing Health Expend. https://ukhsa.blog.gov.uk/2019/01/29/ageing-and-health-expenditure/ (accessed 9.25.21).

[3] Myers, R., Anderson, M., Korba, C., 2020. Striving to Become More Patient-Centric in Life Sciences. What It Really Takes to Optimise Patient Trust and Health Outcomes.

[4] Meinert, E., Alturkistani, A., Brindley, D., Knight, P., Wells, G., de Pennington, N., 2018. Weighing Benefits and Risks in Aspects of Security, Privacy and Adoption of Technology in a Value-Based Healthcare System. *BMC Med. Inform. Decis. Mak.* 18, 100. https://doi.org/10.1186/s12911-018-0700-0.

[5] Ferlie, E.B., Shortell, S.M., 2001. Improving the Quality of Health Care in the United Kingdom and the United States: A Framework for Change. *Milbank Q.* 79, 281–315. https://doi.org/10.1111/1468-0009.00206.

[6] NHS Providers, 2021. The NHS Provider Sector [WWW Document]. NHS Provid. https://nhsproviders.org/topics/delivery-and-performance/the-nhs-provider-sector (accessed 9.25.21).

[7] Jarvis, T., Scott, F., El-Jardali, F., Alvarez, E., 2020. Defining and Classifying Public Health Systems: A Critical Interpretive Synthesis. *Health Res. Policy Syst.* 18, 68. https://doi.org/10.1186/s12961-020-00583-z.

[8] Simonse, L., Albayrak, A., Starre, S., 2019. Patient Journey Method for Integrated Service Design. *Des. Health* 3, 82–97. https://doi.org/10.1080/24735132.2019.1582741.

[9] Meskó, B., Drobni, Z., Bényei, É., Gergely, B., Győrffy, Z., 2017. Digital Health is a Cultural Transformation of Traditional Healthcare. *mHealth* 3, 38. https://doi.org/10.21037/mhealth.2017.08.07.

[10] Hermes, S., Riasanow, T., Clemons, E.K., Böhm, M., Krcmar, H., 2020. The Digital Transformation of the Healthcare Industry: Exploring the Rise of Emerging Platform Ecosystems and Their Influence on the Role of Patients. *Bus. Res.* 13, 1033–1069. https://doi.org/10.1007/s40685-020-00125-x.

[11] Lin, S. Y., Mahoney, M. R., & Sinsky, C. A. 2019. Ten Ways Artificial Intelligence Will Transform Primary Care. *J. General Gen. Internal Intern.*

MedicineMed., 34(8), 1626–1630. https://doi.org/10.1007/s11606-019-05035-1 1.

[12] Gualandi, R., Masella, C., Viglione, D., Tartaglini, D., 2019. Exploring the Hospital Patient Journey: What Does the Patient Experience? *PLoS One* 14, e0224899. https://doi.org/10.1371/journal.pone.0224899.

[13] Marques, I.C.P., Ferreira, J.J.M., 2020. Digital Transformation in the Area of Health: Systematic Review of 45 Years of Evolution. *Health Technol.* 10, 575–586. https://doi.org/10.1007/s12553-019-00402-8.

[14] Kraus, S., Schiavone, F., Pluzhnikova, A., & Invernizzi, A. C. (2021). Digital transformation Transformation in healthcareHealthcare: Analyzing the current Current stateState-of-researchResearch. *Journal of. Business Research*,. 123, 557–567. https://doi.org/10.1016/j.jbusres.2020.10.030.

[15] Doarn, C.R., 2021. Technological Advances Making Telemedicine and Telepresence Possible, in: Latifi, R., Doarn, C.R., Merrell, R.C. (Eds.), *Telemedicine, Telehealth and Telepresence: Principles, Strategies, Applications, and New Directions.* Springer International Publishing, Cham, pp. 257–271. https://doi.org/10.1007/978-3-030-56917-4_16.

[16] The Lancet. (2021). Can Digital Technologies Improve Health? *The Lancet*, 398(10312), 1663. https://doi.org/10.1016/S0140-6736(21)02219-4 4.

[17] Deloitte, U. 2015. Connected health | Deloitte UK. Deloitte United Kingdom. https://www2.deloitte.com/uk/en/pages/life-sciences-and-health-care/articles/connected-health.html.

[18] Budd, J., Miller, B.S., Manning, E. M., Lampos, V., Zhuang, M., Edelstein, M., Rees, G., Emery, V. C., Stevens, M. M., Keegan, N., Short, M. J., Pillay, D., Manley, E., Cox, I. J., Heymann, D., Johnson, A. M., & McKendry, R. A. *et al.* (2020). Digital technologies Technologies in the Public-Health Response to COVID-19. *Nature Medicine*,. 26(8), 1183–1192. https://doi.org/10.1038/s41591-020-1011-4.

[19] World Health Organization, G. 2021. Global strategy Strategy on Digital Health 2020--2025. WHO Licence: CC BY-NC-SA 3.0 IGO. https://www.who.int/docs/default-source/documents/gs4dhdaa2a9f352b0445bafbc79ca799dce4d.pdf pdf.

[20] US Food and Drug Administration. 2020. What is Digital Health? FDA Digital Health Center for Devices and Radiological. https://www.fda.gov/medical-devices/digital-health-center-excellence/what-digital-health.

[21] Ćwiklicki, M., Klich, J., Chen, J. 2020. The adaptiveness Adaptiveness of the Healthcare System to the Fourth Industrial Revolution: A Preliminary Analysis. *Futures*, 122, 102602. https://doi.org/10.1016/j.futures.2020.102602.

[22] Health, C. for D. and R., 2020. What is Digital Health? FDA.

[23] Davies, L. *2020.* Closing the Gap: The Rise of Prescription Digital Therapeutics [WWW Document]. OxfordSM. URL https://oxfordsm.com/events-articles/2020/the-rise-of-prescription-digital-therapeutics (accessed 10.9.21).

[24] reSET® & reSET-O® [WWW Document], n.d. Pear Ther. https://peartherapeutics.com/products/reset-reset-o/ (accessed 10.9.21).

[25] Kollins, S.H., DeLoss, D.J., Cañadas, E., *et al.*, 2020. A Novel Digital Intervention for Actively Reducing Severity of Paediatric ADHD (STARS-ADHD): A Randomised Controlled Trial. *Lancet Digit. Health* 2, e168–e178. https://doi.org/10.1016/S2589-7500(20)30017-0.

[26] Lu, L., Zhang, J., Xie, Y., *et al.*, 2020. Wearable Health Devices in Health Care: Narrative Systematic Review. *JMIR MHealth UHealth* 8, e18907. https://doi.org/10.2196/18907.

[27] Gao, Y., Yu, L., Yeo, J.C., Lim, C.T., 2020. Flexible Hybrid Sensors for Health Monitoring: Materials and Mechanisms to Render Wearability. *Advanced Materials* 32(15), 1902133. https://doi.org/10.1002/adma.201902133.

[28] Dong, B., Yang, Y., Shi, Q., *et al.* 2020. Wearable Triboelectric–Human–Machine Interface (THMI) Using Robust Nanophotonic Readout. *ACS Nano* 14(7), 8915–8930. https://doi.org/10.1021/acsnano.0c03728.

[29] Khan, Z.F., Alotaibi, S.R., 2020. Applications of Artificial Intelligence and Big Data Analytics in m-Health: A Healthcare System Perspective. *J. Healthc. Eng.* 2020, e8894694. https://doi.org/10.1155/2020/8894694.

[30] Davenport, T., Kalakota, R., 2019. The Potential for Artificial Intelligence in Healthcare. *Future Healthc. J.* 6, 94–98. https://doi.org/10.7861/futurehosp.6-2-94.

[31] Babic, B., Gerke, S., Evgeniou, T., Cohen, I.G., 2021. Beware Explanations from A.I. in Health Care. *Science* 373, 284–286. https://doi.org/10.1126/science.abg1834.

[32] Aggarwal, N., Ahmed, M., Basu, S., *et al.*, 2020. Artificial Intelligence in Health Care: Benefits and Challenges of Technologies to Augment Patient Care. *GAO Rep. i*–91. https://doi.org/10.31478/202011f.

[33] Reddy, S., Allan, S., Coghlan, S., Cooper, P., 2020. A Governance Model for the Application of A.I. in Health Care. *J. Am. Med. Inform. Assoc.* 27, 491–497. https://doi.org/10.1093/jamia/ocz192.

[34] Flavián, C., Ibáñez-Sánchez, S., Orús, C., 2019. The Impact of Virtual, Augmented and Mixed Reality Technologies on the Customer Experience. *J. Bus. Res.* 100, 547–560. https://doi.org/10.1016/j.jbusres.2018.10.050.

[35] Sutherland, J., Belec, J., Sheikh, A., *et al.*, 2019. Applying Modern Virtual and Augmented Reality Technologies to Medical Images and Models. *J. Digit. Imaging* 32, 38–53. https://doi.org/10.1007/s10278-018-0122-7.

[36] Lewis, S., 2021. The Importance of Maintaining the Essence of Relational Medicine in a Technological Age. *J. Gen. Fam. Med.* 22, 65–66. https://doi.org/10.1002/jgf2.369.

[37] Paul, S., Riffat, M., Yasir, A., *et al.* 2021. Industry 4.0 Applications for Medical/Healthcare Services. *J. Sensor Actuat. Netw.* 10(43), 43. https://doi.org/10.3390/jsan10030043

[38] Metcalf, D., Milliard, S.T.J., Gomez, M., Schwartz, M., 2016. Wearables and the Internet of Things for Health: Wearable, Interconnected Devices Promise More Efficient and Comprehensive Health Care. *IEEE Pulse* 7, 35–39. https://doi.org/10.1109/MPUL.2016.2592260.

[39] Greco, L., Percannella, G., Ritrovato, P., Tortorella, F., Vento, M., 2020. Trends in IoT Based Solutions for Health Care: Moving A.I. to the Edge. *Pattern Recognit. Lett.* 135, 346–353. https://doi.org/10.1016/j.patrec.2020.05.016.

[40] Kumar, S., Arora, A.K., Gupta, P., Saini, B.S., 2021. A Review of Applications, Security and Challenges of Internet of Medical Things, in: Hassanien, A.E., Khamparia, A., Gupta, D., Shankar, K., Slowik, A. (Eds.), *Cognitive Internet of Medical Things for Smart Healthcare: Services and Applications, Studies in Systems, Decision and Control. Springer International Publishing, Cham*, pp. 1–23. https://doi.org/10.1007/978-3-030-55833-8_1

[41] Mahmoud, N.M., Fouad, H., Soliman, A.M. 2021. Smart Healthcare Solutions using the Internet of Medical Things for Hand Gesture Recognition System. *Complex Intell. Syst.* 7(3), 1253. https://doi.org/10.1007/s40747-020-00194-9

[42] Tsai, J.-M., Cheng, M.-J., Tsai, H.-H., Hung, S.-W., Chen, Y.-L. 2019. Acceptance and Resistance of Telehealth: The Perspective of Dual-Factor Concepts in Technology Adoption. *Int. J. Inf. Manag.* 49, 34–44. https://doi.org/10.1016/j.ijinfomgt.2019.03.003.

[43] Verma, R. 2021. Smart City Healthcare Cyber Physical System: Characteristics, *Technologies and Challenges. Wireless Personal Commun.: Int. J.* 1. https://doi.org/10.1007/s11277-021-08955-6.

[44] Jiang, Y., Reimer, V., Schossig, T., Angelmahr, M., Schade, W. 2020. Fiber Optical Multifunctional Human-Machine Interface for Motion Capture, Temperature, and Contact Force Monitoring. *Optics Lasers Eng.* 128, 106018. https://doi.org/10.1016/j.optlaseng.2020.106018

[45] Zhu, M., Sun, Z., Zhang, Z., *et al.* n.d. Haptic-Feedback Smart Glove as a Creative Human-Machine Interface (HMI) for Virtual/Augmented Reality Applications. *Sci. Adv.* 6(19), eaaz8693. https://doi.org/10.1126/sciadv.aaz8693.

[46] Brunete, A., Gambao, E., Hernando, M., Cedazo, R. 2021. Smart Assistive Architecture for the Integration of IoT Devices, Robotic Systems, and Multimodal Interfaces in Healthcare Environments. *Sensors* 21(6), 2212. https://doi.org/10.3390/s21062212.

[47] Mavropoulos, T., Symeonidis, S., Tsanousa, A., *et al.* 2021. Smart Integration of Sensors, Computer Vision and Knowledge Representation for Intelligent Monitoring and Verbal Human-Computer Interaction. *J. Intell. Inf. Syst.* 57(2), 321–345. https://doi.org/10.1007/s10844-021-00648-7.

[48] Zhang, S., Wei, Z., Nie, J., Huang, L., Wang, S., Li, Z. 2017. A Review on Human Activity Recognition Using Vision-Based Method. *J. Healthcare Eng.* 2017, e3090343. https://doi.org/10.1155/2017/3090343.

[49] Liu, R., Ramli, A. A., Zhang, H., Henricson, E., Liu, X. 2021. An Overview of Human Activity Recognition Using Wearable Sensors: Healthcare and Artificial Intelligence. ArXiv:2103.15990 [Cs, Eess]. http://arxiv.org/abs/2103.15990.

[50] Liao, J., Jia, Y., Wu, Y., Shi, K., Yang, D., Li, P., Qian, Z. 2020. Physical-, chemical-, and biological-responsive nanomedicine for cancer therapy. *WIREs Nanomedicine and Nanobiotechnology* 12(1), e1581. https://doi.org/10.1002/wnan.1581.

[51] Martins, J.P., das Neves, J., de la Fuente, M., Celia, C., Florindo, H., Günday-Türeli, N., Popat, A., Santos, J.L., Sousa, F., Schmid, R., Wolfram, J., Sarmento, B., Santos, H.A., 2020. The solid progress of nanomedicine. *Drug Delivery and Translational Research* 10(3), 726–729. https://doi.org/10.1007/s13346-020-00743-2.

[52] Soares, S., Sousa, J., Pais, A., Vitorino, C., 2018. Nanomedicine: Principles, Properties, and Regulatory Issues. *Frontiers in Chemistry* 6, 360. https://doi.org/10.3389/fchem.2018.00360.

[53] Genome U.K.: The Future of Healthcare, 2020.

[54] Aslam, B., Basit, M., Nisar, M.A., Khurshid, M., Rasool, M.H., 2017. Proteomics: Technologies and Their Applications. *J. Chromatogr. Sci.* 55, 182–196. https://doi.org/10.1093/chromsci/bmw167.

[55] Roth, S.C., 2019. What is Genomic Medicine? *J. Med. Libr. Assoc. JMLA* 107, 442–448. https://doi.org/10.5195/jmla.2019.604.

[56] Troccaz, J., Dagnino, G., Yang, G.-Z., 2019. Frontiers of Medical Robotics: From Concept to Systems to Clinical Translation. *Annu. Rev. Biomed. Eng.* 21, 193–218. https://doi.org/10.1146/annurev-bioeng-060418-052502.

[57] Gassert, R., Dietz, V., 2018. Rehabilitation Robots for the Treatment Of Sensorimotor Deficits: A Neurophysiological Perspective. J. NeuroEngineering Rehabil. 15, 46. https://doi.org/10.1186/s12984-018-0383-x.

[58] Thai, M.T., Phan, P.T., Hoang, T.T., Wong, S., Lovell, N.H., Do, T.N., 2020. Advanced Intelligent Systems for Surgical Robotics. *Adv. Intell. Syst.* 2, 1900138. https://doi.org/10.1002/aisy.201900138.

[59] Choi, P.J., Oskouian, R.J., Tubbs, R.S., 2018. Telesurgery: Past, Present, and Future. *Cureus* 10. https://doi.org/10.7759/cureus.2716.

[60] Zhu, J., Lyu, L., Xu, Y. *et al.*, 2021. Intelligent Soft Surgical Robots for Next-Generation Minimally Invasive Surgery. *Adv. Intell. Syst.* 3, 2100011. https://doi.org/10.1002/aisy.202100011.

[61] Beuss, F., Schmatz, F., Stepputat, M., Nokodian, F., Fluegge, W., Frerich, B. 2021. Cobots in Maxillofacial Surgery – Challenges for Workplace Design and the Human-Machine-Interface. *Procedia CIRP* 100, 488–493. https://doi.org/10.1016/j.procir.2021.05.108.

[62] He, Y., Eguren, D., Luu, T.P., Contreras-Vidal, J.L., 2017. Risk Management and Regulations for Lower Limb Medical Exoskeletons: A Review. *Med. Devices Auckl. N.Z.* 10, 89–107. https://doi.org/10.2147/MDER.S107134

[63] Sirlantzis, K., Larsen, L.B., Kanumuru, L.K., Oprea, P., 2019. 11 – Robotics, in: Cowan, D., Najafi, L. (Eds.), *Handbook of Electronic Assistive Technology*. Academic Press, London, pp. 311–345. https://doi.org/10.1016/B978-0-12-812487-1.00011-9.

[64] Duret, C., Grosmaire, A.-G., Krebs, H.I., 2019. Robot-Assisted Therapy in Upper Extremity Hemiparesis: Overview of an Evidence-Based Approach. *Front. Neurol.* 10, 412. https://doi.org/10.3389/fneur.2019.00412.

[65] Dodziuk, H., 2016. Applications of 3D Printing in Healthcare. Kardiochirurgia Torakochirurgia Pol. *Pol. J. Cardio-Thorac. Surg.* 13, 283–293. https://doi.org/10.5114/kitp.2016.62625.

[66] Trenfield, S.J., Awad, A., Madla, C.M., *et al.*, 2019. Shaping the Future: Recent Advances of 3D Printing in Drug Delivery and Healthcare. *Expert Opin. Drug Deliv.* 16, 1081–1094. https://doi.org/10.1080/17425247.2019.1660318.

[67] Manero, A., Smith, P., Sparkman, J., *et al.*, 2019. Implementation of 3D Printing Technology in the Field of Prosthetics: Past, Present, and Future. *Int. J. Environ. Res. Public. Health* 16, 1641. https://doi.org/10.3390/ijerph16091641.

[68] Blockchain in Healthcare | HIMSS [WWW Document], 2020. URL https://www.himss.org/resources/blockchain-healthcare (accessed 10.9.21).

[69] Bodó, B., Brekke, J.K., Hoepman, J.-H., 2021. Decentralisation in the Blockchain Space. *Internet Policy Rev.* 10.

[70] Agbo, C.C., Mahmoud, Q.H., Eklund, J.M., 2019. Blockchain Technology in Healthcare: A Systematic Review. *Healthcare* 7, 56. https://doi.org/10.3390/healthcare7020056.

[71] Reiff, N., 2020. How Blockchain Can Protect the Global Economy [WWW Document]. Investopedia. https://www.investopedia.com/tech/how-blockchain-can-protect-global-economy/ (accessed 10.9.21).

[72] Tandon, A., Dhir, A., Islam, A.K.M.N., Mäntymäki, M., 2020. Blockchain in Healthcare: A Systematic Literature Review, Synthesising Framework and Future Research Agenda. *Comput. Ind.* 122, 103290. https://doi.org/10.1016/j.compind.2020.103290.

[73] Agbo, C.C., Mahmoud, Q.H., 2020. Blockchain in Healthcare: Opportunities, Challenges, and Possible Solutions. *Int. J. Healthc. Inf. Syst. Inform.* 15, 82–97. https://doi.org/10.4018/IJHISI.2020070105.

[74] Roman-Belmonte, J.M., De la Corte-Rodriguez, H., Rodriguez-Merchan, E. C., 2018. How Blockchain Technology Can Change Medicine. *Postgrad. Med.* 130, 420–427. https://doi.org/10.1080/00325481.2018.1472996.

[75] Ometov, A., Shubina, V., Klus, L., *et al.*, 2021. A Survey on Wearable Technology: History, State-of-the-Art and Current Challenges. *Comput. Netw.* 193, 108074. https://doi.org/10.1016/j.comnet.2021.108074.

[76] Yaqoob, I., Salah, K., Jayaraman, R., Al-Hammadi, Y., 2021. Blockchain for Healthcare Data Management: Opportunities, Challenges, and Future Recommendations. *Neural Comput. Appl.* https://doi.org/10.1007/s00521-020-05519-w.

[77] Ben Fekih, R., Lahami, M., 2020. Application of Blockchain Technology in Healthcare: A Comprehensive Study, in: Jmaiel, M., Mokhtari, M., Abdulrazak, B., Aloulou, H., Kallel, S. (Eds.), *The Impact of Digital Technologies on Public Health in Developed and Developing Countries, Lecture Notes in Computer Science. Springer International Publishing, Cham*, pp. 268–276. https://doi.org/10.1007/978-3-030-51517-1_23.

[78] Stafford, T.F., Treiblmaier, H., 2020. Characteristics of a Blockchain Ecosystem for Secure and Sharable Electronic Medical Records. *IEEE Trans. Eng. Manag.* 67, 1340–1362. https://doi.org/10.1109/TEM.2020.2973095.

[79] Khezr, S., Moniruzzaman, M., Yassine, A., Benlamri, R., 2019. Blockchain Technology in Healthcare: A Comprehensive Review and Directions for Future Research. *Appl. Sci.* 9, 1736. https://doi.org/10.3390/app9091736.

[80] Gordon, W., Wright, A., Landman, A., 2017. Blockchain in Health Care: Decoding the Hype. *NEJM Catal.*

[81] Ben Fekih, R., Lahami, M., 2020. Application of Blockchain Technology in Healthcare: A Comprehensive Study. In M. Jmaiel, M. Mokhtari, B. Abdulrazak, H. Aloulou, & S. Kallel (Eds.), *The Impact of Digital Technologies on Public Health in Developed and Developing Countries* (pp. 268–276). Springer International Publishing. https://doi.org/10.1007/978-3-030-51517-1_23.

[82] Azaria, A., Ekblaw, A., Vieira, T., Lippman, A., 2016. MedRec: Using Blockchain for Medical Data Access and Permission Management, in: 2016 2nd International Conference on Open and Big Data (OBD). Presented at the 2016 2nd International Conference on Open and Big Data (OBD), IEEE, Vienna, Austria, pp. 25–30. https://doi.org/10.1109/OBD.2016.11.

[83] Xia, Q., Sifah, E.B., Asamoah, K.O., Gao, J., Du, X., Guizani, M., 2017. MeDShare: Trust-Less Medical Data Sharing Among Cloud Service Providers via Blockchain. *IEEE Access* 5, 14757–14767. https://doi.org/10.1109/ACCESS.2017.2730843.

[84] Hyperledger, 2020. Introduction — hyperledger-fabricdocs master documentation [WWW Document]. Hyperledger Fabr. URL https://hyperledger-fabric.readthedocs.io/en/release-2.2/blockchain.html (accessed 10.11.21).

[85] Ray, P.P., Dash, D., Salah, K., Kumar, N., 2021. Blockchain for IoT-Based Healthcare: Background, Consensus, Platforms, and Use Cases. *IEEE Syst. J.* 15, 85–94. https://doi.org/10.1109/JSYST.2020.2963840.

[86] Yuan, B., Li, J., 2019. The Policy Effect of the General Data Protection Regulation (GDPR) on the Digital Public Health Sector in the European Union: An Empirical Investigation. *Int. J. Environ. Res. Public. Health* 16, 1070. https://doi.org/10.3390/ijerph16061070.

[87] Betz, U.A.K., Betz, F., Kim, R., Monks, B., Phillips, F., 2019. Surveying the Future of Science, Technology and Business – A 35 Year Perspective. *Technol. Forecast. Soc. Change* 144, 137–147. https://doi.org/10.1016/j.techfore.2019.04.005.

Chapter 17

Smart manufacturing workplace safety with virtual training, AR and haptic technologies

Bhakti Yudho Suprapto[1], Ahmad Farhan Aristz[1], Eric Sean Kesuma[1] and Suci Dwijayanti[1]

Accidents due to undetected fires have caused huge losses in various contexts in the world, such as in offices, manufacturing workplaces, residential areas, even in forest areas. The evacuation process in the firefighting system requires an effective and fast response. The source of fire must be found quickly to prevent it from spreading. Image processing based on an object detection system, it is believed, can circumvent this problem. The method for detecting objects is performed by the You Only Look Once (YOLO) algorithm in real-time. The most suitable YOLO model for the system is the Tiny YOLO VOC model which was built with the Darkflow framework. Besides detecting the fires, the robot can be used to ease the search process and firefighting efforts. However, it needs planning to fit the robots with the system optimized to do the best in the minimum time possible. In robotics, finding the fastest path can be solved by embedding the A* algorithm in the robot. A* algorithm is one of the artificial intelligence methods for finding the fastest path. Not many studies have applied this algorithm to the wheel robots to find the shortest path. Therefore, this A* method will be used for finding the fastest path to find the fire with accuracy by finding the best distance and route with the coordinate system. In this study, the wheel robots had three missions, finding the fire, extinguishing the fire, and returning to the starting points. The system accuracy for those missions was 100%, 83.33%, and 50%, respectively.

17.1 Introduction

Many accidents can occur from a human being's carelessness. Fire is one of the accidents that often occur in the factory. Thus, the manufacturing process needs to be smart to handle various types of accidents, including fire. The need for a fire detection system to be incorporated as a part of the process is growing rapidly [1]. There is a need for not only the evacuation process to be fast and effective, but also the search process to find the source of the fire. The firefighting system is efficient

[1]Department of Electrical Engineering, Universitas Sriwijaya, Palembang, Indonesia

provided a robot is there to find the fire and extinguish it. In addition, a detection system is also important to find the source of the fire. Automatic systems of fire detection can utilize and detect temperature, smoke, and fire. Each sensor operates using different principles and gives different responses.

A robot can be utilized to search and detect fire by finding the best route. The procedure of finding the best route can be developed using artificial intelligence. The shortest path algorithm which is embedded in the robot may help the robot to complete its mission. However, the process of firefighting becomes a problem, especially for a wheeled robot [2]. In the previous research, some methods have been developed to find the shortest path, such as by Dijkstra [3], by Bellman-Ford [4], and research on the D* algorithm [5], and A* algorithm [6]. D* and A* algorithms have been implemented by Takayuki Goto and Takeshi in a robot and they showed that A* algorithm had better performances than D* algorithm [7]. Nevertheless, the A* algorithm has not been applied to the wheeled robot to extinguish the fire. Thus, this research utilized A* algorithm to find the best and shortest path to find the source of the fire by a robot. To do so, the robot needs to detect the fire. This can be implemented by using the image processing approach. The image processing method is applied to object detection by finding the characteristics of fire such as smoke; while fire detection is accomplished through images or real-time video. Various methods have been developed for object detection using image processing, such as R-CNN using PASCAL VOC [8], fast R-CNN for PASCAL VOC dataset and MS COCO, single shot multibox detector using PASCAL VOC, imageNet, and MS COCO [9], and YOLO using PASCAL VOC and ImageNet [10].

Some studies have shown that YOLO gives better performance in the frames per second, which is 91 fps faster than other methods often below 59 fps [11]. Even, PASCAL VOC 2007 Fast YOLO has the highest speed up to 155 fps which is twice the accuracy for real-time detector [10].

Thus, this study proposes an integrated system to detect and extinguish the fire. This system combines image processing and the shortest path algorithm. Here, YOLO is implemented to detect fire and some wheeled robots are used to extinguish the fire in a labyrinth which is assumed a factory. A* algorithm is then used to determine the shortest path passed by the robots based on the information given by the YOLO. This chapter is organized as follows: Section 17.2 describes the robot design and Section 17.3 explains the fire detection using YOLO, followed by the shortest path using A* algorithm in Section 17.4. Results and discussion are presented in Section 17.5. Finally, this chapter is summarized in Section 17.6.

17.2 Robot design

The design of the robot was divided into two parts, namely the mechanical part and the wiring. Acrylic was chosen as the material for the upper and lower part of robots. The sensors, components, and motors were put in place based on the design. The designed robot can be seen in Figure 17.1.

Figure 17.1 Firefighter robot

As shown in Figure 17.1, Robot_1 is yellow and Robot_2 is red. Both robots have the same sensors despite having different designs. The design was aimed to be distinguished by the name of the robot and by the sending of position data of the robot easily.

The hardware needed to build this robot is diverse such as a microcontroller, wireless module nRF24L01, flame sensor, driver motor DCMP H-bridge MOSFET, ultrasonic sensor, compass magnetometer 3 axis HMC5883L, relay, and LCD OLED. Each component has a different function.

Arduino Uno is the microcontroller used for radio communication with the robot. The robot receives the data sent by YOLO through wireless module nRF24L01. A flame sensor finds the location of fire or flame, driver motor DCMP H-bridge MOSFET drives the DC motor, and PING sensor is an ultrasonic sensor to detect the distance of an object in front of the sensor which beams an ultrasonic wave with a frequency of 40 kHz. The sensor detects the value of reflection received by it.

17.3 Fire detection using YOLO

A flowchart of the designed system for fire detection using YOLO can be seen in Figure 17.2. The data used for training were fire images and firefighter robots' images. In total, there were 1000 data where the fire and robot were placed randomly in the labyrinth. The data were taken above the path of the labyrinth using the external camera Logitech C525. Later, the image data were bounded box as a dataset for training YOLO.

The YOLO frames object detection as a regression problem to spatially separated bounding boxes and associated class probabilities. It trains on full images and directly optimizes detection performances [12]. YOLO detects the object by

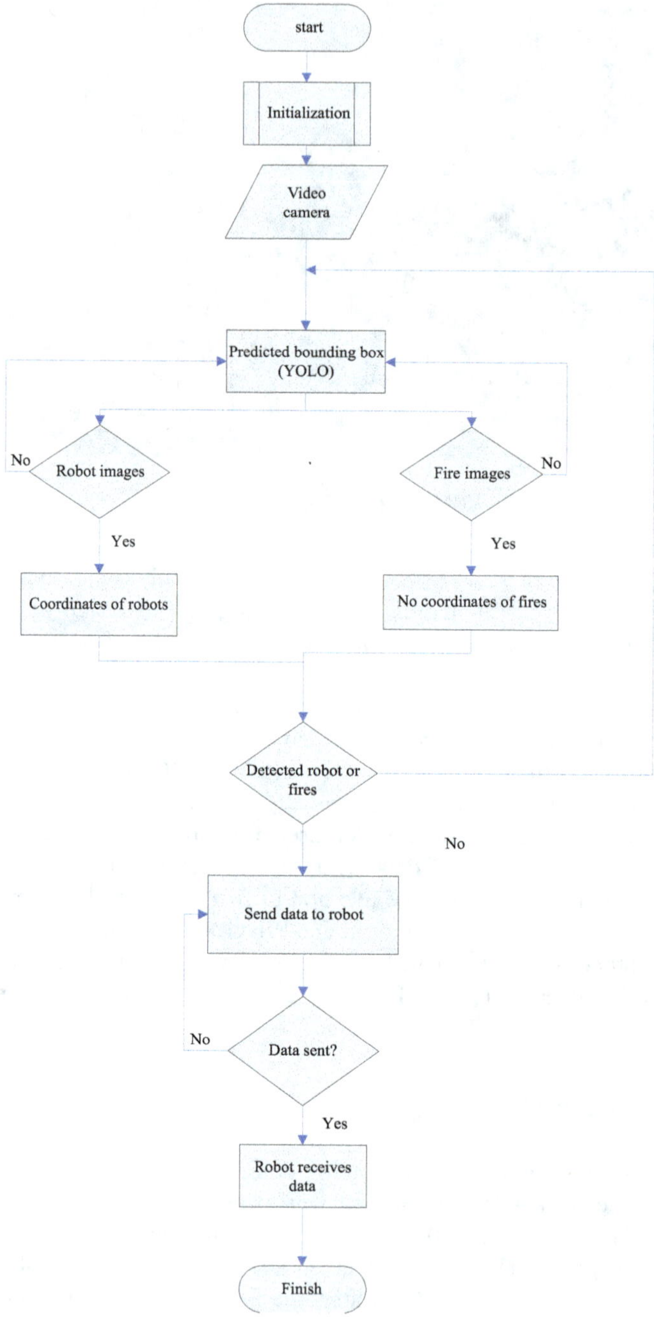

Figure 17.2 Flowchart of the designed system for fire detection using YOLO

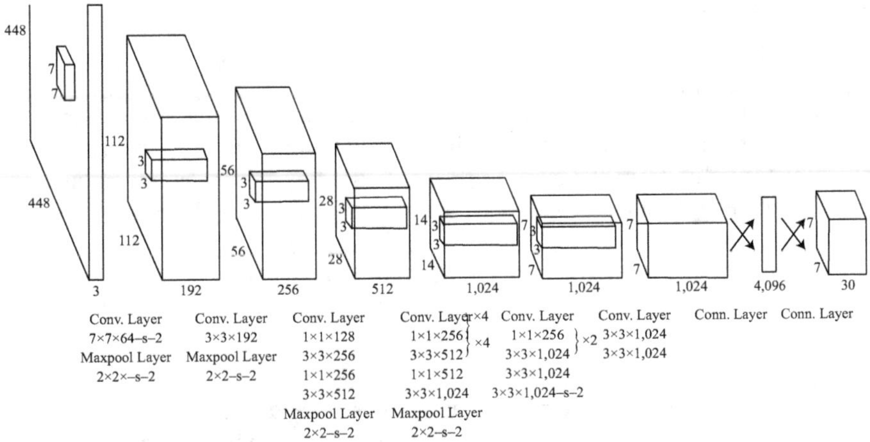

Figure 17.3 The architecture of YOLO [12]

dividing the image into some regions and predicts every bounding box and probabilities for each region. Those bounding boxes are then compared to every possibility to be predicted. YOLO has benefits compared to other classifiers since it can see the entire collection of images during training and test time. Thus, YOLO outperforms R-CNN by making less than half the number of background errors. The architecture of YOLO can be seen in Figure 17.3.

Some structures of YOLO were used in this study since the training process aims to get a model with appropriate architecture. Some models were compared to obtain the model with the smallest loss and shortest computation time. The most optimal model was then used to test the system.

Testing was performed to evaluate the designed system. In this study, two evaluation processes were done, including the detection test to find the robot and fire, and the sending data from the system to the firefighter robot.

17.4 Shortest path using A* algorithm

17.4.1 A algorithm*

A* algorithm is a computation algorithm to find the path by giving plots that can be passed from a point to several destinations. A map with several points may complicate the process to find the best path. Hence, the shortest path algorithm is needed to be embedded in a robot so it can find the best route to its destination. A* algorithm introduces heuristic in the algorithm. Hence, each step must be planned well to get the optimal decision.

A* algorithm works based on the shortest distance from the starting point to the destination point. It uses function $f(n)$ which approximates the distance from one point to another point. The distance traveled from one point to another point is calculated as follows:

$$f(n) = g(n) + h(n) \tag{1.1}$$

where $f(n)$ is the total estimated distance from the starting to the destination point, $g(n)$ is the distance value to get the destination point, and $h(n)$ is the heuristic value that estimates value from the starting to the destination. Here, $h(n)$ is obtained by using a Euclidean distance heuristic which is formulated as follows

$$h(n) = \sqrt{(x_{destination} - x_{start})^2 + (y_{destination} - y_{start})^2} \tag{1.2}$$

17.4.2 Map and node design

In this study, the map of the labyrinth and nodes were designed to embed the A* algorithm in the robots. Each node has a code in letter form to get the position. The point was determined to ease the finding path algorithm. Coordinates serve to label positions were placed in the map so the A* algorithm can be embedded in the robots to navigate from the start point to the destination. Maps and nodes can be seen in Figure 17.4.

As shown in Figure 17.4, location points that represent the doors are A, C, F, G, and I. Besides, this map has four rooms, where room 1 has two doors C and I, room 2 with door A, room 3 has door F, and the last is room 4 with a door of G. Meanwhile, the paths are shown in the points of B, D, E, H, and J. These paths are routes which are passed by the robots to the rooms with fire inside. Coordinates in the map consist of 14 points on the x-axis and the y-axis. The coordinate points can be found using the Euclidian distance heuristic approach.

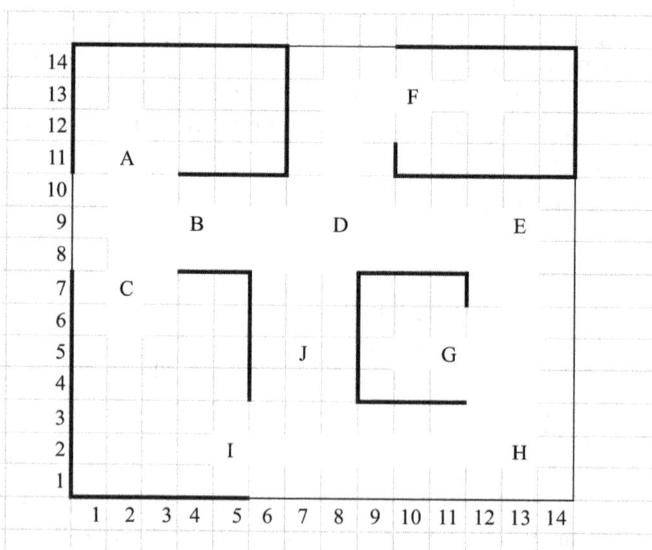

Figure 17.4 Map and obstacle nodes for firefighter robots

17.4.3 How the robots work

Figure 17.5 shows how the robot works. In the first stage, the system is initialized to find and use variable values given by the sensors and other components. Then,

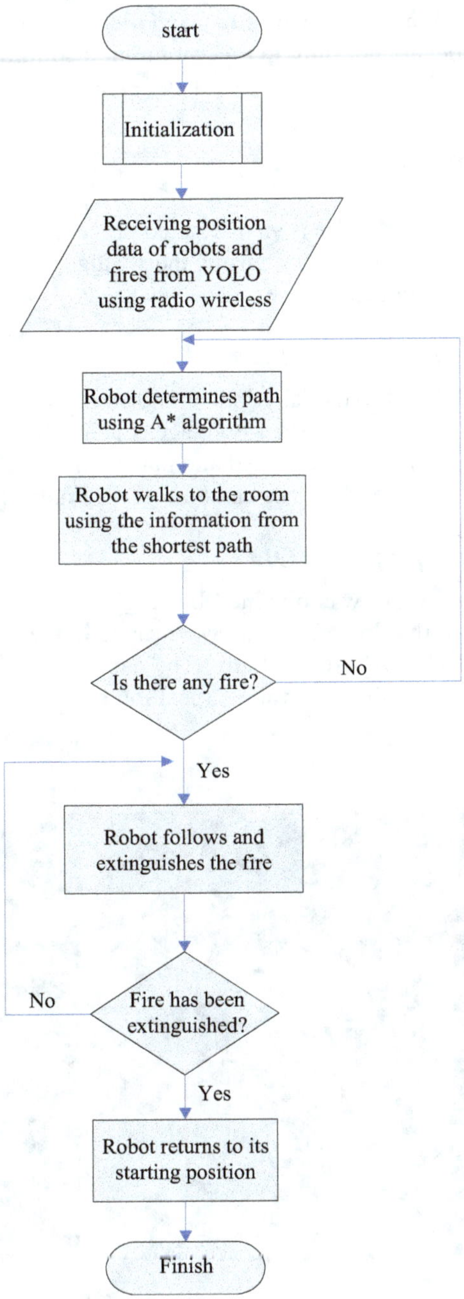

Figure 17.5 How the robot works

the robot receives the data which consists of its start position and position of flames using radio wireless 24L01. After receiving the codes, such as room and flames, both robots walk to the fire point with the shortest path given by the A* algorithm. The robot extinguishes the fire after finding the flames. Those two robots divide their work according to the rooms with fire using the shortest path given by the A* algorithm if the fire occurs not only in one room but also in other rooms. Finally, the robots return to their position once they have extinguished the fire.

17.5 Results and discussions

The workplace safety from the fire is an integrated process of object detection using YOLO and the shortest path calculation using the A* algorithm. Thus, both systems are examined in the experiment.

17.5.1 Position of robots and fire

Robots and fire were placed in the labyrinth as shown in Figure 17.6. The labyrinth is assumed as the real condition of a workplace. Each robot was placed in a different room. Then, three flames were used and placed in the different rooms. Some samples of robots' position and flames can be seen in Figure 17.7.

17.5.2 Collecting data for YOLO

The data for training YOLO was obtained by Logitech C525 camera which was placed above 3 m from the floor. The data were saved in the .PNG format with the file name 0.PNG to 1023.PNG sequentially. The data obtained were the combination taken of several possibilities of robots' position as well as flames' position in

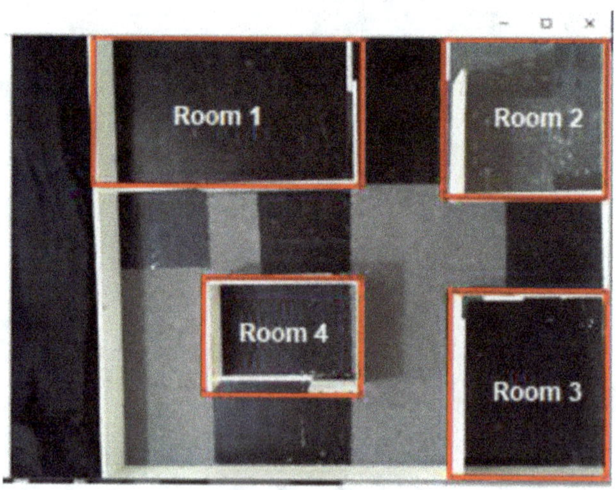

Figure 17.6 Room position in labyrinth

(a) (b) (c)

Figure 17.7 Some samples of robots' position (a) 2 robots and 1 fire (b) 2 robots and 2 fires, and (c) 2 robots and 3 fires

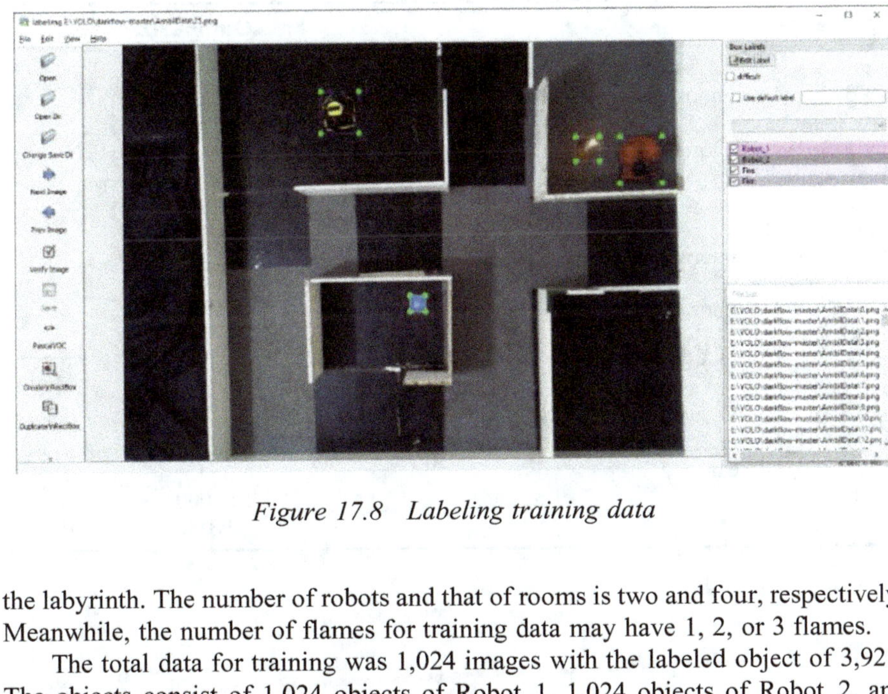

Figure 17.8 Labeling training data

the labyrinth. The number of robots and that of rooms is two and four, respectively. Meanwhile, the number of flames for training data may have 1, 2, or 3 flames.

The total data for training was 1,024 images with the labeled object of 3,921. The objects consist of 1,024 objects of Robot_1, 1,024 objects of Robot_2, and 1,873 objects of fire. The objects were labeled with a labeling program as shown in Figure 17.8.

The results of annotations consist of three classes, namely Robot_1, Robot_2, and Fire which were saved in XML file format.

17.5.3 Training YOLO

The training was performed using the YOLO algorithm utilizing Darkflow repository. Here, CUDA 9.0, Cuddn 7.0.5, OpenCV 3.4.6, Anaconda3 with Python 3.6, and TensorFlow 1.5 were used in the training process.

Two models were used in the training process, namely Tiny YOLOv2 and Tiny YOLO VOC. The number of epochs was set to 20, 50, and 100 epochs.

17.5.3.1 Tiny YOLOv2

The structure of Tiny YOLOv2 is shown in Table 17.1. Tiny YOLOv2 was a modified set of YOLOv2.

The training result using Tiny YOLOv2, an epoch of 20, and iteration of 1280 is shown in Figure 17.9.

As shown in Figure 17.9, the training results have large loss values. Meanwhile, a good model must have a loss value that approaches zero. Thus, the number of epochs was improved to 50 as shown in Figure 17.10.

Table 17.1 Structure of Tiny YOLOv2

Layer description	Output size
Input	(416, 416, 3)
conv 3x3p1_1 + bnorm leaky	(416, 416, 16)
maxp 2x2p0_2	(208, 208, 16)
conv 3x3p1_1 + bnorm leaky	(208, 208, 32)
maxp 2x2p0_2	(104, 104, 32)
conv 3x3p1_1 + bnorm leaky	(104, 104, 64)
maxp 2x2p0_2	(52, 52, 64)
conv 3x3p1_1 + bnorm leaky	(52, 52, 128)
maxp 2x2p0_2	(26, 26, 128)
conv 3x3p1_1 + bnorm leaky	(26, 26, 256)
maxp 2x2p0_2	(13, 13, 256)
conv 3x3p1_1 + bnorm leaky	(13, 13, 512)
maxp 2x2p0_2	(13, 13, 512)
conv 3x3p1_1 + bnorm leaky	(13, 13, 1024)
conv 3x3p1_1 + bnorm leaky	(13, 13, 1024)
conv 1x1p0_1 linear	(13, 13, 40)
max_batches	120,000
Step	−1,100, 80,000, 100,000

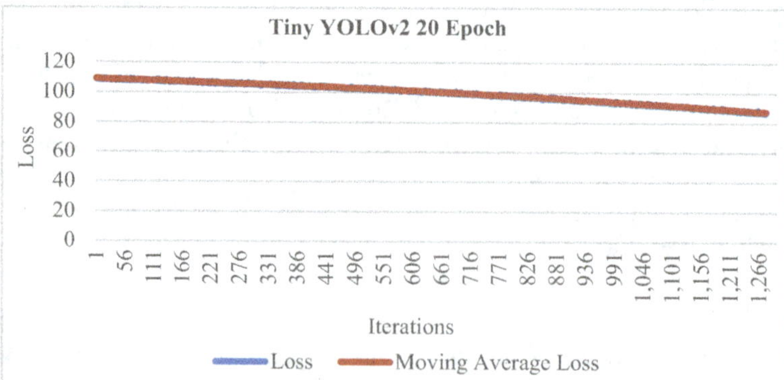

Figure 17.9 Training graph using Tiny YOLOv2 with 20 epochs

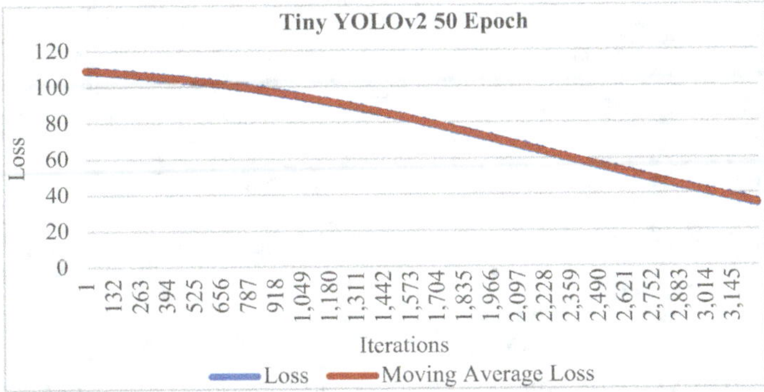

Figure 17.10 Training graph using Tiny YOLOv2 with 50 epochs

Figure 17.11 Some samples of testing results using Tiny YOLOv2 50 epochs model

Figure 17.10 shows an improvement of loss values with the final value of the loss being 34.84 in the 3,200th iteration. This value is lower than the 20 epochs. However, the model was not able to detect the robot as shown in Figure 17.11. Thus, the number of epochs is increased to 100 epochs.

The training result using 100 epochs can be seen in Figure 17.12. As shown in the figure, the performance of the model is much better than the previous two models. However, the model is not able to detect the fires and robots as shown in Figure 17.12.

These results indicate that the model is not able to detect the object. Given confidence has been decreased to 1%, the system was still not able to detect the fire or robot as shown in Figure 17.13. Thus, another structure of YOLO has to be implemented since the Tiny YOLOv2 is not able to detect the object well.

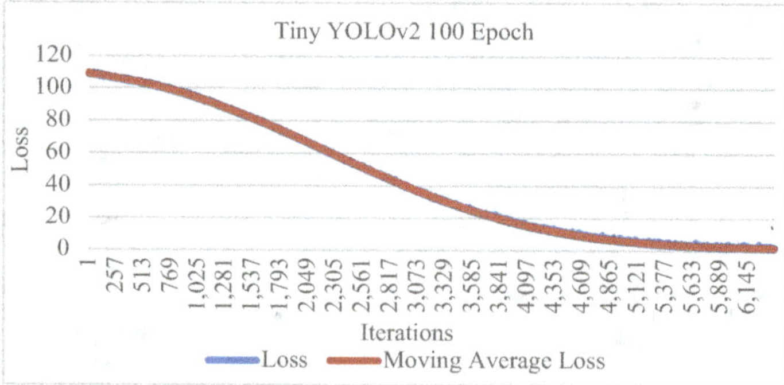

Figure 17.12 Training graph using Tiny YOLOv2 with 100 epochs

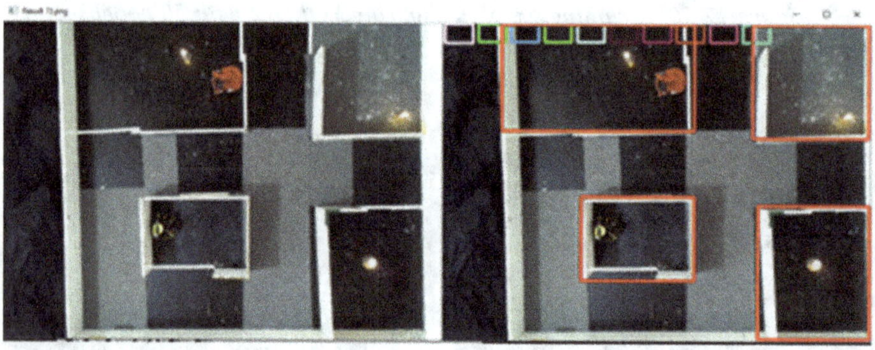

Figure 17.13 Some samples of testing results using Tiny YOLOv2 50 epochs model

17.5.3.2 Tiny YOLO VOC

Tiny YOLO VOC is a modified model of YOLOv2 and the structure of Tiny YOLO VOC can be seen in Table 17.2.

Here, three epochs were used, including 20, 50, and 100 epochs. The results of the training process using those epochs can be seen in Figures 17.14–17.16, respectively. As shown in Figure 17.14, the training result is much better than Tiny YOLOv2. The final loss value is 1.93. To validate this model, the testing was performed in 15 images as shown in Table 17.3.

As shown in Table 17.3, the class of Robot_1 has average confidence of 86.75%, Robot_2 has 76.28%, and Fire has 84.4%. These results indicate the Tiny YOLO VOC can be used to detect the object of fire and robots. However, the performance may improve by increasing the number of epochs. Figure 17.15 shows the results using 50 epochs. As shown in the figure, the loss value has decreased to 0.23 which indicates this model is good enough to be used in the

Table 17.2 Structure of Tiny YOLO VOC

Layer description	Output size
Input	(416, 416, 3)
conv 3x3p1_1 + bnorm leaky	(416, 416, 16)
maxp 2x2p0_2	(208, 208, 16)
conv 3x3p1_1 + bnorm leaky	(208, 208, 32)
maxp 2x2p0_2	(104, 104, 32)
conv 3x3p1_1 + bnorm leaky	(104, 104, 64)
maxp 2x2p0_2	(52, 52, 64)
conv 3x3p1_1 + bnorm leaky	(52, 52, 128)
maxp 2x2p0_2	(26, 26, 128)
conv 3x3p1_1 + bnorm leaky	(26, 26, 256)
maxp 2x2p0_2	(13, 13, 256)
conv 3x3p1_1 + bnorm leaky	(13, 13, 512)
maxp 2x2p0_2	(13, 13, 512)
conv 3x3p1_1 + bnorm leaky	(13, 13, 1,024)
conv 3x3p1_1 + bnorm leaky	(13, 13, 1,024)
conv 1x1p0_1 linear	(13, 13, 40)
max_batches	40,100
Step	1,100, 20,000, 30,000

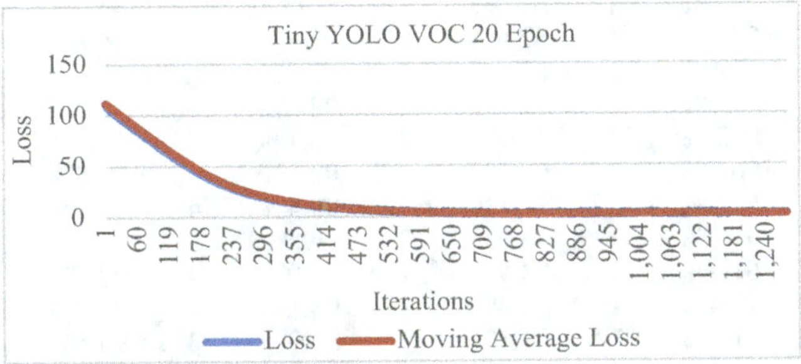

Figure 17.14 Training graph using Tiny YOLO VOC with 20 epochs

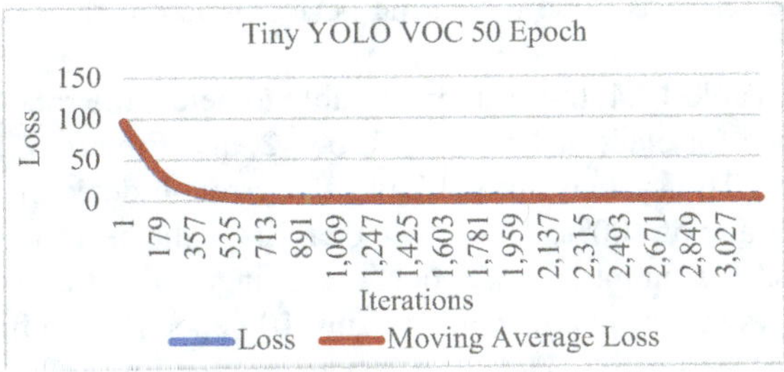

Figure 17.15 Training graph using Tiny YOLO VOC with 50 epochs

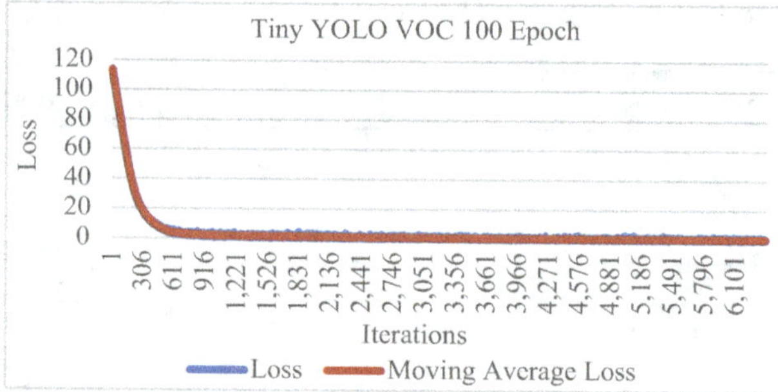

Figure 17.16 Training graph using Tiny YOLO VOC with 100 epochs

Table 17.3 Testing result using Tiny YOLO VOC 20 Epoch, threshold = 20%

Image	Robot_1		Robot_2		Fire_1		Fire_2		Fire_3	
	Loc.	(%)	Loc.	(%)	Loc.	(%)	Loc.	(%)	Loc.	(%)
1.png	1	99.1%	2	43.8%	4	96.1%	3	63.1%	3	58.3%
2.png	1	96.1%	2	76.4%	2	60.4%	3	99.5%	4	43.7%
3.png	1	79.5%	4	79.1%	–	–	–	–	–	–
4.png	1	97.3%	4	67.7%	4	27%	4	45.6%	3	97.7%
5.png	1	97.2%	4	81.9%	1	93.8%	1	26.8%	4	56.4%
6.png	3	94%	2	88.4%	2	98.8%	–	–	–	–
7.png	4	99.2%	3	91.9%	2	98.6%	–	–	–	–
8.png	4	89.8%	2	41.7%	2	86.1%	4	97.3%	–	–
9.png	4	96.7%	2	68.3%	2	92.8%	3	99.5%	–	–
10.png	4	67.6%	2	92.8%	1	99.4%	4	82,5%	–	–
11.png	4	96.9%	3	84.2%	2	99.%	3	99.3%	–	–
12.png	3	61.4%	4	73.9%	1	97.9%	4	97.7%	3	98.8%
13.png	4	49.4%	1	86.4%	1	98.5%	4	88.9%	3	98.8%
14.png	3	91.8%	2	93.1%	1	87.9%	2	90.5%	4	99.3%
15.png	4	85.2%	1	74.6%	1	95.7%	2	97.7%	3	98.6%

testing. Thus, the test was done to see the performance of the model as given in Table 17.4.

As shown in Table 17.4, the system was able to detect the robots and fires. The average confidence for class of Robot_1, Robot_2, and fire are 87.06%, 87%, and 91%, respectively. The system did not have any error in detecting the robots and fires with an accuracy of 100%. To see whether this is the best model for detecting fires and robots, the number of epochs was increased to 100 epochs. The figure below shows the results of training using 100 epochs. The final loss value is 0.12 which approached zeros. Thus, this model has the lowest loss value. To test this model, 15 testing data were used as shown in Table 17.5.

Table 17.4 Testing result using Tiny YOLO VOC 50 Epoch, threshold = 20%

Image	Robot_1		Robot_2		Fire_1		Fire_2		Fire_3	
	Loc.	**(%)**	**Loc.**	**(%)**	**Loc.**	**(%)**	**Loc.**	**(%)**	**Loc.**	**(%)**
1.png	1	99.4%	2	53%	4	98.1%	3	93.0%	–	–
2.png	1	99.2%	2	92.1%	2	78.3%	3	99.5%	4	28%
3.png	1	89.9%	4	93.4%	–	–	–	–	–	–
4.png	1	98.9%	4	92.5%	4	70.7%	3	97.2%	–	–
5.png	1	95.8%	4	98.3%	1	95.1%	4	64.3%	–	–
6.png	3	98.1%	2	95.3%	2	99.4%	–	–	–	–
7.png	4	98.8%	3	99.2%	2	99.3%	–	–	–	–
8.png	4	89.5%	2	87.7%	2	85.8%	4	99.3%		
9.png	4	97.4%	2	56%	2	93.7%	3	99.4%		
10.png	4	61.2%	2	99.4%	1	99.7%	4	80.3%		
11.png	4	99.4%	3	96%	2	99.8%	3	99.7%		
12.png	3	65.85%	4	64%	1	98.6%	4	99.5%	3	99.8%
13.png	4	76.8%	1	73.1%	1	99.2%	4	98.8%	3	99.3%
14.png	3	81.9%	2	98.7%	1	98.7%	2	98.2%	4	99.5%
15.png	4	53.8%	1	99%	1	97.7%	2	98.9%	3	99.1%

Table 17.5 Testing result using Tiny YOLO VOC 100 Epoch, threshold = 20%

Image	Robot_1		Robot_2		Fire_1		Fire_2		Fire_3	
	Loc.	**(%)**	**Loc.**	**(%)**	**Loc.**	**(%)**	**Loc.**	**(%)**	**Loc.**	**(%)**
1.png	1	99.9%	2	22.7%	4	97.8%	3	87.3%	–	–
2.png	1	99.9%	2	92.9%	2	66.8%	3	99.8%	fail	fail
3.png	1	73.2%	4	95.7%	–	–	–	–	–	–
4.png	1	99.9%	4	96.4%	4	33.3%	3	96.9%	–	–
5.png	1	99.3%	4	99.8%	1	99.8%	4	30.6%	–	–
6.png	3	99.8%	2	98.7%	2	99.6%	–	–	–	–
7.png	4	99.9%	3	99.8%	2	99.8%	–	–	–	–
8.png	4	93.2%	2	84.6%	2	78.8%	4	99.8%	–	–
9.png	4	99.6%	2	17.6%	2	89.1%	3	99.8%	–	–
10.png	4	75.3%	2	99.8%	1	99.9%	4	42.8%		
11.png	4	99.9%	3	99.1%	2	99.9%	3	99.9%		
12.png	3	13%	4	29%	1	99.2%	4	99.7%	3	99.9%
13.png	4	60.2%	1	73.1%	1	99.3%	4	99.2%	3	99.7%
14.png	3	95.3%	2	99.8%	1	99.9%	2	98.6%	4	99.7%
15.png	Fail	Fail	1	99.8%	1	95.7%	2	99.7%	3	99.6%

As shown in Table 17.5, the model of Tiny YOLO VOC with 100 epochs has shown instability compared to the model with 50 epochs. For example, for image 1, the model with 100 epochs has a confidence of 22.7% which is lower than 50 epochs. Besides that, a few errors occurred in detecting the fires as in images 2 and

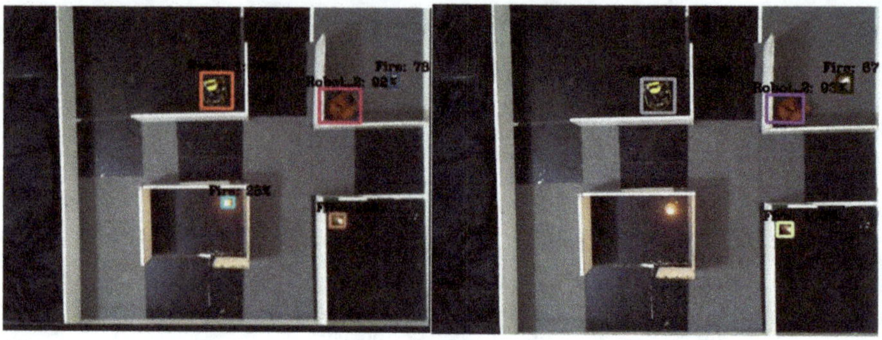

(a) (b)

Figure 17.17 Testing results using Tiny YOLO VOC (a) 50 epochs and (b) 100 epochs

15. Comparison of testing results using Tiny YOLO VOC with 50 epochs and 100 epochs can be seen in Figure 17.17.

As shown in Figure 17.17, the Tiny YOLO VOC with 50 epochs can detect the fire in room 4. This result may be improved by decreasing the confidence threshold. However, it may influence the performance of the system as a whole. Thus, this study shows that Tiny YOLO VOC with 50 epochs is the most appropriate model to detect fire and robots.

Based on the structure of Tiny YOLOv2 and Tiny YOLO VOC as shown in Tables 17.14 and 17.15, the difference between them is the batch number. The Tiny YOLOv2 has 120,000 batches, meanwhile, the Tiny YOLO VOC has 40,100 batches. Batch shows the maximal value of sample data can be processed before the YOLO model is updated. Thus, the Tiny YOLOv2 has more cost computation than the Tiny YOLO VOC. Hence, this study utilized the model obtained by the Tiny YOLO VOC with 50 epochs.

17.5.4 Evaluation of the shortest path

Evaluation of the performance of A* algorithm was done manually by finding the shortest path. The results can be seen in Table 17.6.

As shown in Table 17.6, the A* algorithm can find the shortest path from the start point to the destination. The distance obtained by the A* algorithm is close to the actual distance which was calculated using a ruler. The distance in the A* for each box is 10.21341. The shortest distance was from A to C which is 85 cm.

An evaluation was also performed by taking the combination of the location of fire and robots. The robot used the A* algorithm to find the shortest path with some random combinations. There are some possibilities of 1 fire, 2 fires, and/or 4 fires.

17.5.4.1 Testing a robot with 1 fire

Here, the robots and fire were placed randomly. A sample of the position of robots and a fire can be seen in Figure 17.18. Green represents the start point of the robot

Table 17.6 Evaluation of the shortest path using A algorithm*

No.	View in OLED	Input location	Input destination	Route	Distance with A*	Actual distance
1	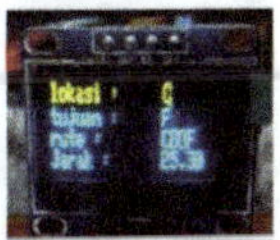	1	3	CBDF	25.30	259.7
2		2	4	ABDEG	35,30	348,8
3		3	1	FDBC	25.30	259.7
4		4	3	GEDF	30.94	325.9
5		1	4	IHG	24.61	239

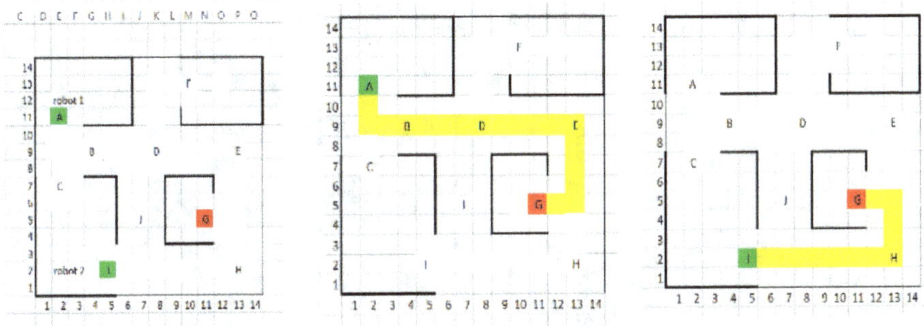

(a) (b) (c)

Figure 17.18 Samples of robot evaluation with 1 fire (a) initial position of the robot, (b) Robot_1 walks to room 4, and (c) Robot_2 walks to room 4

and red is the door where the fire takes place. The results of using 2 robots and 1 fire as the means of obtaining the shortest path can be seen in Table 17.7. As shown in the table, both robots tried to reach the same room since there was only one source of the fire. The robot determined the shortest path to the fire.

Table 17.7 Sample of evaluation results using 2 robots and 1 fire

No	Location Robot_1	Destination Robot_1	Route Robot_1	Distance 1 (A*)	Location Robot_2	Destination Robot_2	Route Robot_2	Distance 2 (A*)
1.	A	G	ABDEG	35.30	I	G	IHG	24.61
2.	C	A	CA	8	G	A	GEDBA	35.30
3.	F	C	FDBC	25.30	A	C	AC	8
4.	G	F	GEDF	30.94	A	F	ABDF	25.30

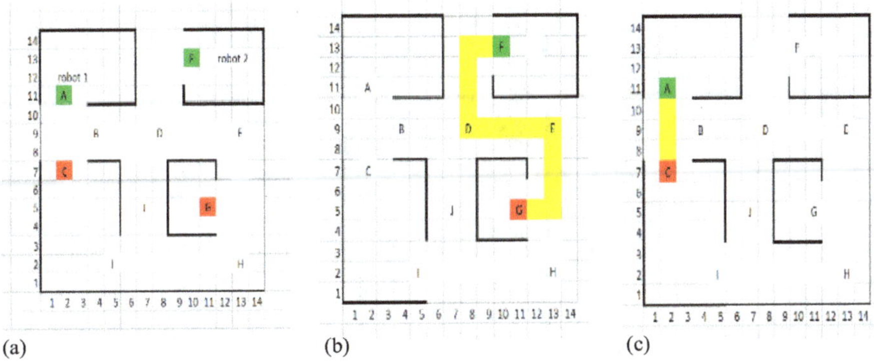

Figure 17.19 Samples of robot evaluation with 2 fires: (a) initial position of the robot, (b) Robot_1 walks to room 2, and (c) Robot_2 walks to room 4

17.5.4.2 Testing 2 robots with 2 fires

This test was performed using two robots and two sources of fire. The sample position of robots and fires can be seen in Figure 17.19. Robots and fires were placed in random locations. A sample of the position of robots and a fire can be seen in Figure 17.19.

The results for all the possible positions of robots and fires can be seen in Table 17.8. As shown in the table, two robots divided their duty to tend to the fire. The robot with the shortest position to the fire based on the A* algorithm looked for the fire, and the other robot found the other fire. The distance obtained by the A* algorithm became the information used by the robot to determine their location to the fire.

17.5.4.3 Testing 2 robots with 3 fires

An evaluation was also performed with 3 fires and the two robots placed randomly. The results can be seen in Table 17.8. As shown in the table, the robot with the shortest distance to the fire extinguished the fire. Then, both robots head to the farther fire. The distance calculated by the A* algorithm was used by robots to determine the shortest path to the fire so both robots can extinguish the fire efficiently. Besides, the A* algorithm was used to compare the distance to the fire between the robots.

17.5.5 Receiving data

Receiving data is performed by module NRF24L0, the data was received in the format data type of unsigned int [robot1, robot2, fire1, fire2, fire3]. As an example, if robot1 is in room 1, robot2 is in room 2, and the two fires are in rooms 3 and 4, the system may receive the data as a code of [12340]. The last digit is "0" since there is no fire in room 3. The result of receiving data can be seen in Table 17.9.

Table 17.8 Sample of evaluation results using 2 robots and 2 fires

No	Location Robot_1	Destination Robot_1	Route Robot_1	Distance 1 (A*)	Location Robot_2	Destination Robot_2	Route Robot_2	Distance 2 (A*)
1.	C	F	CBDF	25.30	A	A	A	0
2.	F	F	F	0	G	I	GHI	24.61
3.	G	I	GHI	24.61	A	F	ABDF	25.30
4.	G	F	GEDF	30.94	C	A	CA	8

Table 17.9 Sample of evaluation results using 2 robots and 3 fires

No.	Route Robot_1	Distance 1 (A*)	Route Robot_2	Distance 2 (A*)
1	G	0	AC	8
	GEDF	30.94	CBDF	25.30
2	C	0	F	0
	CHG	24.61	FDEG	30.94
3	GEDF	30.94	C	0
	FDBA	25.30	CA	8
4	A	0	G	0
	ABDF	25.30	GEDF	30.94
5	FDEG	30.94	A	0
	GHI	24.61	AC	8

As shown in Table 17.9, the module radio NRF24L01 succeeded to send and receive the data since the sent code is the same as the received code. Received code is processed as the location of robot position and fire position so the robot knows its position and the location of the fire. After sending the data, the robot calculates the distance using the A* algorithm and walks toward the fire's location.

17.5.6 Evaluation of the entire system

The evaluation of the system as the complete system of safety using the combination of the YOLO algorithm and the A* algorithm can be seen in Table 17.10. The robots tried to find the location of the fire and the fire was extinguished. This testing was done in ten experiments, where the fire and robots were placed randomly in the labyrinth. The number of fires for this testing was 1, 2, and 3 fires. Testing aims to see the performance of the YOLO model and A* algorithm.

As shown in Table 17.10, YOLO can detect the location of the fire and the robot's as well. The robot succeeds in finding the location of the fire. The whole system can send the code of robots' position and the location of fires using the A* algorithm. The result of detection and localization can be seen in Table 17.11.

In the evaluation of the entire system, the robots succeeded in finishing their mission to look for the fire. However, some experiments showed the robot failed to extinguish the fire because the fan installed on the robot was not strong enough to extinguish the fire. The percentage for the robot to detect and find the fire using the YOLO algorithm and shortest path of the A* algorithm is 100%. Meanwhile, the success of extinguishing the fire was 83.33%.

Analysis of the confidence value of YOLO to detect objects was also performed. This data can be seen in Table 17.13. It shows that the image processing approach can be utilized to detect fire. The average confidence value for Robot_1,

Table 17.10 Evaluation of receiving data

Experiment no,	Detection image by YOLO	Data location sent	Data location received	Success
1		 [12430]	 [12430]	✓
2		 [12243]	 [12243]	✓
3	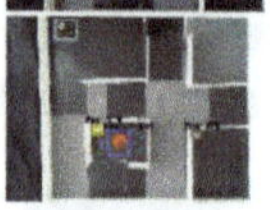	 [14430]	 [14430]	✓
4	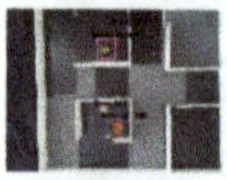	 [14140]	 [14140]	✓
5	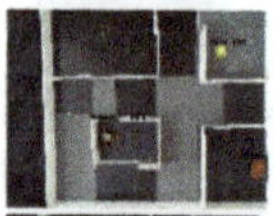	 [43200]	 [43200]	✓
6		 [42240]	 [42240]	✓
7		 [42230]	 [42230]	✓
8		 [42140]	 [42140]	✓
9		 [43230]	 [43230]	✓
10		 [34143]	 [34143]	✓

Robot_2, and Fire are 84%, 92%, and 88%, respectively, using the model given by Tiny YOLO VOC.

The results indicate that the combination of YOLO – to detect the fire and robot location – and the A* algorithm – to find the shortest path based on the location given by YOLO – is good enough to be implemented in various contexts,

Table 17.11 Evaluation of the entire system

Experiment no.	Detected images by Tiny YOLO VOC	Data sent	Remarks
1		21,400	Robot finished the missionRobot_1 failed to return to the start position
2		32,100	Robot finished the missionRobot_2 failed to return to the start position
3	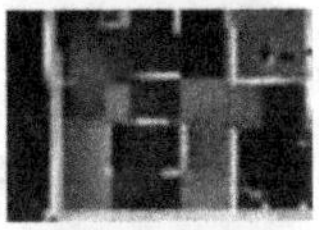	12,230	Robot finished the mission
4		41,320	Robot_1 finished the missionRobot_2 found the fire
5		13,430	Robot finished the mission
6		23,140	Robot finished the mission
7		13,123	Robot finished the mission to fire_2Fire_3 was found
8	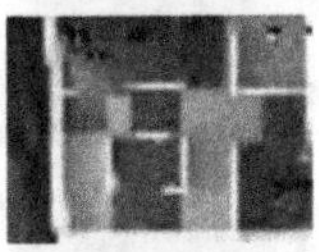	13,200	Robot finished the mission
9		21,400	Robot finished the missionRobot_2 failed to return to the start position
10	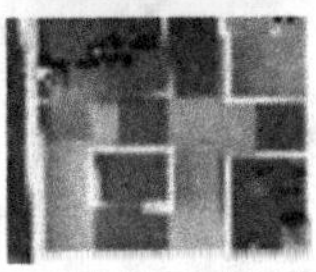	13,123	Robot finished the mission

Table 17.12 Evaluation of the entire system in real-time

Experiment no.	Robot_1		Robot_2		Fire	
	Location	Remark	Location	Remark	Location	Remark
1	2	True	1	True	4	True
2	3	True	2	True	1	True
3	1	True	2	True	2,3	True
4	4	True	1	True	3,2	True
5	1	True	3	True	4,3	True
6	2	True	3	True	1,4	True
7	1	True	3	True	1,2,3	True
8	1	True	3	True	2	True
9	2	True	1	True	4	True
10	1	True	3	True	1,2,3	True
Percentage of success		100%	–	100%	–	100%

Table 17.13 Confidence value of system evaluation

Experiment no.	Confidence value Robot_1	Confidence value Robot_2	Confidence value Fire
1	96%	99%	93%
2	99%	90%	99%
3	91%	93%	95%, 99%
4	100%	100%	98%, 99%
5	90%	91%	99%, 98%
6	90%	93%	94%, 85%
7	85%	90%	85%, 90%, 90%
8	36%	88%	27%
9	85%	90%	79%
10	67%	90%	85%, 86%, 77%
Average	84%	92%	88%

including manufacturing. The proposed method is effective in extinguishing the fire which may take place in the manufacturing with various rooms.

17.6 Conclusions

Image processing can be used to detect fire using a deep learning approach, namely the YOLO algorithm. Tiny YOLO VOC shows better performance than Tiny YOLOv2 to detect the object with an accuracy of 100%. The A* algorithm can be implemented in a multi-robot scenario to determine and compare the distance from the start point to the destination (fire location) by calculating the shortest distance. The system to detect and extinguish the fire can find the shortest path to the fire with a success rate of 100%. A combination of YOLO and A* algorithm can be implemented in real-time conditions, such as manufacturing workplace.

However, further study is still needed to improve the capability of the robot to extinguish the fire and return to the start position since the current system has an 83.33% success rate to extinguish the fire and 50% to return to the start point. Besides, various data for training based on the environment of the robot's acting out to extinguish the fire can be improved.

References

[1] S. Wilson, S.P. Varghese, G.A. Nikhil, I. Manolekshmi, and P.G. Raji, "A comprehensive study on fire detection," *Proceedings of IEEE Conference on Emerging Devices and Smart Systems ICEDSS 2018*, no. March, pp. 242–246, 2018.

[2] F. Duchon, A. Babinec, M. Kajan, *et al.*, "Pathplanning with modified A star algorithm for a mobile robot," *Procedia Eng.*, vol. 96, pp. 59–69, 2014.

[3] M.A. Javaid, "Understanding Dijkstra's Algorithm Abstract: Introduction," pp. 1–27.

[4] D. Walden, "The Bellman-Ford Algorithm and 'Distributed Bellman-Ford,'" pp. 1–12, 2003.

[5] D. Ferguson and A. Stentz, "The Field D * Algorithm for Improved Path Planning and Replanning in Uniform and Non-Uniform Cost Environments."

[6] T.H. Cormen, C.E. Leiserson, and R.L. Rivest, Introduction to Algorithms, 2nd ed., vol. 7, 2001.

[7] T. Goto, T. Kosaka, and H. Noborio, "On the Heuristics of A* or A Algorithm in ITS and robot path-planning, *Proc. 2003 IEEE/RSJ International Conference on Intelligent Robots and System (IROS 2003)*, Vol. 2, pp. 1159–1166, 2003.

[8] R. Girshick, J. Donahue, T. Darrell, J. Malik, U. C. Berkeley, and J. Malik,"R-CNN: region based convolutional neural networks," *Proc. IEEE Comput. Soc. Conf. Comput. Vis. Pattern Recognit.*, vol. 1, p. 2–9, 2014.

[9] R. Girshick, "Fast R-CNN," Proc. I*EEE* Int. C*onf. Comput. Vis.*, vol. 2015 Inter, pp. 1440–1448, 2015.

[10] S. Ren, K. He, R. Girshick, and J. Sun, "Faster R-CNN: Towards Real-Time Object Detection with Region Proposal Networks," *IEEE Trans. Pattern Anal. Mach. Intell.*, vol. 39, no. 6, pp. 1137–1149, 2017.

[11] W. Liu *et al.*, "SSD: Single shot multibox detector," *Lect. Notes Comput. Sci. (including Subser. Lect. Notes Artif. Intell. Lect. Notes Bioinformatics)*, vol. 9905 LNCS, pp. 21–37, 2016.

[12] J. Redmon, S. Divvala, R. Girshick, and A. Farhadi, "You only look once: Unified, real-time object detection," *Proc. IEEE Comput. Soc. Conf. Comput. Vis. Pattern Recognit.*, vol. 2016-Decem, pp. 779–788, 2016.

Chapter 18

Blockchain technology in smart manufacturing

Braimah Sumaina[1] *and Wai Yie Leong*[1]

18.1 Blockchain technology

Blockchain is a system of recording information in a way that makes it difficult or impossible to change, hack, or cheat the system. Blockchain technology stores value-based records called the block, of the activities carried out by human and stored in distributed databases, with blocks linked into the "chain" in a system made up of shared hubs. Incidentally, this stockpiling of blocks is alluded to as a computerized record [1].

Each transaction in this record is approved by the advanced approval of the proprietor, which verifies the transaction and protects it from alterations. Consequently, the data making up the computerized record is inherently secure [1]. Blockchain technology is shown in Figure 18.1.

Simply put, the computerized record resembles a Google spreadsheet shared among various PCs in a system, in which the value-based records are dependent on real buys. The interesting point is that anyone can see the information; however, they cannot spoil it [1].

18.2 Why is blockchain popular?

Assume you are moving cash to your family or companions from your ledger. You would sign in to your internet banking account and move the sum to the next individual utilizing their record number. At the point when the transaction is done, your bank refreshes the transaction records and the change is updated in the bank's records. It appears to be sufficiently straightforward, isn't that so? However, there is a potential issue here that the majority of us disregard [1].

These sorts of transactions can be altered rapidly with the mediation of a trusted third party, like the bank. Individuals who know this are frequently careful about utilizing these sorts of transactions; subsequently, the advancement of third-party applications of late. Be that as it may, this weakness is basically why Blockchain technology was created [1].

In fact, Blockchain is an advanced digital ledger that is attracting a great deal of attention and footing as of late. In any case, why has it gotten so mainstream? All things considered, we should delve into it to understand the entire idea [1].

[1]The Institution of Engineering and Technology, Malaysia

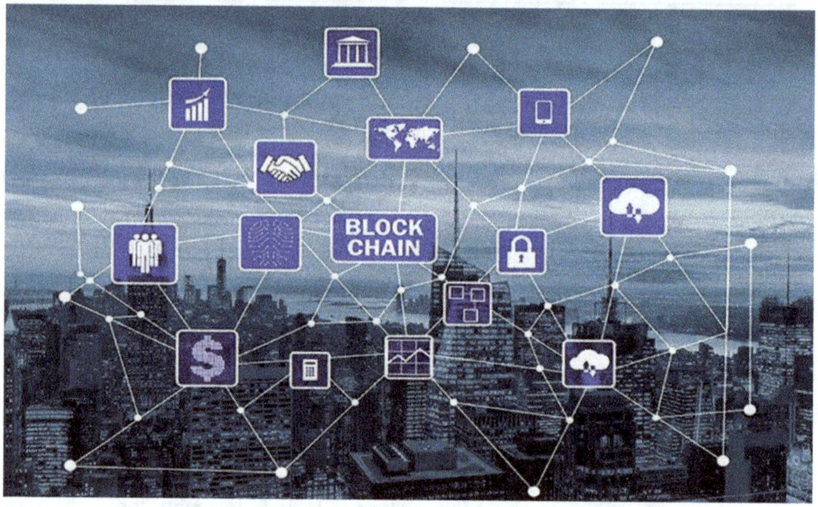

Figure 18.1 Blockchain technology [2]

Record keeping is an essential part of the business' daily activities. Regularly, transaction data is taken care of in-house or with the help of an outsider like intermediaries, investors, or legal advisors increasing time, cost, or both for the business. Luckily, Blockchain does away with this long procedure and encourages the transaction to execute quicker and therefore to save both time and money [1].

The vast majority expect Blockchain and Bitcoin can be utilized reciprocally, however, in actuality that is not the situation. Blockchain is the technology equipped to support different applications identified with various enterprises like fund management, supply chain, production, and so forth. However, Bitcoin is electronic cash that depends on Blockchain technology to be secure [1].

18.3 Advantages

- Blockchain technology takes into consideration verification without being subjected to third-party interactions [3].
- Within a Blockchain, the information structure is connected so that a record is linked to the record before it. The details in a record cannot be changed or deleted unauthorized without breaking the link [3].
- It uses assured cryptography to ensure the fidelity of records of the information. Likewise, to ensure the cryptography process is final, the current record is linked to its contiguous finished record [3].
- After the most intense confidence test procedure, all the transactions and details are affixed to the block. There is consensus about what is to be reported in the block amongst the record members [3].
- Transactions are documented in a sequential order. All the blocks in the Blockchain are spaced along these lines [3].

- The record is distributed among the Blockchain hubs that are the record members. It's represented in this way [3].
- The transactions put away in the blocks are located in a large number of PCs that have an interest in the chain. This is why it is decentralized. However, there is no chance that the information will not be recovered if it is lost [3].

18.4 Disadvantages

It is significantly more difficult to plan and build a stable Blockchain system than an almost identical centralized system [4].

Running the Blockchain section would typically be a lot more difficult than an analogous non-Blockchain program. You can ease this by shifting items away from the Blockchain component, but this will generally nullify a large number of focal points [4].

The data in the Blockchain network is more difficult to compartmentalize. For a non-blockchain scheme, for the most part, everything's a mystery even if you offer a way to get to it. Blockchain systems rely on individuals to know what they are doing with whom they can trust [4].

Blockchain involves the use of public and private keys, and reportedly, there have been problems with private keys. If a user loses their private key, they face numerous challenges, making this one disadvantage of blockchains. On certain issues, this is a beneficial situation, because it is important to rely on them to carry on the equivalent later than they do today. In any case, some of the time is a little less [4].

Another disadvantage is the scalability restrictions, as the number of transactions per node is limited. Because of this, it can take several hours to finish multiple transactions and other tasks. It can also be difficult to change or add information after it is recorded, which is another significant disadvantage of blockchain [4].

18.5 Blockchain technology's possibilities

Many organizations are located in areas where resources are scarce, and corruption is widespread. In such cases, Blockchain renders a significant advantage to these affected people and organizations, allowing them to escape the tricks of unreliable third-party intermediaries [5]. Numerous new computerized monetary forms have endeavored to update their blockchains to accommodate these issues; however, with differing margins of success. Later, one of the most significant advancements preparatory to Blockchain technology, going ahead, will probably have to do with adaptability [5].

Apart from this, new utilizations of Blockchain technology are being found and put into execution constantly. It's hard to state precisely where these advancements will lead the technology and the cryptocurrency business, overall. Supporters of Blockchain are probably going to discover this unimaginably energizing; from their point of view, we are living in a moment with an epoch-making technology that is proceeding to develop and unfurl [5].

18.6 Statistics of Blockchain technology

Blockchain is a technology that has taken the business world by storm. Blockchain technology deals with a collection of linked records that use cryptography to restrict access to the records and to prevent unauthorized edits on the records. These records contain transactional data. Blockchain data is intended to be safe from unauthorized changes and safe as a transaction record [6]. The intricacy of the technology itself may be difficult to understand even for tech enthusiasts, but the increasing importance and usefulness of the technology are clear to all. Blockchain systems can be used as a public means of making transactions, as well as private links for inter-company transactions and record keeping [6]. Worldwide spending on Blockchain solutions is projected to rise from USD1.5 billion in 2016 to an estimated USD15.9 billion by 2022, as shown in Figure 18.2 [6]. Companies in almost every sector are scrambling to take advantage of the potential advantages of the Blockchain revolution and others are trying to implement their private implementations of Blockchain and cryptocurrency [6]. The financial industry accounts for more than 60% of the market value of Blockchain spending worldwide in 2018, and it has spread to almost every field from healthcare to agriculture

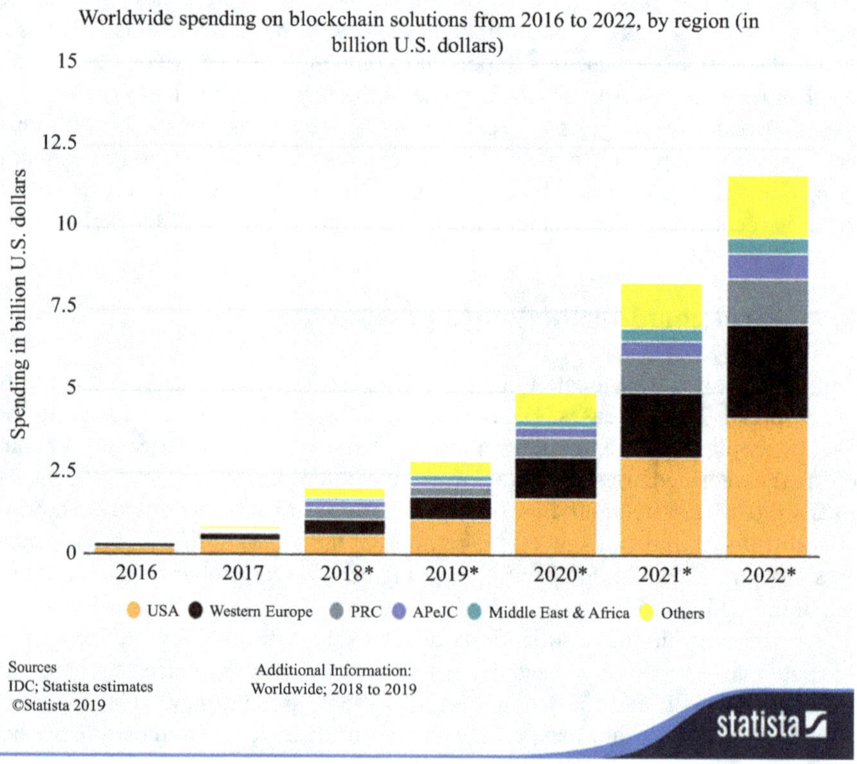

Figure 18.2 Blockchain technology worldwide spending [11]

[6]. The Blockchain technology in Healthcare is shown in Figure 18.3, statistics of Blockchain technology in the retail sector is shown in Figure 18.4, statistics of Blockchain technology business projections is shown in Figure 18.5, statistics of Blockchain technology in the agricultural sector is shown in Figure 18.6.

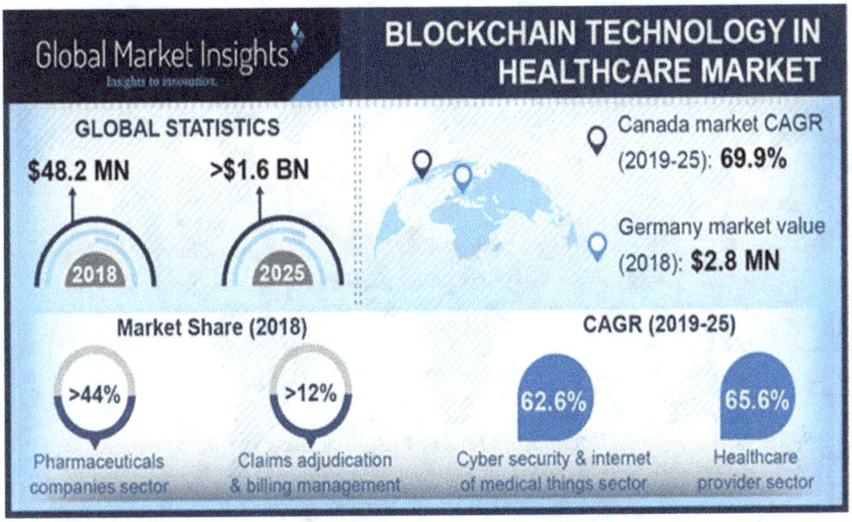

Figure 18.3 Blockchain technology in healthcare [7]

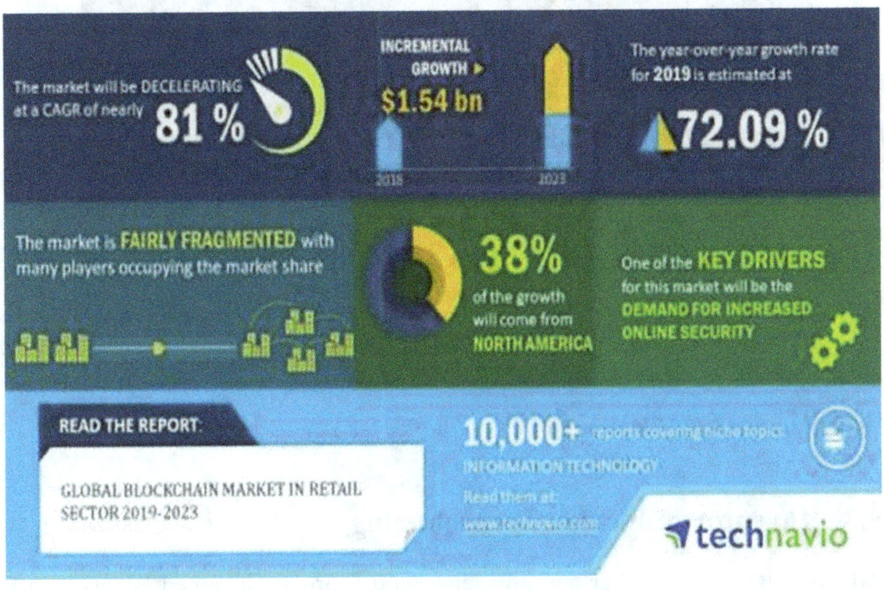

Figure 18.4 Blockchain technology in the retail sector [8]

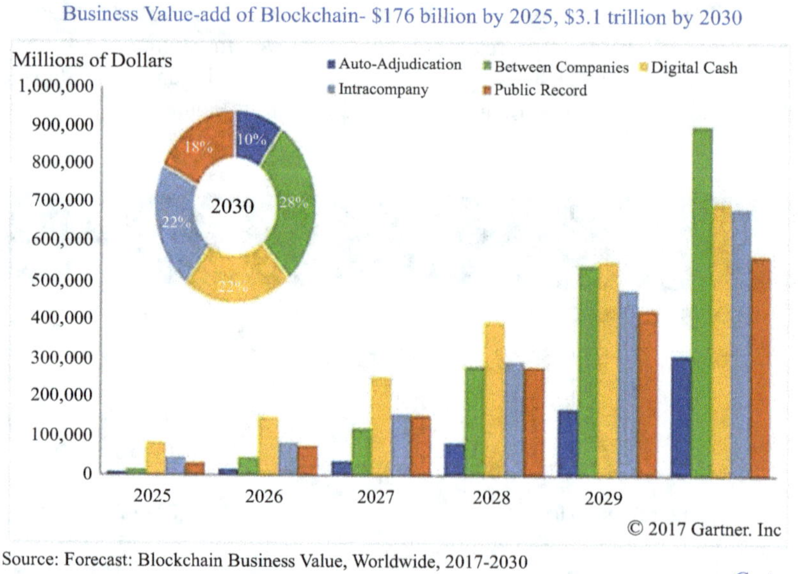

Figure 18.5 Blockchain technology business projections [9]

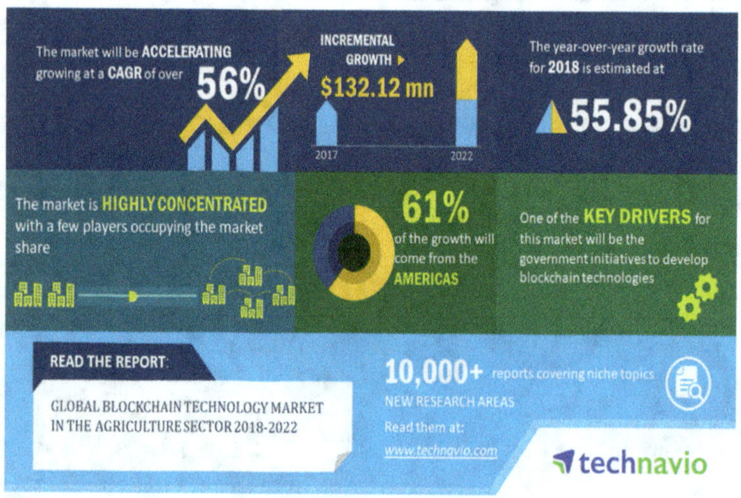

Figure 18.6 Blockchain technology in the agricultural sector [10]

18.7 Inventor of blockchain technology

Mathematician Stuart Haber and physicist Scott Stornetta invented Blockchain in 1991. Stuart Haber is often called the "Father of Blockchain" [12]. Some people say it's Satoshi Nakamoto who is the founder of Blockchain technology, although the thought behind the technology had started much earlier in Haber and Stornetta. Bitcoin was also invented independently by the anonymous Satoshi Nakamoto several years later. Bitcoin is an implementation of the Blockchain framework [12].

Blockchain technology is segregated from both the currency as well as discusses its potential for other cross-organizational financial contracts. Blockchain 2.0 was therefore created, referring to non-monetary implementations [13].

The Ethereum Blockchain infrastructure integrates programming trust programs into the blocks, reflecting financial assets such as bonds. They are widely recognized as smart contracts [13]. Ethereum is a decentralized blockchain platform that establishes a peer-to-peer network that securely executes and verifies application code, called smart contracts. Smart contracts allow participants to transact with each other without a trusted central authority.

18.8 The role of bitcoin

Posting their pioneering white paper in 2008 and releasing the initial code in 2009, Nakamoto developed Bitcoin to be a form of cash that could be transmitted on a peer-to-peer network with no need for a central bank or any other entity to regulate and preserve the ledger, as much as physical cash could have been [13]. Although it wasn't the first electronic currency to be proposed, the proposal for Bitcoin addressed many issues in the digital field and was by far the most popular edition [13].

The engine running the Bitcoin ledger developed by Nakamoto is called the Blockchain; the original and greatest Blockchain is the one that still orchestrates today's Bitcoin transactions [13].

Many blockchains include those running the several hundred "altcoins" – many related currency projects with different rules – as well as completely different implementations, such as:

Ethereum: the cryptocurrency spurring on the second-biggest Blockchain development after Bitcoin. Ethereum disburses a currency called ether, which further allows multiple codes to be stored and run so that smart contracts can be made [13].

Ripple: a gross settlement mechanism in real-time, a network of currency trade and remittances, dependent on a public ledger [13].

18.9 How blockchain technology works?

One of the dominant and misleading views of financial specialists unacquainted with both the recession environment and prosperity is mostly Blockchain. Blockchain technology drives and promotes innovative money space, and multiple experts agree that it requires a variety of reasonable implementations and incorporates previous physical sources of money as well [14]. You have also recognized how banking technology is organized in your usual activities by large and wide-ranging firms and even ordinary firms. However, it continues to be a little bit of a puzzle about what the Blockchain is and how it works [14]. Below, we will analyze the technical aspects of

Blockchain, including an overall idea of its kind of technology, how it operates with digital money types as well as other future applications, and why this could become one of the revolutionary inventions on the Internet [14].

18.10 Important points

Technology from Blockchain is underlying crypto-monetary structures and could be used in many applications [14].

Blockchain networks combine key private infrastructure, centralized entities and collective records [14].

Exchange confirmation and approval is a vital capacity for the crypto-monetary Blockchain [14].

18.11 The Blockchain three principal components

The synthesis of a few distinctive developments will potentially be seen as Blockchain. While these innovations are not recent, they are consolidated and implemented accordingly [14]. The following three key components are:

- Crystallization of private keys.
- A Centralized network with a common registry.
- Means of transactions transparency and network information.

(I) **Private keys–secret keys**

It supports the creativity of two individuals wishing to execute a web transaction to demonstrate the technology of private cryptographic keys [14]. There are two keys to each of these people: one is private and the other is publicly available. This cryptographic feature enables individuals to create a secure digital reference point by consolidating individual's signatures in public and private keys [14]. This secure nature is an important Blockchain technology element [14]. Digital signatures are a powerful means of obtaining, maintaining and preserving possession, together with private and public keys.

(II) **Centralized network**

The digital signature is then paired with both the cryptography section in the distributed network technology component. Blockchain technology serves as a large network of people that may act as verifiers to achieve a consensus on different topics, such as transactions. This approach is supported by statistical tests which are used to guarantee that the network is integrated. Blockchain allows different forms of digital transactions by integrating the use of cryptographic keys on a distributed network.

(III) **Method of confirmation**

One of the significant characteristics of Blockchain technology has been the extent to which it monitors and affirms transactions. In the example in the previous chapter, in which two parties wish to perform a web transaction, each with his own private and public key, Blockchain allows the digital signature (individual A) to include A's private key to be appended to the transaction details towards the next public key (individual B) [14]. This information is part of a block that includes a digital signature, a time signature and other significant transaction details but not, however, the characters of the persons involved [14]. Digital signatures are a fundamental building block in blockchains, used mainly to authenticate transactions. When users submit transactions, they must prove to every node in the system that they are authorized to spend those funds, while preventing other users from spending them. This block is transmitted to all hubs or segment parts of the network through the Blockchain network, which then acts as validators [14].

All this information transmission and transaction validation need huge quantities of registration power. In practical terms, a huge number of PCs across the world can seem unrealistic to expect all of them to devote registration power and other assets to this task [14]. One solution to this problem is mining for the Blockchain network [14]. In essence, this concept sums up a situation in which people who act individually freely in their circumstances usually do not act in the best interests of all customers because their activity drains an asset at an aggregate level [14]. The mining is linked to the conventional money problem known as the "tragedy of the lodge." A person who surrounds a little of his or her computing, this individual was encouraged to help meet the requirements of the broader Blockchain network by using the personal circumvent (in this case, a small number of cryptocurrencies) [14].

18.12 Chain of blocks

Why should this complicated validation procedure be undertaken? This is one crucial step forward for Blockchain networks to ensure that cryptocurrencies cannot be spent in numerous transactions simultaneously, an idea known as double expenditure's power to help the network gain a reward point for the time spent in Blockchain validation [14]. Blockchain networks must ensure that cryptocurrencies are uniquely owned and instilled with interest to ensure that they prevent double expenditure [14]. One way of providing such an assurance is by having the hubs inside the Blockchain network as parts of the ledger frame itself and by attempting to resolve complex mathematical issues with each coin [14]. These hubs help to confirm or deny places with bits of transaction information. If the majority of hub operators reach the same response to a problem, the block will be confirmed and added to the preceding block in the chain [14]. This new block is timestamped and will possibly contain details on different aspects of previous transactions [14]. There is space for variation depending on the particular network here: certain Blockchain networks recognize some types of information for their places while others integrate different information arrangements [14].

It is this last aspect of Blockchain that a few accept gives future applications the greatest potential later [14]. For instance, the data that obstructs a Blockchain is connected with previous transactions between various people, acting as an open record for all previous transactions, in comparison to the Bitcoin network [14]. Yet, the recorded data can be essentially anything for blocks. For example, for governments, aspects of

Blockchain technology may prove useful in relation to transactions that are normally performed through compliance systems [14]. Blockchain technology may be helpful in auditing trails or in cultivating new linkages between various financial institutions and potential partners [14]. For other parts of the financial sector, Blockchain will streamline the conventional days of clearing and settlement. It may also help to simplify regulatory enforcement by, for example, translation of legal disclosure into code or by authorizing certain types of transactions and blocking others. In and outside the financial world, there are wide-ranging chances for Blockchain technology [14].

As with any modern technology, however, the best way to use the incredible capabilities of Blockchain is not obvious [14]. On a long-term basis, all things considered and experimented will divulge new ways to use Blockchain for a variety of different ends and new methods to using Blockchain, to make them gradually sustainable, competent, safe and wonderful. In the meantime, the biggest Blockchain networks, such as those for digital currency forms such as Bitcoin, are developing [14].

18.13 Literature review

18.13.1 Introduction

Blockchain technology is wholly new. We will identify a concise summary of current developments in Blockchain research in this paper and explore future consequences and recommendations. Although Blockchain is not well known, it is increasingly becoming a topic of conversation, and in current media, it is a genuinely hot subject [15]. However, media trends are often not associated with research trends, so it is also a very good exercise to see how trends in scholarly, peer-reviewed science publications cover a trendy subject. Not too long ago, there were hardly any scientific papers on Blockchain at all, but this is rapidly changing [15].

In this article, we will summarize current theories in academic publications and consider three key issues related to the Blockchain. This is followed by a discussion of why Blockchain is important and how it is currently being used, including our specific suggestions on future uses. We end with a review and our reflections on the potential of the technology and the need for future Blockchain studies [15].

18.13.2 Methodology

We started our inquiry with a query on the online UT library server. Searching for the keyword ‘Blockchain’ resulted in a solid mix of media sources [15]. Newspaper articles were the highest with more than 23,000 entries. We counted the total newspaper articles about one-tenth of that for newspapers, coming in at just over 2,000. About as many magazine articles have been registered as newspaper articles, although this is not representative of the actual magazine articles. Newsletters only had under 800 records and there were around 150 books on the topic of Blockchain. Although peer-reviewed journal articles have been our target, it was intriguing to see the figures leaping out as just a comparison figure. Upon looking over the total number of submissions, we modified our search queries to show us only journal articles [15]. We worked our way down from the top and picked journals that appeared to be both important and useful. When the articles were downloaded, we glanced through them for the narratives. We then put together a list of narratives and avoided downloading over one article for each narrative. We compiled about 20 articles from peer-reviewed journals that we believe reflect the existing literature on the topic of Blockchain. From these posts, we have been able to find some common trends in the existing literature [15]. Although these theories do not represent all developments in current Blockchain literature, they provide a very representative overview [15].

Blockchain technology uses the concept of a peer-to-peer network in which every node gets the copy of another node, and in this chain, there is no central authority, and every node of a chain or network performs as a client or server [16]. The Blockchain transactions must be validated and authorized. Every node of the network must verify the other node, and due to this, the presence of the trusted middleman is removed from this technology. Blockchain is also known as a decentralized database that stores confidential records of customers and also manages them efficiently [16]. Whenever a new transaction is initiated inside this network, an existing node validates it and holds a record of a new transaction; hence, a new block is created in a chain that is mostly known as a mined transaction. After generating a new block for formal verification, it immediately performs in a way in which a block hash is generated to define a block, then the subsequent hash is also evolved to link a block with a chain, per block. And, in this way, the chain of blocks is formed, which is now recognized as the Blockchain [16]. Blockchain technology is however completely secure, for when someone tries to tamper with the code, the block hash automatically switches and the next block breaks from the previous hash, so that a defective block is created due to tampering in either block and therefore the block chain gets disrupted [16]. If anyone wishes to make a legitimate change, then they will follow the correct process known as proof of operation, where any node in the network will verify the node and have a copy of the node, and the change will then occur. So, in short, in data tampering, this technology is not easy to manipulate [16]. The ledger in this system is used to eliminate impurity or accountability because the core principle of the ledger is open access to the chain for everyone. Invalid operation is immediately declined and does not establish a single block because it is not checked or accepted by certain network nodes [16]. All data or information on the network is encrypted [16].

18.13.3 Blockchain in education

Blockchain is a technology that transforms the educational system too given it is the sole connection to a chain as proof of digital activity, generating duplication on every network node [15]. The entire database is spread collectively to node networks [15] and it stores the attainment of degrees, credit hours completed which are also known as degree certificates in the education sector. Every graduation credential produces blocks in the chain, and approved industrialists can have access to it, students will even exchange it with respected companies and can put it on their curriculum vitae [15]. The Blockchain platform offers university degrees for publicly awarded students and it is authenticated and approved [15]. Blockchain in education is shown in Figures 18.7 and 18.8.

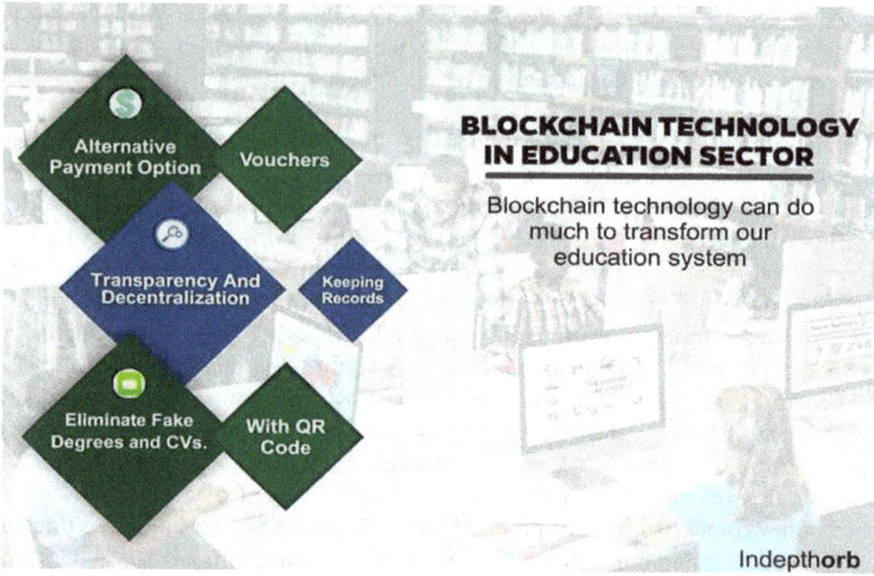

Figure 18.7 Blockchain technology in education sector [17]

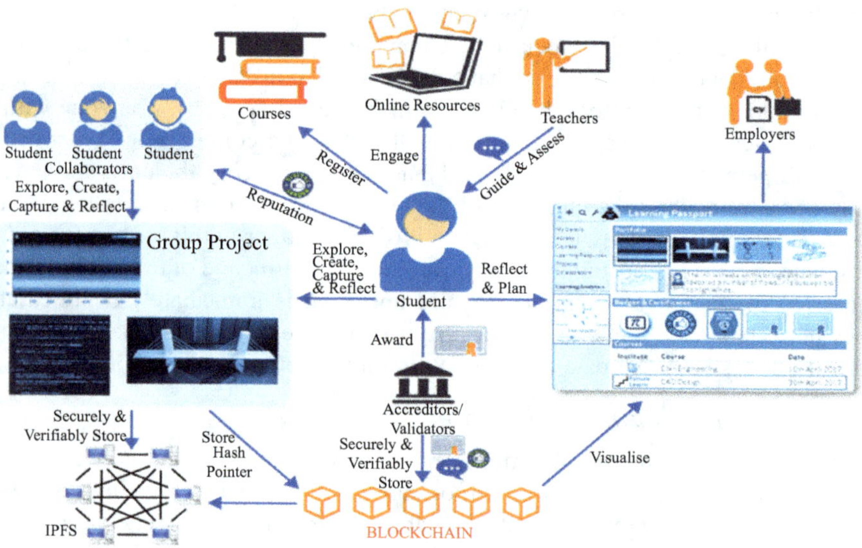

Figure 18.8 Blockchain for teaching and learning [18]

18.13.4 Blockchain in cryptocurrency

This technology is recognizable to everyone and is rising in use day by day. There are massive numbers of Blockchain developed in the world but Bitcoin and

Ethereum are by far the most popular of these. It has also functioned as a traditional currency; consumers can purchase products and services conveniently using it [15]. Cryptocurrency is a digital currency designed to work as a medium of exchange through a computer network that is not reliant on any central authority, such as a government or bank, to uphold or maintain it [15]. Cryptocurrency is an electronic communication medium established and processed mechanically in a chain, where all those transactions are properly secured due to the use of the encryption and decryption technique of this system to analyze financial segments and monitor all transactions safely [15]. This technology is not physically acclimatized for use, but it is only used in the digital networking chain. This cryptocurrency is not owned or controlled by any bank, because the Blockchain network which is exceedingly decentralized handles it. In the Blockchain, blocks collect information relevant to consumer transactions, when any transaction is carried out in a chain, where the transaction must be checked against another block, if it is true or not, until the new block is established in a chain and the transaction record is placed in a new block. Once the construction of a new block is performed for specific identification, the block hash is generated as proof of its existence [15]. While, in some Asian and European nations, it is considered legal, but most of the time, owing to certain state problems, it is banned by the government in a variety of countries. Cryptocurrency has many benefits and improves the openness of any transaction reported permanently, leading to online exposure where anyone can see and experience trust [15]. There is efficient monitoring of the transaction due to the correct block hash and the previous hash, and no one can reverse the transaction, which is often called a forward transaction. A definitive ledger, in this system, is used as a definitive record of transactions. Transaction costs are high due to direct selling, which saves both time and monetary expense. Faster and cheaper, the deal saves an enormous amount of money [15]. Voting can also be carried out using this technology via smartphones and tablets with correct and instantaneous results [15]. Blockchain cryptocurrency is shown in Figures 18.9 and 18.10.

18.13.5 Blockchain in medical care

Big data and Blockchain in the field of health may come from distinct practices such as insurance information, doctors' consultations, scientific, research studies, policies, and many more. Data comes in various quantities and forms for evaluation and study [15]. Hospitals require efficient research tools to carry out analysis of these data. In medical care, we have also found that often sensitive and secure data is required, for which reason Blockchain technology is used in this field to make data safer, open to hospital personnel so that no one can alter or compromise data integrity [15]. Health records are often considered critical that no one should handle in the wrong way. Blockchain can solve this kind of big data problem. In a literature review, we have found that Blockchain technology has made it possible to safely pass data to its users and even to keep the information safe. In Blockchain, no duplication of data is possible, where each transaction has a special hash for the recognition of blocks and user accounts [15]. Health files are encrypted so that one

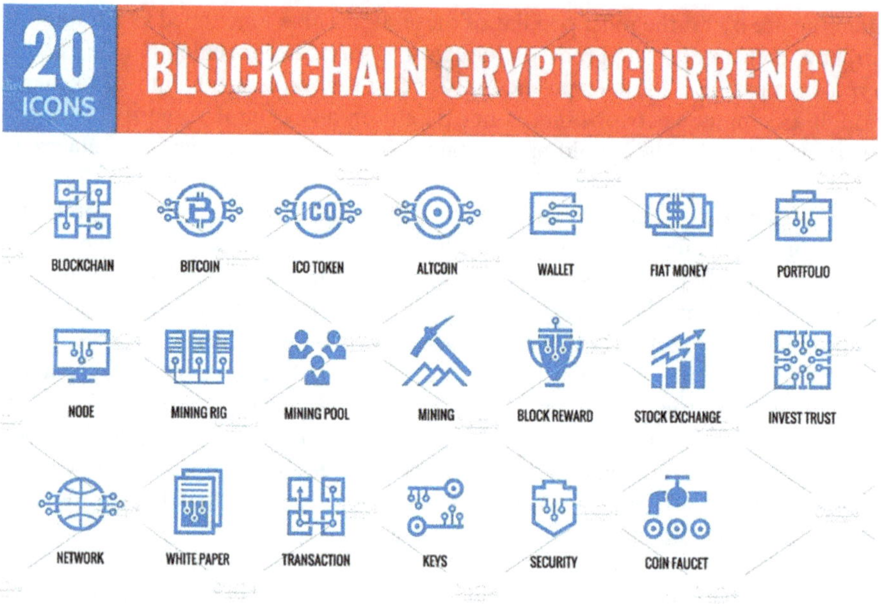

Figure 18.9 The 20 icons of Blockchain in cryptocurrency [19]

Figure 18.10 The features of Blockchain in cryptocurrency [19]

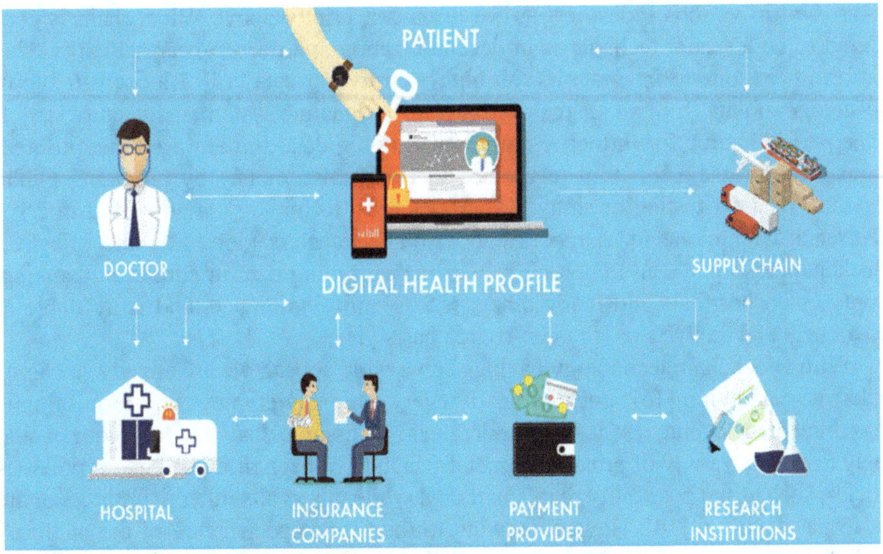

Figure 18.11 Blockchain in medical care [20]

can properly interpret and interact with data and can even be digitally signed. In medical care, we can easily store vast amounts of data using the "MedRec" approach in Blockchain in three different contracts. The first is the "Registrar Agreement." It holds only the specific name of the patient, the simple definition of the patient and allows entry to the only approved institution [15]. The second one is "Patient Provider Relationship Contractor." It gives the necessary details to other nodes and is also known as the patient-custodian relationship [15]. The third is the "Review Agreement." This maintains proper patient medical records, which also contain a brief description of the patient's disease [15]. In addition, this contract makes all transactions validated and authorized by the patient on the basis of refusing access from unauthorized persons. The MedRec approach allows patients to view their data and also enables mining or analysis techniques in medical research to discover new illnesses solutions [15]. Blockchain in healthcare is shown in Figure 18.11.

18.13.6 Conclusion

In this article, we addressed Blockchain and big data, which are two separate technologies, but currently both function together in the industry and play an increasingly significant role in data mining [15]. We reviewed separate research papers from different researchers, collected data from well-known sources, and reviewed Blockchain and big data. Blockchain is a decentralized database with a network of multiple computers around the world, so it's very complicated to tamper or remove data from this network [15]. Blockchain also guarantees that a copy of the data is stored on all network computers in one second. Blockchain is the most advanced technology that is being used in various industries, as we discussed briefly above. Blockchain is considered to be a secure technology in decentralization and peer-to-peer networks. This platform helps one to integrate it with high-security big data [15]. Blockchain allows the storage of data from all sources attributable to dynamic system architecture [15]. It is the perfect source for further research utilizing big data. In the meantime, different companies are using this tool to protect data and raise money by providing public access to data for testing or study [15]. Blockchain technology is now used in cryptocurrencies for monetary purposes and to make transfers more safe and effective. Blockchain also plays an important role in managing big data in education, banking, supply chain hospitals and many others with security assurance, and is available to researchers and analysts on an open-source basis [15].

With Blockchain technology having such a wide appeal, we are already seeing its widespread adoption [16]. Because nearly every company employs any form of agile record-keeping techniques, it is not unrealistic to assume that this technique would apply to a wide variety of projects, some of which have been discussed in our previous articles, such as the possibility for a smart city, while others are either in progress or are yet to be identified [16]. However, owing to the peer-to-peer nature of the platform, this infrastructure and the fact that any stakeholder has access to their block of books, falsification of data has never been more complicated. That alone has the potential to increase consumer confidence in these new technological disruptions. Like for all modern technology, the underpinnings are not fully known and, for this reason, it is impossible to tell how broadly the technology should be implemented [16].

Future researchers will investigate these issues and the latest technologies introduced, as well as review the pace of acceptance of technology. To those that do adopt Blockchain, further analysis will provide us with insight into what improvements (if any) in efficiency have been reported. Studies can also reflect on roadblocks as to why this technology has not been embraced, as well as contemplate developments for customer interest. In addition, as technology advances, potential studies may help to shed light on any safety concerns not initially found [16].

From what began as some coded anonymous programmer to develop a digital currency network, Blockchain has risen in prominence, in almost every sector from finance and healthcare, all the way to education and community planning [16].

In addition, Blockchain technology tends not only to enhance activities in current industries but also to have the ability to revolutionize processes that keep track of asset history into a dramatically enhanced, fully accessible ledger system [16].

18.14 Comparison

We will be comparing Blockchain technology, IoT(Internet of Things) and Network technology.

(I) **Blockchain technology**

Blockchain is a distributed ledger technology (DLT). We understand that to be a structure for the data contained in the blocks. As compared to only one, data is exchanged or distributed between various computers. This system records any transaction occurring in the data; hence it is known as a ledger [21].

The reason it is called Blockchain is that every transaction is represented by blocks. All the blocks are intertwined, in chains [21].

(II) **Internet of Things (IoT)**

The IoT is a simple concept whose implications are vast. The Internet of Things connects anything to the internet, at its most basic, with a power button. All can be connected to the internet like your toaster oven, your TV or even your home [21].

The IoT does come with crucial security issues, though. This means a network can mainly be compromised by the data it transmits. Blockchain comes into play here, and that's how both go hand in hand. A decentralized solution like Blockchain is something that is being applied to IoT [21].

Its peer-to-peer format has proved useful, rather than a centralized server. This principle has been extended to its applications by other organizations. They found a peer-to-peer format offering a high level of security [21].

It is because copies of the same data are transferred to all nodes within a server in a peer-to-peer format, it means that if one node is tampered with, then all other nodes that are interconnected reflect a discrepancy. So, if data has to be hacked, then all nodes on a server have to be hacked, which is virtually an impossible task [21].

The IoT using Blockchain as its security is therefore perfect [21].

(III) **Network technology**

Network technology is the technology that helps us connect with other people through a computer network and electrical devices connected [22]. The major things covered by network technology are as follows:

Communication devices: these comprise network components, such as router, modem, adapter, etc., from cable to wireless technology [22].

Intranet and Extranet: private networks and public networks constitute the two core areas of network technology studies [22].

Communication tools: From messaging to video conferencing, they are part of network infrastructure. Virtual reality communication is also part of network technology which is said to be the next big thing [22].

18.15 Gap analysis and benchmarking of Blockchain technology

Blockchain has been envisioned to be a disruptive technology with potential for applications in various industries. As more and more different blockchain platforms have emerged, it is essential to assess their performance in different use cases and scenarios. In Table 18.1, researchers have conducted gap analysis and benchmarking on the blockchain performance evaluation by categorizing all reviewed solutions into two general categories, namely, empirical analysis and analytical modelling. In the empirical analysis, researchers comparatively reviewed the

Table 18.1 Gap analysis and benchmarking of Blockchain technology

References	Authors	Designs	Gap analysis	Issues addressed
[15]	Umair Aslam, Saad Manzoor, Muhammad Imran Babar.	Blockchain In Requisites Of Big Data And Its Applications	Discussed about Blockchain and big data, both are two distinguishing technologies but nowaday's both are working concurrently in industry and playing important role in data mining	Discussed and addressed the issues with big data and their applications with different technologies
[22]	Piers Lissette	What-Is-Network-Technology	Discussed about network technology, how it works and later compared to Blockchain technology	Addressed the issues on how both network technology and Blockchain technology work together by research and comparison of results
[21]	Sahu Mayank	What-is-the-difference-between-Blockchain-and-IoT (Internet of Things.)	Discussed about the meaning of IoT and compared with Blockchain technology and how the IoT uses Blockchain technology as its security.	Addressed issues arising with IoT and Blockchain technology and how both technologies work together.
[20]	Davis, Robert Anascavage, Nathan	Blockchain Technology	Discussed what Blockchain technology means and what it does.	Addressed the issues on how Blockchain technology works and its application.
[18]	Universa	Blockchain in education	Discussed Blockchain technology impact in the education sector	Addressed the issues Blockchain technology is facing in the education sector
[23]	Jaideep Ghosh	The Blockchain: Opportunities for Research in Information Systems and Information Technology	Discussed how Blockchain technology is used in information systems and how it is integrated with information technology	Addressed the issues of Blockchain technology in information systems and information technology

(Continues)

[24]	Juho Lindman,Matti Rossi, Virpi Kristiina Tuunainen	Opportunities and risks of BlockchainTechnologies in payments– a research agenda	Discussed the opportunities and risk of using Blockchain technology	Addressed the issues of using Blockchain technology in payments and addressed the risk involved with Blockchain technology
[25]	Rosanna Cole, Mark Stevenson, James Aitken	Blockchain technology: operational and supply chain implications	Discussed about Blockchain technology implications and operations of supply chain management	Addressed the issues Blockchain technology faces in supply chain management

current empirical blockchain evaluation methodologies, including benchmarking, monitoring, experimental analysis and simulation. In analytical modelling, they investigated the stochastic models applied to performance evaluation of mainstream blockchain consensus algorithms.

18.16 Industry Revolution 5.0

Industry 5.0 is a re-humanization of the human race towards automation. The acknowledgement that both robotic/automatic/digital developments and the questioning/inspection/innovation/creativity that comes from human beings are of equal importance in the production phase [26] is gaining acceptance among people.

Since Industry 5.0 is like the Jaguar F type; symbolic of the great era of Grand Touring and just one letter from the new classic iteration, but filled with creativity and automation; always still – and here's the point – with the human being at the epicenter of expertise [26].

What is perhaps more fascinating is that the human element of Industry 5.0 is at both ends of the cycle. A great sandwich, maybe [26].

18.16.1 Blockchain technology Industry Revolution 5.0

Although AI and IoT are revolutionary innovations in their own right, it is the facilitating position that Blockchain can perform that makes them a "revolution." Blockchain provides a modern method of processing and exchanging data, as well as allowing transactions that are more reliable and permanent than the existing internet. Blockchain utilizes cryptographic authentication to encrypt data and provide more protection than current computer programming and applications do [27]. The time and date details that cannot be modified are authenticated with this data, resulting in documents that cannot be challenged [27].

Blockchain technology is however decentralized. Data is stored on multiple nodes, or computers, throughout the Blockchain network, instead of being in the enterprise-owned computer system or data farm [27]. Decentralization shifts authority over data away from government entities and large enterprises. It renders a technology hacker-safe for information exchange [27]. A unified network often allows easier for users utilizing a Blockchain network to obtain the knowledge they need [27].

Blockchain technology has amassed a good amount of awareness and it has shifted mostly from controversy to early technological innovations, proof-of-concept and pilot initiatives over the last 12 months [27].

The distributed ledger is being studied in basically almost any industry, but mainly in finance, logistics, trade, healthcare and governments. Blockchain is also being considered as a potential technology mechanism for political voting that could significantly profit from the encryption and tamper-proof performance of Blockchain [27]. IT giant IBM was a trendsetter in engaging in Blockchain technology. Based on the current data presented on blocklr.com, IBM has registered the second-largest rate of Blockchain patents in the world [27]. IBM maintains that it has filed 9,043 patents [27].

Blockchain, IoT and AI are ideally designed for working together [27]. As a result, the compatibility and synchronization of these technologies are fast becoming a major concern for development companies. The three technologies and many more will potentially work together smoothly [27]. For example, IoT will capture information on everything that's going on around us. In the absence of human guidance, AI may make a judgment based on new knowledge given by IoT [27]. Undergirding all this, Blockchain networks will hold and protect this content, distribute it to humans and control centers for more informed decision-making [27].

The very first industrial revolution became an age of steam power, the second was heralded with the discovery of electricity. The third industrial revolution was focused on electronics, technological innovations and the Internet [27].

Today, the fourth industrial revolution is a Blockchain, IoT and AI-based revolution, and it is already here. This 'trinity' of emerging technology would be critical for both customers and the industry [27]. Every company must shift to the modernization and digitization that comes with the fourth industrial revolution [27].

18.16.2 Features

From a technical perspective, Blockchain technology has four attributes: decentralization, traceability, immutability and currency properties [28].

Decentralization focuses on data authentication, storage, maintenance and transmitting processes on Blockchain premised on a distributed network architecture. In this system, trust amongst distributed nodes is established through mathematical techniques rather than through centralized organizations [28].

Traceability indicates that almost all transactions on Blockchain are organized in a sequence, and a block is attached to various adjacent blocks by a cryptographic hash function. Consequently, each transaction can be tracked by analyzing the block data connected to the hash keys [28].

There are two reasons for the immutability of Blockchain technology. On the one hand, all transaction records are in blocks with one hash key linked from the prior block and one hash key referring to the next block. Meddling with any transaction will also lead to different hash values and would, therefore, be discovered by all other nodes running the same validation algorithm [28]. On the other hand, Blockchain is a decentralized distributed database held on thousands of computers, where both ledgers tend to synchronize in real-time. Effective tampering will entail a modification to more than 51% of the network-stored ledgers (Tschorsch and Scheuermann 2016) [28].

Blockchain technology and cryptocurrency are inseparable, i.e., every Blockchain network has a type of cryptocurrency ownership. The essence of Blockchain technology is point-to-point transactions, with no third-party involved, which implies that all transactions do not require the involvement of third parties [28]. Digital currency distribution based on Blockchain technologies is fixed [28].

In particular, the currency base in Bitcoin is set at 21 million caps, so the electronic currency production is generated using a different mineral algorithm and is pre-defined in a formula [28]. There will, therefore, be no inflation problem, collapse, etc. [28]. These non-financial operations will fall under currency in Blockchain 2.0 and 3.0 implementations by integrating other practices, such as political programs, educational activities and finance [28].

18.16.3 Characteristics

- Services and business models [29]
- Profitability of business [29]
- Reliability and continuous increase of efficient productivity [29]
- IT security [29]
- Machine and human safety [29]
- Product life cycles increase [29]
- Industry value chain environmental impact decreases [29]
- Workers' education, skills and involvement increase [29]
- Socio-economic factors improvement [29]

18.16.4 Architecture of blockchain technology

A Blockchain is essentially a distributed data system designed linearly, over time, and verified and audited by all network actors independently [30].

Generally speaking, blockchains comprise transactions bundled into blocks that are mined utilizing considerable capital and generating new mining tokens [30].

The network-wide encryption means that all transactions are legal and utilizes consensus rules to decide what is found in a secure Blockchain [30].

Consequently, blockchains are the implementation of an innovative new way of building networks outside of any single trustworthy agency to determine reality [30], as shown in Figure 18.12.

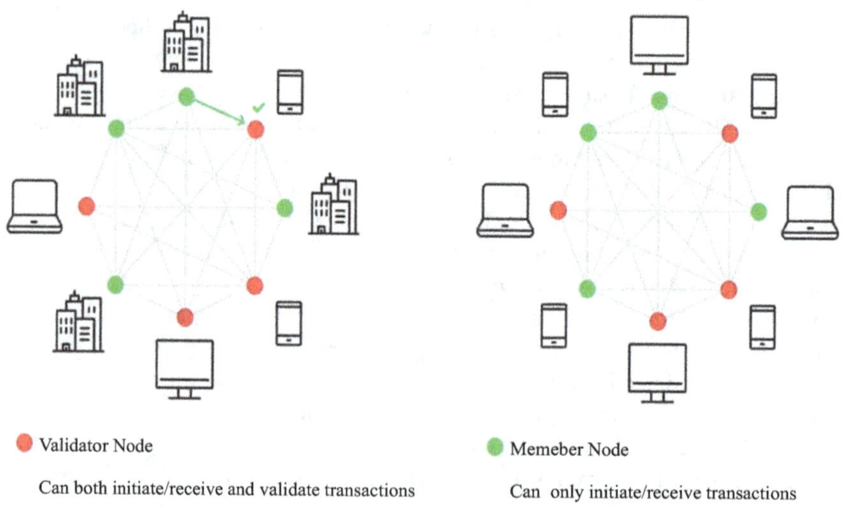

Figure 18.12 The architecture of Blockchain technology [31]

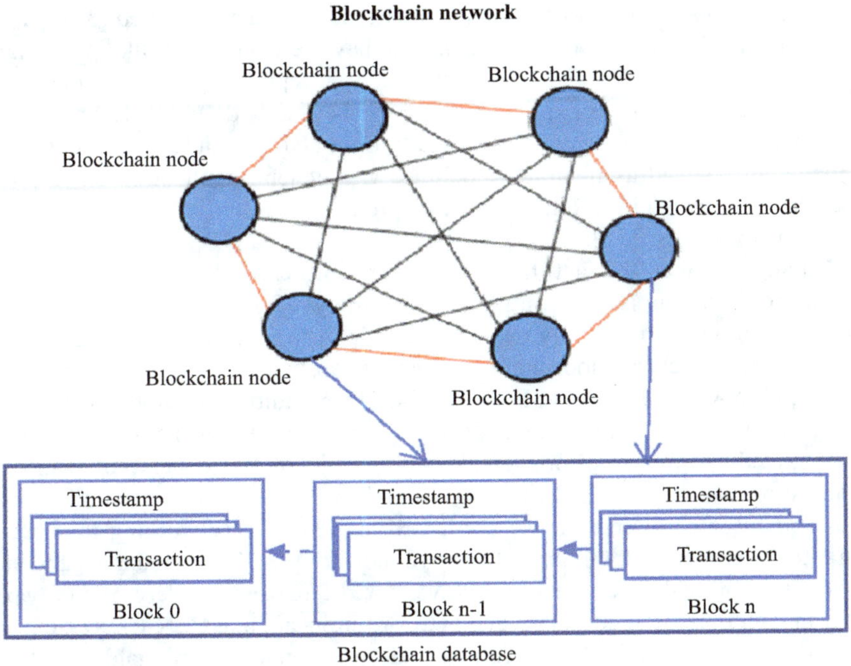

Figure 18.13 The architecture of Blockchain technology [32]

18.17 Characteristics of Blockchain technology

- Cryptography
- Immutability
- Provenance
- Decentralization
- Anonymity
- Transparency

18.18 Professional standards of Blockchain technology

Business, industry and government actors focus on and underline the value of Blockchain standardization, whigham, 2017 by referencing Alexander Peter, Head of Data Communication Agency [33]. Standards have been established as practical standards for multiple sectors, influencing and impacting different stakeholders participating in the businesses, such as regulators and consumers [33]. In relation to new technologies, standards may also be produced [33]. Standards are the result of lengthy debates, hearing the other party and final agreement by professionals who engage in a specific area and, although not legally binding in itself, have a significant impact on the way the subject matter is treated and applied [33]. Part of their influence derives from the mechanisms they are exposed to, like peer examination, that are rigorous and based on widely accepted norms. In some cases, regulations and legislations explicitly recognize the standards and therefore provide it its legal status [33]. In contrast to normally rigid and inflexible legal regulations, standards are viewed as flexible guidelines with general application on different concepts and issues [33]. Naturally, when produced carefully, this does not take away their robustness [33].

Having said that, the fact that many groups influenced by Blockchain ask for their standardization is shocking [33]. A variety of respected standardization organizations, who are actively engaged in updating and/or developing guidelines that deal with aspects in Blockchain, are liable to answer. The leaders in this aspect are ISO, CEN-CENELEC, ETSI, and ITU [33]. In addition to widely known and respected institutions, the impetus for implementing Blockchain principles is demonstrated in private efforts conducted outside of the context of standards organizations. In the so-called Mobility Open Blockchain Initiative (MOBI), three renowned automobile enterprises, Ford, GM and BMW, are working together to guarantee interoperability in the automotive industry [33]. The standards generated at the supranational level are informative, for the reasons entirely of the paper, either internationally, such as ISO and ITU, or European, like CEN-CENELEC and ETSI [33]. These standards are valued and are intended to mitigate social and market issues [33].

Unless addressing the key principles on which Blockchain specifications will evolve, standardization is a highly challenging undertaking in an ever-evolving technology, such as Blockchain. It is often challenging to find the right balance with respect to the pace of the production of standards. Standards should be produced in a way that increases blockchain's potential implementation [33]. The premature standardization of Blockchain is of concern. Standardization could be implemented rapidly enough to prevent insufficient, though not hurried, usage of Blockchain that would damage innovation [33]. The interconnection of these groups, which implies that a specific category of criteria presupposes or is rendered simpler by the advancement of other sections and the advancement of Blockchain technologies during the growth, is to be viewed within a spread-out time frame by evaluating the nature of the questions involved [33], as shown in Figure 18.14.

18.19 Applications of Blockchain technology

In recent years, Blockchain technology has obtained considerable digital ink. This is as big as the internet was in the nineties. That is a game-changer [35]. Nevertheless, we often interpret the terms "Blockchain" with "Bitcoin." What many don't realize is that Bitcoin or cryptocurrency, as an extension of Blockchain technology, has been the core of many sectors in this open and distributed digital archive [35].

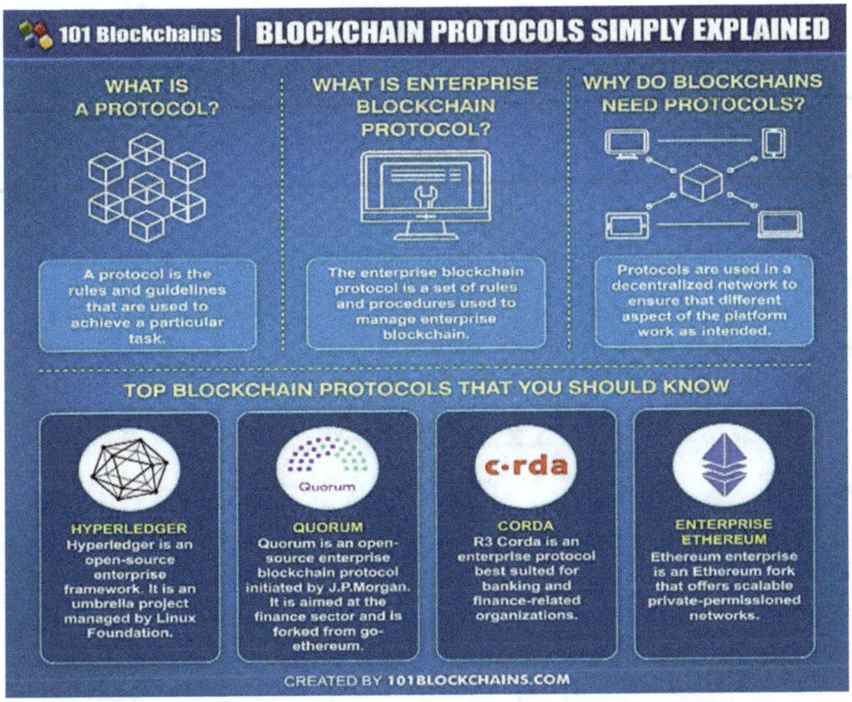

Figure 18.14 Professional standards of Blockchain technology [34]

(I) Healthcare

Blockchain technology has heavily influenced the healthcare sector in clever yet beneficial ways [35]. With the exponential development of the Internet of Medical Things (IoMT) and medical test technologies incorporating artificial intelligence, Blockchain technology implementations will quickly, vastly improve healthcare with interoperability, data portability, transparency and stability [35]. The vast potential of this distributed ledger technology is suitable to address the challenges affiliated with the decentralization of health information [35] In addition, three main fields are anticipated to benefit from Blockchain platform technologies, including clinical research, enforcement and legislation, medical, healthcare and engineering [35], as shown in Figure 18.15.

(II) Cybersecurity

The wide variety of information security breaches is soaring, posing a consistent risk to consumers [35]. According to the Identity Theft Resource Center report, "668 breaches compromised 22,408,258 files between January 1 and July 2, 2018." The 2018 Ponemon file states that "the common complete value of an information breach is $3.86 million and the common breach fee discount for groups the usage of protection automation is $1.55 million [35]." This worrying upward trend in security breaches and hacks has pressured companies to improve their safety

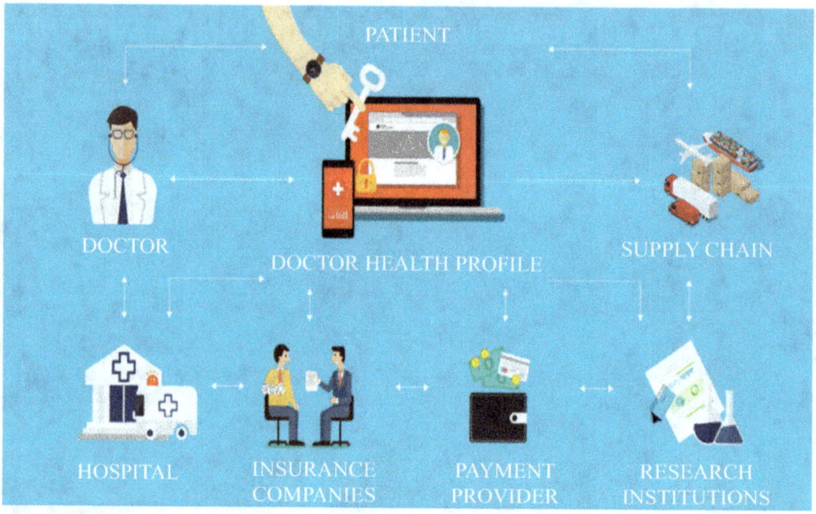

Figure 18.15 Blockchain technology applications in healthcare [36]

options to foresee what the malicious actor uses in methodologies [35]. But are these investments worth it? Are they successful in preventing a counter-attack? It does not seem like that because the attackers are winning and the threats are constantly changing [35]. So how can Blockchain preclude these attacks and thefts from happening? This technology has three key features that enhance cyber security – decentralization, cryptographic protection, and tamper-proof ledgers. Comprehensive encryption and data sharing protocols over a peer-to-peer network guarantee the data stays reliably secure and out of the reach of hackers [35], as shown in Figure 18.16.

(III) **Smart contracts**

According to Nick Szabo, the American computer scientist who had first introduced the idea of smart contracts, it is a "computerized transaction method that implements the contract terms [35]." When you hear "contract," you think of a signed piece of paper that outlines the details of a legal agreement among various parties [35]. Blockchain-based smart contracts are software contracts operating on a computerized transaction protocol controlled by many Blockchain-connected nodes [35]. Although Blockchain contracts operate on a shared network among all approved parties, they can self-execute and manage themselves, thereby removing the need for mediators such as lawyers or notaries [35]. Unlike conventional systems, smart contracts are transparent, traceable, safe, and irrevocable, and that's why many industries are switching over to them [35], as shown in Figure 18.17.

(IV) **Food safety**

A food health Blockchain? Is this a surprise? Well, technology experts think that Blockchain technology can restructure entire business processes of data

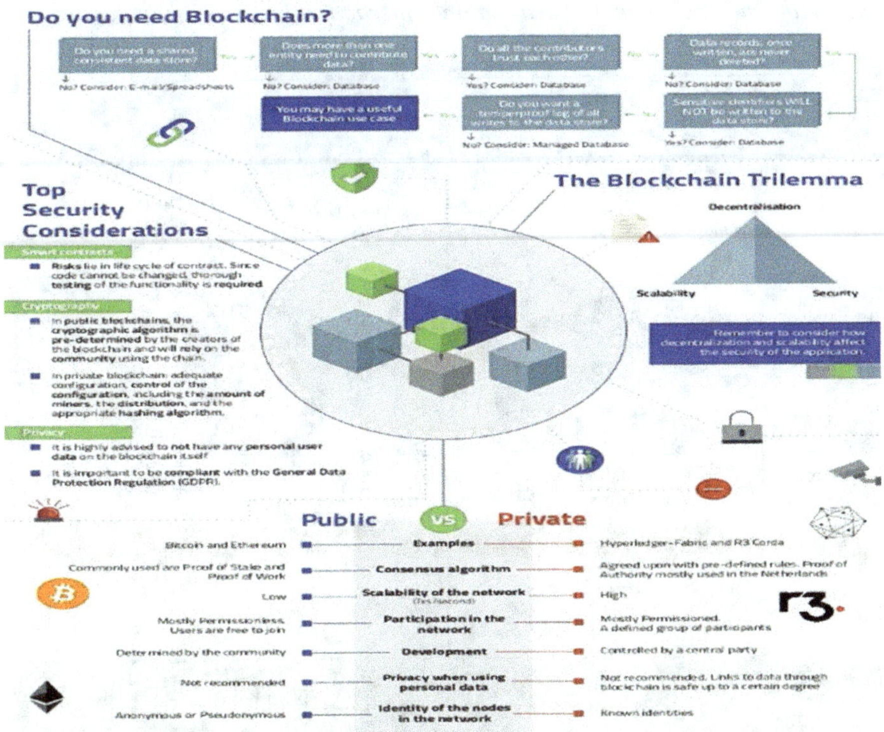

Figure 18.16 Blockchain technology applications in cybersecurity [37]

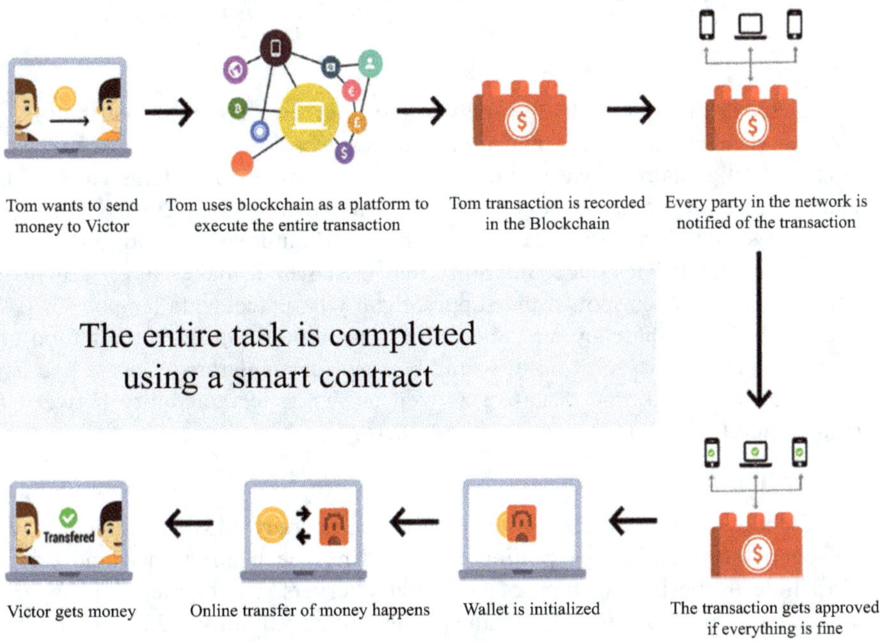

Figure 18.17 Blockchain technology applications in smart contracts [38]

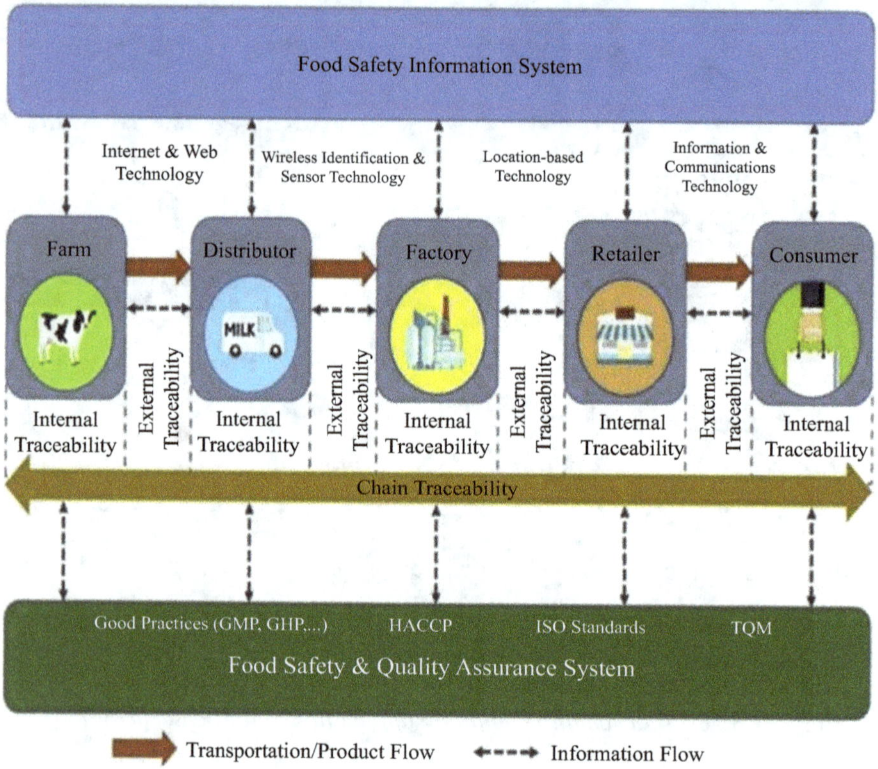

Figure 18.18 Blockchain technology applications in food safety [39]

management through a diverse network of farmers, brokers, distributors, processors, retailers, regulators and users. Looking back at the background, we can quickly find a litany of cases when a bad shipment of food items caused supermarket stores and other companies to be shut down [35]. In 2008, China's child milk crisis, and North America's *Escherichia coli* outbreak in 2006 [35]. The food industry currently lacks accountability and trust which makes it very difficult to determine the proper spot in the supply chain where accidents happen [35]. With Blockchain's transparency and traceability, customers can trace their food easily from farm to fork. In addition, it would be easy for regulators to locate and handle the origin of food-borne diseases, as well as foster accountability between food manufacturers [35], as shown in Figure 18.18.

(V) **Voting**

Elections serve as the cornerstone of democracy [35]. Through the years, however, the reliability of polling processes has been questioned and criticized [35]. Indeed, the involvement of centralized servers, vulnerable to infringements and hacks, has begun to diminish election credibility [35]. Through the

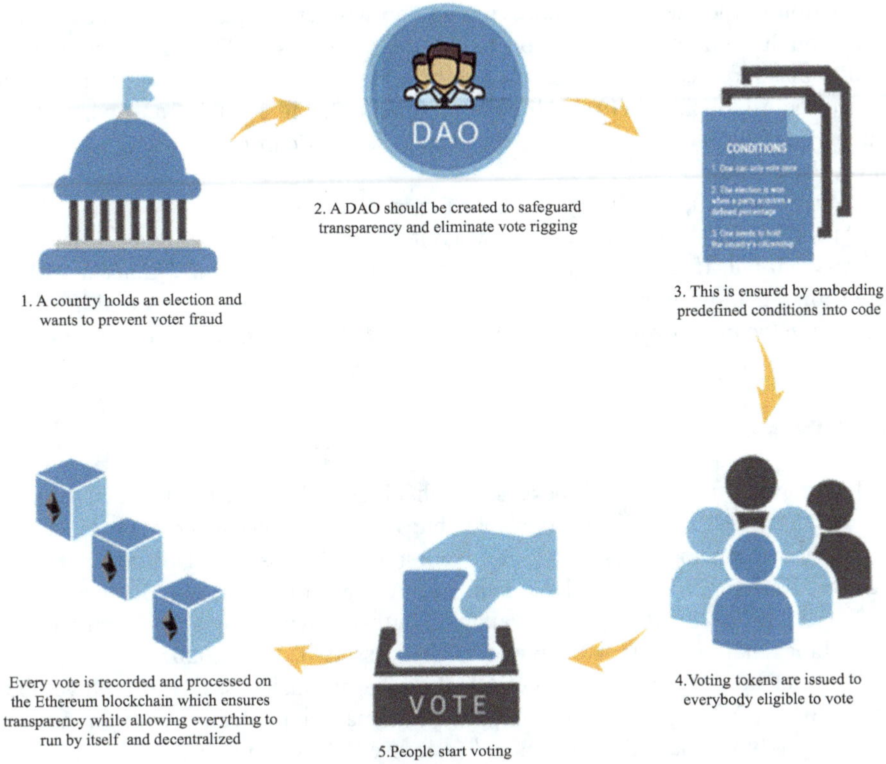

Figure 18.19 Blockchain technology applications in voting [40]

implementation of Blockchain technologies, the whole election cycle may be protected and made transparent [35]. The ease of automated voting is paired with the immutability of Blockchain to remove problems such as double voting, changing of vote and deletion of ballots [35], as shown in Figure 18.19.

18.20 Conclusion

Technology advancements as well as transformations are happening faster and faster, which is why an organization needs strong goals for the company innovations as well as a consistent mentality for the transition [41]. The organization must be able to handle the unexpected and take action to ensure the firm is future-proof. Big businesses in Europe are now focused on the introduction of Industry 4.0; they have not yet understood, as the study discussed underlines, the next industrialization to Industry 5.0 [41].

To obtain quality services and products on customer demands, the business must acknowledge that success will come to those that are creative and adaptable to market changes [41]. Industry 5.0 emphasizes a clear shift from widespread automation to the method of improving human workers' capacity to achieve skill-sets through product customization to the next level [41]. Future questions then need to be addressed as follows: what abilities need to improve, what kind of rules need to be established between human and computer interaction, and what effect can AI have on worker upskilling and what disputes can occur between humans and AI?

Stuart Scanlon, managing director of epic ERP says:

"people have distinct viewpoints when it comes to evaluating the beginnings of Industry 5.0 but, if you look at the pace of technology evolution, I think it will be here sooner than what most people think. The future is happening now and if we are to thrive in the next revolution we need to face the challenges." (Scanlon, 2018) [41].

References

[1] Simplilearn, "What-Is-Blockchain-Technology-and-How-Does-It-Work-Article," 7 May 2020. [Online]. Available: https://www.simplilearn.com/what-is-blockchain-technology-and-how-does-it-work-article (Accessed 10 November 2021).

[2] P. Kashyap, "Why Is Blockchain Technology Important For Business?," 27 March 2018. [Online]. Available: https://www.beingcrypto.com/why-is-blockchain-important-for-business/ (Accessed 10 May 2020).

[3] A. Patel, "Top-Advantages-Blockchain-for-Businesses," 17 November 2018. [Online]. Available: https://www.smartdatacollective.com/top-advantages-blockchain-for-businesses/ (Accessed 10 November 2021).

[4] D. Schwartz, "What-Are-the-Disadvantages-of-Blockchain-Technology," 21 November 2017. [Online]. Available: https://www.quora.com/What-are-the-disadvantages-of-blockchain-technology (Accessed 10 November 2021).

[5] N. Reiff, "Blockchain-Technologys-Three-Generations," 29 February 2020. [Online]. Available: https://www.investopedia.com/tech/blockchain-technologys-three-generations/ (Accessed 10 Novemebr 2021).

[6] S. Liu, "Blockchain – Statistics & Facts," 13 March 2020. [Online]. Available: https://www.statista.com/topics/5122/blockchain/ (Accessed 10 Novemeber 2021).

[7] Scitech Europa, "Blockchain in the Healthcare Market to Hit €1.4 bn by 2025," 12 September 2019. [Online]. Available: https://www.scitecheuropa.eu/blockchain-in-the-healthcare-market-to-hit-e1-4bn-by-2025/97036/ (Accessed 10 May 2020).

[8] Technavio, "Global Blockchain Market in Retail Sector 2019–2023," 12 April 2019. [Online]. Available: https://www.technavio.com/report/global-blockchain-market-in-retail-sector-industry-analysis (Accessed 10 May 2020).

[9] ConsenSys, "Blockchain Will Deliver $3.1 Trillion Dollars in Value by 2030," 6 June 2019. [Online]. Available: https://media.consensys.net/gartner-blockchain-will-deliver-3-1-trillion-dollars-in-value-by-2030-d32b79c4c560 (Accessed 11 May 2020).

[10] Technavio, "Global Blockchain Technology Market in the Agriculture Sector 2018–2022," 13 September 2018. [Online]. Available: https://www.technavio.com/report/global-blockchain-technology-market-in-the-agriculture-sector-analysis-share-2018 (Accessed 11 May 2020).

[11] C. Shalchi, "Ecommerce Shops Take Advantage of Blockchain Rewards Programs," 20 September 2019. [Online]. Available: https://www.bigcommerce.com/blog/author/chris-shalchi/ (Accessed 11 May 2020).

[12] B. Ross, "Who-Invented-Blockchain," 26 April 2019. [Online]. Available: https://www.quora.com/Who-invented-blockchain (Accessed 10 November 2021).

[13] The Institute of Chartered Accountants in England and Wales, "What-Is-Blockchain/History," [Online]. Available: https://www.icaew.com/technical/technology/blockchain/blockchain-articles/what-is-blockchain/history (Accessed 10 November 2021).

[14] N. Reiff, "How-Does-Blockchain-Work," 8 February 2020. [Online]. Available: https://www.investopedia.com/tech/how-does-blockchain-work/ (Accessed 10 November 2021).

[15] S.M.M.I.B. Umair Aslam, "Systematic literature review of blockchain in requisites of big data and its applications," *J. Softw. Eng. Intell. Syst.*, vol. 4, no. 2019, 2019 .

[16] R. A. a. N. Davis, "Blockchain technology: a literature review," *Blockchain Technology: A Literature Review,* no. May 17, 2018, May 3, 2018.

[17] C. Jack, "https://www.indepthorb.com/potential-impact-of-blockchain-technology-in-the-education-sector/," 25 June 2019. [Online]. Available: www.indepthorb.com (Accessed May 2020).

[18] Universa, "https://medium.com/universablockchain/blockchain-in-education-49ad413b9e12," May 2018. [Online]. Available: medium.com (Accessed May 2020).

[19] Enterprise Management 360, "Top-10-Cryptocurrencies-Predictions," 18 Feburary 2018. [Online]. Available: https://www.em360tech.com/tech-news/top-10-cryptocurrencies-predictions/ (Accessed May 2020).

[20] MarksMan Healthcare Communications, "Application of Blockchain Technology in Healthcare," 28 March 2019. [Online]. Available: https://marksmanhealthcare.com/application-of-blockchain-technology-in-healthcare-hope-or-hype/ (Accessed 25 May 2020).

[21] M. Sahu, "https://www.quora.com/What-is-the-difference-between-Blockchain-and-IoT?share=1," 14 April 2020. [Online]. Available: www.quora.com (Accessed 28 May 2020).

[22] L. Piers, "https://www.quora.com/What-is-network-technology," 17 April 2017. [Online]. Available: www.quora.com (Accessed 26 May 2020).

[23] J. Ghosh, "The blockchain: opportunities for research in information systems and information technology," *J. Global Inf. Technol. Manag.*, vol. 22, no. 4, pp. 235–242, 2019.

[24] M. R. V. K. T. Juho Lindman, "Opportunities and risks of Blockchain Technologies in payments – a research agenda," In *Proceedings of the 50th Hawaii International Conference on System Sciences*, Hawaii, 2017.

[25] M. S. J. A. Rosanna Cole, "Blockchain technology: implications for operations and supply chain management," *Supply Chain Manag.*, vol. 24, no. 4, pp. 469–483, 2019.

[26] A. Cooper, "Industry 5.0 – Cobots and Personalisation at Scale," 20 January 2020. [Online]. Available: https://www.freestyle.agency/blog/industry-5.0-cobots-personalisation/ (Accessed 9 June 2020).

[27] M. Down, "How Blockchain is Driving the Fourth Industrial Revolution," 2 March 2019. [Online]. Available: https://hackernoon.com/how-blockchain-is-driving-the-fourth-industrial-revolution-1abdc0535691 (Accessed 10 June 2020).

[28] Chen, G., Xu, B., Lu, M. *et al.* Exploring blockchain technology and its potential applications for education. *Smart Learn. Environ.* vol. 5, no. 1 (2018). https://doi.org/10.1186/s40561-017-0050-x.

[29] M. Rada, "Industry 5.0 Definition," 22 January 2018. [Online]. Available: https://medium.com/@michael.rada/industry-5-0-definition-6a2f9922dc48 (Accessed 9 June 2020).

[30] Pluralsight, "Blockchain Architecture," 10 January 2019. [Online]. Available: https://www.pluralsight.com/guides/blockchain-architecture (Accessed 9 June 2020).

[31] MLSDev, "Blockchain Architecture Basics: Components, Structure, Benefits & Creation," 7 March 2019. [Online]. Available:https://medium.com/@MLSDevCom/blockchain-architecture-basics-components-structure-benefits-creation-beace17c8e77 (Accessed 9 June 2020).

[32] T. Salman, "Probabilistic Blockchains: A Blockchain Paradigm for Collaborative Decision-Making," 20 November 2018. [Online]. Available: https://www.researchgate.net/figure/Blockchain-architecture_fig333486562 (Accessed 9 June 2020).

[33] A. Christiana and M. Evdokia, "Blockchain standards and government applications," *J. ICT Standard.*, vol. 7, no. 3, pp. 287–312, 2019.

[34] N. Singh, "Top 5 Blockchain Protocols That You Should Know," 17 December 2019. [Online]. Available: https://101blockchains.com/blockchain-protocol/ (Accessed 9 June 2020).

[35] N. Ravindran, 5 Real-World Applications of Blockchain Technology, Friday, January 4th, 2019, 5 Real-World Applications of Blockchain Technology - Logiticks].

[36] Market Research Future, "Blockchain Technology in Healthcare Market Expected to Reach at CAGR 71.8% | Industry Insights, Size, Share, Trends, Outlook Analysis 2023," 14 March 2019. [Online]. Available: https://www.medgadget.com/2019/03/blockchain-technology-in-healthcare-market-expected-to-reach-at-cagr-71-8-industry-insights-size-share-trends-outlook-analysis-2023.html (Accessed 10 June 2020).

[37] dutchblockchaincoalition, "Cyber Security Framework Assists in Safe Use of Blockchain Technology," 07 December 2018. [Online]. Available: https://dutchblockchaincoalition.org/en/news/cyber-security-framework-helpt-organisaties-met-veilig-toepassen-blockchains-1 (Accessed 10 June 2020).

[38] S. Mehrotra, "Why Are Smart Contracts so Important," 8 July 2018. [Online]. Available: https://medium.com/acycliclabs/why-are-smart-contracts-so-important-81883d93a0cc (Accessed 10 June 2020).
[39] M. J. Kay Behnke, "Boundary conditions for traceability in food supply chains using blockchain technology," *Int. J. Inf. Manag.*, vol. 52, no. 1, pp. 12–17, 2020.
[40] G. T. Athena, "How Will Voting be Affected by Blockchain in the Future?,"10 August 2018. [Online]. Available: https://www.quora.com/How-will-voting-be-affected-by-blockchain-in-the-future (Accessed 10 June 2020).
[41] A. M. A. D. Daniel Paschek, "Industry 5.0 – The Expected Impact of Next Industrial Revolution," *Innovation and Industrial Management International conference*, Piran, Slovenia, May 2019.

Chapter 19

Reducing Waste and Pollution with Automation and CPS in Manufacturing

Shuh Huey Ho[1], *Nurul Izzatul Akma Katim*[2], *Mohd Sabirin Rahmat*[1] *and Muhammad Sufyan Safwan Mohamad Basir*[3]

This chapter of the book presents an overview of the cyber physical systems (CPSs) of reducing waste and pollution in the manufacturing process. The conceptual model of a CPS is elaborated in detail for the concept model of CPS in the manufacturing process. The role of CPSs is introduced along with manufacturing applications. It is worth mentioning that the major concern of cyber security issues of the CPS, however, was not discussed in this paper. The authors mentioned an overview and elements for the manufacturing process such as new evolving technology. Today, new business models are emerging that will alter how goods are created, moved, and consumed. The system-on-chip (SoC), Internet of things (IoT), and cloud computing platforms are the enabling computer technologies for such change (CCPs). Now, sensors, actuators, and SoCs with WiFi and/or Ethernet are integrated at the production line's edge. The SoC, Internet of things (IoT), and cloud computing platforms are the enabling computer technologies for such change (CCPs).

19.1 Introduction

The fourth industrial revolution has resulted from the integration of the physical, digital, and biological worlds (Industry 4.0). One of the primary elements of Industry 4.0 is the transition from mass and semi-customized production to mass, fully customized, and variant product manufacturing [1]. Today, new business models are surfacing that will alter the way goods are created, delivered, and consumed. Sensors, actuators, and SoC equipped with Ethernet or WiFi are integrated at the manufacturing line's edge to transform the existing embedded systems into cyber-physical systems (CPSs). The CPS is used to remotely build, monitor, run, and control the cyber physical products (CPPs). The mechanical part of the

[1]Department of Mechanical Engineering, MAHSA University, Selangor, Malaysia
[2]Department of Electrical and Electronic Engineering, MAHSA University, Selangor, Malaysia
[3]Department of Electrical Engineering, Politeknik Mukah Sarawak, Sarawak, Malaysia

product occupies physical space, while the product digital twin, which contains the product data electronic sheet and dynamic data, occupies virtual space. Most goods will have their own identification number, IP address, and access code, allowing them to be accessed through Ethernet or WiFi from any location at any time. It covers the product's history from manufacture to recycling, and the state of the product's dynamic variables [4].

Nowadays, CPS has a wide range of applications. It may be applied in smart grid, water networks, healthcare wearable devices, factory, manufacturing, monitoring and control of gas and oil pipelines, unmanned aerial and autonomous underwater vehicles, greenhouse control, and hybrid electric vehicles [5].

19.2 Waste

Waste is defined as any activity that uses resources but does not add any value to the products and therefore can be eliminated and may affect the living [6]. Waste can be in the form of solid, semi-solid, liquid or gas. Waste can be categorized into hazardous and non-hazardous waste. Hazardous waste is toxic, reactive, corrosive, and expensive such as solvent waste including Endrin, Acetone, and Creosote whereas non-hazardous waste is any trash that did not harm to environment and human health such as agricultural waste, municipal solid waste. Chemical waste, solid waste and toxic waste are the three common types of manufacturing waste. Waste can occur through the defects in the raw materials or during processing or during transporting. Figure 19.1 shows the typical manufacturing facility. In Figure 19.1, air, water, solvents, catalysts, energy, and raw materials will be used to produce products while waste energy, gaseous wastes, water or liquid wastes, and solid wastes will be released.

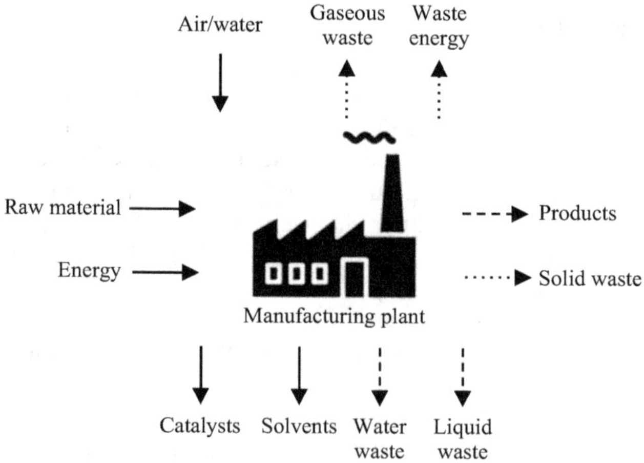

Figure 19.1 Typical manufacturing facility with pollution [7]

Improper disposal of the industrial waste will lead to pollution issues. The environmental pollution issues in Malaysia are increasing with the economic growth. Air pollution, water pollution, and waste have become obvious and serious problems in Malaysia with the economic development. Currently, the major contributor in the waste generation is the manufacturing sector, followed by agriculture, medical, and household. Pollution from factory wastes cause serious landfill issues and health issues when the wastes travel to the air and water.

According to the Malaysia Environment Statistics, 4.0 million tonnes of the scheduled wastes were reported in 2019. 57.1% (2.3 million tonnes) of the total scheduled wastes were the wastes for metal refinery, electrical & electronics, power plant, and chemical industry. In addition, a total of 4,932 factories were reported as the resources of water pollution. According to the statistics, Selangor shows the highest number of industrial sources of pollution (1,668 factories), followed by Johor (945) and Negeri Sembilan (371). These create challenges for the waste management system.

19.2.1 Types of waste in manufacturing

In manufacturing operations, defects, excess processing, waiting, overproduction, inventory, and transportation are common production process-oriented wastes [8]. Waste in manufacturing can also take the form of time and resources wastage.

Overproduction: In manufacturing, overproduction occurs when components are produced before the next downstream process and leading to the storage of too much unused products. Overproduction can be caused by poor automation, unstable production schedules, inaccurate information, long set-up times, and unreliable processes. Overproduction is costly and affects the productivity and quality of the products.

Defects: Defects refer to products that did not fulfill the customer's expectation. Defects can be caused by the poor machine repair, inaccurate inventory, poor quality control at production, poor documentation, and failure in understanding customers' requirements.

Transportation: Waste from transportation is the waste produced during the unnecessary transportation from one position to another position. Excessive handling and movement can affect the quality of the products. Large batch sizes, multiple storage facilities, and poor layouts are common types of transportation waste.

Motion: Motion defects refer to any unnecessary motion by person or machine which can cause wear and tear on machines or products. The health and safety of the workers can be affected when there are additional delays in the production. Additionally, physical motion such as lifting, reaching, or bending is also considered as motion waste if it is unnecessary. Lack of production standards, process design, production plan are common examples of motion waste.

Inventory: Inventory waste refers to the waste produced by unprocessed inventory. Inventory waste can be caused by the inventory defects, delays in production, overproduction and excessive transportation. Inventory waste can be in raw materials or finished products. High costs and large spaces will be needed if the items are holding in the inventory.

Waiting: Waiting waste refers to a waste produced when the item is not being processed. Waiting waste can be materials, manpower, or equipment. Waiting waste can be caused by delays in the set-up, unplanned downtime, and lack of process control.

Excessive processing: Excessive processing refers to additional activities that do not add any value to the products like re-inspecting, documentation, and re-counting. Excessive processing can be related to poor communication and human error.

19.3 CPS

CPS is an important component in Industrial Revolution 4.0 (IR 4.0) and received great attention recently as it can be applied in many applications such as manufacturing, process control, environment, and transportation. Three phases involved in the development of CPS. The first development involved the identification technologies, followed by the development of CPS equipped with actuators and sensors. CPS can store and analyze data in the third development [9].

CPS is characterized by the integration of cyber and physical parts. Cyber parts refer to any software or hardware resources that achieve computation, communication and control purposes whereas physical parts refer to any system that operates in continuous time [10]. The difference between the properties of the cyber part and physical parts in CPS is tabulated in Table 19.1. Communication, computing and physical elements must exist in CPS. Controllers and sensors are used to control the physical dynamics and are connected via a communication network [11].

CPS can be described using a corresponding set of characteristics: technical emphasis, cross-cutting aspects, level of automation, and life-cycle integration. The individual system in CPS is connected according to the requirements and all the data are accessible. Hence, activities such as optimization, parameterization, and maintenance can be automated. Connection, cyber level, cognition, and configuration must be considered when implementing CPS in manufacturing.

19.3.1 Concept of CPS

The concept of CPS consists of five layers for completing the system of manufacturing. The first layer is the physical layer that contains sensors, actuators,

Table 19.1 The properties of cyber and physical parts in CPS

	Cyber	Physical
Structure	Computing	Physical law
Method of ensuring proper order	Sequences	Real time
Time properties	Discrete	Continuous
Event synchronization	Synchronous	Asynchronous

tracking devices, and computing elements. The mapping of the CPS concept is shown in Figure 19.2.

The sensors record the data for sending the information to the controller then the obtained data is processed and transmitted to the cloud for further processing. Based on the process algorithm of the system embedded, the signal received by the actuator from the controller can be executed remotely or locally. The sensors convert analog or digital signals from the appliances, tools, and surrounding devices into electrical signals. The sensors are used to collect the data and processed locally under control algorithms.

Actuators are an important component in CPS physical layer (see Figure 19.3). The microcontroller actuated the actuators to alter the status or the condition of the system under production or test based on the application process algorithm. The actuator receives the control signal (an analogue or digital signal with a maximum voltage of 5 V and a maximum driving current of 25 mAmps) from microcontrollers. Driver circuits, such as relays and power amplifiers, should be employed between the actuating device and microcontroller output, depending on the application, to deliver higher voltage and current as needed.

Using multiple networking protocols such as GPRS, 3G/4G/LTE technology WiMAX, and WiFi, CPSs can access the cloud or cyber environment. Other lightweight IoT message-oriented data protocols used to send data from edge devices to the cloud for data processing and storage include Websocket, MQTT, AMQP, CoAP, and Node. In terms of bandwidth, reliability, speed, latency, security, and scalability, each protocol has its own benefits over the others. Medical

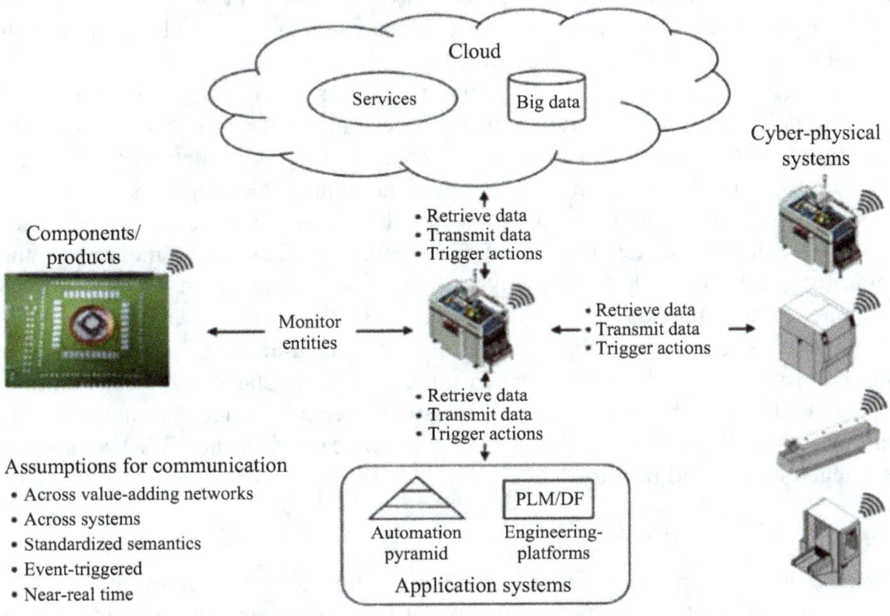

Figure 19.2 Interaction of the entities and system [12]

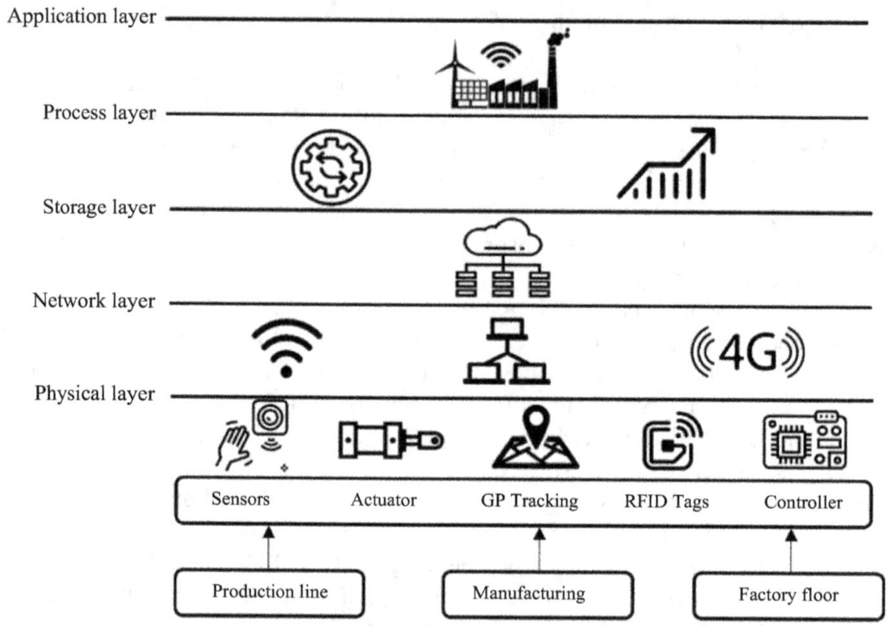

Figure 19.3 Interaction of the entities and system [12]

CPSs, factory control, smart water networks, smart buildings, self-driving cars, hydropower plant and monitoring, greenhouse efficient control, and wind power plant are just a few of the CPS applications that use these protocols. Other details related to these protocols can be obtained in Ref. [13].

The data is processed using simulation models such as map reduction and dimensional modelling techniques in the processing and analytic layer. Graphs, reports, and dashboard visualization can be created using SQL queries for near real-time monitoring. Spot pricing, planning, data mining techniques such as data clustering, regression, and classification may be applied in order to monetize the data for predictive maintenance. Besides, this layer can also convey monitoring and controlling actions back to the physical layer, allowing some equipment and machines to be actuated.

Operators, consumers, third-party suppliers, manufacturers, as well as other service providers will be interacting in this layer. Smart factories, grid, building, transportation, and healthcare are all possible users. It includes a user-friendly interface that allows the stakeholders to communicate with the CPS levels using privileged access and priority.

19.3.2 Benefit of CPS

The advantages of implementing CPS in manufacturing are to monitor the status of the equipment easily, control the function of the equipment, minimize downtime during the manufacturing process, and reduce the manufacturing waste [14]. CPSs are a set of interactive systems with the support of intelligent machines which, coordinated and controlled by a central entity, deliver operational details or useful information to qualified workers. Thus, operations during the production can be self-control while potential issues and possible solutions directly in this cybernetic replica can be proposed by the engineers.

Innovation in CPS is based on the input from many disciplines and needs a skilled workforce to sustain and support its development. It has the great potential to improve and change the industry while addressing many critical challenges. The role of CPS has changed and evolved with the increase in the number of devices connected to the Internet of Things (IoT). Increasing productivity, efficiency, and economic benefits will be the role of CPS. This integration includes advances in real-time processing, detection, and actuation between IoT systems and physical domains provides capabilities for system analysis of the cybernetic and physical structures involved. Nowadays, the key to these systems is artificial cognition, defined as artificial intelligence in the networking of people, processes, data, and things. From this integration comes all the added value that CPS possesses.

19.4 Architecture CPS for manufacturing

Integration of CPS in the manufacturing process requires further developments. Based on Figure 19.4, CPS architecture is divided into five tiers. The first step in

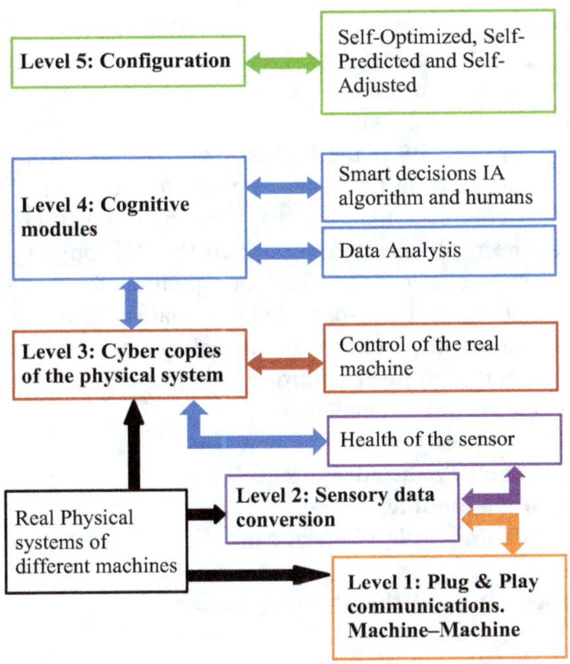

Figure 19.4 Interaction of the entities and system [12]

acquiring reliable and accurate data among the various machines is the level of connection. The communication employs machine–machine (MM) protocols and necessitates plug-and-produce and plug-and-play connection that can be automatically connected. Other methods to identify people for human collaboration can be used at this level. This communication can be employed in the date conversion process at the second level. At this level, several processes can be utilized at this level to turn the date into information to perform intelligent data analysis. Some processes are designed to forecast and manage the machine's health, while the others are used to analyze performance of the degradation and predict the performance as well as correlate the multidimensional data.

The third level involves the creation of virtual copies of cyber copies of physical systems. This level serves as a hub for details and information that is sent from all the connected machines. This data should be saved and analyzed in order to extract additional information, thus enhancing the understanding of the status or condition of each fleet machine. These analyses enable the machines to self-compare, allowing the performance of a single machine to be compared and qualified across the fleet. Besides that, comparison on the performance of the machine and the previous assets can be identified, hence future behavior can be predicted. The generation of historical data is used to identify any changes or variation of the machine.

Next, level of intelligence and decision-making that allows expert humans and algorithms to be viewed remotely by providing an appropriate presentation of analytical information. The operator or engineer can obtain the analyses and make appropriate decisions by using human–machine interfaces (HMI) for industry 4.0. Collaborative diagnosis for maintenance processes can be performed, which can be easily determined because of the individual machine's status information and availability of comparative information.

The fifth level of the CPS is the highest level. In this level, configuration is made to provide feedback from the cybernetic part to the physical part. This level provides supervisory control to the machine, allowing it to self-configure and self-adapt. It serves as the resilience control system (SCR), applying the appropriate controls to cognition-level decisions. Self-configuration for resilience, automatic adjustment for variation, and self-optimization against disturbances are typical functions. In order to avoid perturbations in real-time control applications, low-level, middle-level, and high-level control are proposed. Levels four and five is where high-level control can be implemented.

Level 5: Configuration
Self-Optimized, Self-Predicted and Self-Adjusted
Level 4: Cognitive modules
Smart decisions IA algorithm and humans
Data Analysis
Level 3: Cyber copies of the physical system
Control of the real machine
Level 2: Sensory data conversion
Real Physical systems of different machines

Level 1: Plug & Play communications. Machine-Machine
Health of the sensor

19.5 Implementation of CPS

CPS should be implemented in ensuring the objectives of IR 4.0 in which information and production systems are integrated [15]. For a smart industry, the 9 pillars of IR 4.0 should be compromised, and CPR will contribute to the IoT and Web of Things (WoB) part. The outcome may benefit in full control over production processes, increase efficiency as well as operational transparency [16]. The added values of CPS in manufacturing are characterized into four. First, integrated system data should be protected, works at continuous connectivity using industrial grade 5G network, and a high-speed data transfer rate. Second, the interconnected objects include the algorithm used, data processing, and parameters displayed must be implemented in a studied way, hence the weakness of the system can be determined for improvisations. In addition, results generated should be benchmarked with other recent systems. From this approach, the system is able to determine the potential problem and faults can be easily diagnosed. Lastly, the machine should be configured under closed loop operation. The input of the system is varied based on the information from the output there the system will work accordingly to the information served or isolate any misbehavior immediately.

The integration of two CPS components, namely complex physical systems and a cyber core is crucial [17]. The cyber core becomes a part to automatically monitor, control, and estimate the physical system condition. The information gathered will be processed and displayed in real-time. For a future smart manufacturing, CPS is one of the options in ensuring the production is maximized [18].

19.6 CPS in waste management

The final goal has been outlined in circular economy principles is to transform into a closed loop the traditional economic linear approach known as take, make, and dispose. The goal can be performed through different objectives such as reuse of the end-of-life products, reduction in waste, and reinsertion of the waste as secondary raw materials. It is a good opportunity to use new adaptable and resilient manufacturing and remanufacturing processes in this field. Waste reduction is one of the circular economy pillars. It could be accomplished through different approaches as shown in Figure 19.5. Elements in the upper part of the triangle (which considers end of life products as a resource) are preferred compared to that in the lower part.

Lately, the growth rate of waste flows (10%) has brought new challenges to the industrial world [20]. Increasing the adaptability of a recycling system using smart lines is the best solution to face the singular variabilities. In this work, CPS for the feed-forward control of recycling processes will be introduced and applied, through a two-step procedure architecture, to size reduction.

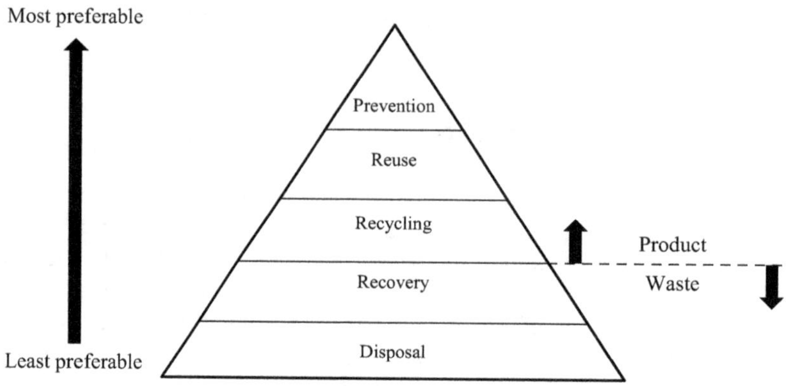

Figure 19.5 Interaction of the entities and system [12]

CPS has received attention in daily human life as it is widely applied to education, medical devices and systems, agriculture, process control, energy management, security, environment, smart manufacturing, and intelligent transportation. In CPS, sensors, which monitor the indicators of cyber and physic and actuators are used, coordinated, and communicated with the elements of computation, to control the environment [21,22]. CPS is characterized into cyber capability in physical components, complex network, computational elements, technology sensors, high automation level, technology actuator, and human system interaction which are required in the current waste management system [22,23]. Waste collection, processing, and disposal activities are the existing CPS adoptions and practices in waste management from waste users to final disposal [24,25].

According to other research, sensors, RFID, GPS, and cloud can be used to improve the performance of waste management and efficiency. Sensors are introduced at the waste sorting stage and enable the recognition of the characteristic of different waste and properties of the material such as texture and shape [26–28]. Global position system (GPS) is usually adopted for tracking vehicle and bin positions. While in wireless technology, RFID is commonly used to monitor transportation processes of the waste [29,30]. CPS and Cloud have been applied into some treatment processes for example recycling and remanufacturing to support waste recovery processes [31,32]. Other than that, smart waste management systems using IoT implementation have been also convinced to manage the waste problem globally [33–35].

19.7 Conclusion

Cyber physical production networks operate smoothly and automatically with IoT-based real-time production logistics and artificial intelligence-based decision-making algorithms. The manufacturing of CPSs possesses progressively automated linkages and networking, leading to big data-driven innovation in sustainable Industry 4.0. The integration of IoT smart devices, cognitive decision-making algorithms, industrial artificial intelligence, and digitized mass production furthers the design of cutting-edge monitoring strategies across Industry 4.0 wireless networks. Industry 4.0 generates cutting-edge developments in sustainable operations and may attain uniform production and consumption practices through cognitive decision-making algorithms across artificial intelligence data-driven IoT systems. An overview of the CPSs of reducing waste and pollution in the manufacturing process was presented in this paper. The conceptual model of a CPS was elaborated in detail for the concept model of CPS in the manufacturing process. The role of CPSs along with manufacturing applications was introduced. It is worth mentioning the major concern of cyber security issues of the CPS, and it was not discussed in this paper. The authors mentioned an overview method and element for manufacturing processes of such new evolving technology.

References

[1] Wurm J, Jin Y, Liu Y, et al. Introduction to cyberphysical system security: a cross-layer perspective. IEEE Trans *Multi-Scale Comput Syst* 3, pp. 215–227, 2017.

[2] Bergera C, Heesa A, Braunreuthera S and Reinharta G. Characterization of cyber-physical sensor systems. In Proceedings of the 48th CIRP Conference on Manufacturing Systems (Naples), 2015, pp. 638–642.

[3] Ashibani Y and Mahmoud QH. Cyber physical systems security: analysis, challenges and solutions. *Comput Secur* 68, pp. 81–97, 2017.

[4] Uhlemanna THJ, Schocka C, Lehmanna C, Freibergera S and Steinhilpera R. Digital twin: demonstrating the potential of real time data acquisition in production systems. In Proceedings of the 7th Conference on Learning Factories (Darmstadt), 2017, pp. 113–120.

[5] Jawhar I, Al-Jaroodi J, Noura H and Mohamed N. Networking and communication in cyber physical systems. In Proceedings of the IEEE 37th International Conference on Distributed Computing Systems Workshops (Atlanta), 2017, pp. 75–82.

[6] Ashok KR. *Zero Waste: Management Practices for Environmental Sustainability*, London: CRC Press, 2019.

[7] Kenneth LM and James AD. *Pollution Prevention: Methodology, Technologies and Practices*, New York, NY: John Wiley & Sons, 2010.

[8] Lonnie W. *How To Implement Lean Manufacturing*, New York, NY: McGraw Hill Professional, 2009.

[9] Fei H. *Cyber-Physical Systems: Integrated Computing and Engineering Design*, London: CRC Press, 2010.

[10] Diego Galar P, Pasquale D and Uday K. *Handbook of Industry 4.0 and SMART Systems*, London: CRC Press, 2019.

[11] Husheng L. *Communications for Control in Cyber Physical Systems: Theory, Design and Applications in Smart Grids,* Burlington, MA: Morgan Kaufmann, 2016.

[12] Sabina J, Christian B, Houbing S and Danda BR. *Industrial Internet of Things: Cybermanufacturing Systems*, New York, NY: Springer, 2016.
[13] Jawhar I, Al-Jaroodi J, Noura H and Mohamed N. Networking and communication in cyber physical systems. In Proceedings of the IEEE 37th International Conference on Distributed Computing Systems Workshops (Atlanta), 2017, pp. 75–82.
[14] Yun Arifatul F, Andi W and Muhtar H. Cyber-physical System Enabled in sustainable waste management 4.0. A smart waste collection System for Indonesian Semi-Urban Cities. *Proc Manuf* 42, pp. 535–542, 2020.
[15] Dalenogare LS, Benitez GB, Ayala NF and Frank AG. The expected contribution of Industry 4.0 technologies for industrial performance. *Int J Prod Econ* 204, pp. 383–394, 2018.
[16] Ghobakhloo M. The future of manufacturing industry: a strategic roadmap toward Industry 4.0. *J Manuf Technol Manag* 29, pp. 910–936, 2018.
[17] Antsaklis PJ, Goodwine B, Gupta V, et al . Control of cyberphysical systems using passivity and dissipativity based methods. *Eur J Control* 19, pp. 379–388, 2013.
[18] Herrmann C, Schmidt C, Kurle D, Blume S and Thiede S. Sustainability in manufacturing and factories of the future. *Int J Precis Eng Manuf Technol* 1, pp. 283–292, 2014.
[19] European Union Directive 2008/98/EC of the European Parliament and of the Council of 19 November 2008 on waste and repealing certain directives. Off J Eur Union 51, pp. 3–30, 2008.
[20] Probst L, Frideres L, Pedersen and Demetri D. Clean technologies: closed loop waste management. EU: Business Innovation Observatory Contract No. 190/PP/ENT/CIP/12/C/N03C01, 2014.
[21] Gelenbe E and Fang-Jing W. Sensors in cyber-physical emergency systems, 2012, pp. 1A0-1A0.
[22] Berger C, Hees A, Braunreuther S and Reinhart G. Characterization of cyber-physical sensor systems. *Proc CIRP* 41, pp. 638–643, 2016.
[23] Letichevsky AA, Letychevskyi OO, Skobelev VG and Volkov VA. Cyber-physical systems, *Cybern Syst Anal* 53, no. 6, pp. 821–834, 2017.
[24] Chandrappa R and Das DB. Solid waste management, environmental science and engineering. In *Sustainable Solid Waste Management*, 2012.
[25] Tchobanoglous G. Solid waste management. *In Environmental Engineering: Environmental Health and Safety for Municipal Infrastructure, Land Use and Planning, and Industry*, Hoboken: John Wiley & Sons, Inc., 2009, pp. 177–308.
[26] Huang J, Pretz T and Bian Z. Intelligent solid waste processing using optical sensor based sorting technology. In Proceedings – 2010 3rd International Congress on Image and Signal Processing, CISP 2010, 2010, vol. 4, pp. 1657–1661.
[27] Killmann D, Scharrenbach T and Pretz T. Perspectives of sensor based sorting for the processing of solid waste material. In 2nd International Symposium Mechanical-Biological Treatment and Automatic Waste Sorting Technology, MBT, 2007, pp. 296–307.

[28] Chandini M. Design of lignin sensor for identification of waste paper grades for an automatic waste paper sorting system. Master Dissertation, NC State University, Raleigh, 2001. Retrieved from https://repository.lib.ncsu.edu/handle/1840.16/18.

[29] Hannan MA, Arebey M, Basri H and Begum RA. RFID and integrated technologies for solid waste bin monitoring system. *Aust J Basic Appl Sci* 4(10), pp. 5314–5319, 2010.

[30] Islam MS, Hannan MA, Arebey M and Basri H. An overview for solid waste bin monitoring system. *J Appl Sci Res* 8(2), pp. 879–886, 2012.

[31] King A, Burgess S, Ijomah W and McMahon C. Design for end-of-life: repair, recondition, remanufacture or recycle? Proceedings of DETC'04 ASME 2004 Design Engineering Technical Conferences, 2004, pp. 745–754.

[32] Wang XV and Wang L. Digital twin-based WEEE recycling, recovery and remanufacturing in the background of Industry 4.0. *Int J Prod Res*, pp. 3892–3902, 2018.

[33] Bharadwaj AS, Rego R and Chowdhury A. IoT based solid waste management system: a conceptual approach with an architectural solution as a smart city application. In 2016 IEEE Annual India Conference (INDICON), 2016, pp. 1–6.

[34] Misra D, Das G, Chakrabortty T and Das D. An IoT-based waste management system monitored by cloud. *J Mater Cycles Waste Manag* 20(3), pp. 1574–1582, 2018.

[35] Abdullah N, Alwesabi OA and Abdullah R. IoT-based smart waste management system in a smart city. *Adv Intell Syst Comput* 843, pp. 364–371, 2019.

Chapter 20

Smart manufacturing workplace safety with virtual training, AR, MR and haptic technologies

Lee Poh Foong[1]

The modern era than any other time known has involved more human-computer interaction technologies including virtual reality, augmented reality (AR), and/or mixed reality (MR) to bring better safety measures to handling the dangerous task, for instance, sending material to melt in a hot furnace or other such risky tasks to be completed with this technology. The subject matter of this chapter is focused on three types of virtual world technologies, and their respective applications, from the past, present, and future. This chapter aims to highlight the technology's use to enhance the safety of the smart manufacturing system.

20.1 What is virtual reality (VR)?

VR provides an immersive environment for the user to have a feeling of 'just be there', temporarily out of touch with the real world by taking part in a simulated experience [1]. They base the modern design of VR headset displays on the technology developed for smartphones, which is incorporated with gyroscopes and motion sensors [2]. Hence, the VR simulated scene depends for its real-life experience on the tracking system in the smartphones and on the movements of the head, hand, and body positions. The components for VR nowadays are relatively cost-effective to be used by independent VR developers.

20.2 Principle of VR

The first step to create a virtual world is to employ software – to draw graphics or video – and interfaces to hardware used – to enable the user wearing a headset to immerse into the virtual environment – where the entire structure of VR is shown in Figure 20.1. The hardware involves sensory interfaces with images and sounds, whereas the haptic technology uses the software to design the scene and interface with the graphic. In the virtual world, the virtual objects are interfaced with

[1]Universiti Tunku Abdul Rahman, Petaling Jaya, Malaysia

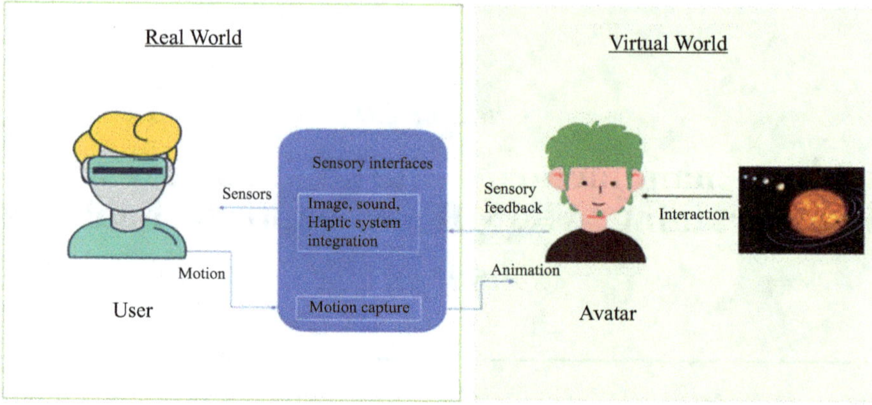

Figure 20.1 Principle of virtual reality

technologies in response to stimulation provided to the human senses – see, hear, taste, smell and touch – by the visualization devices, sound and reinforced feedback, motion platform and tactile experiences.

VR creates the awareness of an interaction with the virtual environment between virtual scenes and virtual objects. Real-time interaction between the user and the virtual environment requires capturing the motion or intention of the user to act in the virtual world [3]. What the virtual reality projection displays to the computer, the third party in the real world can observe in the form of a scene and the actions of the avatar (the virtual persona of the VR user) in the virtual world. The VR software developed for displaying in the smartphone is simpler and allows one to replace the VR headset for the smartphone. For this setting, the virtual scene and action are to be observed by the user himself. The lenses in the VR head-mounted display (HMD) bend the light that enables the user to look at VR [4]. This resolves the limitation of human vision that can see front objects at a distance of 3–7 cm only [5].

20.3 Application of VR

In current years, VR has been developed for use in many domains. This includes the areas of education, automotive industry, healthcare, retail, tourism, construction, entertainment, military, and different product types of manufacturing.

The application of VR in education allows students to take themselves to a place that does not require traveling to physically [6]. They can travel to the virtual world from their desk chairs. From the perspective of dispersing education, the VR education apps put into use VR for the students to visualize the concepts from the book using this platform to do tasks and practice what may be considered dangerous for them to do in the physical world. In a virtual world, the students can interact with the objects and practice the required skills to manage complex equipment. Taking a virtual tour of the ancient world for a history lesson is possible for education and learning with VR. For the biology class, the students can travel even inside the human body to study its structure. The VR learning platform in education has helped to bring great student participation in a lesson. It is believed that student engagement in learning can be fetched with tremendous benefits, for instance, increased student distraction that uses smartphones for playing games and neglecting to learn lessons can be curbed with VR education.

The application of VR in the automotive industry saves financial resources spent on manufacturing and managing physical assets. The tedious steps involved in vehicle design and prototyping can be resolved through the VR platform. The vehicle design efforts require continuous incremental improvement, reviews, and testing, where the constant modifications during the mock-up sections have been reported to be cost-incurring and time-consuming stages in car manufacturing [7]. Therefore, the use of VR has enabled car designers to visualize the exterior vehicles virtually to carry out amendments rather than do the new physical mock-up repeatedly. The design team can carry out real-time car exterior modifications during the mock-ups. Besides, the fabulous application of VR in this area is the VR platform can bring in different talented car designers from different countries to work together on the same mock-up session without moving out of their countries. Moving to the application in the automotive industry for training in assembling the car parts is conventionally carried out through the observation approach. After the workers learn the theoretical part of their training process, they practice under experts' supervision and the mistakes that can happen at this stage can bring damage to a particular vehicle model. The advantage of VR technology has to do with allowing the workers to obtain sufficient training in the virtual environments and make mistakes safely to gain experience. Meanwhile, VR brings safety to the workers, allowing them to practise in any dangerous situation and teaching them how to behave in a particular situation that seems real in the virtual world, minimizing stress and life-threatening situations for car specialists [8].

VR application in healthcare is claimed to help physicians empathize with their patients. For instance, it is challenging for a physician to feel what it's like to have dementia and what their caregivers go through when caring for the patient. If such situations can be simulated by VR to give a real-life experience to the physician, he will understand the situation of the patients in a more constructive way [9]. Besides, VR is evidenced to distract the patients to focus on the virtual world and hence reduces chronic pain. This helps patients who rely on too many painkillers by reducing stress with the VR programme. One of the companies, namely Karuna Labs, has worked in treating chronic pain through creating softwares to provide motion and visual-based experiences [10] to patients. Training with VR in healthcare is a primary aim of disbursing knowledge without any of the costly apparatus. For example, doctors can observe the details of any part of the body in stunning 360° images and create training scenarios that are similar to surgical procedures to be practised for multiple times and without a drop of blood shed. VR has another good use by improving patient response during rehabilitation, especially for patients with neurologic conditions in which treatment VR-based interventions attempt to promote neuroplasticity and motor learning [11].

Applications of VR in retail have been new but are challenging the *status quo* by requiring the customer to have the VR equipment at home. The actual tactile feeling of the product that they intend to buy, such as shoes, clothing, towel, gadgets, furniture, and many more types of products, is a need for the customer which the VR can resolve, especially, where the pandemic has taken away the opportunities for customers to visit the store and so VR has to step in to end the crisis [12]. COVID pandemic has lessened physical shopping and encouraged online purchases, where VR has given the customer an improved way to engage in the purchasing experience of the items not by physically selecting the item but by viewing the preferred items through the photos or pictures. IKEA came up with the ARKit app that lets users have a 3D preview of the chosen furniture to be placed at their home before making the purchasing decision [13]. But among other applications, retails through VR are still early as the customers have to own the VR sets of equipment, which is expensive at the moment. Another challenge for VR in retail is the feeling of touching the products which still remains to be true to the actual tactile feeling and gives less confidence to the customer to make a purchasing decision. If the VR experience for a brand is not satisfying to the customer, the reputation loss of a brand and loss in customers because of the unrealistic interaction with VR with the related retail can happen.

The application of VR in tourism astoundingly transforms the travel and tourism industry in a way to create more marketing chances for hotel rooms, flights, and places to travel [14]. For hotel room booking, normally customers require lots of research on information about the environments and room sizes of the hotels. This technology has speeded up the decision-making as the customers have a chance to immerse in the virtual environment to see the room and scene of stay. Besides, VR helps to attract the visitors to see the destination and convince themselves of the worthiness to visit the actual tourist spot, conventionally with video, documentary, or photos. With VR, potential tourists may gain the experience of the expensive tourist place and be willing to spend money to visit the place after engaging in the virtual experience made with video, sound, and images.

The application of VR in construction and architecture is popular especially in providing the client with a three-dimensional model of the project before the first purchase decision is made and to enable clients to walk and explore the building closely, with decisions on which changes to be made and checked in real-time [15]. The advantage for the builders in this is to create better construction projects and spot problems before the construction starts. Building information modelling (BIM) has been supported well by the VR platform where the architect can save time on designing something that might not work in the real world, saving the builder useless efforts and the client money for the construction project [16]. It is challenging for the verification of a building project from architectural 2D plans but 3D models can show the design in a more realistic fashion increasing design accuracy. VR transforms the conventional construction procedures from having the scale models on the computer screen and printing them on paper as physical scale models to shifting towards working with full-size structures in the VR environment. Another great advantage of VR in construction is the virtual space sharing enabling collaboration with more people such as the designer and builder inviting comments from the doctors and nurses by theirs taking a visit to the virtual construction site of a hospital. Professional training with VR in construction allows real-life experiences to be gained by engineers and architects by working with VR of a tower crane; furthermore, experience working with the designs in a virtual world with dangerous equipment helps to rectify mistakes without adverse fallouts on real-world apparatus where mistakes can cause life to be put to risk.

Many more areas are made beneficial by the VR technology; besides the concept of VR is good for gaming and entertainment too. The future of VR is uncertain and hard to predict as to how far this technology will take us. There is a lot of research and effort in process for the development of VR in all areas of endeavor and, someday, it will become an alternative option to creating real-world experiences in our daily life for different areas.

20.4 Application of VR in safety of smart manufacturing

The high number of injuries and fatalities in mining sector in US has been reported yearly, and VR is poised to reduce the injury rate in the manufacturing sector [17]. VR can simulate the production process, assembly line configurations, and highlight potentially dangerous situations. The virtual environments engage and bring effective training for the safety of the workers. How does this work? In smart manufacturing, the data of the machines, tools, environments having different sensitivity indices towards temperature, pressure, humidity and many more important parameters will be gathered automatically in the big data system. The data from different aspects of the manufacturing process can be developed into VR and provide training to the workers where they are trained on using a tool, procedure, or machines in a particular environment. There are several companies that have been working with VR for their company employees' safety in mind. For instance, the manufacturing of Brightline in supplementing the worker training at refinery and petrochemical facilities [18] is one example of VR in manufacturing. A total of 130 tasks in simulation is helping to ensure the workers obtained sufficient training to minimize injury or death. The main arena of activity, operations, for hands-on training in the field is challenging due to the poor potential outcomes, in the case where there are any inaccurate executions of a process. However, to provide training through simulation of the operation process in manufacturing is now possible with VR. Furthermore, the VR training to improve the safety of smart industry is enabling the workers to consult experts in the field, in real-time, regardless of location and distance, especially helpful in conducting highly complex tasks.

Smart manufacturing may remain bogged with heavy equipment, vehicles and hazardous substances that bring negative effects on the workers' wellness and reduce in-site work performance [19]. Besides, facility layout planning in VR allows the workers to gain a circumspective idea of the workplace to provide a better ergonomic effect and healthy distancing for the positioning of the equipment and workspaces. Moreover, the manufacturer can simulate the production rate by deciding on the VR-supported layout to predict the scenario when to alter the production plan and evaluate the impact of every decision in optimizing the production rate and minimizing the injury risk. This is essential for the industry that deals with high temperature or highly inflammable products.

In a nutshell, VR technology in smart manufacturing has enabled the worker training for complex tasks and minimized injuries and fatality to be possible. This is achieved through the dangerous scene simulation into the VR system and training in different settings with the confirmed actual scene data. The facility and layout of the floorplan in the VR system is providing the workers with a more ergonomic way of reducing stress in working with heavy machinery or any facility. VR training can provide great safety, impart skill training and minimize errors in real operation [20].

20.5 What is AR?

AR is an interactive experience of a real-world environment. This technology is defined as a system that incorporates three basic features which are combining the real and virtual worlds, real-time interaction, and the exact 3D location of virtual and real objects to produce an imaginative reality rivalling the actual world in experiencing it. AR is one of the biggest technology trends right now as AR-ready smartphones are becoming more accessible around the world. The most famous example of AR technology is the mobile app, Pokemon Go, that was released in 2016 and has been receiving astounding feedback from different age range users. For this game, the user can locate and capture Pokemon characters that pop up in the real world in your garden, near the fountain, in the shop, in condos or anywhere that is possible for the characters to pop up [21]. AR can alter one's perception of the real-world environment while VR immerses the user in the virtual world completely.

20.6 Principle of AR

AR interfaces convert the surrounding environment into a digital world by placing virtual objects in the real world and in real-time. There are three different categories of AR tools: games such as Pokemon Go, for example; AR browsers that aim at enriching the camera display with contextual information like the label pops from the phone while you point to the object in the shopping mall to display its name, ingredient or calories depending on the parameter that is programmed into the software; and 3D viewers allowing the users to place life-size 3D models in our environment with or without the use of trackers [22].

AR is available to be used on mobile devices, with which the viewer can see the 3D models from smartphones and tablets. Another display AR method is on PC or connected TV players where the AR is incorporated through the webcam and relayed through the screen. There is an option on using the AR with head-mounted displays, glasses, and lenses [23].

20.7 Application of AR

AR has highly advanced applications in creating real-world experiences and obtains its high demand in the market on its different applications. This interactive technology which can be used with smartphones has become an impressive tool for the industry and public. The thriving world of virtual real-life experiences in AR is a significant trend that can be traced to the areas of education, automotive industry, healthcare, retail, tourism, construction, entertainment, and different product types of manufacturing.

Application of AR in education and e-learning applications has been changing the traditional classroom with AR-supported workbooks, textbooks, story books and many more different book types to be found in the bookstore at a reasonable price. The benefits taking effect from AR in the educational setting provide learners with opportunities to enhance their understanding of different knowledge bases through the AR-endowed interactivity with and engagement in a virtual world which enhances the recollection and remembering of newly learned knowledge. In the digital world, the focus of the mind of children revolving around games and entertainment more than on studies has been a great concern for the parents. Therefore, AR technology with its capability to recreate pages from lessons in a way that fosters creativity and knowledge assimilation can minimize their mind's distraction and increase concentration and faster pickup of learning. For example, AR apps for learning animals from the book enable displaying of 3D graphic images of animals appear on top of the smartphone camera while it points to a page related to the animal from the book [24]. Some of the Zoo Apps have integrated the actual animal sound to make the experience more life-like. Else, the detailed information can appear from the camera screen about the animal that is shown in a 2D-page image from a book. Multisensory learning through AR by seeing and hearing at the same time is stated to improve the learning capacity of the students [25]. However, the limitation in the wide use of AR in education is the internet connectivity issue because AR requires a stable internet connection. Another concern is about the memory capacity and battery usage of the phone, which is limited; therefore, a made-for-AR phone for AR learning is required.

Application of AR in the automotive industry has transformed the way we traditionally do car prototyping and car assembly in car manufacturing and conduct business in showrooms [26]. The first car company to employ AR technology in business is Mercedes, which launched its first virtual showroom in the year 2017 [27]. The AR app available on smartphones enables the customers to take a closer look at the car model in the 3D presentation showing them the capabilities of the car. On the other hand, Volkswagen, another branded car company, has utilized AR technology in the service departments to help technical workers to speed up the car maintenance performance. The high demand for AR technology in the automotive industry has given the customer the expectation of increased safety through the addition of features on assisting navigation and more technical functions to minimize human mistakes while driving. The windshield AR projected through the heads-up display (HUD) can project the AR images for the car as well as the connected sensors so that the driver can see all information he needs, like speed limit or slow-moving vehicles, for a safe ride without taking his eyes off the road. With a browser – that is being incorporated in the car industry – the customer can view the car in 3D which experience the buyer uses to enrich his shopping behavior in the car showroom. Given car manufacturers are commonly applying AR to optimize the assembly process [28], where the AR integrates the necessary information for the assembly line workers in a step-by-step instruction on the screen of AR projection glasses, it follows that the process of assembling can be faster while minimizing human errors.

Application of AR in healthcare has combined the virtual and physical worlds to enrich the patient's hospitalization experience with the working of medical devices and drugs interacting with the patient body [29]. AR technology has helped in increasing the accuracy and efficiency of surgery given the surgeons can augment the scan results from the patient's body with the AR capabilities to assist the surgeons in real-time when performing the surgery [30]. Besides, AR has improved the accuracy with which nurses and surgeons push injections by showing the exact location of the vein by connecting to a handheld scanner projecting over the skin showing images where the vein is [31]. Furthermore, AR can assist the doctor in determining the symptoms accurately. Another application of AR can project the data of the patient during their visits.

Application of AR in retail gets its popularity in checking out merchant products. The customer can interact with the products virtually getting exact information on products and choosing those with better chances of matching with their requirements as merchants get to know their customers better and become more capable of making suggestions for buying [32]. Besides, another advantage with AR retail is helping the merchant to reduce the overhead and shopping costs with AR, where the merchants can permit the customer to engage with an infinite amount of digital products and their variations. This can reduce the storage space and cost to the merchant. The customer can do away with the long time spent in manually selecting products through the faster AR procedure of selecting products as the merchant gives vouchers or discounts to appreciate and encourage more purchases using AR. This helps in increasing the sales of the merchants. Further, fitting rooms for clothes shoppers are now on AR. The virtual fitting room for customers shows him his looks by trying on the selected clothing on them without physically having to put it on, which saves time and manpower in collecting the clothes outside the fitting room [33]. Some retail stores have applied AR into innovative and modern ways to build brand awareness with the customer by projecting the AR graphic in front of their shop.

The application of AR has become popular in tourism where the benefits of AR and VR in the travel industry are more or less the same [34]. The interactive experience for the tourists to view the 3D image of the hotel room before booking and estimating its size and environment increases the satisfaction of the tourists and with theirs least likely to complain about the difference between the photos from the internet and the actual environment of the hotel room, amenities and many more aspects. Some of the hotels set the hotel lock into the AR for the customer with Bluetooth connection, providing maps, reviews and vouchers to the tourists. The AR even innovates by adding animals and fishes in the AR app at the hotel where the tourists can take part in family games, taking photos with the AR avatar by adding to the actual hotel-stay experience that brings more fun to the vacation.

The application of AR in construction has helped by constructing 3D models combining building data where the data is displayed digitally over the user's physical surroundings in real-time which the user sees with the AR device or special helmet [35]. With similar purposes of AR in construction as VR, enabling the designers and clients to preview the buildings based on its 3D models, AR lets clients get a realistic view of the project outcomes and make changes before the construction begins. AR gives a chance to the teams to conduct the examination using the details and components of the structure before it is executed, prevent over-expending on the budget and identify mistakes before it happens preventing delays in schedule. Besides, a lot of machinery is involved in carrying out building projects that may require hours of training to come to a level where they can be used without accidents or erroneous operations can be avoided. Sometimes, wasteful expenditures in waiting for an expert to come to resolve some issues or investigate an accident may cause safety issues or damages whereas AR assists in resolving this matter in no time by projecting the necessary information and steps that are prerecorded or by collaborating with the expert real-time in another place to repair the machine [36]. Consequently, AR technology is lowering the training cost, increasing efficiency and perhaps minimizing downtime.

20.8 Application of AR in safety of smart manufacturing

Smart manufacturing becomes essential where safety is a prime concern and is forcing businesses to transform into the digital era through implementing a combination of various technologies to improve productivity, safety and even increase profit. AR as an immersive technology plays a part in bringing a significant change to companies combining artificial intelligence (AI) and Internet of Things (IoT) [37]. AR has advanced the smart manufacturing in production, maintenance, training and streamlined operations. Product development with AR in different areas has been discussed in which the designer uses AR to make realistic ideation from the concept stage through collaboration with the clients and fast modification through this immersive technology to product development. The complex tasks in the assembly process are now incorporated into AR to provide information to the workers whenever there is a requirement to correct a mistake or carry out the procedures without misfires; and in training, AR lets one practice multi-times over without the machine giving out [38]. The AI may help to predict and prevent most of the accidents through data collection and to project future development efforts with instructions to the AR to reduce the downtime with real-time monitoring and ongoing analytics. Quick customer support without waiting for a long time has shed light on AR technology in providing real-time support instructions with a wearable camera pointing to the instrument.

AR is increasing the safety of smart manufacturing as it has been shown to do effectively by helping the workers to retain more knowledge through informative information projected with the physical instrument [38]. This makes them less likely to make dangerous mistakes. The analytical data of the machinery can be relayed to the AR and connected to an alert-giving sensor to trigger the real-time alert when procedural steps go wrong. The safety real-time alerts potentially increase the safety of the workplace. Besides, AR has also increased the accessibility of the workers to the remote experts to resolve some unique safety risks that might need experienced workers who may not be working in the industry at that moment. Although the safety of the smart manufacturing is well protected with AR technology, however, the investment required to build AR in manufacturing is still entangled with the issue of whether the technology is worthwhile to pursue given the cost.

20.9 What is MR?

AR and MR are both considered immersive technologies, but the difference between both is that the MR extends AR that allows real and virtual elements to interact in an environment. MR is the intermediate range from total immersive VR to partial integration of the real world with virtual graphic display. These configure the extreme of the virtuality continuum from the proper environment and VR in Figure 20.2. The technology has started its progress in early 1990s, was developed at the US Air Force's Armstrong Laboratories and the term was officially coined in the year of 1994 by Paul Milgram and Fumio Kishino [39,40].

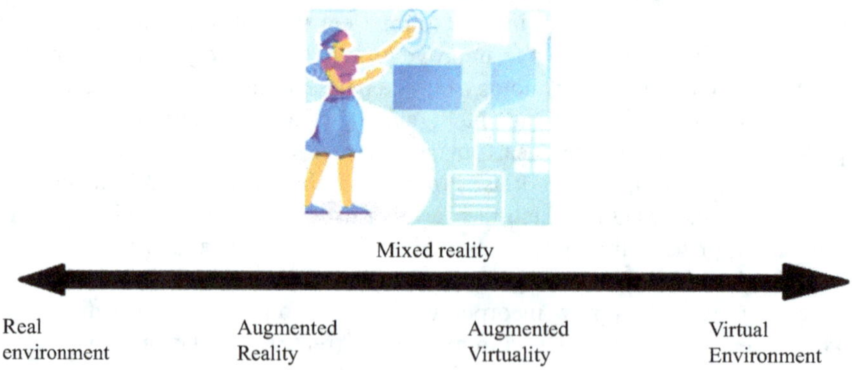

Figure 20.2 The extrema of the virtuality continuum

20.10 Principle of MR

The development of the devices that contain sensors to build the MR is focused on capturing information about the physical world around the user which is beyond the sensors in the mobile phone. MR gives a model of the world to integrate the device and place the content on surfaces and hide content under tables or warn users if they're about to walk into a wall in VR. Eye-tracking can support natural interaction and allow disabled people to navigate by using just their eyes with MR. Besides, MR includes access to camera data that allows applications to detect and track objects in the world, like equipment being repaired or props being used in a game [41].

20.11 Application of MR

MR has given a promising future application to the community, with its assimilation of experiences between total virtual and partial physical world engagement. MR has its application quite like the AR; however, MR is a better application in education at the moment.

MR application in education provides an immersive and effective learning experience to children and young learners. The learners can interact with the environment produced by a mix of real and VR by MR requiring the learners to wear a head-mounted display and motion controller [42]. The interaction of the human–computer interface from MR allows learners to touch and manipulate objects which make learning easier with abstract concepts. Besides, MR enhances the collaboration between students, teachers, and experts to work together in a piece of art or workshop, which can guide thousands of learners in a museum, besides its ability to engage learners from all over the world at the same time.

MR technology is working along with the HoloLens device that has gained appreciation in the automotive industry. MR is always the key tool for the designer or the methodology for training new staff to view the 3D design, naming the parts, and more details about the new model on its aerodynamics flow. Meanwhile, this technology provides better support to technical staff in engaging the service professional. The challenges that are faced by the automotive industry cover the inadequate decision support for engineering design, where the physical mock-ups are time-consuming and costly to undertake to do the projects according to norms. Besides, the inconsistency in the complex automotive assembly can result in missed processes and bring warranty issues to this industry [43]. MR can reduce the stress and provide sufficient training or on-the-spot requesting of help from the expert in calibrating or resolving a challenge in a complex situation. MR can show the real-time aerodynamic model, which optimizes the efficiency of the design on its wind drag and fuel consumption efficiency. MR technology has stepped permanently into the automotive industry as a cost-effective tool of manufacturing automobiles [44].

The application of MR in healthcare has made the amazing journey from enabling training nurses and medical students to engage more deeply into the module through immersive learning [45] to simulate multiple rare situations in MR for real clinical settings. Besides, the MR with patients' hologram helps in training healthcare professionals to diagnose and give to professional's medical issues [46]. A programme that was developed by SphereGen technology namely, 'The Learning Heart' enables users to view the hologram of the heart from all directions and examine its functions, providing learning immersive environment for medical students [47]. MR also helps surgeons locate and reconnect major blood vessels to optimize the accuracy of the surgical outcome that brings benefit to the patients. Another application of MR in healthcare is the revolutionizing surgery, where it supports surgeons to get patients' information on anatomical pictures in 3D during the surgery where HoloLens is a standalone computer worn like a helmet. MR is constantly improving its quality in healthcare and medical training as it positively affects the medical industry [48].

The application of MR in retail has grown in its popularity in recent years because this immersive technology has provided a high visual and realistic customer experience [49]. The technology allows customers to e-try to feel the product before buying with ease and convenience characteristic to online stores. Another application of MR in retail is mixed commerce kiosks which can be placed in areas visited by many visitors and allow them to select, view and try the products. For instance, Ralph Lauren [50] has employed this technology in their business where the customers will purchase through texting messages. In the rapidly changing world, businesses must be extremely flexible to survive including retail; however, there are still challenges for retail to employ this technology which is discussed in VR and MR such as the costliness of the apparatus that may burden the business owner and may not be effective for all the customers who still prefer the online to test and try the product [51].

The application of MR in tourism has provided good information and an immersive environment to the tourism industry. Most of the tours are built based on VR and AR into MR. The technology has provided good information to the visitors and even provided education on the history of the place while visiting the real scene or virtual scene. Besides, tourists can have better planning on the hotel booking and achieve a higher satisfactory quotient of room booking during the tour visit where the MR allows the virtual scene to be projected with the detailed information on it [52]. However, this technology is still a great investment for the tourism industry.

Application of MR and holographic technology has a high growth market in coming years in the construction industry. The MR provides the platform to the team to go through the entire project from idea to implementation together, viewing the 3D models which can prevent schedule delays, reduce overrun of cost, and identify errors before they happen [53]. Further, the MR has provided a comprehensive on-job inspection facility by projecting detailed information on the helmet while walking through the construction sites [54]. This optimizes the project monitoring and its successful completion that comes along with the other benefit of higher safety during operation offered to the industry.

The application of MR in multiple types of manufacturing has generally brought in advantages by creating virtual instruction through data collection. Training and education about the operational procedures or complex assembling tasks may be accessible by the workers through training [55]. Meanwhile, MR is not solely a passive observer of the data in the workplace but is also acquiring data through the sensors that connect to Bluetooth devices and IoT sensors. The treasure trove of data, with data collected from time to time, can merge with the predictive model, through AI, and hence provide a more promising production and quality to the product [56]. Alongside this is the increase in safety levels like AR and VR.

20.12 Application of MR in the safety of smart manufacturing

Industry 4.0 encourages the rapid development of MR has contributed to the safety of smart manufacturing through identifying the fault and by alerting the operators with an alarm fitted on a machine to give notifications [57]. In addition, the MR can provide the immediate remedy in resolving the issue, perhaps including calling the expert to come into the scene through MR in their place. Besides, training the workers to familiarize themselves with the operation procedures and floor map through MR, the training can be taken multiple times until the workers are masters of the necessary skill to reduce mistakes and preserve a higher safety rate on the factory floor. The real-time data projection and acquisition must allow the predictive model of the MR in the industry to be more accurate and safer.

20.13 What is haptic technology and its importance in VR/AR/MR in smart manufacturing?

Haptic technology has integrated and equipped the VR/AR/MR with senses, which includes the perception of touch and different digital touch sensations [58]. The purpose of haptic technology is to enhance user engagement, immersion, and to enhance the sense of presence. Besides, the haptic technology can reduce cybersickness which can be fatigue or dizziness caused by an imbalance of perceptions between reality and the virtual world [59]. For user interface feedback and confirmation, the user gives tactile feedback to confirm the interaction. Recoiling weapon shots in the extended reality (XR) game to illustrate user immersion when they feel the haptic feedback can enrich haptic textures with specific perceptions. Further, another important feature of haptics for the VR/AR/MR technology is to bring in the remote presence such as hugging and shaking hands [60]. In increasing the awareness of the user in the virtual world are the context and virtual environment that gives directions to the user or warns him about something that might happen next which can be notified to the user through the sense of touch.

Conclusion

The future direction for haptic technology might aim at the development of multimodal haptic stimuli to accommodate the perceptual capabilities of humans and hence bring up the economy in design and manufacturing the haptic technology.

References

[1] T. Baumgartner, D. Speck, D. Wettstein, O. Masnari, G. Beeli, and L. Jäncke, Feeling present in arousing virtual reality worlds: prefrontal brain regions differentially orchestrate presence experience in adults and children, *Front. Hum. Neurosci.* 0 (2008) 8. https://doi.org/10.3389/NEURO.09.008.2008.

[2] E. Tsekleves, A. Warland, C. Kilbride, I. Paraskevopoulos, and D. Skordoulis, The use of the Nintendo Wii in motor rehabilitation for virtual reality interventions: a literature review, *Intell. Syst. Ref. Libr.* 68 (2014) 321–344. https://doi.org/10.1007/978-3-642-54816-1_17.

[3] F. Mérienne, Virtual reality: principles and applications, in Encyclopedia of Computer Science and Technology, 2nd ed., 2017, pp. 1–11. https://doi.org/10.1081/E-ECST2-140000194.

[4] J. Grubert, L. Witzani, E. Ofek, M. Pahud, M. Kranz, and P.O. Kristensson, Text entry in immersive head-mounted display-based virtual reality using standard keyboards, in 25th IEEE Conference on Virtual Reality and 3D User Interfaces, 2018, pp. 159–166. https://doi.org/10.1109/VR.2018.8446059.

[5] L. Sidenmark, C. Clarke, X. Zhang, J. Phu, and H. Gellersen, Outline pursuits: gaze-assisted selection of occluded objects in virtual reality, in Conference on Human Factors in Computing Systems – Proceedings, 2020. https://doi.org/10.1145/3313831.3376438.

[6] Y. Slavova and M. Mu, A comparative study of the learning outcomes and experience of VR in education, in 25th IEEE Conference on Virtual Reality and 3D User Interfaces, VR 2018 – Proceedings, 2018, pp. 685–686. https://doi.org/10.1109/VR.2018.8446486.

[7] G. Lawson, D. Salanitri, B. Waterfield, Future directions for the development of virtual reality within an automotive manufacturer, *Appl. Ergon.* 53 (2016) 323–330. https://doi.org/10.1016/J.APERGO.2015.06.024.

[8] The Thousand Faces of Virtual Reality – Google Books, (n.d.). https://books.google.com.my/books?hl=en&lr=&id=FU-QDwAAQBAJ&oi=fnd&pg=PA153&dq=VR+technology+has+to+allow+the+workers+to+obtain+sufficient+training+in+the+virtual+environments+and+make+mistake+safely+to+gain+experience.+Meanwhile,+VR+brings+safety+to+the+workers,+allowing+the+to+practise+in+any+dangerous+situation&ots=52k8WbplE5&sig=qSwqh2v0YtHJ_baghF4A7DDZqYY&redir_esc=y#v=onepage&q&f=false (accessed September 30, 2021).

[9] M. Javaid, A. Haleem, Virtual reality applications toward medical field, *Clin. Epidemiol. Glob. Heal.* 8 (2020) 600–605. https://doi.org/10.1016/J.CEGH.2019.12.010.

[10] Karuna Labs | VentureRadar, (n.d.). https://www.ventureradar.com/organisation/KarunaLabs/acc51f44-4d2d-4b26-b912-e15b6aaecc0e (accessed September 30, 2021).

[11] D.E. Levac and H. Sveistrup, Motor learning and virtual reality, *in Virtual Reality for Physical and Motor Rehabilitation*, 2014, pp. 25–46. https://doi.org/10.1007/978-1-4939-0968-1_3.

[12] V.K. Ketoma, P. Schäfer, and G. Meixner, Development and evaluation of a virtual reality grocery shopping application using a multi-kinect walking-in-place approach, *Adv. Intell. Syst. Comput.* 722 (2018) 368–374. https://doi.org/10.1007/978-3-319-73888-8_57.

[13] R.L. Seligmann, Creating a Mobile VR Interactive Tour Guide, 2018. http://www.theseus.fi/handle/10024/144759 (accessed September 30, 2021).

[14] S.M. Brondoni, Global tourism management. Mass, experience and sensations tourism, Symphonya. *Emerg. Issues Manag.* 0 (2016) 7–24. https://doi.org/10.4468/2016.1.02BRONDONI.

[15] R. de Klerk, A.M. Duarte, D.P. Medeiros, J.P. Duarte, J. Jorge, and D.S. Lopes, Usability studies on building early stage architectural models in virtual reality, *Autom. Constr.* 103 (2019) 104–116. https://doi.org/10.1016/J.AUTCON.2019.03.009.

[16] S. Chien, S. Choo, M.A. Schnabel, and W. Nakapan, Immersive visualization of building infor-mation models, in *Usage and Future Possibilities During Design and Construction*, 2016, pp. 673–682.

[17] M.T. Filigenzi, T.J. Orr, and T.M. Ruff, Virtual reality for mine safety training, *Appl. Occup. Environ. Hygiene*. 15 (2010) 465–469. https://doi.org/10.1080/104732200301232.

[18] Sponsored Content | Game On: Refinery and Petrochemical Workers Train with AR & VR - POLITICO, (n.d.). https://www.politico.com/sponsored-content/2020/01/safety-with-virtual-reality (accessed September 30, 2021).

[19] A. Badri, B. Boudreau-Trudel, and A.S. Souissi, Occupational health and safety in the industry 4.0 era: a cause for major concern?, *Saf. Sci.* 109 (2018) 403–411. https://doi.org/10.1016/J.SSCI.2018.06.012.

[20] E. Matsas and G.-C. Vosniakos, Design of a virtual reality training system for human–robot collaboration in manufacturing tasks, *Int. J. Interact. Des. Manuf.* 11 (2015) 139–153. https://doi.org/10.1007/S12008-015-0259-2.

[21] J. Paavilainen, J. Stenros, H. Korhonen, E. Koskinen, K. Alha, and F. Mäyrä, The Pokémon GO experience: a location-based augmented reality mobile game goes mainstream, in Proceedings of the 2017 CHI Conference on Human Factors in Computing Systems, 2017. https://doi.org/10.1145/3025453.

[22] J. Carmigniani, B. Furht, M. Anisetti, P. Ceravolo, E. Damiani, and M. Ivkovic, Augmented reality technologies, systems and applications, *Multimed. Tools Appl.* 2010 511. 51 (2010) 341–377. https://doi.org/10.1007/S11042-010-0660-6.

[23] C. Geiger, B. Kleinnjohann, C. Reimann, and D. Stichling, Mobile AR4ALL, in Proceedings IEEE and ACM International Symposium on Augmented Reality, 2001, pp. 181–182. https://doi.org/10.1109/ISAR.2001.970532.

[24] I. Alakärppä, E. Jaakkola, J. Väyrynen, and J. Häkkilä, Using nature elements in mobile ar for education with children, in Proceedings of the 19th International Conference on Human-Computer Interaction with Mobile Devices and Services, 2017. https://doi.org/10.1145/3098279.

[25] S. Yuen, G. Yaoyuneyong, and E. Johnson, Augmented reality: an overview and five directions for AR in education, *J. Educ. Technol. Dev. Exch.* 4 (2011) 11. https://doi.org/10.18785/jetde.0401.10.

[26] A.Z.A. Halim, Applications of augmented reality for inspection and maintenance process in automotive industry, *J. Fundam. Appl. Sci.* 10 (2018) 412–421. https://doi.org/10.4314/jfas.v10i3S.35.

[27] AR in the Automotive Industry — Jasoren, (n.d.). https://jasoren.com/ar-in-automotive/ (accessed September 30, 2021).

[28] J.L. Gabbard, G.M. Fitch, and H. Kim, Behind the glass: driver challenges and opportunities for AR automotive applications, *Proc. IEEE.* 102 (2014) 124–136. https://doi.org/10.1109/JPROC.2013.2294642.

[29] R. Kohli and E.D. Hoadley, Healthcare, *Inf. Syst. Action Res.* (2007) 241–253. https://doi.org/10.1007/978-0-387-36060-7_10.

[30] L.T. De Paolis and V. De Luca, Augmented visualization with depth perception cues to improve the surgeon's performance in minimally invasive surgery, *Med. Biol. Eng. Comput.* 57 (2018) 995–1013. https://doi.org/10.1007/S11517-018-1929-6.

[31] E. Watanabe, M. Satoh, T. Konno, M. Hirai, and T. Yamaguchi, The trans-visible navigator: a see-through neuronavigation system using augmented reality, *World Neurosurg.* 87 (2016) 399–405. https://doi.org/10.1016/J.WNEU.2015.11.084.

[32] E. Cruz, S. Orts-Escolano, F. Gomez-Donoso, et al., An augmented reality application for improving shopping experience in large retail stores, *Virtual Real.* 23 (2018) 281–291. https://doi.org/10.1007/S10055-018-0338-3.

[33] D. Protopsaltou, C. Luible, M. Arevalo, and N. Magnenat-Thalmann, A body and garment creation method for an internet based virtual fitting room., *Adv. Model. Animat. Render.* (2002) 105–122. https://doi.org/10.1007/978-1-4471-0103-1_7.

[34] D.-I. Han, T. Jung, A. Gibson, Dublin AR: implementing augmented reality in tourism, *Inf. Commun. Technol. Tour.* 2014. (2013) 511–523. https://doi.org/10.1007/978-3-319-03973-2_37.

[35] C. Kwiatek, M. Sharif, S. Li, C. Haas, and S. Walbridge, Impact of augmented reality and spatial cognition on assembly in construction, *Autom. Constr.* 108 (2019) 102935. https://doi.org/10.1016/J.AUTCON.2019.102935.

[36] B. Bhattacharya, J. Oliver, R. Radkowski, S. Gilbert, and M. Dorneich, Automatic Generation of Augmented Reality Guided Assembly Instructions Using Expert Demonstration, (n.d.).

[37] F.R. Ramalho, A.L. Soares, and A.H. Almeida, Immersive systems in human-centered manufacturing: the informational dimension, *IFIP Adv. Inf. Commun. Technol.* 598 (2020) 297–307. https://doi.org/10.1007/978-3-030-62412-5_25.

[38] U.H. Govindarajan, A.J.C. Trappey, and C. V. Trappey, Immersive technology for human-centric cyberphysical systems in complex manufacturing processes: a comprehensive overview of the global patent profile using

collective intelligence, *Complexity*. 2018 (2018). https://doi.org/10.1155/2018/4283634.

[39] S. Vi, T.S. da Silva, and F. Maurer, User experience guidelines for designing HMD extended reality applications, *Lect. Notes Comput. Sci. (Including Subser. Lect. Notes Artif. Intell. Lect. Notes Bioinformatics)*. 11749 LNCS (2019) 319–341. https://doi.org/10.1007/978-3-030-29390-1_18.

[40] (13) (PDF) New Poetics of Augmented Space: Re-conceptualise Augmented Space Based on Fundamentals of Augmented Reality, (n.d.). https://www.researchgate.net/publication/319472206_New_Poetics_of_Augmented_Space_Re-conceptualise_Augmented_Space_Based_on_Fundamentals_of_Augmented_Reality (accessed September 30, 2021).

[41] E. Barba and R.Z. Marroquin, A primer on spatial scale and its application to mixed reality, in *Proceedings of the 2017 IEEE International Symposium on Mixed and Augmented Reality*, ISMAR 2017, 2017, pp. 100–110. https://doi.org/10.1109/ISMAR.2017.27.

[42] Y.M. Tang, K.M. Au, H.C.W. Lau, G.T.S. Ho, and C.H. Wu, Evaluating the effectiveness of learning design with mixed reality (MR) in higher education, *Virtual Real.* 24 (2020) 797–807. https://doi.org/10.1007/S10055-020-00427-9.

[43] S. Borsci, G. Lawson, and S. Broome, Empirical evidence, evaluation criteria and challenges for the effectiveness of virtual and mixed reality tools for training operators of car service maintenance, *Comput. Ind.* 67 (2015) 17–26. https://doi.org/10.1016/J.COMPIND.2014.12.002.

[44] M. Bordegoni and G. Caruso, Mixed reality distributed platform for collaborative design review of automotive interiors, Virt. Phys. Prototyp. 8 (2012) 243–259. https://doi.org/10.1080/17452759.2012.721605.

[45] T. Stretton, T. Cochrane, and V. Narayan, Exploring mobile mixed reality in healthcare higher education: a systematic review, *Res. Learn. Technol.* 26 (2018) 2131–2131. https://doi.org/10.25304/RLT.V26.21301.

[46] E. Bottani, F. Longo, L. Nicoletti, et al., Wearable and interactive mixed reality solutions for fault diagnosis and assistance in manufacturing systems: implementation and testing in an aseptic bottling line, *Comput. Ind.* 128 (2021) 103429. https://doi.org/10.1016/J.COMPIND.2021.103429.

[47] SphereGen Technologies – Making Technology Relevant, (n.d.). https://www.spheregen.com/ (accessed October 1, 2021).

[48] J.T. Verhey, J.M. Haglin, E.M. Verhey, and D.E. Hartigan, Virtual, augmented, and mixed reality applications in orthopedic surgery, *Int. J. Med. Robot. Comput. Assist. Surg.* 16 (2020) e2067. https://doi.org/10.1002/RCS.2067.

[49] S. Jain and D. Werth, Current state of mixed reality technology for digital retail: a literature review, *Lect. Notes Comput. Sci. (Including Subser. Lect. Notes Artif. Intell. Lect. Notes Bioinformatics)*. 11588 LNCS, 2019, pp. 22–37. https://doi.org/10.1007/978-3-030-22335-9_2.

[50] Apple's mixed reality headset might require an iPhone to work | Australia Daily News, (n.d.). https://ralph-lauren.net.au/apples-mixed-reality-headset-might-require-an-iphone-to-work/ (accessed October 1, 2021).

[51] N. Vaidyanathan, Augmented reality in retail – a case study: technology implications to utilitarian, aesthetic and enjoyment values, in Proceedings of the 2020 4th International Conference on Virtual and Augmented Reality Simulations, 2020. https://doi.org/10.1145/3385378.

[52] F. Debandi, R. Iacoviello, A. Messina, et al., Enhancing cultural tourism by a mixed reality application for outdoor navigation and information browsing using immersive devices, *IOP Conf. Ser. Mater. Sci. Eng.* 364 (2018) 012048. https://doi.org/10.1088/1757-899X/364/1/012048.

[53] Wang, Xiangyu, and Marc Aurel Schnabel, eds. Mixed reality in architecture, design, and construction. Springer Science & Business Media, 2008.

[54] P. Raimbaud, R. Lou, F. Merienne, F. Danglade, P. Figueroa, and J.T. Hernandez, BIM-based mixed reality application for supervision of construction, in IEEE Conference on Virtual Reality and 3D User Interfaces, VR 2019 – Proceedings, 2019, pp. 1903–1907. https://doi.org/10.1109/VR.2019.8797784.

[55] M. Juraschek, L. Büth, G. Posselt, and C. Herrmann, Mixed reality in learning factories, *Procedia Manuf.* 23 (2018) 153–158. https://doi.org/10.1016/J.PROMFG.2018.04.009.

[56] S. Doolani, C. Wessels, V. Kanal, et al., A review of extended reality (xr) technologies for manufacturing training, *Technology* 8 (2020) 77 (Page 77, 78). https://doi.org/10.3390/TECHNOLOGIES8040077.

[57] D.B. Espíndola, L. Fumagalli, M. Garetti, et al., A model-based approach for data integration to improve maintenance management by mixed reality, *Comput. Ind.* 64 (2013) 376–391. https://doi.org/10.1016/J.COMPIND.2013.01.002.

[58] P. Wang, X. Bai, M. Billinghurst, et al., Haptic feedback helps me? in *A VR-SAR Remote Collaborative System with Tangible Interaction*, vol. 36, 2020, pp. 1242–1257. https://doi.org/10.1080/10447318.2020.1732140.

[59] J. Plouzeau, D. Paillot, J.-R. Chardonnet, and F. Merienne, Effect of Proprioceptive Vibrations on Simulator Sickness During Navigation Task in Virtual Environment, 2015. https://sam.ensam.eu/handle/10985/10422 (Accessed October 1, 2021).

[60] J.B.F. van Erp and A. Toet, Social touch in human–computer interaction, *Front. Digit. Humanit.* 0 (2015) 2. https://doi.org/10.3389/FDIGH.2015.00002.

Chapter 21

Conserving environment using resources wisely with reduction of waste and pollution: exemplary initiatives for Education 4.0

Ng Khar Thoe[1], Masanori Fukui[2], Corrienna Abdul Talib[3], Tairo Nomura[4], Peter Chew Ee Teik[5] and Rajendra Kumar[6]

As the global attention gears towards the fourth Industrial Revolution (IR4.0), there is increasing awareness of siloed educational settings giving the push to revise curriculum focusing on Education 4.0 as a desired pedagogical approach that is aligned with IR4.0. Among the initiatives taken are the emphasis on the teaching of computational thinking (CT) skills, development as well as wise use of Science, Technology, Engineering, Arts, Mathematics (STEAM)-based projects and artificial intelligence (AI) age visualization tools for monitoring and evaluating activities. This chapter discusses the importance of raising awareness of conservation and wise use of resources (Conservation and Wise Use of Resources (ConWUR)) in the educational and industrial settings in achieving 'United Nation Sustainable Development Goals' (UN SDGs). It outlines exemplary practices in Japanese and Malaysian educational institutions that promote the wise use of resources by educational stakeholders and computational thinking skills in students aiming at bridging the education–industrial gap to make students future-ready in facing the increasingly challenging world. In this chapter, the exemplary Southeast Asia Minister of Education Organisation (SEAMEO) Learning Science and Mathematics Together (LeSMaT) Education 4.0 project initiative is elaborated. This chapter also discusses the development and evaluation for the future Human-Machine Interface (HMI) educational system promoting computational thinking (CT). In this chapter, the Scratchtopia Challenge as an exemplary initiative to promote CT at the elementary

[1]Training and Research Division, SEAMEO RECSAM, Penang, Malaysia
[2]Center for University Education, Tokushima University, Japan
[3]School of Education, Faculty of Social Sciences and Humanities, Universiti Teknologi Malaysia, Johore, Malaysia
[4]Faculty of Education, Clinical Study of Education, Saitama University, Japan
[5]PCET Ventures, Bukit Mertajam, Malaysia
[6]School of Engineering and Technology, Sharda University, India

level will be deliberated on. The challenging Internet of Things (IoT) with Arduino for data processing from the sensor as well as how IoT can be used for Sustainable Energy for All (SE4ALL) and reduction of waste are also illustrated in this chapter. The discussion on future directions to incorporate the use of AI age tools for monitoring and evaluation activities ends this chapter, where the development of apps, use of digital tools as well as illustration of exemplary modules that are part of the initiatives of the author(s) towards sustainable living through ConWUR and waste reduction are some of the future efforts or conservation activities to promote sustainable living through LeSMaT Education 4.0 project initiative.

21.1 Introduction and SEAMEO LeSMaT Education 4.0 project initiative

SEAMEO 'Learning Science and Mathematics Together' in a borderless world [abbreviated as LeSMaT (Borderless)] took its name from the SEAMEO Borderless School project that was an area identified under the vision of Golden SEAMEO 'Basic Education and Student Networking' that was endorsed by HOM in 2012. This initiative is in line with SEAMEO Priority Area No. 7 (adopting a twenty-first-century curriculum), in addition to No. 5 (revitalising teacher education) and No. 3 (resiliency in the face of emergencies). In the SEAMEO RECSAM's 11th Five-Year Development Plan (July 2020–June 2025), this project has started piloting tools related to Education 4.0 with schools in both urban and rural areas to reach out to a wider audience in fulfilling Priority Area No. 2 (addressing barriers to inclusion) as well as SDGs No. 4 (quality education), No. 10 (reduced inequalities), No. 13 (climate action) and No. 17 (partnership for the goals). UN SDGs emphasize the conservation of resources for sustainability as reflected in No. 6 (clean water and sanitation), No. 11 (sustainable cities and community), and No. 12 (responsible consumption and production). The UN Decade of 'Sustainable Energy for All' (UN-SE4ALL) 2014 to 2024 also emphasizes SDGs No. 7 (affordable and clean energy).

From 2021 to 2025, R&D activities in line with current trends are invested with the need to and will be conducted to inform pedagogies that promote 'Sustainable Energy for All' (*SE4ALL*), 'Conservation and Wise Use of Resources' (*ConWUR*) (in which 'Water and Values Education' was also adapted), 'Telecare and Healthy Lifestyle' (*TeleHeal*) as well as 'Climate Awareness and Disaster Risk Reduction EDucation' (*CADRRED*) as sub-themes of LeSMaT project initiatives. Awareness will be raised on SDG issues to use low-cost materials through learner-centred teaching, learning and assessment techniques. Apart from involving the SEAMEO countries already participated during the 10th Five-Year plan, i.e., Malaysia, Indonesia, Thailand, Philippines, Vietnam, Lao PDR, Cambodia, Myanmar and Singapore, various blended-mode activities including digital tools and hand-held technology to bridge the digital divide will be explored for the rest of the SEAMEO countries, i.e. Brunei Darussalam and Timor Leste. In line with governmental aspirations to integrate Education 4.0 in Science/Mathematics education, R&D activities were conducted with evidence-based output/deliverables in the form of revised curriculum materials that are shared as OERs, lesson examples, framework, presentations and publications.

The R&D activities are also in line with the concerns raised globally and nationally on 'Science, Technology, Reading, Engineering, Arts, Mathematics' (STREAM) education using digital tools and e-platform. The enhancement of science/mathematics/reading literacy and improved STREAM competency from an early age are crucial towards building a science and technology culture via self-directed lifelong learning important to promote human resource development in line with SEAMEO Priority No. 7, also SDGs Nos. 4 and 10. Literacy (an action with components being embedded into how information is consumed) is always a part of every content area, namely, mathematics, science, social studies, arts, music and reading, i.e., STREAM is the intentional alignment of standards within the identified areas for teaching and assessment activities focusing on application, creation and evaluation.

STREAM competency standards including research into current trends on gamification, edu-/entrepreneurship, financial literacy, statistical thinking and assessment techniques are also areas of concern in the new era of digital literacy. LeSMaT (Borderless) resources [with more Education 4.0 focused curriculum for future integrating technology such as Internet of Things (IoT), artificial intelligence (AI), computational thinking (CT), coding, design-based learning and engineering processes in project-based activities (PBA), augmented/virtual reality (AR/VR) for deeper understanding of abstract scientific and mathematical concepts] were revised for blended training programmes.

Digital tools including Youtube videos, e-books as well as e-platforms such as Edmodo, WebEx and Google Classroom web-conferencing tools have been used since 2017 when AR was first introduced through the e-course series to promote 'Learning Transdisciplinary Science Integrating Mathematics, Arts-Reading-Language-Culture, Engineering-Environmental Education-Entrepreneurship, Technology' (*LearnT-SMArET*), i.e., another sub-theme of LeSMaT (Borderless). Experts from India, Indonesia, Japan, Malaysia, Pakistan, Philippines, Singapore and Thailand, to name a few, with diverse fields of expertise related to LearnT-SMArET as especially technology-focused tools, were involved in delivering topics as aforementioned in the Centre's 11th Five-Year Development Plan. Emphasis was also given on revised versions of the curriculum prepared in presentation videos accessible from LearnT-SMArET e-course series. Prior to the lockdown due to pandemic outbreak, physical workshops were conducted to prepare e-modules and low-cost prototypes of the programmes. After participation in blended-mode input, participants were requested to invent artifacts by preparing innovative outputs to the learning process with sharing on social learning platforms such as Telegram and Edmodo. Competitions such as 'Heritage Immortalised: A nation building Minecraft championship 2021' that were launched by third co-author (URL: https://youtu.be/u7jwSugOAB8) was also conducted from the period of May–August 2021 organised in partnership with the collaborator. It is expected LeSMaT (Borderless) Education 4.0 initiative will promote thinking, technology and life (work/entrepreneurial/survival) skills by preparing students to become global players who are creative/innovative, enterprising, equipped

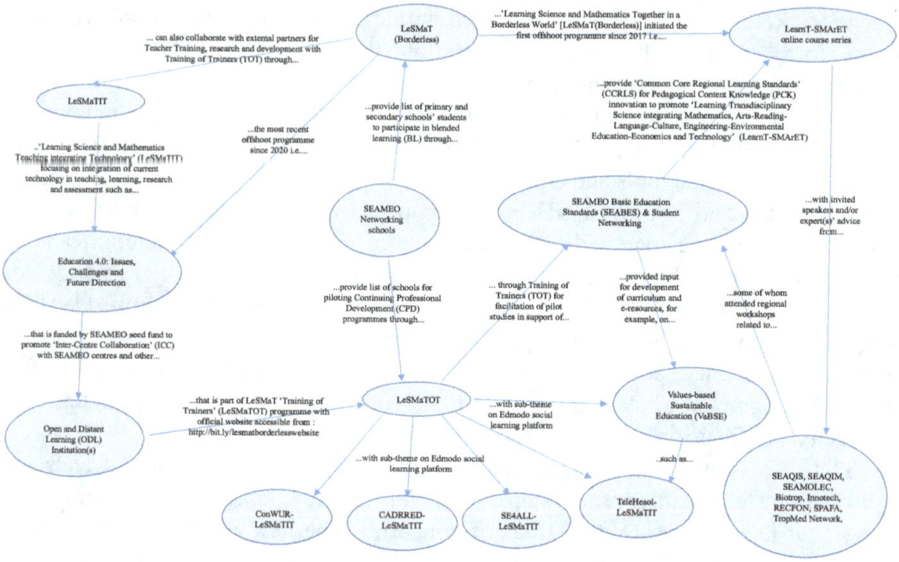

Figure 21.1 Conceptual framework of LeSMaT (Borderless) and its offshooting project-based programmes

with twenty-first-century skills in cross-cultural education and having been inculcated with ethics, good moral values with behavioural changes towards achieving Society 5.0 and Global Citizenship through blended-mode activities as well as technology-enhanced investigative project, inquiry-based and problem-based learning.

LeSMaT (Borderless) framework and some learning output supported by evidence-based findings to promote STREAM literacy are exemplified in this book chapter as summary in the format of infographic or elaborated in Sections 21.2–21.8. Discussions will also be made in delineating future directions of the initiative Education 4.0. Subsequently, exemplary practices are also reported from LeSMaT (Borderless) educational partners in Japan and Malaysian educational institutions that promote wise use of resources and CT skills aiming at bridging the education-industrial gap to make students future-ready in facing the increasingly challenging world. The development and evaluation for future human–machine interaction/interface (HMI) educational systems promoting CT will be discussed in Section 21.2. Scratchtopia Challenge as an exemplary initiative to promote CT at elementary level will be deliberated in Section 21.3. The challenging IoT with Arduino for data processing from the sensor, with discussion on how IoT and other technological tools can be used for Sustainable Energy for All (SE4ALL), conservation of resources and reduction of waste, are illustrated in Section 21.4. Figure 21.1 is the conceptual framework of LeSMaT (Borderless) and its offshooting project-based programmes through Inter-Centre Collaboration (ICC). Figure 21.2 is the infographic summary of the recent activities conducted.

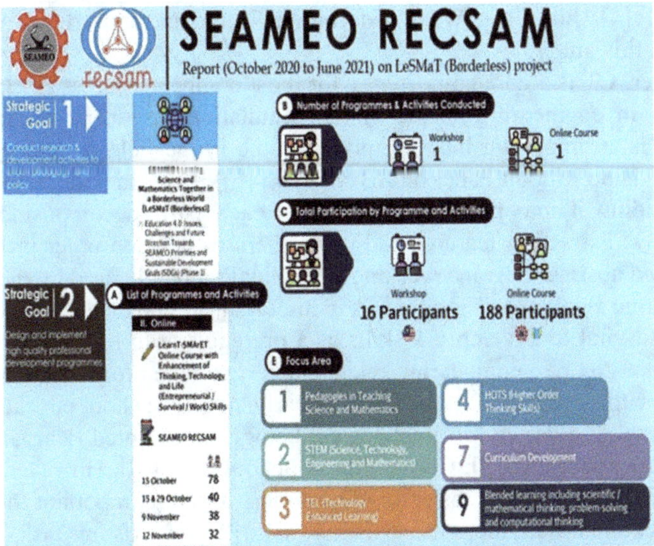

Figure 21.2 The infographic summary for the recent activities conducted

21.2 Development and evaluation for future HMI educational system to promote CT

The importance of fostering CT has been pointed out worldwide, and various methods for fostering CT, such as programming and unplugged computing, have been studied. It is also important to acquire CT in engineering education. On the other hand, it is essential to use the existing learning contents for changing the way content is understood by burying beneath the top soil to understand the existing contents at a deeper level. Furthermore, it is necessary to connect these activities to the development of core skills. Of course, this is equally necessary for engineering education.

This chapter describes the possibility of deepening the understanding of the existing learning contents by using the existing learning contents as CT tasks, illustrating a CT training learning support system using a figure classification task as an example. Then we describe a future HMI educational system.

21.2.1 Background

The information environment in which people live is developing further, with the rise of advanced information technology and AI, and the Singularity's predicted arrival (Kurzweil, 2005). In education, many projects are being implemented, including programming education and training of AI engineers. In this context, the importance of fostering CT has been increasingly emphasized (Wing, 2006 [1]). In addition, CT is composed of several subconcepts (Gov.UK: Department for Education, 2013 [2]; BBC Bitesize, 2020 [3]), and it is important to focus on these concepts to foster CT. There are many other definitions of CT (e.g. ISTE and

CSTA, 2011 [4]) but since Wing's definition is the most used, we use Wing's definition in this study.

Moreover, CT is a core competency for most engineers (Gross *et al.*, 2014 [5]). Besides, CT in engineering is one of the Simulation-based engineering sciences (SBES) that is an important interdisciplinary topic in the field of science and technology (NSF Blue Ribbon Panel on SBES, 2006 [6]). For example, when creating a product, dividing it into parts and examining each part (decomposition), finding commonalities, extracting features and using them to develop other products (generalization and abstraction), and working procedurally are all based on the concept of CT (algorithmic thinking). Therefore, it is important to enhance CT in engineering education and vital to develop CT skills as a prerequisite to higher studies, even if students do not go on to engineering-related undergraduate programs. Consequently, it is necessary to develop CT skills from an early stage of school education.

To foster CT, a variety of practices are being implemented (Harvard Graduate School of Education, n.d. [7]; Computing at School, n.d. [8]). However, at present, most of them prepare special tasks for CT training, and it is important to train CT in existing educational activities, i.e., existing tasks. As an existing task, we use the figure classification task. Classification tasks can also be very useful in fostering CT, and many examples of this are found in unplugged computing (Bell *et al.*, 2007 [9]).

In this study, we develop a learning support system that aims to foster individualized CT. The system gives feedback to the learner on right and wrong errors, and the learner can learn from his or her errors and perform error-based simulation. We define this interactive system as the future HMI educational system and proceed with the following discussion. In this chapter, we describe a model for fostering CT in figure classification tasks and a system for supporting these activities.

21.2.2 Design of this activity

21.2.2.1 Venn diagram and Yes/No chart

We will focus on the use of Venn diagrams and yes/no charts to develop CT in a figure classification task. Venn diagrams and Yes/No charts are used to train CT in the figure classification task. The Venn diagram is used to classify the figures based on the conditions such as 'Is there a pair of parallel sides?,' 'Does it have three vertices?,' and more.

21.2.2.2 Experiential learning

Experiential learning is expected to foster the development of decomposition, generalization, abstraction, and algorithmic thinking of CT in the figure classification task using Venn diagrams and the yes/no chart. However, it is impossible to completely separate CT subconcepts (e.g., it is not easy to distinguish between generalization and abstraction). Therefore, to evaluate whether the figure classification tasks can foster CT, we design a practice task based on experiential learning and we evaluate whether CT as a whole has been enhanced or not. The experiential model involves a process of experiencing a learning phenomenon and applying the experience to the next phase of education (Kolb, 1984 [10]). The process consists of a four-stage learning cycle of 'concrete experience', 'reflection', 'conceptualization', and 'trial (practice)'.

First, the students are asked to classify the given figures uniquely. The students are asked to consider the classification conditions by themselves (Phase I). Then, the students are asked to work on a Yes/No chart. (Phase II) And again, have them classify the shapes uniquely (Phase III). If the score of Phase III is higher than that of Phase I through the experience of Phase II, we can say that the CT training effect of Phase II was practical.

21.2.2.3 Evaluation criteria

I have already pointed out that we evaluate whether CT as a whole has been enhanced or not. We will discuss this in detail. For example, it is difficult to completely discriminate between generalization and abstraction in the figure classification task, and it is thought that they are related in some way. We mentioned the use of Venn diagrams and Yes/No charts in figure classification. In the Venn diagram, the learner selects and places figures that match the given conditions, but procedurally the figures cannot be uniquely classified. On the other hand, in Yes/No charts, learners can procedurally classify figures uniquely. In other words, the Yes/No chart is expected to foster algorithmic thinking. Therefore, one of the most important factors in the development of CT is whether or not the students have acquired this algorithmic thinking. Based on this, we developed the following evaluation criteria, which are shown in Table 21.1.

However, the conditional statement used in the Yes/No chart varies from person to person, even when classifying the same figure. Therefore, there are many different answers, and it is not easy to evaluate the solutions. Consequently, we developed a learning support system that provides feedback on whether the Yes/No chart created by the learner is correct or incorrect. An overview of the system is given in the next chapter.

21.2.3 System overview

21.2.3.1 About future HMI educational system

In this study, we develop a future HMI educational system, which consists of a Venn diagram learning system and Yes/No chart learning system. Both systems allow for

Table 21.1 Evaluation criteria of CT in figure classification

Score	Classification using Venn diagram	Classification using Yes/No chart or similar method	Achievement of procedure
5	○	○	Over 80% correct answers
4	○	○	Correct answers more than 3-step
3	○	△	2-step correct answer and is aware of proceduralization
2	○	×	×
1	×	×	×

Note: ○: Correct, △: partially correct, ×: incorrect.

easy drag-and-drop placement of figures and conditional blocks, and both can return correct or incorrect answers for individual answers created by the user. In other words, the system behaves like an AI in learning where various solutions are assumed.

By using these systems, users can learn from their own mistakes. That is, they can perform Error-based simulation. Error-based simulation means to follow the learner's idea and to lead to an answer that the learner can judge as an error (Horiguchi *et al.*, 2014 [11]). We consider this interaction between the learner and the system to be one of the HMI educational systems. In addition, it is assumed that the HMI educational system will be effective in fostering CT, which is composed of multiple concepts, in the existing subject content.

21.2.3.2 Venn diagram learning system

A Venn diagram learning system was developed to check whether learners can perform simple figure classification when using the Yes/No chart learning system. This screen of the system is shown in Figure 21.3.

21.2.3.3 Yes/No chart learning system

We have developed a system that can give feedback to learners on whether they have answered correctly or not on a Yes/No chart that they have created. The screen of the system is shown in Figure 21.4.

As a result of the implementation of this system based on Sections 21.2.3.2 and 21.2.3.3, it was found that the CT scores of the learners improved significantly. Therefore, the usefulness of the method and the system described in this paper was confirmed (Fukui *et al.*, 2020 [12]).

21.2.4 Conclusion

In this paper, we propose a practice based on experiential learning as an example for fostering CT and describing a future HMI educational system to support the practice. The usefulness of the system was clarified.

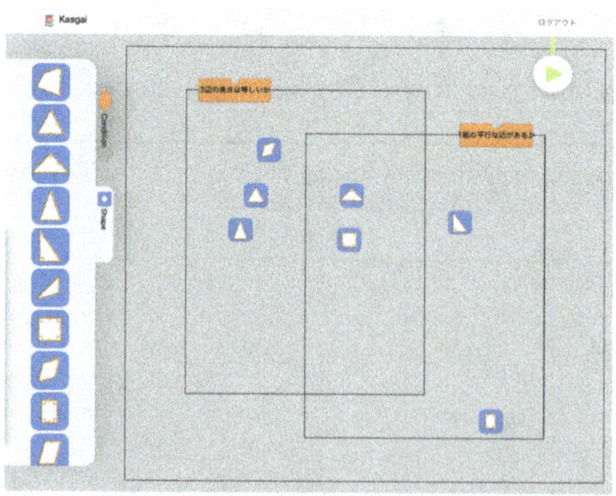

Figure 21.3 The screen of the Venn diagram learning system

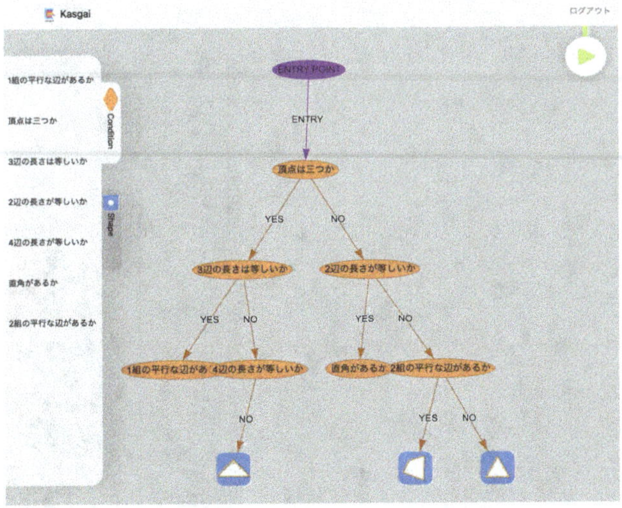

Figure 21.4 The screen of Yes/No chart learning system

In the future society, it is expected that the development of CT will become more critical. In addition, CT as a fundamental skill in engineering education will become even more critical. Therefore, it is essential to consider how to foster CTs, focusing on promoting CTs in programming education and fostering CTs in contexts other than programming education.

21.3 Scratchtopia Challenge as an exemplary initiative to promote CT at elementary level

This sub-section exemplifies Scratchtopia Challenge as an exemplary initiative to promote CT at the upper primary school level. Scratchtopia programming is integrated into a science experiment of primary science curriculum so that opportunities are given to students to take control over their self-directed/self-assessed learning of abstract or difficult concepts when science lessons were to be taught online as well as students have to study from home regardless of location and time within the constraints faced during the global pandemic.

Illustrations will be made on the development of e-module anchored to an instructional design model namely 'Analyzing, Designing, Developing, Implementing and Evaluating' (ADDIE) based on the concepts of science experiments that are aligned with the objectives of the primary science standards-based curriculum (Kurikulum Sains Sekolah Rendah or henceforth abbreviated as KSSR). The following Table 21.2 depicts the phases of modules anchored on the ADDIE model with sample printscreens of Scratch programmes.

In a nutshell, Scratchtopia Challenge can assist students in learning by applying the principles of coding to other subjects such as science. Through learning the

Table 21.2 Development of e-module anchored on ADDIE model integrating Scratchtopia programming into a science experiment

Phases	Description	Integrating Scratch and experiment
Analyzing	The need for an alternative model to introduce science experiment	Due to Covid-19 pandemic, all schools were closed leading to educational disruption and instability. Education assistance using visual programming tools. 
Designing	Designing learning content aligned with objectives of KSSR	E-Modules consist of guidelines and introduction on Scratch programming science experiments that align with the KSSR Science Curriculum addressing waste and pollution 

Developing	Developing a prototype namely 'Scratchtopia challenge: From science experiment to coding'. Developing survey instruments to elicit views of students and teachers	Scratchtopia challenge creates a computational environment that helps improve understanding of a concept, free open courseware. 
Implementing	Facilitating online learning to make science learning interesting and easier to master abstract concepts	Scratchoptia Challenge can be used with or without an internet connection when carrying out science activities. Complements science experiments in the classroom 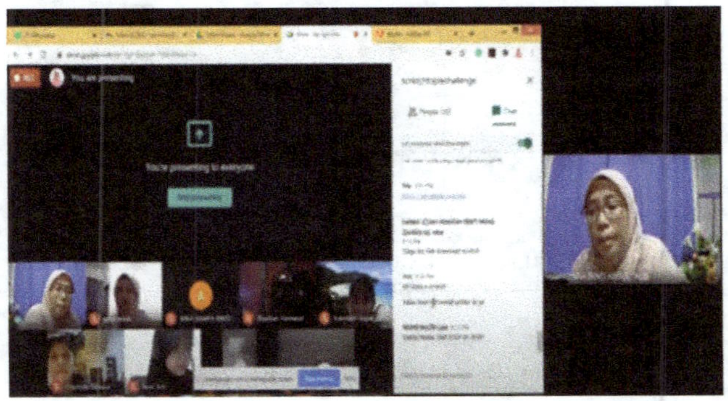

(Continues)

Table 21.2 (Continued)

Phases	Description	Integrating Scratch and experiment
Evaluating	Verification of e-module by technical/market experts in the field and by researchers. Instruments will be validated, reliability tests conducted and data collected will be analyzed using SPSS software	Questionnaire pre and post-competition on the impact of Scratchtopia Challenge for a science experiment 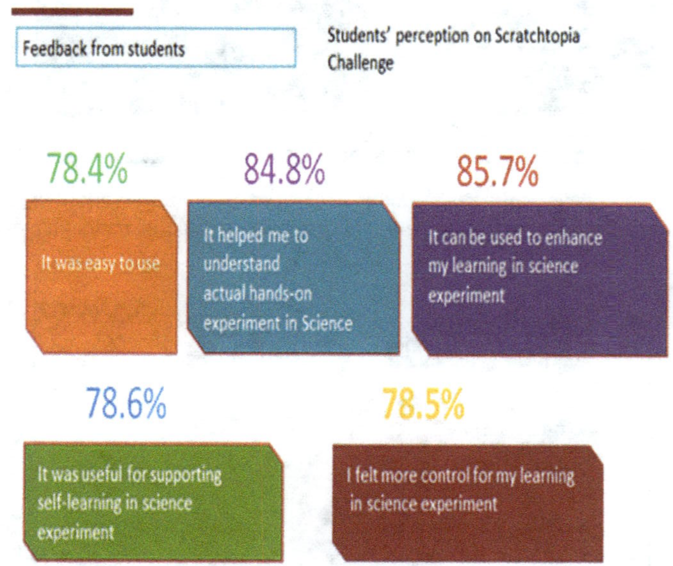

principles of coding with development of CT, students can think analytically and critically, hence facilitating their ability to master skills in science experiments and investigative research. They are also able to make connections for problem-solving in real-life situations.

The analysis of survey data revealed that students were found to have changed in learning attitude as they became independent and were more able to understand the procedure of experiment as well as able to answer questions confidently and no longer felt bored to learn science experiments.

21.4 Exemplary projects on conservation of energy and/or other resources as well as waste reduction integrating IoT concept and technological tool(s)

Managing energy resources sustainably is important to enhance the resilience of living ecosystems. There are 10 alternative sources of energy that can be broken down into renewable and non-renewable types for power generation, which is primarily used to produce electricity to help modern life function and to carry out daily activities of human beings. Due to the availability of resources and the logistics aspects involved to generate electricity, the cost to consume electricity that is normally based on the number of kilowatt-hours (kWh) (occasionally known as the domestic units of electricity) of electrical energy used may vary also based on the device used by an individual or the industry since kWh is the measure of the energy used by a device at a rate of 1,000 W in an hour.

This sub-section illustrates the recent Education 4.0 initiative to promote energy conservation and waste reduction incorporating the concept of IoT using Arduino for data processing from the sensor during a blended-mode regional workshop conducted from 15 to 19 March 2021. Subsequently, the prototype produced was pilot-tested among secondary school learners attending the class on the topic 'Technological Design' [or Reka Bentuk Teknologi (RBK)] with draft lesson ideas prepared for the LeSMaT-Sustainable Energy for All (SE4ALL) e-module.

During the blended-mode regional workshop, the participants were first introduced to the basic components of the robotics kit (Figure 21.5) and guided to download (Figure 21.6) information posted on a whiteboard by the first author of sub-section 21.4 or third author of this chapter.

The participants were then guided to role play as engineers to build an electric wind fan with sample programmes (Figure 21.7).

Upon completion of the prototype, the participants were given another challenging task to build a better fan (Figure 21.8).

They were also presented with scenarios to prepare (Figure 21.9) and displayed (Figure 21.10) prototypes that conserve energy.

Discussions are made after programming using Arduino software with brainstorming of lesson ideas (Figure 21.11).

After the prototypes were built, a draft plan was made to connect all devices through 'Internet of Things' (IoT) (Figure 21.12).

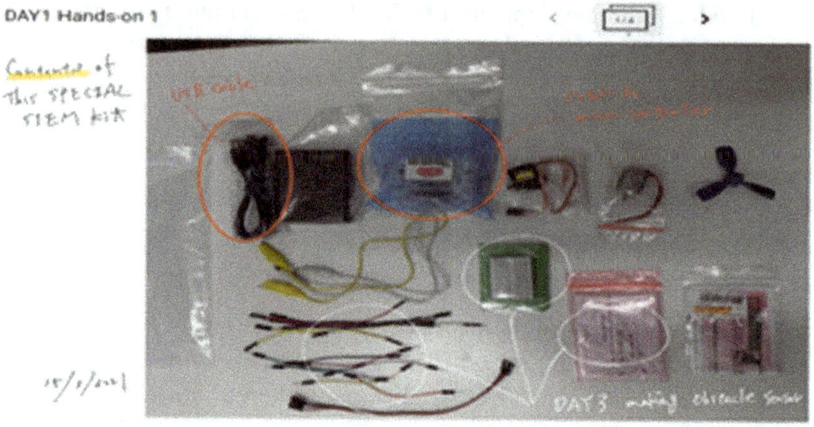

Figure 21.5 Basic components of a STEM Du robotic kit

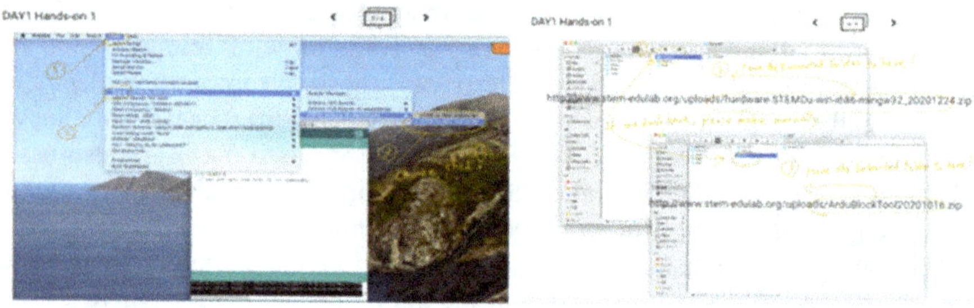

Figure 21.6 Guide to download

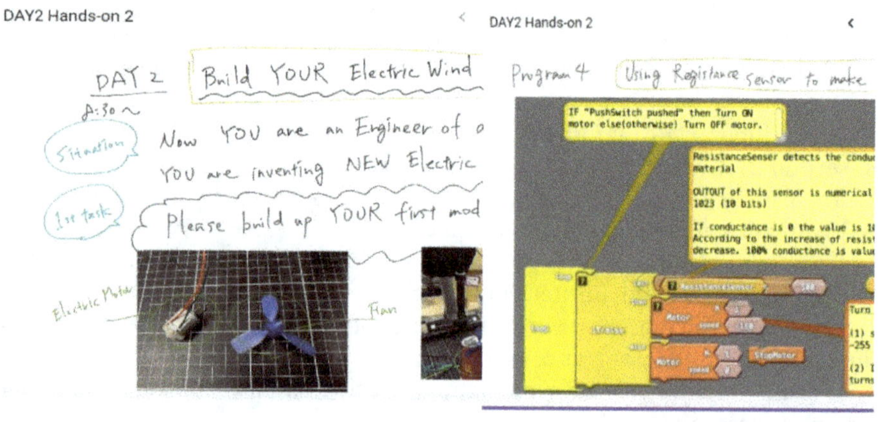

Figure 21.7 Building electric fan role-playing engineer with guided programmes using Arduino software

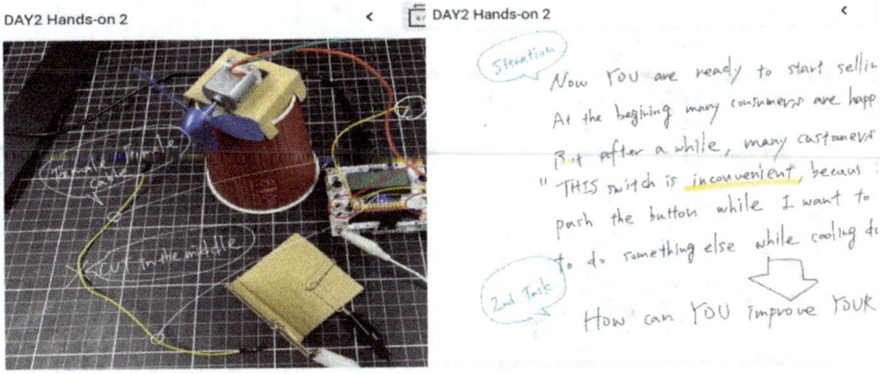

Figure 21.8 Exemplary prototype and more challenging task to build better fans

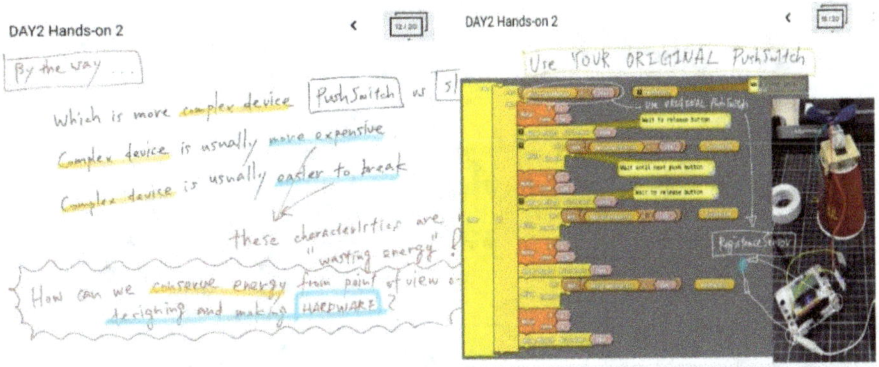

Figure 21.9 Scenario-based learning to challenge the design of prototypes by the need to conserve energy

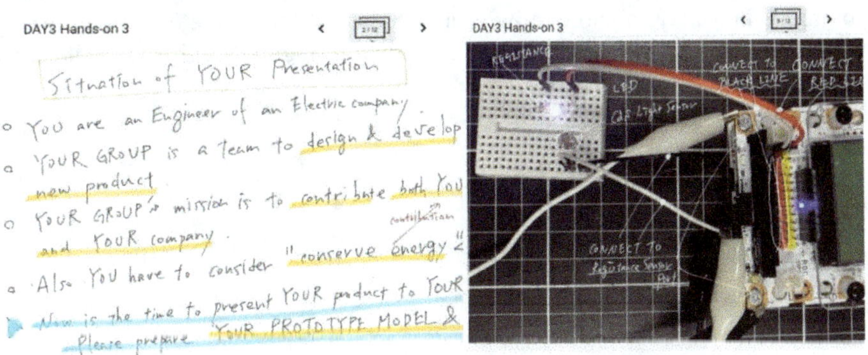

Figure 21.10 Guided presentation on a prototype to conserve energy

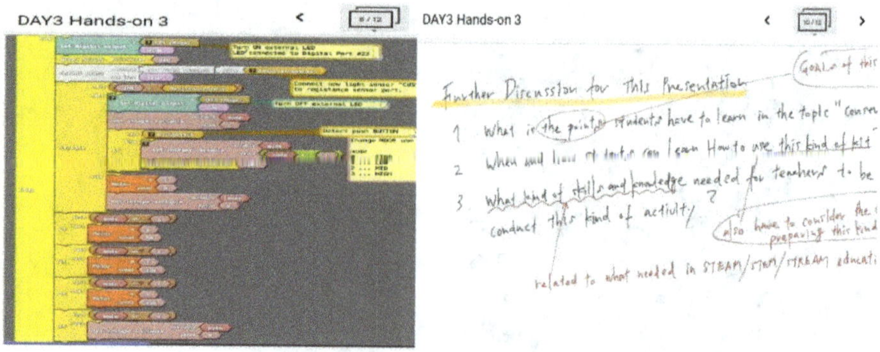

Figure 21.11 Programming in Arduino to prepare prototype and lesson ideas to conserve energy

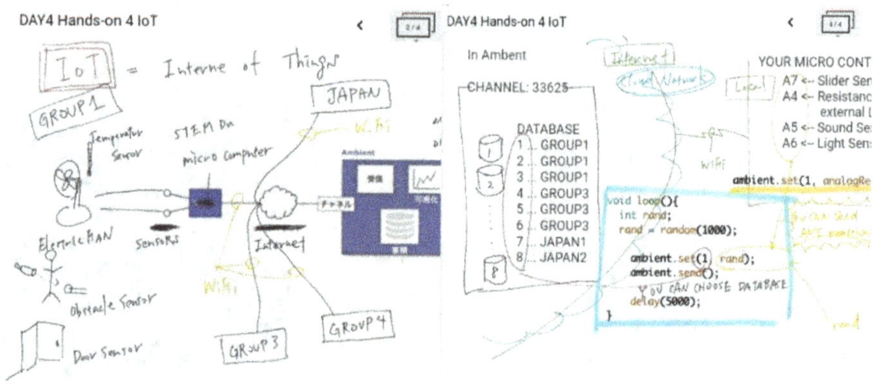

Figure 21.12 Draft plans to connect all devices through 'Internet of Things' (IoT)

Finally, the prototypes illustrating the electric equipment (e.g. fan) prepared by two groups in Malaysia and one by a Japanese consultant were controlled through IoT using STEM Du Micro Controller as shown in Figure 21.13 for monitoring of electrical usage.

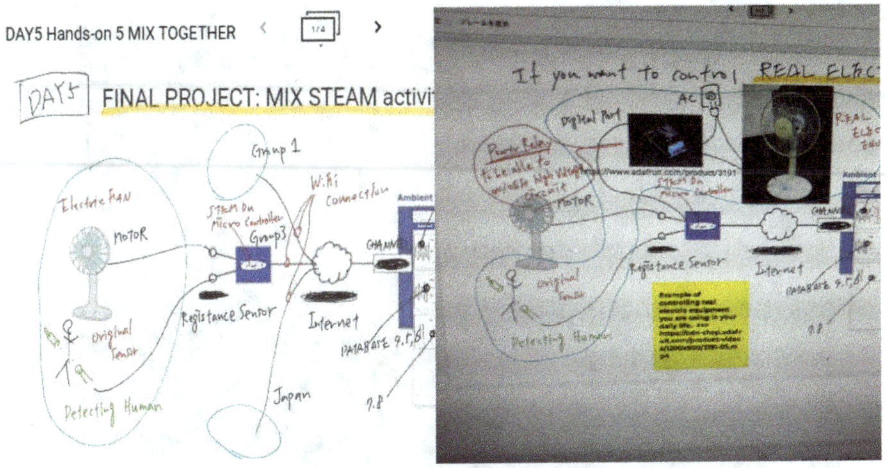

Figure 21.13 Using the concept of IoT to connect electrical devices through STEM Du Micro Controller

The following are exemplary projects integrating technological tools by two participants who attended the workshop.

1. Smart home

Smart homes are designed to be eco-friendly because they can save energy and heat, and can even be cheaper to live in. The basic principle of energy conservation in smart homes has to do with keeping track of which space a person is using and when he is using it. The prototype (Figure 21.14) shows the design of a smart home that uses the Arduino Uno R3 as the main controller (red), with the aid of a digital temperature and humidity sensor (green). The electrical devices built in this smart home are ceiling fans (purple), air conditioners (yellow) and LED lights.

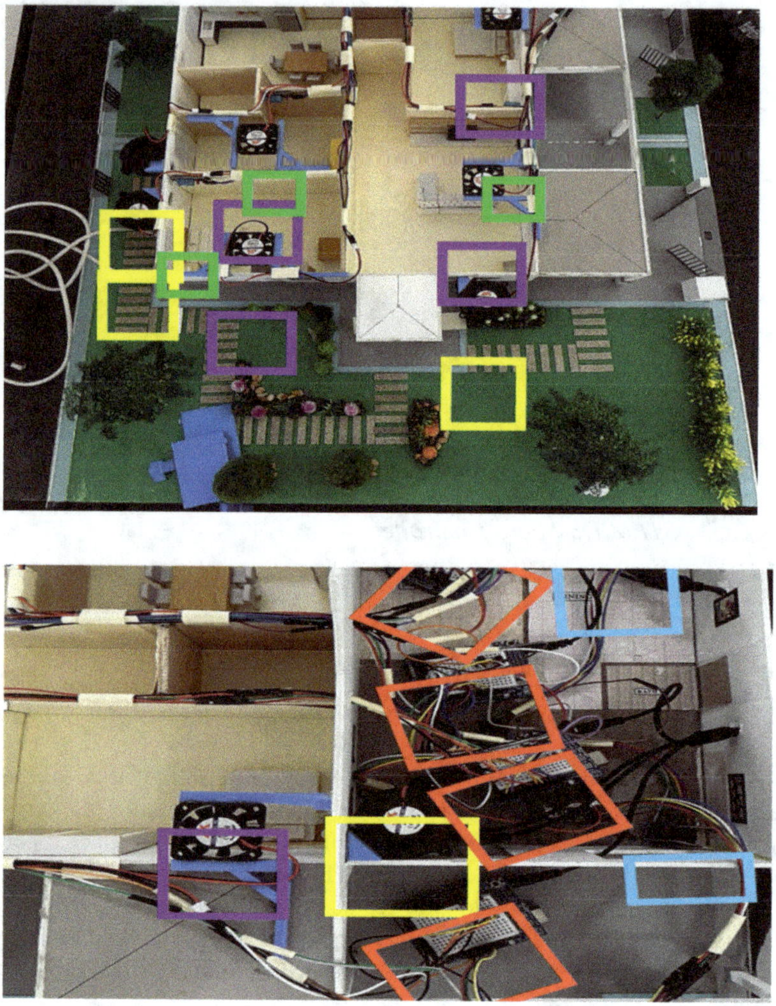

Figure 21.14 The smart home control system prototype

The first way in which smart homes save energy is by using a thermal sensor to conduct heat monitoring. The motion sensor will detect which rooms are occupied and make sure that the temperature in the rooms is sufficiently warm or cool by being programmed to maintain a temperature. A radio receiver is integrated with this system to enable the homeowner to monitor the house remotely, by using a handphone application. The working principle of this IoT controller is to await the homeowner to send a certain command using a phone application, to forward the command to the microcontroller at home, and then execute the command based on the settings that have been configured.

To demonstrate the feasibility and effectiveness of this system, devices such as light switches, power plugs, temperature sensors and current sensors have been integrated with the C++ programming (Figure 21.15)

```
#include "DHT.h"
#include <SPI.h>
#include <nRF24L01.h>
#include <RF24.h>

#define dht_apin 8
#define DHTTYPE DHT11
DHT dht(dht_apin,DHTTYPE);

RF24 radio(10,9);

const byte address[6] = "00001" ;

void setup() {
  // dht
  Serial.begin(9600);
  delay(500);
  Serial.println("dht is online");
  delay(1000);
|
  // nrf
  radio.begin();
  radio.setPALevel(RF24_PA_MAX);
  radio.setChannel(0x76);
  radio.openWritingPipe(0xF0F0F0F0E1LL);
  const uint64_t pipe = (0xE8E8F0F0E1LL);
```

Figure 21.15 Programming demonstrates the configuration of the radio receiver

2. Rainwater reminder incorporating Bluetooth module

Rainwater Reminder is the prototype to remind school workers when the rainwater tank's level is enough for watering purposes. This prototype is to reduce the water wastage in watering plants. The concept of the prototype is based on the principle of an ultarasonic sensor detecting water levels. An ultrasonic sensor is placed in the rainwater tank to measure the water level. When the water level is low, the red LED will light up while the green LED will light up when the water level is high. When the workers see the green LED, they can control the water pump using the apps created to water the plants. The app is controlled by a Bluetooth module (Figure 21.16). This could reduce the social contract deemed as one factor in the spread of COVID-19 pandemic.

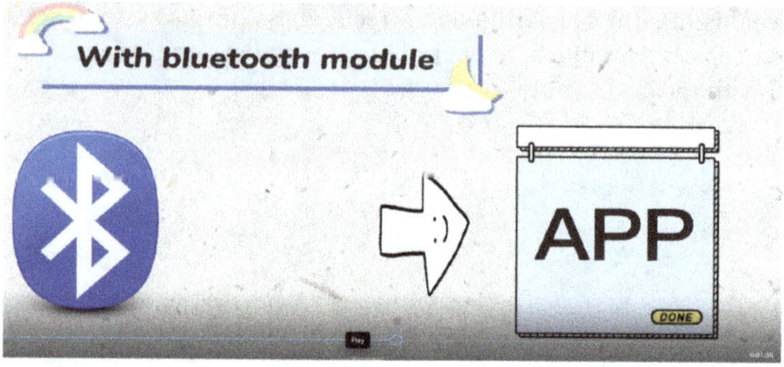

Figure 21.16 Bluetooth module [13]

The projects in this sub-section illustrates R&D activities with evidence on conservation of energy and other resources in line with SDGs especially No. 4 (Quality Education), No. 11 (Sustainable Cities and Communities) as well as No.12 (Responsible Consumption and Production). With the availability of digital tools in the era of Industrial Revolution (IR) 4.0, many efforts can be made to enhance sustainable living and this article concludes with suggestions for the way forward towards Education 4.0.

21.5 Development and wise use of tools for monitoring, evaluation and research activities

This sub-section elaborates recent initiatives for monitoring and evaluation activities with discussion on future directions incorporating the use of AI age tools, digital tools as well as an illustration of exemplary modules to minimize pollution that are part of the initiatives of the author(s) towards sustainable living through ConWUR and waste reduction to reach out to wider audience including the marginalized groups.

Figure 21.17 illustrates the print screens of guided activities posted on the apps developed by the second co-author to facilitate the monitoring and evaluation of activities for the Education 4.0 project initiative [https://bit.ly/lesmated40info].

Figure 21.18 shows the print screens of e-surveys entitled Attitudes towards Conservation of Energy and other Resources (AToCONEoR)[https://bit.ly/atoconeor] and Motivation towards STREAM education (MoToS) [https://bit.ly/motivationtowardsstreameducation] developed for monitoring and evaluation activities for pilot studies of Education 4.0 project initiative.

Figure 21.19 are selected print screens for analysis through the PLS-SEM tool on e-survey entitled Attitudes towards Conservation of Energy and other Resources (AToCONEoR) [https://bit.ly/atoconeor] using data collected from pilot studies of Education 4.0 project initiative.

Technology-enhanced monitoring and evaluation activities [such as use of apps (Figure 21.17), e-surveys on Google forms (Figure 21.18), PLS-SEM (Figure 21.19), to name a few] as aforementioned are conducted. The Education 4.0 project initiative also planned to reach out to a wider audience including respondents from the far-flung countries of the SEAMEO region (including Cambodia, Lao PDR and Vietnam as in Figure 21.20) with pilot studies conducted among samples from educational institutions of both urban and rural areas as shown in Figure 21.21.

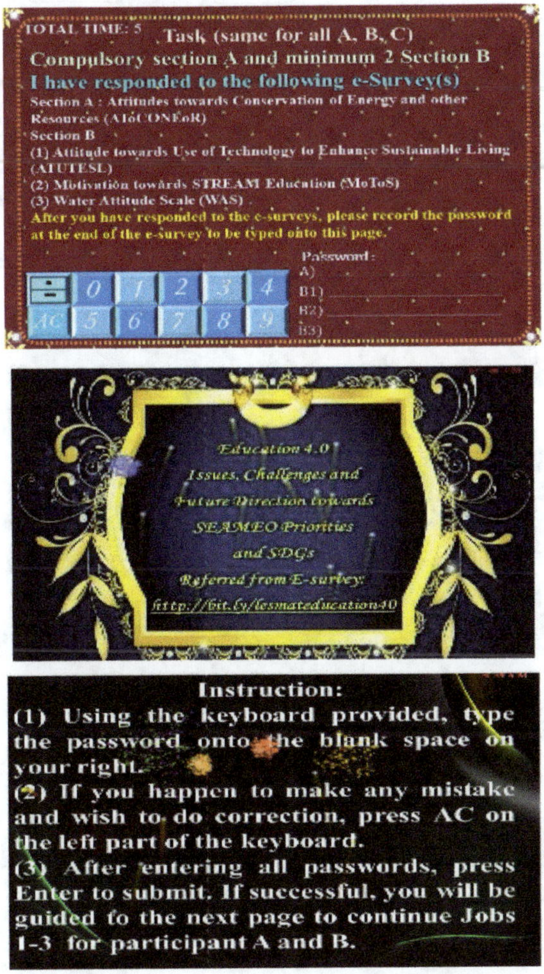

Figure 21.17 Printscreens of selected guided activities posted on the Apps for monitoring and evaluation activities

To conclude, in the advent of the technological era with much evidence of the digital divide especially during the pandemic, the initiatives of Education 4.0 should also aim at preventing drop-outs as reported by Zainal *et al.* [14] and be able to reach out to a wider audience including the marginalized groups in SEAMEO region and beyond. For example, other tools such as Edmodo social learning platforms and Wolfram programming (as experienced and reported by Chin *et al.* [15]) that are also planned in the implementation of further studies should also incorporate human values-based education as reported by Ng [16]. Academics as well as experts from both educational and industrial settings should work collaboratively to bridge the education-industrial gap towards producing younger generations who are future-ready to face the increasingly challenging world in the IR 4.0 era using the most current technological tools in line with SDGs with conservation of resources for sustainable living [17].

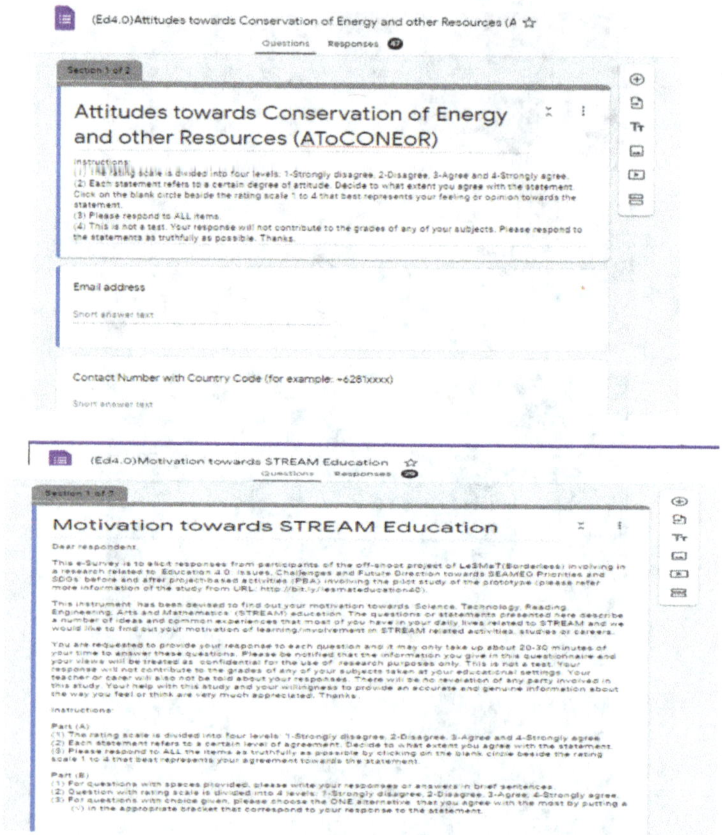

Figure 21.18 Printscreens of two e-surveys developed for monitoring and evaluation activities for pilot studies of the project

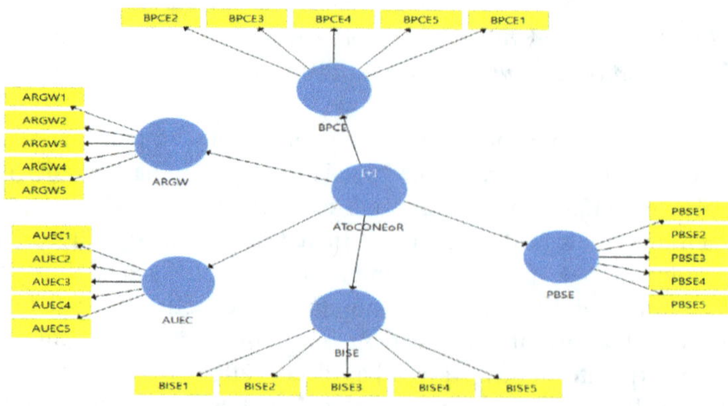

Figure 21.19 Printscreens of draft data analysis using PLS-SEM on e-survey AToCONEoR using data collected from pilot studies

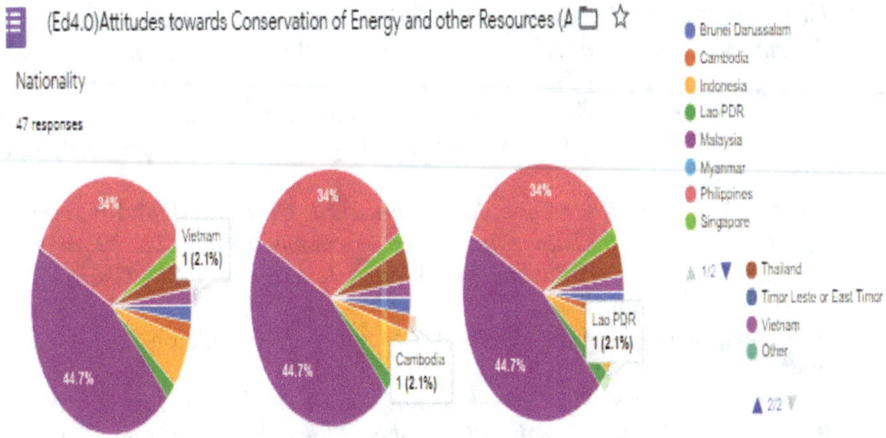

Figure 21.20 *Pie chart summarizing SEAMEO countries involved in the pilot study including Vietnam, Cambodia and Lao PDR responded by pilot study samples*

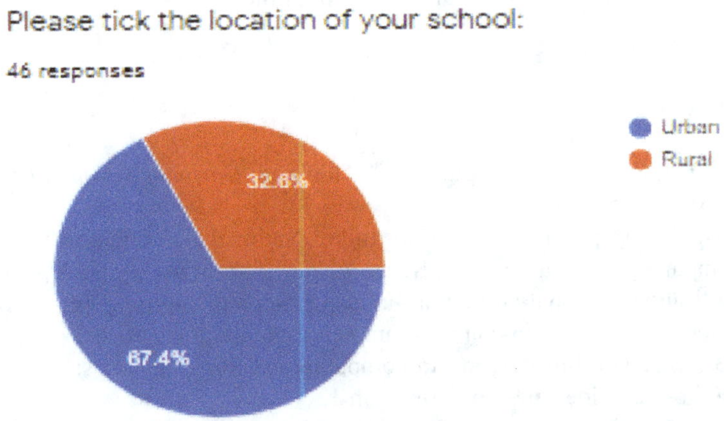

Figure 21.21 *Pie chart summarizing location of school from both urban (67.4%) and rural (32.6%) areas responded by pilot study samples*

Special mention

All authors who contributed to this chapter receive a special mention.

1. Introduction and SEAMEO LeSMaT Education 4.0 project initiative
   Ng Khar Thoe, Mariam binti Othman, Shah Jahan bin Assanarkutty, Wan Noor Adzmin binti Mohd Sabri, Wong Lai Cheng, Baharulnizam Baharum and Mohd Faizal Zainoldin
2. Development and evaluation for future Human Machine Interaction (HMI) educational system to promote Computational Thinking (CT)
   Masanori Fukui, Yuji Sasaki, Jo Hagikura, Jun Moriyama, Yusuke Hayashi and Tsukasa Hirashima

3. Scratchtopia Challenge as exemplary initiative to promote CT at elementary level
   Corrienna Abdul Talib, Tang Kuen Shuen, Nur Wahidah Abd. Hakim and Ng Khar Thoe
4. Exemplary projects on conservation of energy and/or other resources as well as waste reduction incorporating Internet of Things (IoT) concept and technological tool(s)
   Tairo Nomura, Ng Khar Thoe, Nelson a/l Cyril, Deva Nanthini a/p Sinniah, Pang Yee Jiea, Cherry Chung Yin Hoong, Muhammad Faiz Abdul Kuthoose, Sivaranjini a/p Sinnih, Asmahani Abu Samah, Mellisa Reeve, Loh Chee Hoong and Norazlina bt. Mohamad
5. Development and wise use of tools for monitoring, evaluation and research activities
   Peter Chew Ee Teik, Ng Khar Thoe, Lay Yoon Fah, Adeline Leong Suk Yee, Corrienna Abdul Talib, Wahyudi, Ryohei Miyadera, Ong Eng Tek and Rajendra Kumar
6. Smart home
   by Pang Yee Jiea
7. Rainwater reminder incorporating Bluetooth module
   by Cherry Chung Yin Hoong

References

[1] Wing, J.M. (2006). Computational Thinking. *Communications of the ACM*, 49(3): 33–35.
[2] Gov.UK: Department for Education, National Curriculum in England, Computing Programmes of Study (2013) https://www.gov.uk/government/publications/national-curriculum-in-england-computing-programmes-of-study/national-curriculum-in-england-computing-programmes-of-study.
[3] BBC Bitesize, Introduction to computational thinking, https://www.bbc.co.uk/bitesize/guides/zp92mp3/revision/1.
[4] International Society for Technology in Education (ISTE) and the Computer Science Teachers Association (CSTA). (2011) Operational Definition of Computational Thinking for K–12 Education. https://cdn.iste.org/www-root/ct-documents/computational-thinking-operational-definition-flyer.pdf.
[5] Gross, S., Kim, M., Schlosser, J., Lluch, D., Mohtadi, C., and Schneider, D. (2014). Fostering computational thinking in engineering education: Challenges, examples, and best practices, *IEEE Global Engineering Education Conference*, EDUCON, 450–459.
[6] NSF Blue Ribbon Panel on SBES. (2000). Simulation-based engineering science: Revolutionizing engineering science through simulation, NSF, Washington, DC.
[7] Harvard University, Computational Thinking with Scratch, http://scratched.gse.harvard.edu/ct/index.html.
[8] Computing at School, https://www.computingatschool.org.uk/.

[9] Bell, T., Witten, I.H., and Fellows, M. (1998). Computer Science Unplugged: off-line activities and games for all ages, *Lulu* (1998) https://jmvidal.cse.sc.edu/library/bell98a.pdf.

[10] Kolb, D. A. (1984). *Experiential Learning: Experience as the Source of Learning and Development*, Prentice Hall.

[11] Horiguchi, T., Imai, I., Toumoto, T., and Hirashima, T. (2014). Error based simulation for error-awareness in learning mechanics: An evaluation. *Journal of Educational Technology and Society*, 17(3), (2014).

[12] Fukui, M., Sasaki, Y., Hagikura, J., Hayashi, Y., and Hirashima, T. (2020). Development and experimental use of assembling Venn Diagram and Yes/No chart exercise system cultivating computational thinking. *The Journal of the Japanese Society for Artificial Intelligence*, 35(6), pp. D-JA7_1–13, (in Japanese).

[13] Bluetooth Module. [URL: https://new.edmodo.com/groups/borderlesslesmat-13656669/ small_groups/3conservationwiseuseofresources-conwur-13668649/posts] (Accessed 13 Novemeber 2021).

[14] Zainal G., Haris M.J. and Ng K.T. (1991). *The Malaysian Drop Out Study Revisited*. Penang, Malaysia: Unit Penyelidikan Pendidikan Asas, Universiti Sains Malaysia.

[15] Chin, C.K., Hairiah M., Miyadera, R., Ng, K.T., Ch'ng, Y.S. and Promsing, N. (2019). Promoting education for sustainable development in teacher education integrating blended learning and digital tools: an evaluation with exemplary cases. *EURASIA Journal of Mathematics, Science and Technology Education*, 15(1): 1–17. Retrieved from https://www.ejmste.com/download/promoting-education-for-sustainable-development-in-teacher-education-integrating-blended-learning-5613.pdf

[16] Ng, K.T. (2007). Integration of human values in HVWSHE curriculum incorporating teaching strategies/ approaches and assessment/evaluation techniques. In P.Pannen, K.T.Ng, Ikhsan, D. Mustafa & Herawati (Eds.), *SEAMEO resource package: Human Values-based Water, Sanitation and Hygiene Education (HVWSHE)*. SEAMEO SEAMOLEC and UN-HABITAT. Retrieved from http://scholar.google.com/scholar?cluster=6622390137260114690&hl=en&oi=scholarr

[17] Conserve Energy Future (2021). *What are different alternative sources of energy?* Retrieved https://www.conserve-energy-future.com/different-energy-sources.php (Accessed 13 November 2021).

Chapter 22

Conserving cultural heritage, monitoring health and safety in the environment integrating technology: issues, challenges and the way forward

Awangku Hassanal Bahar Pengiran Bagul[1], Khoo Nee Kah[2], Ng Jing Hang[3], Pang Yee Jiea[2] and Ng Khar Thoe[4]

Promoting sustainable living through conservation, health education, as well as preventive healthcare, is the aspiration of all academics and industrialists. This chapter will discuss various issues and challenges related to this aspiration, especially focusing on the importance of conserving cultural heritage, monitoring health and safety in work, study, recreation and living environments in line with sustainable development goals (SDGs). An example of how occupational health and safety in the small and medium industrial (SMI) manufacturing sector could be promoted will illustrate an initiative of the first author in developing a module for the same in Malacca, Malaysia. The discussion will continue on enhancing awareness towards environmental as well as preventive healthcare through recreational activities and sports science through technological tools. Subsequently, the conservation of cultural heritage through the Minecraft digital tool will be deliberated upon. This chapter will conclude with lessons learnt from the development of digital platforms in managing sustainable ecotourism programmes with suggestions as to the way forward.

22.1 Environmental conservation for sustainability in fulfilling SDGs

The complex environmental, social and economic challenges are plaguing the sustainability and safety of the world today. It is estimated that 1.2 billion people are living in extreme poverty [1] and more than half of the world's arable land is degraded, and this phenomenon is undermining human livelihoods for at least 1.5 billion people [2].

[1]Faculty of Business, Economics and Accountancy, Universiti Malaysia Sabah, Sabah, Malaysia
[2]Institute of Technology Management And Entrepreneurship, Universiti Teknikal Malaysia (UTeM), Melaka, Malaysia
[3]Faculty of Medicine, Bioscience and Nursing, Malaysian Allied Health Sciences Academy (MAHSA) University, Selangor
[4]Training and Research (T&R) Division, SEAMEO RECSAM, Penang

One of the major threats to humans this century is climate change that is mainly caused by human overconsumption, and it poses a serious risk to human survival as habitats are inundated by floodwaters caused by sea-level rise even as population displacements take place from unprecedented drought or shifting precipitation patterns.

In 2015, 17 SDGs were introduced to address the current complex global challenges related to climate change, ocean and land resources, deprivation and hunger, inequalities, clean water, governance, consumption, economic growth, wellbeing and education. The SDGs are the direct successor of the millennium development goals (MDGs) and built on the Rio Declaration on Environment and Development [3] as well as the Rio 20 UN Conference on Sustainable Development (UN General Assembly, 2012) and taking into consideration the different national realities, capacities and level of development while respecting national policies and priorities. The agreement of the SDGs as the universal goals that involve the entire world and their implementation has attracted the interest of many scholars, sparking a whole new narrative on how SDGs should be achieved in an integrated manner by the end of 2030.

It is important to know that the promotion of sustainable living is explicitly embedded across most of the SDGs. Humanity depends more than ever on the sustainability of the environment for survival and better quality of life. The incorporation of sustainable living into the urgent action to combat climate change and its impact is key to achieving the SDGs. The SDG 13 (Combat Climate Change) has a high impact on SDG 6 (Sustainable Management of Water and Sanitation), SDG 7 (Access to Sustainable Modern Energy), SDG 8 (Sustainable Economic Growth), SDG 9 (Sustainable Industrialization), SDG 11 (Sustainable Cities and Human Settlement), SDG 12 (Sustainable Consumption and Production), SDG 14 (Sustainable Use of Oceans, Seas and Marine Resources) and SDG 15 (Sustainable Use of Terrestrial Ecosystem). Human activities such as the overconsumption of any resources will put increasing pressure on the environment that ultimately contributes to the threat of expedited climate change.

The SDGs were ultimately designed to reconcile environmental protection with social–economic development. Despite the advancing progress in addressing social and economic development, the environmental crisis is still unprecedented even with the integration of nature protection into the wider scope of human development. Locke *et al.* [4] propose the 3Cs framework where nature and its vital contributions to humans, which together embody biodiversity and ecosystem functions and services, make progress towards a vision of humanity 'Living in Harmony with Nature' by 2050. If the vision is met, biodiversity will be valued, conserved, restored and used sustainably where the ecosystem service can be maintained to sustain a healthy planet. The proposed 3Cs framework includes:

C1: Increasing the conservation effort in order to secure endangered species while protecting all remaining primary ecosystem fragments. Sustainable practices such as protecting good farmland, practicing productive regenerative agriculture, and keeping nitrogen out of freshwater will be elevated as mainstream and at the same time, increase ecological restoration and maintain pollinators. 'Green' cities with the purpose of reducing carbon emissions, preventing urban sprawl while providing access to nature for urbanites' health and well-being.

C2: Establishing 'ecologically representative and well-connected systems of protected areas (PAs)' while increasing the coverage of key biodiversity areas (KBAs). Viable populations of native species and ecological processes should be restored and maintained.

C3: Ecological integrity and associated global processes such as carbon storage, rainfall generation or fluvial flows will be retained. Large protected areas (PAs) will be established while conserving the indigenous and community areas, hence securing their knowledge and livelihoods.

Ultimately, the 3Cs aim to provide a systematic and consistent framework for countries in committing to the common global goals but with realistic measures that suit their current national realities. The framework not only provides adaptable responsibilities for international cooperation to preserve nature and people but also serves as a guide for the participation of non-state stakeholders to meet the 2050 vision.

One of the greatest challenges in promoting sustainable living through conservation will always be to incorporate all the dimensions of sustainable development i.e., environmental, social and economic, in a balanced way. Each of these dimensions requires mindful evaluation and reasoning on how they will be interlinked while contributing to the SDGs. Environment and socio-economic development are so intricately linked that any socio-economic development would not be sustainable without considering the environmental dimensions.

22.2 Monitoring occupational health and safety in small and medium industrial (SMI) manufacturing sector: challenges and future direction

This subsection illustrates part of a bigger-scale study to showcase an initiative of the first author in developing a module to monitor occupational health and safety in the Small and Medium Industrial (SMI) manufacturing sector in Malacca.

Illustrations are made on the development of an e-module anchored on an instructional design model namely 'Analyzing, Designing, Developing, Implementing and Evaluating' (ADDIE) based on the concepts of 'Occupational Safety, Health, Environmental Management Tools' (OSHEMT) and its seven elements that are aligned with the Occupational Safety and Health (OSH) environment to measure the Compliance to Malaysia OSH and environmental regulations.

The development of an e-module anchored on OSHEMT was found to promote OSH practices towards the OSH environment through non-formal learning (e.g. discussion, experience sharing and meeting) as well as formal training, seminars and workshops. Other proper health and safety practices include managing the schedule of industrial waste to be re-used via the use of Electronic Schedule Waste Inventory System (ESWIS) involving activities to recycle, store and prepare an inventory.

Table 22.1 elaborates on the phases of the module anchored on the ADDIE model with a summary on the descriptions on the phases of Analyzing, Designing, Developing, Implementing and Evaluating in ADDIE as well as challenges faced integrating OSHMT and expected outcome with suggestions for the way forward.

Table 22.1 *Development of e-module anchored on ADDIE model integrating OSHEMT*

Phases	Description	Challenges faced integrating OSHEMT	Expected outcome and/or future directions
Analyzing	Training needs analysis (TNA)	The need to promote technology-enhanced lifelong learning on the use of new tools and technology through online education	Industrial workers who are keen to take self-interest courses through online or blended-mode training
Designing	Designing module that suits training needs of adult workers	The need to incorporate an andragogy (i.e. adult education or any form of adult learning) approach in the design of modules.	Industrial employees at S&M Manufacturing as adult learners motivated for self-accessed/self-directed/self-paced learning and e-module designed for OSHEMT
Developing	Development of modules for adult competency-based learning	The development of training should be blended mode that fulfills requirements for adult learning with certification as well as focusing on competency-based learning. Examples include training of assessor for noise risk and ergonomic risk, boilermen, chargemen, fire manager/officer, certified professional in erosion and sediment control (CPESC), certified environmental professional in scrubber operation (CePSO), certified environmental professional in bag filter operation (CePBFO), certified environmental professional in schedule waste management (CePSWaM), to name a few	Series of online education that can offer professional certificates to produce OSH and OSHEMT coordinator, initial noise assessor and Training of Trainer Sample Certificate for Training of Trainer/ 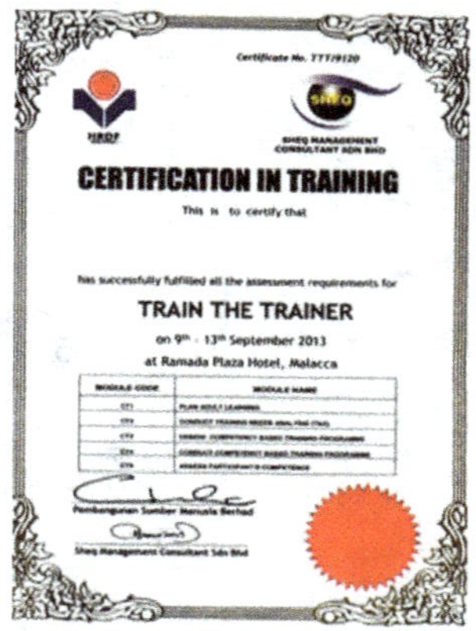

Implementing	Implementation of activities introduced in the e-module	The greatest challenge is to implement Cleaner Production (CP) that can create a healthy environment with good health among workers. Subsequently, a safe environment will result in better efficiency and productivity	A long-term education programme should be implemented to raise awareness on the importance of preventing hearing loss
Evaluating	The e-module serves as a guide for monitoring and evaluation of practices in the industry	Evaluation is to be made on an ongoing basis to inspire workers to implement CP as well as ensure healthy surroundings for the workforce and a secure environment to achieve protection, proper use of resources as well as avoidance of effluents	An engineering study should be implemented to reduce the source of noise exposure with more high-tech machines, noise protector equipment, noise sensors, etc.

22.3 Enhancing awareness on environmental and preventive healthcare for sports science supported by technology: a systematic review and suggested research

Apart from monitoring health and safety in the workplace, health education and preventive healthcare integrating recreational and/or sports activities in various communities of learning as well as educational/industrial settings is also important for the balanced growth of human beings towards achieving SDG No. 3 (Good health and Well-Being) and SDG No. 4 (Quality Education). People with physical fitness and healthy minds are more productive, alert and creative. Hence, efforts in enhancing awareness on environmental and preventive healthcare integrating recreation and sporting activities will indirectly respond to the global call for achieving SDG No. 9 (Industry, Innovation and Infrastructure) as well as No. 11 (Sustainable Çities and Communities).

It is believed that sports as part of physical science/health education are the best mode to produce physically and mentally active learners/workers in the industries that introduce sports and social clubs as part of the social welfare activities for employees. Nevertheless, in the advent of the global pandemic, many sports activities could not be conducted, especially those involving human body touch such as football and rugby, to name a few. This subsection aims at enhancing environmental health awareness with discussions on various issues pertaining to preventive healthcare, especially sports science or physical education, and how technological tools could be optimised to replace physical training, as experienced by the first author as an amateur footballer who explored various ways of improving sports science training in football supported by digital tools [5].

Through the use of search engines, articles became accessible from the Internet, especially from Google Scholar and ResearchGate. The quality literature published between 2000 and 2021 on topics like 'preventative healthcare and medicine, environmental conservation/health/quality, storyboards, sports education/medicine, human's movement in football playing, football pitch and top players, developing characteristics, biostatistical data, functional abilities, physical/mental health psychology, IR4.0, visual learning, e-tools', to name a few, from different reputed journals was systematically reviewed for in-depth discussions in this subsection.

According to the literature, preventive healthcare is dealing with lowering the risk factors that are associated with illnesses. However, preventive medicine is defined as a branch of medical service dealing with avoiding or prevention of diseases through the medical practices that are so designed as to enfold preventive care such as screening or medical tests with possible treatment before further deterioration [6,7]. Both avoiding and preventive care can be achieved if there is proper management of environmental conservation for sustainability as summarised in the following Table 22.2 as well as monitoring of environmental health, security and quality resources such as energy and water (refer to selected summary in Table 22.3).

Table 22.2 Review on managing environmental conservation using the application of Biotechnology concepts

Aim/purpose	Biotechnology concept	Possible impact/benefits
To sense contamination or pathogens in the sewage system	Involvement of microbial communities	Protecting the public from dangerous exposure to pathogens
To ensure it is difficult for insects to develop resistance via bioinsecticides attack at once in various ways	Involvement of biopesticides e.g., bio fungicides (Trchoderma), etc.	Potential for insect control in the long-term and conserve the environment sustainably
To conserve the environment through regular detoxification of contaminated water with heavy metals, wastewater, sediment, sludge of soil	Involving microbial communities such as bioremediation bacteria (e.g., bacillus, alcanivorax, etc.)	Cleaner environment with improved water resources for domestic/ industrial usage

[8–10].

Table 22.3 Review on monitoring environmental health, security and quality resources for sustainability

Aim/purpose	Biotechnology concept	Example/name of the process	Possible benefits
Producing sustainable sources of renewable energy: From biomass/organic waste to electricity	Involvement of biomass materials used for energy from plants such as corn and soy	Conversion of organic wastes directly to electricity or biomass energy	Results in very much less dependence on fossil fuel
Ensuring the production of food for humans and animals with minimal interruption to the environment	Managing symbiotic microbial communities effectively or microbe-based symbioses in plants	Use of molecular approaches based on the continuity of microbial pools which are circulating frequently between animal-provided niches, plant and soil inside the agricultural and natural ecosystems	Effective ways to develop highly productive microbe-based sustainable agricultural systems and, at the same time, to address the ecological and genetic consequences of the large applications of microbes in the agricultural practice

[8–11].

Apart from managing environmental conservation as well as monitoring environmental health, security and quality resources, preventive healthcare can also be achieved through abiding by the 'Standard Operating Procedure' (SOP) during pandemic by wearing a mask, sanitizing oneself or the environment regularly and not participating in sports involving a large number of people possibly with body touch such as football. Managing good mental health, maintaining physical fitness and body immune system such as taking fruits or vegetables that contain natural phenolic compounds due to the antioxidant power that could reduce chances of getting diseases [12] are also important practices during the pandemic era.

One way to manage stress and cope with sudden lifestyle changes is the replacement technique or the building of new habits [13]. Due to physical training becoming out-of-reach, various technological tools that can replace physical training for virtual training were reviewed as well as research to improve sports training, motivation for self-directed learning to enhance performance were undertaken as summarised in Table 22.4.

Case examples are illustrated on selected technological tools available used to guide the first author as an amateur footballer who was involved in the game since 2017. In addition to playing in the school as part of the extra co-curricular activities, the first author was also a member of a local football club. During leisure time, more skills were learnt through playing football using advanced technological gamification tools such as smartphones and television connected to Sony PlayStation than on the field. From playing online games and watching football matches, various observations were made with learning output, demonstrating communication skills and multiple intelligence (visual-spatial) via verbalising understanding of concepts using drawing as part of the learning output.

Table 22.4 Technological tools with biostatistical data to improve sports training and self-directed learning

Features	Research	
	Visual learning and/or Gamification tools	
Enhancing knowledge and skills (e.g., game, research, etc.)	# Huizenga *et al.* [14] for deep learning of concepts	# Li [15] for the concept of game with educational or entertainment purposes# Football gamification playstation [35]
Improving motivation and team spirit	# Buzko *et al.* [17] on improving students' motivation to learn	# Erhel and Jamet [18] on how game-based learning enhances intrinsic and extrinsic motivation
Engaging hands-on real-life experience	# Westhuizen and Haar [19] on fatigue level	#Football gamification through Sony playstation [16] (Sony PlayStation, n.d.)
Influence by motivation theories or sports and health psychology	# Narulita *et al.* [20] on motivation theories in science learning using Augmented Reality VL tool	

An observation made was that there are various ways that football can travel in the air before reaching the goal post on land. This observation was captured in the form of sketched drawing as illustrated in Figure 22.1 and Figure 22.2. The first author was also involved in blended-mode training on the use of digital tools and e-platforms during school co-curriculum activities.

The drawing activities (Figures 22.1 and Figure 22.2) were completed prior to drawing the path of the 3D-model on the Blender tool for Augmented Reality (AR) and reported by the team members as part of the project team's output submitted at the end of the blended-mode training. The AR tool allowed the sketching of paths of the moving ball that was reflected as an animated object such as football before reaching football post [Exemplary AR learning output is accessible from http://bit.ly/footballAR, names of application files Football.apk and Footballpost.apk, to name a few] [21].

This subsection provides environmental health awareness and preventive healthcare with an overview of how visual learning e-tools displaying statistical data were used to replace physical training in sports science (focusing on football) as well as physical health education with elaboration on the author's experience. It is part of an ongoing study to explore various functions of advanced technological tools that support sports training and physical health education, especially in football. Due to the constraints faced especially in terms of time and resources

Figure 22.1 Inside curve

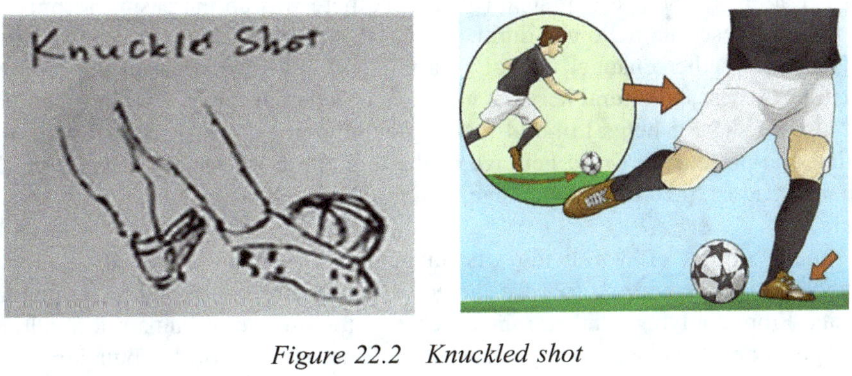

Figure 22.2 Knuckled shot

available as well as the recent pandemic outbreak, the authors were only able to report selected cases as aforementioned.

The authors as teams of educators/ researchers in the fields of 'Science, Mathematics, Engineering and Mathematics' (STEM) education managed to keep abreast with the latest technology for IR4.0 in sports science and physical health education. The second and fourth co-authors are Professors of Engineering education whereas the third and sixth (who is also football fan) co-authors are currently STEM educators majoring in Science and Mathematics integrating technology education. Some of the concepts learnt that related to biostatistical data in sports medicine could also be applied to health education at various levels in line with the current trends and government's aspirations to promote Science, Technology, Reading, Engineering, Arts and Mathematics' (STREAM) education or STEM integrating Arts education. Hence, other areas that could be explored include digital drawing tools such as Wacom Intuos (Creative Pen Tablet) and biometrics tools for data collection through wearable devices that tracked movement during sports activities as reported by [22]. The animation functions in Geogebra are also explored, including the mathematical modelling of soccer dynamics as reported by [23]. Continuous efforts should also be made to incorporate or devise more advanced tools for the implementation of values-based technology-enhanced applications in sports science, sports as well as physical and psychology of health [24] incorporating transdisciplinary studies to facilitate human resource development (HRD) [25].

22.4 Conserving cultural heritage through Minecraft digital tool

The storehouse of our heritage is threatened as it has suffered from weathering, deterioration and destruction, both by human and natural factors. As of June 2020, there are a total of 1,121 UNESCO World Heritage Sites, and only 15 percent of the world's cultural heritage is presently preserved in a digitalised format. A great majority of historic artefacts and sites are certainly at risk of corrosion because of age. Once a heritage site is lost, damaged, or destroyed, the recovery process is slow, if at all. In the wake of such calamities, there was an increasing forming of opinion and resultant action to digitize cultural heritage to keep them away from the risk of weather change, natural disaster, bad coverage or insufficient infrastructure. Access to cultural heritage locations is essential for societal development and also because it helps to build resilient communities. Thus, the effort in protecting and preserving world heritage in the digital era is crucial, in line with the famous adage: "part of our heritage, pillar of our present, and enabler of our future".

In the efforts of worldwide arts and cultural heritage preservation, digital technology is an emerging tool for the collection, maintenance and dissemination of data. From applying synthetic intelligence to the lineage of humanities, cultural heritage is headed closer to a virtual future. The use of world-simulation video games for cultural heritage preservation provides the best alternative for humans in safeguarding cultural heritage resources. However, not all simulation video games have the capability to present how we perceive heritage values efficiently. Based on the evaluation of the effectiveness of video games to motivate, immerse and constitute reality, the game-based educational tool, Minecraft is recognised as one of the most popular tools in constructing the cultural heritage environment.

Minecraft is an online multiplayer game recently gaining much attention especially among game players who would like to build and play in their world. Hence, it has recently been adopted widely as game-based learning activities in some classrooms using Minecraft Education Edition (MEE). With the features available in this MEE such as artefacts, boxes and scenarios, educators as well as students could make real-world connections and develop self-accessed, self-directed and self-paced contextual or problem-based learning. Minecraft, additionally, performs its vital role as a teaching and learning tool which is implemented through worldwide education institutions to deliver digital knowledge and skills effectively. The Minecraft video game shows that nothing, not even war, can destroy worlds built in Minecraft. This digital tool shares the same ideal in mind: to promote the usage of digital media and virtual reality to preserve buildings, especially world heritage sites, such that they can be immortalised.

This subsection illustrates the authors' experiences to introduce Minecraft digital tool among enthusiastic learners to showcase exemplars for 'conservation of cultural heritage' as one of the focus topics of learning output to promote 'Learning Transdisciplinary Science integrating Mathematics, Arts-Culture-Language, Engineering-Environmental Education-Economics-Entrepreneurship, Engineering, Technology' (abbreviated as LearnT-SMArET). While learning to build cultural heritage using Minecraft, students also develop higher-order thinking, technological skills and engineering prowess in construction activities apart from learning other subjects such as 'spatial geometry' in Mathematics, visualizing 3D-animation of scientific concepts/processes, computer art application, digital storytelling and project management, to name a few. Table 22.5 summarises the Minecraft projects related to the conservation of cultural heritage prepared by primary and secondary school students.

22.5 Development of digital platforms to manage sustainable edutourism programmes: lessons learnt and the way forward

Prior to the global pandemic era, tourism was once an industry that provides the essential source of income for a country with natural resources. In fact, a tourist is operationally defined as someone away from their residence for at least 24 h and less than a year. However, over the past more than one year, the tourism industry has been the hardest hit of the sectors of the economy due to the outbreak of coronavirus that affects the movement of global citizens. For the industry to continue to sustain itself, it became very much like a fashion industry and was replaced

Table 22.5 Summary of Titles and URLs of Minecraft projects related to cultural heritage developed by project teams

No.	Project title [URL]	Printscreen
1	LearnT-SMArET – use of Minecraft Digital tool in preserving heritage (SMJK Yok Bin School) [26] [URL: https://youtu.be/bOvs3_4bjRQ]	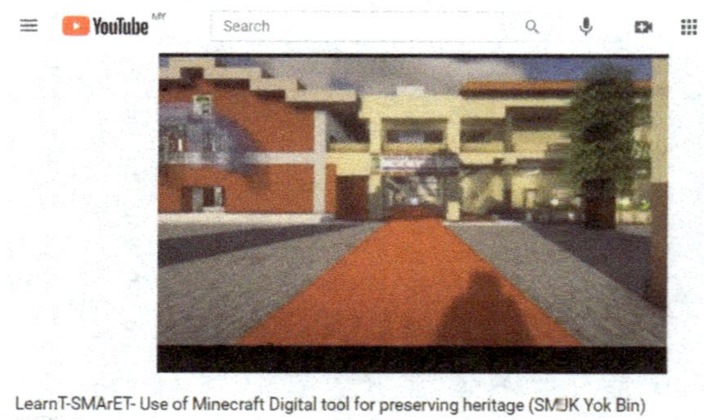
2	LearnT-SMArET – use of Minecraft Digital tool for cultural celebration (CNY Classroom 5 Einstein) [27] [URL: https://youtu.be/Khj0u3HZ5gM]	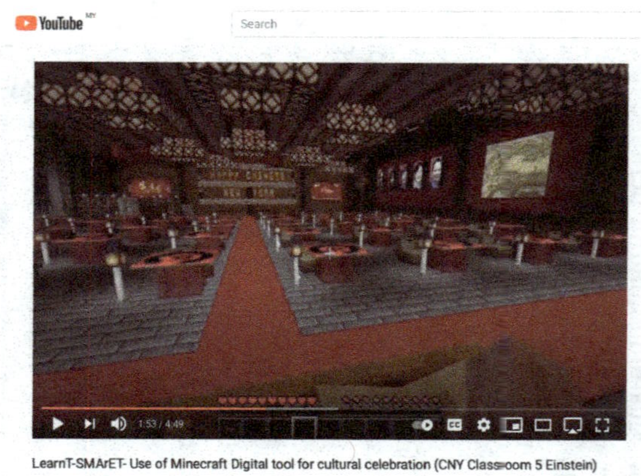

3	LearnT-SMArET – use of Minecraft Digital tool for cultural celebration (CNY Classroom 2 Red) [28] [URL: https://youtu.be/cCpFIBUg4m8]	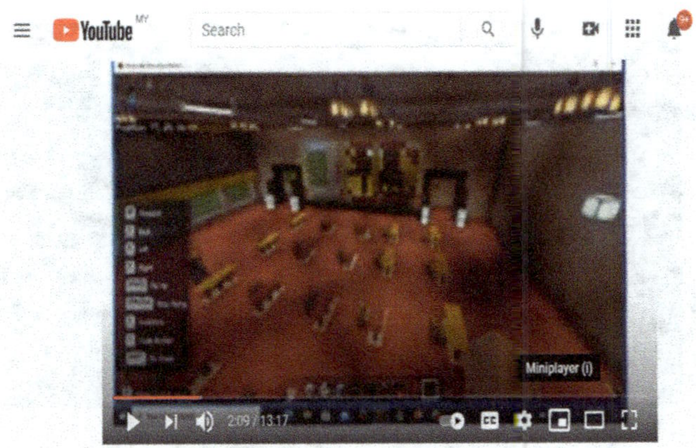 LearnT-SMArET- Use of Minecraft Digital tool for cultural celebration (CNY Classroom 2 Red)
4	LearnT-SMArET – use of Minecraft Digital tool for cultural celebration (CNY Classroom 5 Newton) [29] [URL:https://youtu.be/HKVVm-NdB_s]	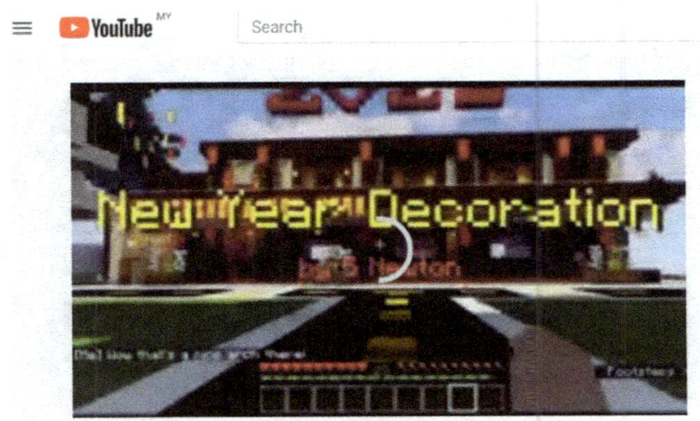 LearnT-SMArET- Use of Minecraft Digital tool for cultural celebration (CNY Classroom 5 Newton)

(Continues)

Table 22.5 (Continued)

No.	Project title [URL]	Printscreen
5	LearnT-SMArET – use of Minecraft Digital tool for cultural celebration (CNY Classroom SMKIJC) [30] [URL: https://youtu.be/FmrV2v2JHFU]	
6	Use of Minecraft Digital tool to preserve Penang Bridge (Heritage Immortalised: a nation building MINECRAFT championship 2021) [31] [URL: https://youtu.be/swPI_7rRmag]	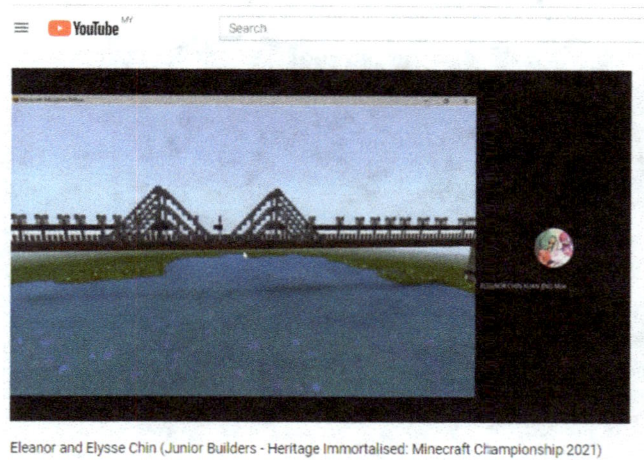

7	Use of Minecraft Digital tool to preserve Melaka- A UNESCO Site (Heritage Immortalised: a nation building MINECRAFT championship 2021) [32] [URL: https://youtu.be/HECD61NnAds]	
8	Use of Minecraft Digital tool to preserve Ipoh Callie's Castle (Heritage Immortalised: a nation building MINECRAFT championship 2021) [33] [URL: https://youtu.be/zILUN3Ejx0A]	

(Continues)

Table 22.5 (Continued)

No.	Project title [URL]	Printscreen
9	Use of Minecraft Digital tool for cultural celebration- Wesak Day (Heritage Immortalised: a nation building MINECRAFT championship 2021) [34] [URL: https://youtu.be/MXDv74HIL1w]	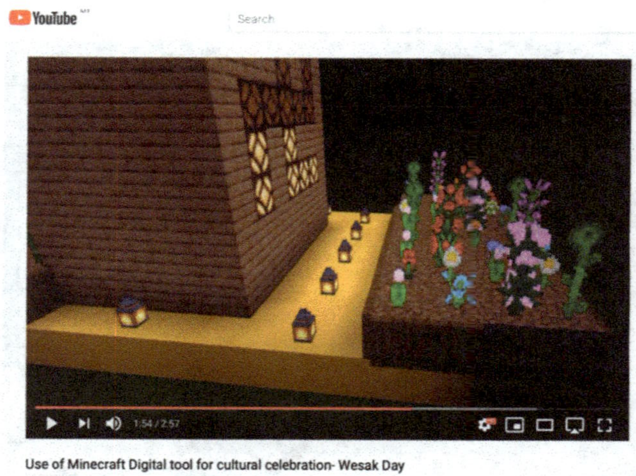
10	Use of Minecraft Digital tool to preserve World Heritage: A Famosa in Malacca (Heritage Immortalised: a nation building MINECRAFT championship 2021) [35] [URL: https://youtu.be/PN8xxnOB36g]	

11 Penang Heritage Tour – Blue Mansion (Heritage Immortalised: a nation building MINECRAFT championship 2021) [36] [URL: https://bit.ly/BlueMansionPenang]

by virtual vacations in some parts of the world. One of the important niches in the ever-growing tourism business is edutourism which incorporates the concepts of tourism with lifelong education or travelling to a unique location for formal and informal learning [37] that is getting popular in both higher learning institutions and industrial settings. However, with the unforeseen circumstances of natural and man-made disasters such as the outbreak of coronavirus pandemic, the income from this industry has been much jeopardised. Hence this subsection explores various alternative modes in which edutourism could be conducted through leveraging on the availability of various blended-mode platforms with evidence on how these tools could be used for green marketing as well as dissemination of information in the IR4.0 era.

Edutourism (or sometimes referred to as educational tours) that blends education with tourism in a way that promotes lifelong learning could be viewed as recreation-based in situ learning. One of the benefits of educational tours is that they can provide interesting ways to explore new things by helping students to learn through visual experience, to have fun and recreation. It can also be the action of travelling with the intended purpose of learning a new skill or engaging in an activity that is otherwise unfamiliar to gain educational and cultural insight into the occupation [37]. Hence while building new knowledge, new skills and a sense of expanded horizons for learners, edutourism provides an enjoyable lifelong learning experience. Given students could not get the full knowledge from published articles or books. As social, cultural, economic, political and human norms environments around the world keep on changing, therefore students cannot just be knowledgeable through only being the arm-chaired travellers. One must personally have the personal touch by visiting the people and place and experience personally its cultural norm and environment to perfect their knowledge of the place. The only way is through a blended-mode of edutourism to fulfill SDG 4 (Quality education).

In line with current trends of Education 4.0, various studies related to sustainable blended-mode edutourism were reviewed. A study on sustainability factors of Private Higher Education Institutions (PHEIs) in Saudi Arabia was conducted. Elaborate discussions were made on how education could be made affordable, accessible anytime anywhere without the restriction of age, gender, nationality and social status in line with the United Nation SDG No. 4 (Quality education), No. 5 (Gender Equality) and No. 8 (Contribution towards decent work and economic growth to expand the nation) [38]. Innovative business models were reviewed [39–42]. Apart from green marketing, suggestions were also made to expand the business into innovative ventures, for example, popularizing edutourism by catering to the Middle Eastern students with block courses during the summer break or the period while they are awaiting their student visa. The business is committed to social causes as it offers employment opportunities to the unemployed and scholarships to students facing financial constraints in the future stages [38].

An online course was conducted titled ‘The Real World of Immersive Augmented Reality (AR)’ (between October and December in 2017) that was completed under the SEAMEO Education Agenda Virtual Learning programme with project output archived on SEAMEO's website. Some of the exemplary projects were also shared in Edmodo social learning platform under sub-themes such as 'Conservation and Wise Use of Resources' (ConWUR) and 'Telecare and Healthy Lifestyle' (TeleHeal). Since the topics for AR projects are diverse to promote LearnT-SMArET, there is some interesting project output related to Environmental Education and Edutourism.

Figure 22.3 (posted onto LeSMaT-ConWUR sub-portal in Edmodo) is an illustration of how a Life Skills School Program was conducted in the northern part of Thailand, a country in which Sufficient Economy is emphasised by the King and worthy to be visited for values-based sustainable edutourism programmes. However, within the constraints of budget, time and space, virtual edutourism could also be conducted through blended-mode platforms as illustrated. A brief description or powerpoint presentation in the form of a video posted on Edmodo is reported in this paper. More elaborate discussions including video uploaded onto YouTube with research/evidence-based findings were presented virtually by Promsing who is the co-author of this paper [43,44].

Figure 22.4 illustrates how Edmodo social learning platform was leveraged (by one student delegate who was a participant of AR online course) to introduce a place of interest in Indonesia, i.e. 'Lubang Jepang' or directly translated as Japanese holes, which are tunnels built by the Japanese army around the second world war in Bukit Tinggi, West Sumatera, Indonesia. Using the same platform, the

Figure 22.3 YouTube link posted on Edmodo LeSMaT-ConWUR sub-portal on Life Skills School programme in Thailand

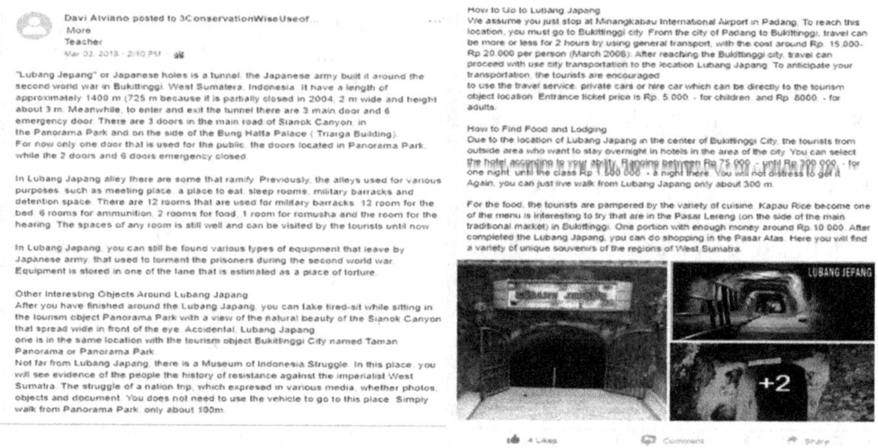

Figure 22.4 Use of Edmodo social learning platform to promote edutourism resort 'Lubang Jepang' in Indonesia

student delegate also extracted his AR project to showcase landmarks of ASEAN countries with brief write-ups and illustrations using 2D and 3D.

Figure 22.5 depicts how Lao PDR could be introduced through posting the landmark of this country onto Edmodo social learning platform.

In Malaysia, an example that facilitated edutourism was the students' societal service for tertiary level in one local university in Johor where the second co-author works. The university students carried out social obligations under their university's corporate social responsibilities to the rural area to conduct social works as well as field trip learning. It is part of edutourism as almost 65% of university students were from cities and urban regions. In the university paper entitled 'Agro Based Marketing', an elective paper for Business Management degree student final semester, a field trip tour was part of their assignment. Therefore, a visit to a rural farm suits them well. Figure 22.6 illustrates the process of planting oil palm in a swampy area. It was part and parcel of students' learning process as, especially, in Malaysian rural vicinities, agrarian societies have existed since independence till now. It will be interesting for students from urban areas to know more about crops the rural folks grow and the process of growing like most of the monoculture in Malaysia. Almost all courses offered in the Malaysian tertiary education of all disciplines could be transformed into edutourism at various levels to benefit the students and society.

A farsighted teacher will organise field trips for his/her students in whatever discipline of studies that would not only enable the students to gain insightful knowledge pertaining to their study but also prepare them for their future undertakings upon graduation. The meaningful adventures through edutourism are of utmost importance in every student's life, in gaining knowledge in socio-cultural, political and economic facets of life.

Figure 22.5 Use of Edmodo social learning platform to promote edutourism resorts in Lao PDR

Figure 22.6 Field trip to study the process of planting oil palm in a swampy area in Malaysia

Knowledge is power, the old saying is very true, unless students are prepared through this means. It is very unlikely that students will gain insightful knowledge with just chalk and talk or just arm-chair traveller syndrome. Therefore, provided one has gone through the edutourism process, only then a student will prepare as a perfect person and equipped with facts and actual scenarios that equate with the old saying that to learn something, one would have to walk a thousand miles. The only way to prepare the human capital of a country is to groom tertiary educational institutions and their students into mature and intelligent people to bring forth the development of the country.

This subsection discusses part of the authors' R&D activities to develop an e-platform to promote transdisciplinary studies integrating life (work, survival and entrepreneurial) skills. Only selected cases presented in Edmodo e-platforms by delegates from Indonesia, Malaysia and Thailand were reported. Upon reflection, the authors felt that the introduction of blended-mode training programmes incorporating transdisciplinary studies has opened up a wide array and diverse curriculum with learning output as aforementioned. For example Indonesian students from Science or Arts streams prepared impressive AR project output during the first (2017-2018) online course training [20,45]. Thai and Malaysian participants also prepared output uploaded onto YouTube for the second (2018–2019) online course training.

The authors managed to keep abreast with the latest technology for Industrial Revolution (IR4.0) by exploring how blended-mode platforms could be used to manage sustainable ecotourism programmes that provide lifelong learning opportunities for work, survival and entrepreneurship ventures. Such exposures are important, especially, for high school students who will be facing the challenges of either to opt for further studies or to stop schooling and start work. Various reasons were reported by researchers (e.g. [46]) to have contributed to students' dropping out from the previous educational system that was not yet supported by technology. But this study implies that wise use of technology in the IR4.0 era will expedite the implementation process of blended-mode programmes.

The lessons learnt from Series 1 online course also made the author become more experienced to train with better management in subsequent online training, e.g. the implementation of an e-survey (http://bit.ly/lesmatesurveyprojectbrief) that was used to guide the preparation project output with various focuses of transdisciplinary education. Hence, continuous efforts should be made to promote the sustainable implementation of values-based technology-enhanced programmes incorporating transdisciplinary education to facilitate human resource development (HRD) [25]. Amid unprecedented global pandemic issues that have impacted tourism, perhaps a safety net should be provided. Lessons learnt from organisers and authorities implementing edutourism programmes could be referred for future practice to ensure sustainable implementation of such programmes. Further, the insightful worldview that students gained from various communities on blended-mode tourism among SEAMEO counterparts would benefit students in gaining an insightful knowledge of the socio-cultural panaroma and beliefs, practices and norms of all the local societies as well as communities in the countries of the SEAMEO region and beyond.

Special mention

All authors who contribute to this chapter receive a special mention.

1. Environmental Conservation for Sustainability in Fulfilling SDGs

Awangku Hassanal Bahar Pengiran Bagul, Ng Khar Thoe, Ong Eng Tek and Careemah Choong Lay Kheng

2. Monitoring occupational health and safety in Small and Medium Industrial (SMI) manufacturing sector: Challenges and future directions
   Khoo Nee Kah, June Chua Huey, Leong Wai Yie, Ng Khar Thoe, Pang Yee Jiea and Miranda Poh Khoon Yeoh
3. Enhancing awareness on environmental and preventive healthcare for sports science supported by technology: A systematic review and suggested research
   Ng Jing Hang, Leong Wai Yie, Ng Khar Thoe, Rajendra Kumar, Miranda Poh Khoon Yeoh and Goh Jia Haur
4. Conserving cultural heritage through Minecraft digital tool
   Pang Yee Jiea, Ng Khar Thoe, Chin Chee Keong, Ng Chin Siang, Ng Yu Yan and Lim Seh Hoe
5. Development of digital platforms in managing sustainable ecotourism programmes: Lessons learnt and the way forward
   Ng Khar Thoe, Lee Tan Luck, Careemah Choong Lay Kheng, Awangku Hassanal Bahar Pengiran Bagul, Raden Indarjani and Natthasurachat Promsing

References

[1] Suresh, P. and Johnson L. R. (2015). The post-2015 agenda: from millennium development goals MDGs to sustainable development goals SDGs. *International Journal of Current Research and Review*, 7(15), 62–67.

[2] UN [United Nations] General Assembly. (2012). Resolution adopted by the General Assembly on 27 July 2012, 66/288. The future we want. http://www.un.org/ga/search/view_doc.asp?symbol¼A/RES/66/288&Lang¼E (accessed on 25 August 2017).

[3] UN [United Nations] General Assembly (2011). High-level Meeting on Addressing Desertification, Land Degradation and Drought in the Context of Sustainable Development and Poverty Eradication. A/65/861. http://www.un.org/en/ga/ president/65/issues/66ga_unccd_hlmtg.pdf (accessed on 23 August 2017).

[4] Locke, H., Ellis, E.C., Venter, O., *et al.* (2019) Three global conditions for biodiversity conservation and sustainable use: an implementation framework. *National Science Review,* 6(6), 1080–1082. Retrieved from https://doi.org/10.1093/nsr/nwz136.

[5] Ng, J.H., Kumar, R., Ng, K.T., Leong, W.Y. & Goh, J.H. (2020). Visual learning tools for sports and physical health education: a reflective study and challenges for the ways forward. Presentation Compiled (p81) in The Proceedings of the 5th International Conference on Management, Engineering, Science, Social Science and Humanities (iCon-MESSSH'20) (Virtual).

[6] Rouse, M. (2020). *Preventive Health.* Retrieved from https://whatis.techtarget.com/definition/preventive-health (accessed on 13 November 2021).

[7] William, C. and Shiel Jr (2021). *Medical Definition of Preventive Medicine.* Retrieved from https://www.medicinenet.com/script/main/art.asp?articlekey=5039 (accessed on 13th November 2021).

[8] Rittmann, B. (n.d.). *Environmental health and microbial biotechnology.* Retrieved from http://www.biodesign.asu.edu (accessed on 13 November 2021).

[9] Mosttafiz, S., Rahman, M. and Rahman, M. (2012). Biotechnology: role of microbes in sustainable agriculture and environmental health. *The Internet Journal of Microbiology*, 10 (1). Retrieved from https://ispub.com/IJMB/10/1/14136 (accessed on 13 November 2021).

[10] Usta, Canan (April 24 2013). Microorganisms in Biological Pest Control—A Review (Bacterial Toxin Application and Effect of Environmental Factors), Current Progress in Biological Research, Marina Silva-Opps, IntechOpen, DOI:10.5772/55786. Retrieved July 20, 2021 from: https://www.intechopen.com/chapters/44118

[11] National Geographic (n.d.). *Biomass Energy*. Washington, DC: National Geographic Headquarters. Retrieved July 20, 2021 from https://www.nationalgeographic.org/encyclopedia/biomass-energy/

[12] Basli, A., Belkacem, N. and Amrani, I. (2017). *Health Benefits of Phenolic Compounds Against Cancers*. Retrieved from https://www.intechopen.com/books/phenolic-compounds-biological-activity/health-benefits-of-phenolic-compounds-against-cancers (accessed on 13 November 2021).

[13] The Meadowglade (2020). *What are coping mechanisms for stress?* Moorpark, Californa. Retrieved from https://themeadowglade.com/what-are-coping-mechanisms-for-stress/

[14] Huizenga, J., Admiraal, W., Dam, G. and Voogt, J. (2019). Mobile game-based learning in secondary education: students' immersion, game activities, team performance, and learning outcomes. *Computers in Human Behavior*, 99 (April), 137–143. Retrieved from https://doi.org/10.1016/j.chb.2019.05.020

[15] Li, M. (2016). Developing skills and disposition for lifelong learning: culture creative issues supervising international doctoral students in New Zealand universities. *Journal of International Students*, 6(3), 740–761.

[16] Sony PlayStation F.C. App. (n.d.). *FootballVvideos onPS4*. Retrieved from https://www.playstation.com/en-gb/explore/ ps4/entertainment/playstation-fc/ (accessed on 13 November 2021).

[17] Buzko, V. L., Bonk, A. V. and Tron, V. (2018). Implementation of gamification and elements of augmented reality during the binary lessons in a secondary school. Paper presented at the Proceedings of the 1st International Workshop on Augmented Reality in Education Kryvyi Rih, Ukraine, October 2, 2018.

[18] Erhel, S. and Jamet, E. (2018). Improving instructions in educational computer games: exploring the relations between goal specificity, flow experience, and learning outcomes. *Computers in Human Behavior*, 91 (September, 2018), 106–114. Retrieved from https://doi.org/10.1016/j.chb.2018.09.020

[19] Westhuizen, v. d. E. J. & Haar, D. v. d. (2018). *A Wearable Device-Based Framework for Determining Player Effectiveness on the Football Pitch*. Retrieved from https://www.researchgate.net/profile/Dustin_Van_der_Haar/publication/ 324814210_A_wearable_device-based_framework_for_determining_player_effectiveness_on_the_football_pitch/ links/5ae94b4f0f7e9b837d3b18b3/A-wearable-device-based-framework-for-determining-player-effectiveness-on-the-football-pitch.pdf (accessed on 13 November 2021)

[20] Narulita, S., Perdana, A. T. W., Annisa Nur F., Darmakusuma, M.D., Indarjani and Ng, K. T. (2018). Motivating secondary science learning through 3D interactive technology: from theory to practice using augmented reality. In K.T. Ng (Ed.), *Learning Science and Mathematics On-line Journal*, 13, December, pp. 38–45. Retrieved from http://myjurnal.my/public/article-view.php?id=135506 (accessed on 13 November 2021).

[21] Soh, J., Khoo, H. and Ng, J.H. (2018). Movement on the earth, water, in the air and space. In Storybook submitted for Augmented Reality online training *(October to December, 2018)*. Penang, Malaysia: Chung Ling Butterworth Secondary School.

[22] Kos, M. Kramberger, I. (2017). *A Wearable Device and System for Movement And Biometric Data Acquisition for Sports Applications*. IEEE Access, March 17, 2017, 99: 1. Retrieved from https://ieeexplore.ieee.org/stamp/stamp.jsp?arnumber=7865994 (accessed on 3 March 2022).

[23] Machado, J.A. Tenreiro & Lopes, A.M. (2017). *Mathematical Modeling of Soccer Dynamics*. NASA Astrophysics Data System (ADS). Retrieved https://www.science.gov/topicpages/d/dynamic+mathematics+software.html# No. 94. Retrieved https://www.science.gov/topicpages/d/dynamic+mathematics+software.html#.

[24] Koumantareas, J. and Oliveira, R. C. D. (2018). Sports and health from the perspective of physical education professionals. Motriz: Revista deEducacao Fisica. Epub *Nov 2018*. Retrieved from http://www.scielo.br/scielo.php?script=sci_arttext&pid=S1980-65742018000400702 (accessed on 13 November 2021).

[25] Ng, K. T. (2017). Development of transdisciplinary models to manage knowledge, skills and innovation processes integrating technology with reflective practices. *International Journal of Computer Applications (IJCA) (0975-8887)*. Scholarly peer-reviewed research publishing journal. No. 2. pp.1-9. Vol. 975, pp. 15–23. Retrieved from https://pdfs.semanticscholar.org/9c1c/f423f6d2c5810c866ffbba64f65832179b5b.pdf (accessed on 13 November 2021).

[26] https://youtu.be/bOvs3_4bjRQ.

[27] https://youtu.be/Khj0u3HZ5gM.

[28] https://youtu.be/cCpFlBUg4m8.

[29] https://youtu.be/HKVVm-NdB_s.

[30] https://youtu.be/FmrV2v2JHFU.

[31] https://youtu.be/swPI_7rRmag.

[32] https://youtu.be/HECD61NnAds.

[33] https://youtu.be/zILUN3Ejx0A.

[34] https://youtu.be/MXDv74HIL1w.

[35] https://youtu.be/PN8xxnOB36g.

[36] https://bit.ly/BlueMansionPenang and https://youtu.be/DPA0Czc5UbA.

[37] ICS (n.d.). *International Career Studies (ICS)*. 2011–2012. Retrieved: http://internationalcareerstudies360.blogspot.com/2012/02/edu what.html

[38] Choong, C. (2020). *Sustainability Factors of Private Higher Education Institution in Saudi Arabia*. Unpublished doctoral thesis, Kuala Lumpur, Malaysia: Asian e-University.

[39] Belmonte and Murray (1993). Getting ready for strategic change. *Information Systems Management*, 10(3), pp. 23–29.

[40] Hannan and Silver (2000). *Innovating in Higher Education Teaching, Learning and Institutional Cultures*. Society for Research into Higher Education Buckingham and Open University Press

[41] Thille, C.M. (2013). *Surfing the Tsunami: Faculty Engagement with the Open Learning Initiative*. University of Pennsylvania. Retrieved from https://repository.upenn.edu/dissertations/AAI3592387/ (accessed on 3 March 2022).

[42] Whitworth, A. (2012). Invisible success: problems with the grand technological innovation in higher education. *Computers & Education*, 59, 145–155.

[43] Promsing, N., Phromsing, K. and Ng, K.T. (2018a). Involving primary students in project-based blended-learning activities: Exemplary practices in a local school at Thailand. Virtual presentation compiled in K.T.Ng. (2018). *Supporting Transdisciplinary Science/Mathematics Education, Management and Social Development Integrating Use of e-Tools and Sustainable Technology-Enhanced Communities*. Co-chair presentation during the International Conference on Recent Trends in Science, Technology, Management and Social Development (ICRTSTMSD-18) conference at Bali, Indonesia, 4–5 August, 2018.

[44] Promsing, N., Phromsing, K., Ng, K.T. and Lee, T.L. (2018b). Integrating indigenous knowledge and skills in technology-enhanced science project-based activities: exemplars, issues and challenges. In K.T. Ng (Ed.), *Learning Science and Mathematics (LSM) Online Journal*, 13, 73–86, December 2018. Retrieved: http://myjurnal.my/public/article-view.php?id=135509 (accessed on 13th NOV 2021)

[45] Wahid, Y., Iyan, S.D., Davi, A.A., Dava, A.A. & Ng, K.T. (2018). Learning mathematical concepts contextually supported by digital tool and e-platform. Published in refereed Proceedings of ICRTSTMSD-18, August 4-5, 2018 at Kuta Central Park Hotel, Bali, Indonesia.

[46] Zainal, G., Haris, M.J. and Ng, K.T. (1991). *The Malaysian Dropout Study Revisited*. Penang, Malaysia: Basic Education Research Unit. Retrieved https://scholar.google.com/scholar?cluster=9786386042006755463&hl=en&oi=scholarr (accessed on 13 November 2021).

Chapter 23

Rethinking and redesigning strategies related to IR4.0 to bridge the gap of human resource development in ICT industries and smart manufacturing

Por Fei Ping[1], Miftahul Hidayah[2] and Ng Khar Thoe[3]

The global landscape is changing rapidly in the fourth industrial revolution (IR4.0). The sustainability of a country's economy is closely linked to its successful strategies to develop existing and future human resources to embrace unprecedented technological advancement in various disciplines, especially ICT-based industries and smart manufacturing. This chapter focuses on the enculturation of lifelong learning to constantly reskill, upskill and cross-skill human resources to surf the waves of technological change. The strategies launched by the government of Malaysia to foster lifelong learning have also been scrutinised in this chapter to ensure there is no gap between stated objectives of the need to produce competent and versatile human resources and actual results. Lifelong learning plays a pivotal role in empowering individuals to keep abreast of the technological shift in smart manufacturing. To develop a relational understanding of geometry concepts required in smart manufacturing, the discussion about designing Techno-based Mathematical Tasks (TbMT) using Cabri 3.3 is included in this chapter. To design TbMT provides opportunities for exploration and construction of tasks independently which help enhance and extend the understanding of geometry concepts. This chapter concludes with the insights gleaned from the competition of building young minds through Minecraft digital tool to embrace innovation and transformation in smart manufacturing. As Minecraft is a 'sandbox game', there is no exact objective or end goal. This seemingly endless creative freedom inculcates creativity and innovation among children since young by using the 1x1 coloured blocks to manipulate the imaginative world. Conclusions are derived and recommendations are made for future work on rethinking and redesigning strategic implementation to bridge the gap of human resource development in ICT-based industries and smart manufacturing.

[1]School of Education, Humanities and Social Sciences, Wawasan Open University, Penang, Malaysia
[2]SEAMEO QITEP in Mathematics. Indonesia
[3]Training and Research Division, SEAMEO RECSAM, Penang, Malaysia

23.1 Lifelong learning for human resource development in the era of IR4.0

Rapid technological change and a fluid future in this uncertain era demand more market-relevant skills and knowledge to remain constantly competitive and sustainable. As observed, the manufacturing industry is undergoing a digital transformation with the integration of IR4.0 technologies. The seamless connection between digitisation and real-world needs allows manufacturing processes to be more flexible, productive, resource-saving and energy-efficient. The internet of things (IoT) solutions and cloud computing enable machines to communicate with production, workpieces and other necessary devices in real-time. Abnormal manufacturing conditions can also alert the system immediately to empower effective action.

With the emergence of this smart manufacturing, lifelong learning is vitally essential to develop and equip existing human resources to meet the needs to move with the waves of change. The concept of lifelong learning existed initially in the 1970s to provide adults the access to formal courses in educational institutions; while in 1996, OECD (Organization for Economic Cooperation and Development) Education Ministers adopted ‘Lifelong Learning for All’ as a policy framework. In essence, a more comprehensive perception of lifelong learning is adopted by covering all purposeful learning activities pursued by all individuals to improve their knowledge and competencies. In a similar move, international organisations such as UNESCO and the European Commission have also adopted the more comprehensive view [1]. People at different stages of life should continually engage in learning. Learning happens in various forms and various settings, from formal education in schools or universities to different types of learning at the workplace, in the community and even at home. Formal, non-formal and informal types of learning contribute to the growth of human resources, especially in mastering new advanced skills in smart manufacturing to strengthen the capacity of individuals to adapt to the scale of current economic change.

In the article written by Klaus Schwab [2], the Founder and Executive Chairman of the World Economic Forum, it is postulated that IR4.0 could result in disruption of labour markets. Human labour is substituted by automation across the entire economy; economic inequality is exacerbated; and the segregation between “low-skill/low-pay” and “high-skill/high-pay” is increasingly widened. Pondering upon the scenario, adopting lifelong learning will help individuals achieve sustainable growth.

According to the Networked Readiness Index 2019 [3], Malaysia is ranked 32nd out of 121 countries in readiness for a networked economy. The Networked Readiness Index 2019 is developed based on four pillars, which are Technology, People, Governance and Impact. As reflected in the overall scores, Malaysia achieved 63.76, leading the group of upper-middle-income countries. Malaysia is considered a consistent performer in all four pillars, ranging from 36th in Technology, 33rd in Impact, 32nd in People to 29th in Governance.

The Malaysian government is a strong advocate of lifelong learning in the era of IR4.0 as evidenced by the formulation of various plans and policies as well as the publication of official documents that bear testimony to its commitment to lifelong learning. In an attempt to address concerns of human resource development, the Ministry of Human Resources in Malaysia is geared to promote lifelong learning by reskilling, upskilling and cross-skilling the workforce. For instance, a human-centred agenda to synergise strategic partnership across ministries and agencies was presented in the 2019 Human Resource Conference: Labour Market Dynamics and Future of Work [4]. The Blueprint on Enculturation of Lifelong Learning for Malaysia 2011–2020 (Ministry of Higher Education Malaysia, 2011) sets out a comprehensive agenda of lifelong learning by recognising Lifelong Learning as the third pillar of the education system and human capital development. The first pillar is School System, while the second pillar is Tertiary Education. In response to meeting the evolving needs due to technology advancement, public health crisis, environmental issues as well as political and economic instability, the education system is to be re-modeled to focus on long-term human capital development by enhancing knowledge and skills obtained from primary and secondary school system, tertiary education in college and university, and then continue the on-going personal and professional development through lifelong learning. The Mid-Term Review (MTR) of the 11th Malaysia Plan (11MP) also focuses on empowering human capital by highlighting it as the fourth pillar of priority [5]. Under the National Workforce Human Capital Development Blueprint 2018–2025, the 5th priority for action is deploying lifelong human capital development solutions (Human Resources Development Fund, 2018). The uncertain and fluid future of IR4.0 requires new approaches to human capital development. However, in reality, the complex phenomenon of IR4.0 is not easily fathomed and translated into educational perspectives. The Ministry of Higher Education Malaysia has taken bold initiatives to redesign the higher education ecosystem in Malaysia with achievable strategies, as articulated in 'Framing Malaysian Higher Education 4.0: Future-Proof Talents' [6] to train and equip the existing talent pool to be future-ready. There are four critical foci and fifteen comprehensive approaches formulated, chief of these are 2u2i, CEO@Faculty Programme, iCGPA, Malaysia MOOCs, APEL, Gap Year, Malaysia English Assessment (MEA) and E-Portfolio [7].

Lifelong learning is a broad and subjective concept. Its interdisciplinary nature penetrates various fields of studies, from manufacturing to healthcare. With a concrete understanding of the concept, a new integrative lifelong learning model can be explored and an assessment tool to measure the attributes of lifelong learning in different disciplines can be developed to determine the preparedness of individuals in embracing IR4.0 to meet the demand of smart manufacturing.>

Acknowledgment

This work was supported by Fundamental Research Grant Scheme (FRGS) by the Ministry of Education Malaysia (FRGS/1/2019/SS109/WOU/03/1).

23.2 Redesigning strategies to enculture lifelong learning for the success of smart manufacturing

Robotic, artificial intelligence, IoT, digitalization and automation are close to people's hearts in the era of the IR4.0. The IR4.0 technologies are intensively deployed in almost every industry – from agriculture and healthcare to education and manufacturing – to save cost, increase productivity and improve efficiency. In the manufacturing industry particularly, the IR4.0 technologies are deployed to boost operational effectiveness, reduce scrap and downtime, heighten quality capabilities and improve safety. The IR4.0 advancement ushers in a revolution of experience. For example, an agriculture company could use IoT technology to monitor and gather data regarding soil moisture contents and nutrient levels to determine water usage for irrigation purposes, customize fertilizer blends and plan for the optimal time to harvest. There is also real-time visibility of farm activities and weather analytics. The statistics presented by Thomson Reuters, CB Insights, and PwC analysis [8] demonstrated a huge increase of USD117.1 billion of IR4.0 technology investments just within 7 years, from 2012 to 2018. In this period, a total of USD646 billion has been invested in IR4.0 related technologies.

With IR4.0 technologies permeating every industry, the global workforce is experiencing the biggest shift in employment history. The analysis by McKinsey Global Institute [9] alerted that approximately 50% of current work activities have been technologically automated. At the fastest speed of adoption of IR4.0 technologies, there will be 375 million of the global workforce that needs to switch occupational categories. In addition, by 2030, 2.66 billion of the global workforce will be in new occupations [9]. In these unpredictable global environments, people face greater threats to survival. They face the threat of being left behind, becoming irrelevant and losing their jobs or their livelihoods. The global workforce is intensely impacted in the era of IR4.0. To be able to surf the waves of unpredictable change, lifelong learning is a must to acquire new skill sets to perform new tasks. Lifelong learning is seen as a continuous process. It is considered as a set of learning activities that aims to improve the skills, knowledge and competence of an individual. In other words, lifelong learning can be defined as giving education to learners throughout their lives, which precisely means education exists within all phases of an individual life cycle or throughout an individual's whole life [10].

In relation to current global scenarios, higher education institutions shoulder great responsibilities to mold human capital that is employable in the IR4.0 smart manufacturing environments. The prosperity of a nation's economy is closely linked to its successful system to develop human capital that can cope with the potential displacement of the workforce or upcoming workforce transitions. The IR4.0 smart manufacturing and its significant implications for the future of human capital present the higher education institutions with a unique opportunity to create a culture of lifelong learning to ensure the continued relevance of the global workforce. This section aims to scrutinise the initiatives launched by the government of Malaysia to develop an education system that is future-ready.

Malaysia is one of the countries forging ahead with IR4.0, which is evidenced by the report of Networked Readiness Index 2019. Of all the 121 countries surveyed, Malaysia is ranked 32nd. This is a remarkable achievement as Malaysia is the only country from the upper-middle-income group that ranked in the top quartile. The rest of the countries in the top quartile are high-income countries, such as Sweden, Singapore, Norway, Germany, and the like [3]. The assessment of the Networked

Readiness Index is based on four pillars, which are technology, people, governance and impact. Firstly, technology is the heart of a networked economy especially in the era of IR4.0, where access to technology, including access to infrastructure and affordability, is evaluated. The content which refers to the type of digital technology provided and the level of future technologies the country is prepared for are also evaluated. Secondly, in terms of people, the application of technologies by individuals, businesses and governments are reviewed. Thirdly, to ensure a positive and inclusive impact on the community and economy, appropriate governance mechanisms need to be implemented. Hence, trust (security), regulation and inclusion are the assessment criteria under governance. Last, given the main concern of using technology is to improve the growth and well-being of the community and economy, therefore, it follows that the impact of technology on the economy, quality of life and contribution to sustainable development goals is included in the assessment [2]. Given the well-structured and comprehensive assessment of Networked Readiness, the ranking of Malaysia indicates the high level of its readiness in embracing IR4.0.

To ensure the nation is ready for IR4.0, the government of Malaysia has launched various measures to mitigate the risks of displacement of the workforce. In the Human Resource Conference held in 2019 with the theme 'Labour Market Dynamics and Future of Work', the Ministry of Human Resources in Malaysia presented the importance of creating skillful, knowledgeable and innovative human capital to meet the requirements of the industry especially in the era of IR4.0. Human capital development initiatives will provide opportunities for quality employment as well as ensure access to quality education and training towards building a more inclusive, equitable and prosperous nation. These initiatives are aligned with Pillar IV: Human Capital Development, Pillar VI: Strengthening Economic Growth and Mid-Term Review of the 11th Malaysia Plan [4].

To be at par with the international standards of Sustainable Development Goal 8 Decent Work and Economic Growth of promoting inclusive and sustainable economic growth, employment and decent work for all, the Ministry of Human Resources is reviewing labour laws: Employment Act 1955, Trade Union Act 1959 and Industrial Relations Act 1967 [4]. The National Policy on Industry 4.0 is formulated to drive the development of an end-to-end Industry 4.0 ecosystem in Malaysia to produce high-skilled employment (Ministry of International Trade and Industry, 2018). Upskilling, reskilling and cross-skilling are critical to power a holistic step up in national productivity across all sectors of the economy.

New priorities and emphases have been tabled in the Mid-Term Review of the Eleventh Malaysia Plan (2016–2020) by outlining six pillars to support inclusive growth and sustainable development [5]. The fourth pillar is to empower human capital. Given knowledgeable and innovative human capital is an important factor to support economic growth [5], therefore, it follows that accelerating human capital development continues to be the priority of the government. Efforts on human capital development by equipping the workforce with relevant skills enhance quality employment towards building a more inclusive, equitable and prosperous nation.

Human capital development is needed to provide the workforce with opportunities to earn high salaries and wages under better working conditions. At present, the composition of jobs in Malaysia is mainly concentrated in semi and low-skilled categories. Various initiatives have been taken by the government to generate skilled jobs which focus on encouraging automation and innovation, adoption of new technologies and commercialization of intellectual property. In this regard, high-quality investment is given priority to create more skilled jobs in the economy. Critical skills required by the industries have been identified by the Industry Skills Committee, and the Critical Occupations List has also been developed to address skills imbalances and job mismatches [5]. The new priorities and emphases are aimed to empower human capital as well as produce competitive and resilient human capital in the era of IR4.0 to support the national aspiration to become an inclusive and developed nation.

The growing prominence of IR4.0 necessitates the urgent call for innovative solutions to equip the current and future workforce to be globally competitive and relevant in the rapidly-developing digital reality. The National Workforce Human Capital Development Blueprint (2018–2025) [6] puts a strong focus on investments in education; and the reskilling, upskilling as well as cross-skilling of local talents are crucial to improving social mobility and the well-being of the society as a whole. Human capital is the backbone of the nation's economy. It is a critical enabler to drive and sustain the nation's economic growth and support the transition of all economic sectors towards knowledge-intensive activities, especially smart manufacturing in this context.

The Blueprint aims to catalyze the upskilling, reskilling and cross-skilling of the Malaysian workforce in line with the national goal of having a 35 per cent skilled workforce by the year 2020. Skilled workforce enhances the potential to drive economic development, social advancement and innovation. The Human Resources Development Fund (HRDF) under the Ministry of Human Resources Malaysia deploys lifelong human capital development solutions by supporting apprenticeships, expanding support for displaced and retrenched workers and promoting recognition of learning done throughout the lifecycle. The 1MALAYSIA Outplacement Centre (1MOC) job portal is ready, and the number of participants under Recognition of Prior Experiential Learning Scheme (RPEL) is under monitoring. RPEL allows employers to claim reimbursement for obtaining recognition of their employees' skills and experience from the Department of Skills Development (Human Resources Development Fund, 2018).

The Ministry of Higher Education's Malaysia Education Blueprint 2015–2025 (Higher Education) includes lifelong learning as one of ten paradigm shifts necessary to create excellence in higher education [11]. The Ministry of Higher Education's Blueprint on Enculturation of Lifelong Learning for Malaysia 2011–2020 also recognises the importance of lifelong learning, calling it the third pillar of human capital development after primary and secondary school and tertiary education [6]. In line with these initiatives, Malaysian Higher Education Framework 4.0 is formulated by the Ministry of Higher Education Malaysia (2018) to cope with the challenges and to welcome the opportunities brought by IR4.0. The Ministry of Higher Education Malaysia is entrusted with the mission to sustain the higher education ecosystem to develop and enhance individual potentials and fulfill the nation's aspirations.

The IR4.0 represents new ways of redesigning the higher education system in Malaysia. This brings education and industry to greater partnership to nurture human capital that is market-relevant. Obsolescence of human capital belongs to the heart of economic challenges. The three types of human capital obsolescence are particularly relevant in the era of IR4.0: job-specific obsolescence due to technological and organisational change; sector-specific obsolescence due to shifts in employment; firm-specific skills obsolescence due to displacement [12]. This obsolescence of human capital addresses the need for higher education to significantly promote lifelong learning to maintain employability of the working population. Firstly, job-specific skills obsolescence occurs due to technological or organisational developments in the production process which changes the skills demanded for a particular kind of job. In this scenario, the skills the workers in that profession possess are probably no longer sufficient to perform their jobs adequately. Secondly, for sector-specific obsolescence, workers' skills may still be adequate for their profession, but the demand for their profession is declining. Shifts in the industry structure of employment can be due to shifts in consumer demand, investments, government purchases and international trade patterns. Due to these demand shifts, part of the workers employed in sectors of industry that face a sharp decline in employment has to find a job in another economic sector to remain employed. Lastly, firm-specific skills are defined as a set of skills that enable an individual to perform highly specialised tasks in a few similar industries with limited mobility, such as agricultural and fishery work, and pilots who work for the Navy or Air force. The obsolescence occurs when workers are laid off in case of company closure or reorganisation. When retrenched, the firm-specific skills of these workers lose their value [12].

In this highly volatile environment, the Ministry of Higher Education Malaysia envisions the need of nurturing lifelong learning of human capital for sustainable development and growth in the era of IR4.0. The Ministry has framed four foci, 15 approaches with detailed initiatives as well as three future-proof attributes and nine Malaysian future-proof skill sets in the book titled "Framing Malaysian Higher Education 4.0: Future-Proof Talents' '. The four foci stressed in the Framework are Future Ready Curriculum, Agile Governance, Research and Innovation, and Talent Planning [7]. The Ministry also encourages all private and public higher education institutions to allocate 30% of its curriculum to floating lessons to equip their students with skills and knowledge so they are well-prepared to face the challenges in the rapidly changing industry.

A future-ready curriculum is a framework that highlights the elements of flexible education, including 2u2i work-based learning degree programs, CEO@Faculty Program, iCGPA (integrated Cumulative Grade Point Average), Malaysia MOOC (Massive Open Online Courses), APEL (Accreditation Prior of Previous Experience-based Learning), Gap Year, MEA and E-Portfolio [7].

To ensure the success of implementing smart manufacturing, developing human capital is a must to accelerate digital transformation. The Smart Manufacturing Industry Report [13] highlighted the priority of promoting a learning culture in the industry to drive optimal results. As reported, one of the top barriers preventing smart manufacturing is people-related: there is a serious lack of required skill sets to manage the implementation of advanced technologies. Rethinking and redesigning strategies to develop current and future human capital is of utmost importance so that training people on the integration and use of smart technologies throughout the entire supply chain becomes imperative. To grow and innovate in what is projected in smart manufacturing, apart from investing in technology and equipment, 'people' play a vital role in transforming the operations. Hence, strategies to bolster training of both the current workforce and new hires with needed competencies to meet the demands of the IR4.0 economy require urgent attention.

This section encapsulates the way forward for Malaysia in embracing IR4.0 by adapting to the ever-changing demands of industry. Regardless of technological advancement, the Malaysian workforce would be able to overcome it through lifelong learning by continuous upskilling, reskilling and cross-skilling. The strategies implemented help build the talent pipeline to ensure current and new workforce are familiar with digital technology to become more productive, competitive and effective in their roles. The philosophy of lifelong learning is to be encultured to train and develop the best people for the realization of smart manufacturing. This approach is good for the economy, the nation and the individuals. Learning should not stop after graduating college/university. Learning continues for the rest of an individual's life for both career and cognitive development. The implementation of the strategies to develop human capital that is market-relevant has to be accelerated in a fast and steady manner; otherwise, the nation will be at risk of falling behind in the stride to development. No institution can succeed in a silo in this fast-changing IR4.0 environment. To be IR4.0 ready, the nation needs to cultivate a new mindset of collaboration across disciplines, organisations and industries to create opportunities, to break barriers and boundaries, and to find progress together.

Acknowledgment

This work was supported by Fundamental Research Grant Scheme (FRGS) by the Ministry of Education Malaysia (FRGS/1/2019/SS109/WOU/03/1).

23.3 Designing techno-based mathematics tasks for learning geometry

To develop a relational understanding of geometry concepts required in smart manufacturing, the discussion about designing TbMT using Cabri 3.3 is included in this sub-section. TbMT design provides opportunities for exploration and construction of tasks independently which help enhance and extend the understanding of geometry concepts.

Tatsuoka and Corter [14] found that there is a high correlation between geometry and the attributes to measure higher-order mathematical thinking skills, which is one of the skills that is required in the IR4.0. However, some research literature revealed two main difficulties that underlie the development of students' spatial skills in geometry, and these are visualization and mental transformation of 3D figures [15,16].

Many studies have shown the significant roles of technologies to support students in learning geometry as reported by National Council of Teachers of Mathematics (NCTM, 17). In some cases, the role of visual function in technology was only for seeing and conjecturing, not for experimenting [18]. However, to develop a relational understanding of geometry concepts, students should be actively involved in the learning process. Ng *et al.* [19] argued that Dynamic Geometry Software (DGS) such as Cabri offered interactive aspects, including dragging, rotating, and changing views dynamically on-screen that can strengthen students' visualization and cognitive processes in learning geometry. Some research also supported that these kinds of activities could positively impact students' learning [20–22].

Therefore, in this sub-section, the authors illustrate how the design of TbMT using Cabri 3.3 might assist the students with the visualization and allow them to explore, construct, observe and manipulate the geometrical figures. The TbMT project began in this study by designing the technology-integrated tasks that embed digital dynamics technology for learning mathematics. The task is defined as mathematical activities that students can manipulate with underlying geometry concepts through tessellation context. The activities have been designed for 45 min which aim to stimulate students in exploring geometry concepts.

This study was conducted to pilot the process using design-based research that consists of four stages: analysis of practical problems, initial development, iterative cycle, and reflection. To provide an appropriate design principle in the use of TbMT in mathematics classroom, we adopted the framework developed by Reeves [23].

In the analysis of practical problems, which was conducted at the beginning of the project, the researchers analysed the mathematics curriculum in Indonesia to find an appropriate theme and context that can support students' understanding of learning geometry by utilising dynamic technologies. Therefore, the context of tessellation was chosen because it can bridge the key ideas of geometry by exploring patterns, angles and similarities.

The second phase of this design-based research is the initial development. The researchers used Cabri 3.3, supported by much research that stated the positive contribution of this DGS to support students' learning [24–27]. Three reviewers have been invited to give feedback and to empower the initial tasks. Some revisions have been made to improve the tasks based on the suggestions from the reviewers. After the revision process finished, the tasks were ready to be implemented.

The iterative cycle was started with the pilot implementation involving the students to determine the effectiveness of the TbMT design. Table 23.1 summarizes the pilot study activities that were conducted involving students in a sequence of tasks related to tessellation as one of the topics chosen under geometry.

Firstly, as an introduction, the students were asked if they had heard about tessellation before. We provided a definition of tessellation and an example to be

Table 23.1 Summary of the activities

Phase	Activities	Diagram/photo
Preliminary	Students were asked to tessellate a triangle and a square based on the given examples	See Figure 23.2
Challenge	Students had to make their own tessellations using the 2D shapes provided	See Figure 23.3
Main	Students investigated four different pentagons to determine which pentagon(s) can be tessellated	See Figure 23.4
Closure	Students determined the pentagon(s) that can be tessellated and explained the reason for their answer	See Figure 23.5

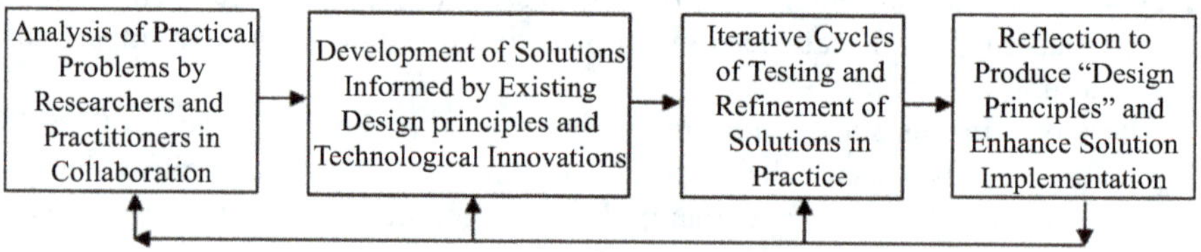

Figure 23.1 Development model [14]

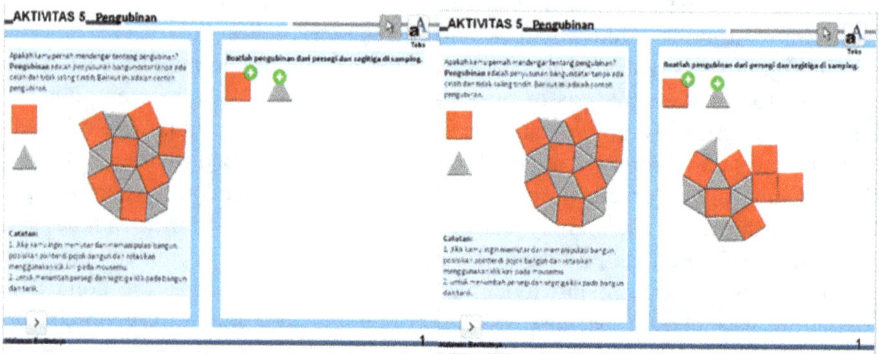

Figure 23.2 Activity 1 – preliminary

observed by the students. For the preliminary activity, the students were expected to create a tessellation based on the given example from a triangle and a square. The researchers thought that if students could not do the rotating and dragging in a right way, they might struggle to finish the whole activity. Therefore, the preliminary activity provided a good opportunity for students to get more familiar with the tools and interactive aspects in Cabri. The activity in the preliminary phase and examples of students' work can be seen in Figure 23.2. Some students created a wrong pattern. However, they were asked to revise their work until getting the correct pattern, as the given example (Figure 23.2).

Turning to the second phase, the students were challenged to create their own tessellations using three shapes provided which are triangle, square, and hexagon. This activity aims to give students the experience to explore and put them at ease to tessellate without any guidance. They were asked to arrange their own pattern. The most interesting thing in this activity was the trial-and-error activities. The phase and an example of students' work can be seen in Figure 23.3. In this phase, students also experienced some errors in the tessellation process, which created a discontinued process (see Figure 23.3)

The third phase of the task, previously mentioned as the main phase, consists of four sub-activities that allow the students to explore and construct tessellation from four different pentagons. In this activity, students investigated and found that there was (were) pentagon(s) that could tile the plane, even though the regular pentagon could not tile the plane. This activity can be seen in Figure 23.4.

Finally, the last phase that took place as the essential part of this task was to allow students to create their hypotheses. Hence, they have to conclude and give a reason for their answers to convince others based on their exploration and experience in the previous main phase. In this closure activity, students were expected to explain the reason for their chosen pentagon(s). From Figure 23.5, there was a student who had only focused on the angle as the reason. This student utilised the angle measurement tool which was provided in this activity. Meanwhile, there was also a student that used the segment tool to measure the length of the side. This condition resulted in the other response dealing with the length of the sides of the pentagon. It supported the argument that the condition of the pentagonal tiling problem can happen not only because of the angles but also the length of the sides of the pentagon.

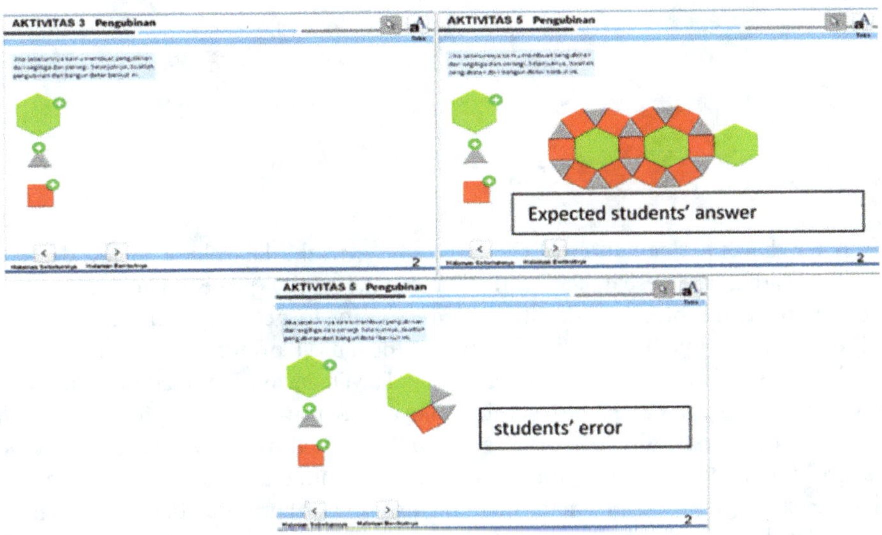

Figure 23.3 Activity 2 – challenge

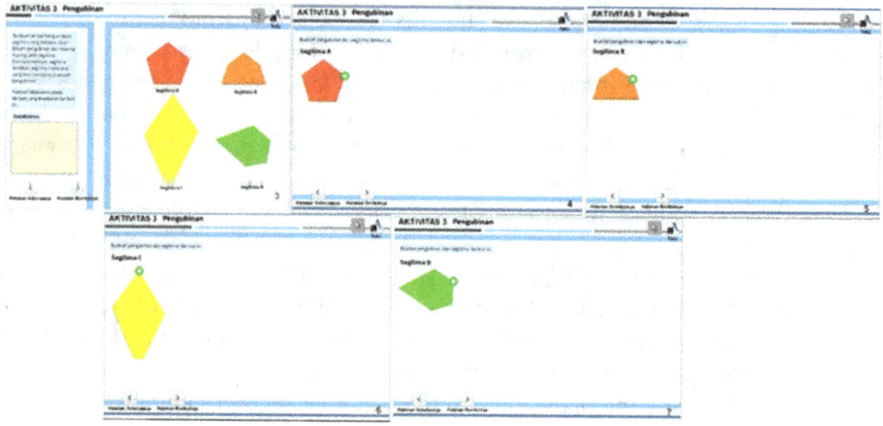

Figure 23.4 Activity 3 – main phase

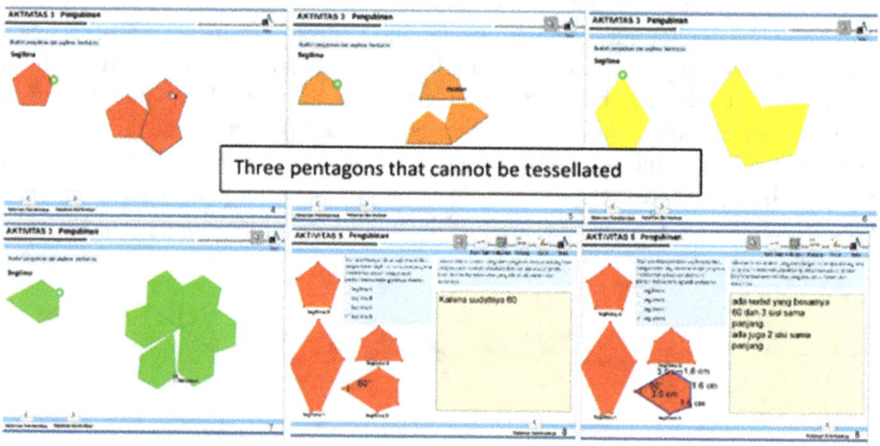

Figure 23.5 Activity 4 – closure

After the implementation of the pilot study, in the last phase of the design-based research, the researchers reflected on the entire project and shared the information to the broader educational community to get suggestions, feedback, and recommendations for the final design of the TbMT project.

The results from this study indicated that TbMT design provided opportunities for students to explore and construct the tasks on their own, which can enhance and extend their understanding of geometry concepts. The designed activities also give a space for students to develop geometric reasoning as one of mathematical thinking skills required in smart manufacturing. Furthermore, the utilisation of technology in the tasks is not only as an amplifier but also as a reorganiser which engages students to think mathematically by predicting, exploring, constructing and solving the given problems. The processes and results in this study are expected to contribute to the knowledge about the optimal use of ICT in mathematics classrooms as one of the ways to develop students' mathematical thinking skills that are essential in the era of IR4.0.

23.4 Building young minds through Minecraft digital tool to embrace smart manufacturing

This chapter provides insights into how young minds could be built through the Minecraft digital tool to embrace innovation and transformation in smart manufacturing. As the saying goes, 'To bend a bamboo, start when it's still a shoot' (as translated from Malay's proverb '*Melentur buluh biarlah dari rebungnya*'), we should start cultivating young minds with good values and creative ideas when they are young.

According to Venkatraman [28], from a longitudinal study on creative potential conducted by NASA, 98% of 1,600 of 4- and 5-year-olds scored at the 'creative genius' level. However, this figure dropped as time goes by. After five years, only 30% of the same group of children scored at the same level. After another five years later, only 12% scored at 'creative genius' level. Eventually, it was revealed that only 2% of the same group of samples scored at this genius level with the same test administered when they were adults.

There are quite a few digital tools that can be leveraged to enhance children's creative potential as reported by Talib *et al.* [29] and Loganathan *et al.* [30]. In this sub-section, the authors will elaborate on Minecraft digital tool as an exemplar to stimulate creative young minds based on the observation findings and experience gained from organising as well as judging (led by the fourth and first authors) the recently completed '*Minecraft Heritage Immortalised: A nation building championship 2021*' event with output accessible from URL: https://www.youtube.com/channel/UC18ynRC45bjbCsAWXsW1MCg/playlists. It was observed that participants of this championship were able to produce creative output even at the age of primary level as contestants in Junior group . Hence, it is also feasible to start exploring this tool that could be leveraged to maximize the potential of preschool children as young as 4- to 5-year-olds since research showed that human's creativity reaches a peak at this level as reported in the abovementioned study.

Minecraft is a 'sandbox game' with no exact objective or end goal. This seemingly endless creative freedom inculcates creativity and innovation among children since young by using the 1×1 coloured blocks to manipulate the imaginative world as shown in Figure 23.6.

Through playing with this tool in the process of developing project output involving design, planning as well as mathematical calculations, the players can develop scientific/mathematical thinking, technology, social and communication skills. These are essential skills required for lifelong learning and that have been evaluated in Programme for International Student Assessment (PISA) organised by OECD as reported by Conner *et al.* [32]. Minecraft can be used to teach science and

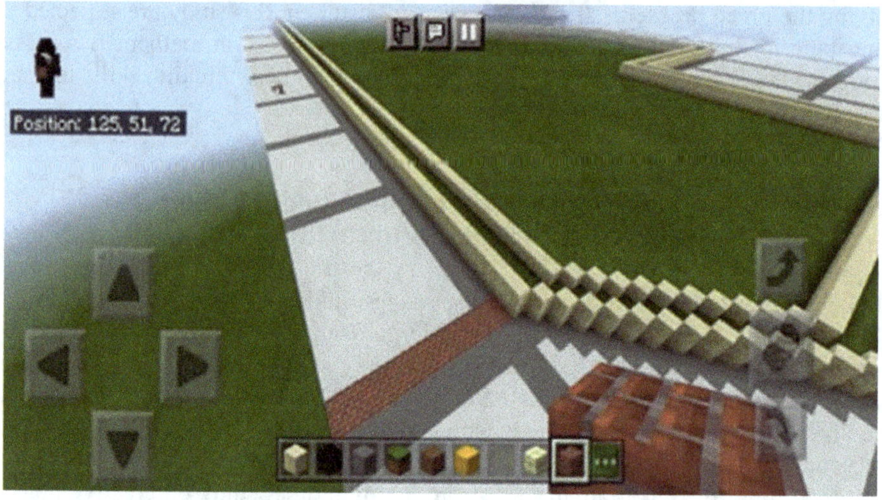

Figure 23.6 View extracted from the project entitled 'Preserving and commemorating Lunet Park' by team members from the Philippines [31]

mathematics concepts through outdoor activities such as a visit to Lily Pond, Penang Botanical Garden [33]. Apart from using Minecraft as a tool to promote outdoor studies, perhaps students at preschool levels can also be introduced to the basic mathematical concepts such as lines, shapes, angles, directions, square, rectangle, 2-dimension, cubes, cuboid, 3-dimension, etc. as well as to the science terminologies such as colours including blue, brown, green, white, to name a few. Once they have a clear understanding of all these concepts in mathematics and science, they could be better equipped with more advanced knowledge and skills to design the world they have in mind using Minecraft digital tool, Thereafter, fast learners could be advanced to acquiring knowledge and skills required for smart manufacturing with the development of higher-level thinking skills including computational thinking (CT) skills, e.g. relating a square (2-D shape) with a cube (3-D shape), matching the concept of 'colour' with 'pixel' to import pictures, creating an aspiring world in Minecraft using Scratch programming language, to name a few.

Due to the rapid development of technology and the emergence of new normal under the constraints faced during the pandemic, the implementation of blended-mode continuing professional development (CPD) courses for educators of young learners is essential to keep abreast with the current trends and meet the needs of children from diverse backgrounds. This effort is also important to echo the emphasis put by the government on the importance of ensuring quality. There is a need for preschool teachers to possess ECE-related qualifications in teaching as stated in the Ministry of Education Blueprint 2015–2025 [11]. To bridge the gap between human resource development in ICT-based industries, on one hand, and smart manufacturing, on the other, the authors would like to make the following recommendations for future work on rethinking and redesigning strategic implementation:

1. Conducting blended-mode in-service training courses/workshops as well as events/activities for science and mathematics educators from primary to secondary school levels to embrace digital tools belonging to the most current trends to enhance creative potentials of young learners in collaboration with regional training centres as reported by Ng [34], Ng [35] and Ng *et al.* [18] as well as primary, secondary and tertiary institutions.
2. Implementing blended-mode CPD programmes for educators from preschool or kindergarten levels to embrace digital tools belonging to the most current trends to enhance the creative potentials of young learners in collaboration with supporting partners or preschool associations.
3. Convening blended-mode meetings/curriculum development workshops to explore topics or niche areas to promote ''Learning Transdisciplinary Science Integrating Mathematics, Arts/Reading/Language/Culture, Engineering/ Environmental Education/ Economics/Entrepreneurship and Technology' (LearnT-SMArET) [36] for the development of spiral curriculum.
4. Carrying out blended-mode research and development (R&D) activities with the development of instruments to examine the motivation levels as reported by Narulita *et al.* [37] and Ng *et al.* [38] as well as fluid intelligence [39] capabilities of preschool young learners participating in developing projects using Minecraft tool by learning from experience in conducting such research at primary/secondary levels.
5. Providing more digital platforms incorporating innovation and best practices in Open and Distance Education [40] to promote lifelong learning for human resource development in the era of IR4.0, an example being the digital library as reported by Alizah *et al.* [41].

23.4.1 Special mention

All authors who contribute to this chapter receive a special mention.

1. Lifelong Learning for Human Resource Development in the Era of IR4.0
   Por Fei Ping, Balakrishnan A/L Muniandy and Fong Soon Fook
2. Redesigning Strategies to Enculture Lifelong Learning for the Success of Smart Manufacturing
   Por Fei Ping, Balakrishnan A/L Muniandy and Fong Soon Fook
3. Designing Techno-based Mathematics Tasks for Learning Geometry
   Miftahul Hidayah, Pasttita Ayu Laksmiwati, Wahid Yunianto, Toh Tin Lam and Emily Chor Wai Si
4. Building Young Minds through Minecraft digital tool to Embrace Smart Manufacturing

Ng Khar Thoe, Por Fei Ping, Ng Chit Peng, Pang Yee Jiea, Edsel Coronado, Subuh Anggoro, Sharfuddin Abdul Shukor, Kamalambal a/p Durairaj and Careemah Choong Lay Kheng

References

[1] OECD. (2001). *Education Policy Analysis 2001*. Paris: OECD Publishing. https://doi.org/10.1787/epa-2001-en.
[2] Schwab, K. (2016, January 14). The Fourth Industrial Revolution: What It Means, How to Respond (edition). [Online]. Available: https://www.weforum.org/agenda/2016/01/the-fourth-industrial-revolution-what-it-means-and-how-to-respond/ (accessed on 13 November 2021).
[3] Dutta, S. and Lanvin, B. (Eds.). (2019). *The Network Readiness Index 2019: Towards a Future-Ready Society*. Washington, DC: Portulans Institute.
[4] Ministry of Human Resources. (2019). *Labour Market Dynamics & Future of Work*. Putrajaya, Malaysia: Kementerian Sumber Malaysia.
[5] Ministry of Economic Affairs. (2018). *Mid-term Review of the Eleventh Malaysia Plan 2016–2020*. New priorities and emphases. Putrajaya, Malysia: Ministry of Economic Affairs.
[6] Human Resources Development Fund. (2018). *National Workforce Human Capital Development Blueprint 2018–2025*. Kuala Lumpur, Malaysia: Human Resources Development Fund.
[7] Ministry of Higher Education Malaysia. (2018). *Framing Malaysian Higher Education 4.0: Future-Proof Talents*. Putrajaya, Malaysia: Ministry of Higher Education Malaysia.
[8] PricewaterhouseCoopers. (2020). *Investing in a 4IR Ecosystem of Connected and Intelligent Things. . . and People*. PwC. https://www.pwc.com/us/en/library/4ir-ready/investment.html
[9] Manyika, J., Lund, S., Chui, M., *et al.* (2021, July 2). *Jobs Lost, Jobs Gained: What the Future of Work Will Mean for Jobs, Skills, and Wages*. Chicago, IL: McKinsey & Company. https://www.mckinsey.com/featured-insights/future-of-work/jobs-lost-jobs-gained-what-the-future-of-work-will-mean-for-jobs-skills-and-wages# (accessed on 13 November 2021).
[10] Laal, M. (2012). Benefits of lifelong learning. Procedia – Social and Behavioral Sciences, 46, 4268–4272. https://doi.org/10.1016/j.sbspro.2012.06.239
[11] Ministry of Education Malaysia. (2015). *Malaysia Education Blueprint 2015–2025 (Higher Education)*. Putrajaya, Malysia: Kementerian Pendidikan Malaysia.
[12] Grip, A. (2006). *Evaluating Human Capital Obsolescence*. Maastricht, Netherlands: Research Centre for Education and the Labour Market.
[13] SME. (2018). Smart manufacturing: building talent to accelerate a digital transformation. *Smart Manufacturing Industry Report*, 1(3).
[14] Tatsuoka, K.K. and Corter, J.E. (2004). Patterns of diagnosed mathematical content and process skills in TIMSS-R across a sample of 20 countries. *American Educational Research Journal*, 41, 901–926.
[15] Ng, Oi-Lam., Shi, L., and Ting, F. (2020). Exploring differences in primary students' geometry learning outcomes in two technology-enhanced environments: dynamic geometry and 3D printing, *International Journal of STEM Education*, 50, 1–13.

[16] Ozdemir, G. (2010). Exploring visuospatial thinking in learning about mineralogy: spatial orientation ability and spatial visualization ability. *International Journal of Science and Mathematics Education*, 8(4), 737–759.

[17] NCTM. (2015). Strategic Use of Technology in Teaching and Learning Mathematics – NCTM Position Statement. [Online]. Available: https://www.nctm.org/Standards-and-Positions/Position-Statements/Strategic-Use-of-Technology-in-Teaching-and-Learning-Mathematics/ (accessed on 13 November 2021).

[18] Pea., R.D. (1985). Beyond amplification: using the computer to reorganize mental functioning. *Educational Psychologist*, 20(4), 167–182.

[19] Ng, K.T., Baharum, B., Othman, M., Tahir, S., and Pang, Y.J. (2020). Managing technology-enhanced innovation programs: framework, exemplars and future directions. *Solid State Technology*, 63(1s), 555–565. http://www.solidstatetechnology.us/index.php/JSST/article/view/741 (accessed on 13 November 2021).

[20] Ghavifekr, S. and Rosdy, W.A.W. (2015). Teaching and learning with technology: effectiveness of ICT integration in schools, *International Journal of Research in Education and Science (IJRES)*, 1(2), 175–191.

[21] Ra, S., Shrestha, U., Khatiwada, S., Yoon, S.W., and Kwon, K. (2019). The rise of technology and impact on skills. *International Journal of Training Research*, 17(s1), 26–40.

[22] Temsah, O.L. and Moukarzel, D.M. (2018). Effect of technology on elementary students' reasoning & communication skills in science at Lebanese private schools: an exploratory study, *European Scientific Journal*, 14, 1857–7881.

[23] Reeves, T.C. (2006). Design research from a technology perspective. In J. van den Akker (Ed.), *Design Methodology and Developmental Research in Education and Training.* The Netherlands: Kluwer.

[24] Klasa, J. (2009). A few pedagogical designs in linear algebra with Cabri and Maple. *Linear Algebra and its Applications*, 432, 2100–2111.

[25] Laksmiwati, P.A. (2015). Developing interactive Cabri 3D assistance media in three-dimensional space for grade X students of senior high school using guided inquiry learning. *Proceedings of the International Conference on Research, Implementation, and Education of Mathematics and Sciences*, Yogyakarta, Indonesia.

[26] Straesser, R. (2001). Cabri-géomètre: Does dynamic geometry software (DGS) change geometry and its teaching and learning? *International Journal of Computers for Mathematical Learning*, 6, 319–333.

[27] Tutak, T., Türkdoğan, A., and Birgin, O. (2009). The effect of geometry teaching with cabri to learning levels of fourth grade students. *E-Journal of New World Sciences Academy 2009*, 4(2), 25–36.

[28] Venkatraman, R. (2018). *You're 96 percent less creative than you were as a child. Here's how to reverse that.* Retrieved https://www.inc.com/rohini-venkatraman/4-ways-to-get-back-creativity-you-had-as-a-kid.html (accessed on 13 November 2021).

[29] Talib, C.A., Ali, M., Zawadzki, R., Baharuddin, N.S., Thoe, N.K., and Aliyu, H. (2017). Video-based learning in chemistry education: exemplars, issues and challenges. *Learning Science and Mathematics (LSM) Online Journal*, 12, 35–51. Retrieved from http://www.recsam.edu.my/sub_LSMJournal/images/docs/2017/(4)CAT%20p35-51_final.pdf (accessed on 13 November 2021).

[30] Loganathan, P., Talib, C., Thoe, N., Aliyu, F., and Zawadski, R. (2019). Implementing technology infused gamification in science classroom: a systematic review and suggestions for future research. *Learning Science and Mathematics (LSM) Online Journal*, 14. Retrieved http://recsam.edu.my/sub_lsmjournal/images/docs/2019/2019_5_PL_6073_Final.pdf (accessed on 13 November 2021).

[31] Satojeto, J., Nim, J., Calumpita, S., Nalupano, A., and Nequia, L. (2021). Preserving and Commemorating Luneta Park.Minecraft project for 'Heritage Immortalised Minecraft Championship 2021' organized by SMJK Yok Bin, UTeM and Ministry of Education in collaboration with SEAMEO RECSAM and Empire Code.

[32] Conner, L., Ng, K.T., Ahmad, N.J., Ab. Bakar, H., Suma, P., and Lay, Y.F. (2013). Evaluating students' performance for scientific literacy, reading and thinking skills in PISA 2009: Lessons learnt from New Zealand and Malaysia. Paper presented and published in the Proceedings (refereed) of 5th International Conference on Science and Mathematics Education (CoSMEd) 2013. Malaysia: SEAMEO RECSAM, pp. 11–14.

[33] Kok, Y.W., Abdul Shukor, S., Sinnih, S., Oh, S.S., and Loh, M.T. (2021). Lily Pond: Penang Botanical Garden. Minecraft Project for 'Heritage Immortalised Minecraft Championship 2021' organized by SMJK Yok Bin, UTeM and Ministry of Education in collaboration with SEAMEO RECSAM and Empire Code.

[34] Ng, K.T. (2005). An Evaluation of the Scientific Creativity and Problem-Solving Behaviours of Young Learners in the Development of Investigative Project Work. Paper published in the Proceedings (refereed) of the 1st International Conference on Science and Mathematics Education (CoSMEd) 2005. Malaysia: SEAMEO RECSAM. Retrieved https://scholar.google.com/citations?view_op=view_citation&hl=en&user=qewEkbgAAAAJ&citation_for_view=qewEkbgAAAAJ:L8Ckcad2t8MC (accessed on 13 November 2021).

[35] Ng, K.T. (2012). *Selected Student Research Projects of the 5th Search for SEAMEO Young Scientists (SSYS) 2006 Congress*. Penang: SEAMEO RECSAM. Retrieved: https://www.researchgate.net/profile/Khar_Ng/publication/292605229_1BookSSYS_2006-Compile-v100712-Final/data/56af75ca08ae9f0ff7b25b2b/1BookSSYS-2006-Compile-v100712-Final.pdf (accessed on 13 November 2021).

[36] Ng, K.T. (2018). Development of transdisciplinary models to manage knowledge, skills and motivation processes integrating technology with reflective practices. Paper presented in ICRDSTHM 2017 and published in *International Journal of Computer Applications* (IJCA) Proceedings of

International Conference on Recent Developments in Science, Technology, Humanities and Management (ICRDSTHM) 2017-Number 2, pp. 15–23. Retrieved May 22, 2019 from https:/research.ijcaonline.org/icrdsthm2017/number2/icrdsthm2017022.pdf OR https://www.ijcaonline.org/proceedings/icrdsthm2017 (accessed on 13 November 2021).

[37] Narulita, S., Perdana, A.T.W., Annisa N.F., Darmakusuma, M.D., Indarjani, and Ng, K.T. (2018). Motivating secondary science learning through 3D interactive technology: from theory to practice using Augmented Reality. *Learning Science and Mathematics (LSM) Online Journal*, 13. Retrieved http://www.recsam.edu.my/sub_lsmjournal/images/docs/2018/(3)Sari%20Narulita%20p38-45_final.pdf (accessed on 13 November 2021).

[38] Ng, K.T., Soon, S.T., and Fong, S.F. (2010). Development of a questionnaire to evaluate students' perceived motivation towards science learning incorporating ICT tools. *Malaysian Journal of Educational Technology*, 10(1). Retrieved https://www.researchgate.net/profile/Khar-Ng/publication/316084831_Development_of_a_Questionnaire_to_Evaluate_Students'_Perceived_Motivation_towards_Science_Learning_Incorporating_ICT_Tool/links/58ef5549a6fdccd5c46d73ae/Development-of-a-Questionnaire-to-Evaluate-Students-Perceived-Motivation-towards-Science-Learning-Incorporating-ICT-Tool.pdf (accessed on 13 November 2021).

[39] Ng, K.T., Fong, F.S., and Soon, S.T. (2010). Design and development of a fluid intelligence instrument for a technology-enhanced PBL programme. In Z. Abas, I. Jung & J. Luca (Eds.), *Proceedings of Global Learn Asia Pacific 2010 – Global Conference on Learning and Technology*. Penang, Malaysia: Association for the Advancement of Computing in Education (AACE), pp. 1047–1052. Retrieved August 30, 2021 from https://www.learntechlib.org/primary/p/34305/ (accessed on 13 November 2021).

[40] Gil-Jaurena, I. and Malik, B. (2013). Innovation and best practice in Open and Distance Education. *Open Praxis*, 5(4), 261–263. Retrieved https://search.informit.org/doi/pdf/10.3316/informit.968812523739055 (accessed on 13 November 2021).

[41] Alizah, A.,Lee, T.L., Ng, K.T., Noraini I., and Siti Zarikh S.A.B. (2019). Transforming pu blic libraries into digital knowledge dissemination centre in supporting lifelong blended learning programmes for rural youths. *Acta Informatica Malaysia (AIM)*, Zibeline International Publishing, 3(1), 16–20. Retrieved https://ideas.repec.org/a/zib/zbnaim/v3y2019i1p16-20.html (accessed on 13 November 2021).

Chapter 24
Summary

Wai Yie Leong[1]

Advanced technologies such as robotics, 5G mobile communications, IoT, cloud computing, and wireless sensor networks have had a huge impact and influence on manufacturing with increased collaboration between humans and smart systems. As the manufacturing process gets more automated, it's becoming more error-free by using real-time data, even as the communication systems, artificial intelligence (AI) techniques, and robotics are feeding data back into the manufacturing process putting in place an ever accurate system. This will enable products that are more customized and personal, and lead to a more competitive, efficient, and value-added production process by reacting quicker to technical or human errors to avoid product and system damage, on one hand; while increasing workplace safety and risk assessment, and reducing waste, pollution, and associated costs, on the other.

24.1 Challenges and technology roadmap for HMI for smart manufacturing

Human–machine interaction (HMI) technology brings robots closer to humans in smart manufacturing environments where computer-integrated manufacturing, high levels of adaptability and rapid design changes, digital information technology, and more flexible technical working training are required in an error-free workplace. Smart manufacturing can also assist in surveying workplace inefficiencies and worker safety. Efficiency optimization is a huge focus for adopters of "smart" systems, which is done through data research and intelligent learning automation. We have discussed the benefits of the integration of lean manufacturing, smart sensors, 5G technology, IoTs, virtual reality, VOIP, 3D printing, robotics, and automation.

24.2 Leading the human intelligence transformation, a technology roadmap for future HMI systems in smart manufacturing

HMI technologies are about reducing complexity and processes and adding value to human intelligence. However, while business processes are changing rapidly, the

[1]The Institution of Engineering and Technology (IET), Malaysia

fact remains that industries and manufacturers are struggling to exploit the full potential of digitization. The roadmap, from both strategic and technological perspectives, to visualize each step on the route towards smart manufacturing was presented. The importance of the human factor has been studied through the implementation of various models. A new framework has been proposed to explain the quality measures to the quality skills, and how it will improve the efficiency and effectiveness of an industrial process.

The challenges and impact of HMI systems in smart manufacturing including legal issues and ethical concerns when using autonomous systems; data compliance and protection; an aging society and a general reluctance to change; training; overproduction; lack of transparency in many processes and industries; and dependency on IT and electricity have been discussed.

24.3 Human–machine interaction (HMI) in smart manufacturing

The development of smart manufacturing has been accompanied by changing tasks and demands for humans in the factory. As the most flexible entity in the cyber-physical production systems, workers will be faced with a large variety of jobs ranging from specification and monitoring to verification of production strategies. Through technological support, it is guaranteed that workers can realize their full potential and adopt the role of strategic decision-makers and flexible problem-solvers. The use of established interaction technologies and metaphors from the consumer goods market seems to be promising. We have discussed HMI systems, architectures, technologies, and professional standards. We have presented cyber-physical-based solutions for the technological assistance of workers via intelligent user interfaces. Besides technological means, we also pointed out the requirements necessary for adequate qualification strategies, which will create the required, inter-disciplinary understanding for smart manufacturing.

24.4 AI implementations in HMI for smart manufacturing

AI is intelligence shown by smart machines in contrast to the natural intelligence displayed by humans and animals. Future smart systems that perceive their environments and take actions can maximize chances of successfully achieving goals. As smart machines become increasingly capable, tasks considered to require "intelligence" are often removed from the definition of AI, a phenomenon known as the AI effect. For instance, optical character recognition is frequently excluded from things considered to be AI, having become a routine technology. Modern machine capabilities generally classified as AI include successfully understanding human speech, autonomously operating machines, and vehicles, delivering intelligent routing in content delivery networks, and performing simulations. This chapter will cover practical case studies, future design, and architecture for AI systems in smart manufacturing.

24.5 Industrial IoT in a 5G and beyond environment for smart manufacturing

Internet of Things has been adopted in smart manufacturing. Along with the permeation and applications of advanced technologies in manufacturing, massive amounts of data have been generated in the manufacturing process. However, the current 3rd generation mobile network (3G), 4G, and other communication technologies cannot meet the demands of cloud management platforms (CMPs) for high data rate, high reliability, high coverage, low latency, etc., which hinders the development and implementation of CPMS. As an advanced wireless transmission technology, 5G and beyond has a significant potential to promote IIoTs and CPMS. Based on the architecture and characteristics of 5G and Beyond wireless communication technology, this chapter proposed an architecture of 5G-based Industrial Internet of Things and Services (IIoTS), and described the implementation methods of different advanced manufacturing scenarios and manufacturing technologies, i.e., enhance mobile broadband (eMBB), massive machine type communication (mMTC), and ultra-reliable and low latency communication (URLLC). Besides their characteristics, we analyzed the key technologies and challenges of the 5G-based IIoT. This chapter has covered interoperability difficulties, network architectures on mitochondrial inner membrane organizing system (MINOS), and security issues within smart manufacturing environments.

24.6 Simulations to support data analytics (DA) applications in smart manufacturing

Modeling and simulation have been used by manufacturers to analyze their operations and support decision-making. We have proposed multiple methods in which simulation can serve as a DA application in a manufacturing environment to address big data issues. An example case will be discussed to demonstrate one use of simulation where a virtual representation of machining operations is used to generate the data required to evaluate manufacturing DA applications.

24.7 Cyber-physical systems engineering for manufacturing

Cyber-physical systems (CPS) are integrated embodiments of cyber systems (consisting of computing, communication, and control elements) and physical systems (consisting of geometrical and material elements). Almost all products used in modern society are CPS. Almost all modern manufacturing systems to produce these products are also CPS. Engineering such complex CPS has reenergized the field of systems engineering, which has been moving steadily away from a document-based practice to a model-based discipline. Success in CPS engineering strongly depends on the proper application of model-based systems engineering

(MBSE). CPS are integrated embodiments of cyber systems (consisting of computing, communication, and control elements) and physical systems (consisting of geometrical and material elements). Almost all products used in modern society are CPS. Almost all modern manufacturing systems to produce these products are also CPS. Engineering such complex CPS has reenergized the field of systems engineering, which has been moving steadily away from a document-based practice to a model-based discipline. Success in CPS engineering strongly depends on the proper application of MBSE.

Manufacturing is a national priority in several countries. These countries are investing heavily in public-private partnerships in what they consider to be strategic manufacturing-technology areas. This chapter described the critical link between modern manufacturing and CPS engineering. The chapter discussed the characteristics of CPS, the evolution of systems engineering as a discipline to meet the challenges of engineering complex systems, and recent efforts to boost research and development to enable manufacturing innovation.

24.8 Industrial cloud-based solutions in smart manufacturing

Modern manufacturing demands are characterized by high fluctuations with a negative impact on resource efficiency. In this framework, Industry Revolution 4.0 key enabling technologies such as cloud manufacturing enable the sharing of distributed resources for effective use at the industrial network level. In this work, an intelligent cloud manufacturing platform is proposed to increase resource efficiency in a manufacturing network through the dynamic sharing of manufacturing services, including computational, software, and physical manufacturing resources that can be offered on-demand according to a service-oriented paradigm. The cloud-based platform includes a database module where user input data are collected, the data is processed by an intelligent module for data processing, optimization and feasible solutions are generated, and analytics are carried out by a decision support module for solutions evaluation and comparison. A case study demonstrates technical and economic advantages for industrial resource efficiency improvement.

24.9 3D processing for human-machine-interaction and additive manufacturing

The use of conventional manufacturing methods is mainly limited by the size of the production run and the geometrical complexity of the component. As a result, we are occasionally forced to use processes and tools that increase the final cost of the element being produced. Additive manufacturing techniques provide major competitive advantages since they adapt to the geometrical complexity and customized design of the part to be manufactured and allow for lighter-weight products, multimaterial products, ergonomic products, efficient short production runs, and fewer assembly errors. This leads to lower associated costs and tool investment costs, more optimized use of materials, and a more sustainable manufacturing process. Additive manufacturing is seen as being one of the major revolutionary industrial processes of the next few years. This chapter has expanded upon the structural design of the machines, history, classification, alternatives existing today, materials used and their characteristics, technology limitations, and the prospects that are opening up for different technologies both in industry innovation and the academic field of research. The choice of technology should be made once the application has been planned.

24.10 Reinforcement learning for human-robot-interaction in smart manufacturing

For many contemporary manufacturing processes, autonomous robotic operators have become ubiquitous. But the number of human operators within these processes remains high and, consequently, the number of interactions between humans and robots has increased. This is a problem, as human beings introduce a source of disturbance and unpredictability into these processes in the form of performance variation. Despite the natural human aptitude for flexibility, their presence makes modeling and optimization of these systems considerably more challenging, and in many cases impossible. The development of a methodology to effectively model these systems and a reinforcement learning agent capable of autonomous decision-making was presented. This decision-making capability provides the robotic operators with greater adaptability by enabling their behavior to change based on observed information of both their environment and human colleagues. The work extends theoretical knowledge on how learning methods can be implemented for robotic control, and how the capabilities that they enable may be leveraged to improve the interaction between robots and their human counterparts. The work further presents a novel methodology for the implementation of a reinforcement learning-based intelligent agent which enables a change in behavioral policy in robotic operators in response to performance variation in their human colleagues. The evaluation demonstrates that the reinforcement agent can effectively learn to adjust its behavior based on the knowledge extracted from observed information and balance the task demands to optimize these adjustments.

24.11 Networked sensing in smart manufacturing

Networked sensing consists of distributed autonomous sensors to monitor physical or environmental conditions, such as temperature, sound, pressure, etc. The more modern networks are bidirectional and can control sensor activity. Today such networks are used in many industrial and consumer applications such as industrial process monitoring. This chapter discussed the next generation of sensing requirements, architectures, technologies, spectrum, and distributed sensing systems for smart manufacturing.

24.12 Intelligent autonomous systems using AI, sensing, and machine cognition in smart manufacturing

AI techniques such as deep learning (classification, regression, and clustering methodologies) allow machines to learn and autonomously execute tasks even under unforeseen circumstances. Machine cognition needs to improve with the development of highly adaptive systems. This includes collaborative robots / cobots who will estimate the needs for assistance, judge risks, watch for safety factors, and offer help. This requires advanced perception, localization, vision, and cognition abilities, along with improvement in computation power in embedded platforms. In this chapter, we presented smart algorithms, AI-based autonomous systems and networks, and sensing integration methods. We addressed the impact of AI on the use of decision-making mechanisms applied to smart manufacturing systems, considering different declinations of AI. There are significant ethical and legal implications related to future decision-making processes for the choice of using force through a system governed by an AI, potentially endowed with a high level of autonomy. It will be essential to apply a paradigm, which defines for each action who is responsible for its implementation, who is associated with its administrative and legal responsibility, who will have to be consulted for further information and possible approval, and, finally, who will have to be informed. All aspects related to international safety certifications that regulate the design, development, and verification of systems whose malfunctions can impact the safety of people are essential. Many of the current reference standards are no longer adequate if we consider the current and anticipated evolution of AI, and therefore they will have to be adjusted accordingly.

24.13 Networked sensors for data interoperability and real-time production tracking in smart manufacturing

Ubiquitous sensing and the collection of big data via networked sensors allow for faster analyses and customization processes to reduce network latency and overload in distributed intelligence networks. Real-time trackers using networked sensors, IoT, and machine learning can be used for efficient resource management, process optimization in the production process, reduction in material wastage, theft prevention, and asset management. This chapter discussed networked sensors for data interoperability and real-time production tracking in smart manufacturing, from architectures to technologies, systems, and professional standards.

24.14 Industrial automation and interoperability

A big part of maximizing the value of industrial automation, and the full realization of the benefits in the industry, is the flow of data and information and human-machine cooperation. This has applications in a lot of environments such as within manufacturing plants, warehouses, and distribution centers. Also known as interoperability, this presents common challenges for companies since overcoming the difficulties of connecting the proprietary machines and the equipment made by different companies is rarely simple. Data exchange has been made easier between equipment from the same company, but the challenge is how products from different companies can communicate and interoperate. This chapter will discuss different types of automation and interoperability systems, architectures, and technologies.

24.15 Smart manufacturing workplace safety with virtual training, AR, and haptic technologies

The haptics, virtual and augmented reality technologies using big data and AI, are used to train drivers and skilled workforce so they will not be exposed to dangerous training scenarios. Due to the characteristics of virtual reality technology itself, it is very convenient, fast, and true to engage in the simulation of the above-mentioned related work. The use of virtual reality technology has been recognized by more and more people. Users can experience the most real feelings in the virtual reality world. The authenticity of experiences of its simulated environment and the real-world experiences are difficult to distinguish between true and false, making the technology truly immersive for the people.

At the same time, virtual reality has all the perceptive functions possessed by human beings, such as hearing, vision, touch, taste, smell, and other perception systems. Finally, it has a super simulation system that truly realizes human-computer interaction and makes people operate in the process. With it, you can operate at will and get the most realistic feedback from the environment. It is the existence, multi-perception, interactivity, and other characteristics of virtual reality technology that make it popular with more and more people.

24.16 Blockchain technology in smart manufacturing

Based on a peer-to-peer topology, blockchain is a distributed ledger technology that allows data to be stored globally on thousands of servers. This makes it difficult for one user to gain control of the network or operation. Subsequently, the data contained in the computerized record is more secure. The next-generation blockchain technology systems, architectures, technologies in smart manufacturing were covered.

24.17 Reducing waste and pollution with automation in manufacturing

Due to volatile and fast-moving markets, increasing competition, or more complex product solutions, industrial companies are facing increasingly intricate challenges. High-energy consumption, emissions output, resource scarcity, and waste are just some of the issues that enterprises must manage. Industry Revolution 4.0 (IR 4.0) focuses on an intelligent and interconnected world in which smart factories represent the connection between digital and physical production networks. This interconnection can lead to a variety of environmental benefits such as consuming fewer resources using intelligent connectivity of objects. The energy consumption of a company can be reduced by using start-stop technologies or big data transparency concerning resource and energy consumption to reach optimization potential in production. We discussed four scenarios: a deployment scenario, an operation scenario, integration and compliance with sustainable development goals, and a long-run scenario. The results indicate that there is a negative relationship related to the flow in the production process from the inputs to the final product, including raw materials, energy requirements, information, waste disposal, and their impacts on the environment. The integration of IR 4.0 and SDGs can enhance environmental sustainability to create ecological support that guarantees high environmental performance with a more positive impact than before. This will help stakeholders and companies to provide solutions to existing environmental challenges that can be mediated through the adoption of these new technologies. The novelty of this research is its depiction of IR 4.0 and its technologies integrated with SDGs to create a sustainable IR 4.0 combining environmental protection and sustainability.

In this edited book, we have covered challenges, new concepts, systems, architectures, new technologies, and design characteristics of human–machine interaction (HMI) systems in smart manufacturing environments using state-of-the-art technologies including AR, 5G, IoTs, VOIP, VR, Blockchains, Sensing, Automation and Robotics in various manufacturing.

Index